CONTEMPORARY ER(

CONTEMPORARY ERGONOMICS 1997

**Proceedings of the Annual Conference
of the Ergonomics Society
Stoke Rochford Hall
15–17 April 1997**

Edited by

S.A. Robertson
University College London

Taylor & Francis
Publishers since 1798

UK   Taylor & Francis Ltd, 1 Gunpowder Square, London EC4A 3DE

USA   Taylor & Francis Inc., 1900 Frost Road, Suite 101, Bristol, PA 19007-1598

A catalogue record for this book is available from the British Library.

ISBN 0-7484-0677-8

Cover design by Hybert Design

Printed in Great Britain by T.J. International (Padstow) Ltd

Preface

Contemporary Ergonomics 1997 are the proceedings of the Annual Conference of the Ergonomics Society, held in April 1997 at Stoke Rochford Hall. The conference is a major international event for Ergonomists and Human Factors Specialists and attracts contributions from around the world.

Papers are chosen by a selection panel from abstracts submitted in the autumn of the previous year and the selected papers have the opportunity to be published in *Contemporary Ergonomics*. Papers are submitted as camera ready copy prior to the conference. Details of the submission procedure may be obtained from the Ergonomics Society.

The Ergonomics Society is the professional body for Ergonomists and Human Factors Specialists, based in the United Kingdom it attracts members throughout the world and is affiliated to the International Ergonomics Association. It provides recognition of competence of its members through the Professional Register. For further details contact:

The Ergonomics Society,
Devonshire House,
Devonshire Square,
Loughborough, Leics.
LE11 3DW
United Kingdom

Tel./Fax. +44 1509 234904

Contents

MEDICAL ERGONOMICS

THE RE-DESIGN OF A HOSPITAL PHARMACY DISPENSARY AREA AND WAITING ROOM

J. May & K. Purdy

AVRU, University of Derby,
Derby, DE3 5GX, UK

An ergonomic perspective was taken to ensure that the re-design of a hospital Pharmacy department was effective in addressing the needs and improvements required by its staff and patients. The assessment techniques consisted of: a task analysis, staff and patient/customer questionnaires; a workflow analysis; measurements of the physical layout of the workstations and room areas; an ergonomic checklist, and measurement of room lighting levels. Several problems were noted regarding: the layout of the existing Pharmacy and waiting room; the dimensions and position of the serving hatches; the physical size and position of the workstations, and the surrounding environmental conditions. Recommendations were made to address these issues.

Introduction

Parts of a hospital Pharmacy department were considered by its staff to be poorly designed for the modern needs of both the patients and the Pharmacy staff. The present reception area was described as giving the appearance of being impersonal and formal. The counters, where prescriptions were issued, also offered no form of privacy, with conversations often heard by other people in the waiting room. The current design of the dispensary where prescriptions are made up for both out-patients and in-patients fails to meet the present day needs and requirements of its staff in terms of workflow patterns, workplace design/layout and environmental factors.

This department is therefore undergoing a major re-design process, and to help ensure that this is successful it is important that ergonomic factors regarding the working environment are considered. Careful assessment of problems and the ergonomic re-design of Pharmacy departments has been successful in other hospitals (Leach, & Pearse, 1994) and has been noted to improve workflow and productivity. Improvements in working conditions, both physical and environmental has also improved staff morale and product storage and contact with patients/customers (Leach, Frosdick, & Friend, 1995).

Methods

The following assessment methods were used to determine the level of satisfaction of both patients and staff with the present design of the dispensary and waiting area, and to highlight particular problems which need to be addressed within the design process:-

Patient questionnaires

A short questionnaire was constructed and administered to the patients visiting the Pharmacy. This determined the users' impression of the Pharmacy, and any problems they encountered while obtaining a prescription. The questionnaire was left in the waiting room and the patient was then invited to complete and post it in an adjacent box. Due to a poor uptake the method of administration was then changed and the questionnaire was given to people by the Pharmacy staff.

Staff questionnaires

A longer questionnaire were also administered to dispensary staff. This recorded their views regarding the existing Pharmacy department and also factors which they would like to see implemented in the design process. All staff members were asked to complete and return this questionnaire.

Workflow study

The work flow through the Pharmacy was studied via informal discussions with staff and direct observation. This involved observation of the three types of prescription processed in the department; 'in-patients', 'out-patients', and 'to take out patients'.

Task Analysis

A task analysis was performed on the major tasks within the Pharmacy. This highlighted possible problems or mismatches between the equipment or furniture provided and the needs of the individuals performing the tasks.

Workplace Measurements

Physical measurements were made of the dispensary and the waiting room. These were then compared to anthropometric data (Pheasant, 1986) to identify inappropriately designed areas.

Measurements of lighting levels

The presence of poor illumination has been directly linked to an increase in errors when dispensing (Buchanan, et al, 1991). The lighting within the Pharmacy was measured by a light meter and compared to established recommended levels.

Results

Patient questionnaire

The questionnaire was administered over the course of a week and 62 completed questionnaires were received. The main results are summarised below:-

- When entering the Pharmacy some 94% of respondents said that they could see clearly where to go to hand in their prescription.
- When handing in their prescription/receiving their medication 91.4% of respondents reported that it was easy to communicate with the Pharmacy staff.
- While waiting for their prescriptions 17% of respondents said they found the waiting room depressing, 37% found it friendly/welcoming, 40% found it pleasant, while 3% found it unpleasant.
- When handing in the prescription/receiving their medication 14% of respondents claimed that the counter was too high. No one reported the counter as being too low.
- 57% of respondents would like more privacy when collecting their prescription.
- Other difficulties reported were that the chairs in the waiting room were positioned too close together and were uncomfortable/cramped when being used. No facilities were provided for wheelchair users when waiting or collecting medication.
- 20% of respondents had hearing problems on the day they collected their prescription, 14% had vision problems and 25% had mobility problems.

Staff Questionnaire

A 100% response rate was achieved. Some of the main findings are:-

- 80% of the staff were unsatisfied/highly unsatisfied with the present design of the Pharmacy and no-one reported being satisfied with the present design.
- 87% of respondents claimed that they had to walk to the opposite side of the room more often than they thought reasonable while working in the dispensary. This was for maintaining the reception desks while dispensing at the same time.
- Tasks identified as problem areas were; serving in the Pharmacy shop, receiving prescriptions, and giving medication to patients. These were due to the design of the serving hatches and shop counter which were too high, narrow and deep. Patients waiting to be served at both the hatches and shop counter could not always be seen from the dispensary.
- When printing labels from the computer 60% of in-patients' staff, and 67% of out-patients' staff rated their workstation as uncomfortable to use. Problems included; the bench height was too low when standing, inadequate leg room under the bench and little space available on the work bench. Glare on VDU screens was also a problem. These are shown in Figure 1 below.
- When dispensing medication 73% of staff found the storage shelves uncomfortable to reach.
- While receiving prescriptions/giving medication, 80% of staff found the counters very high. 73% of staff reported that they did not find it easy to communicate with patients when receiving prescriptions or giving medication. This was due to; the lack of privacy (47% of staff would like more privacy), the bench width at the serving windows, and the overall size of the windows which were small.
- 93% of the staff indicated that the temperature in the Pharmacy was inconsistent, being too hot in the summer and too cold in the winter. Lighting was also a problem for 67% of staff who experienced glare on VDU screens and other areas were reported to be too dull. 33% of staff experienced problems of noise while working. This was reported to originate from other hospital staff who pass through the dispensary area on their way to and from the stores.

- 60% of staff found that the most frequently dispensed medication was not stored conveniently and that the workstations often did not provide adequate storage space or enough room to work efficiently.

Figure 1. The most frequently reported workstation problems

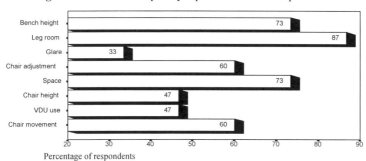

Percentage of respondents

Work flow study

The directional flow of people within the Pharmacy was not structured, resulting in directional conflicts between staff or problems with access. The long benches and narrow width of the dispensary created lengthy, narrow aisles, forcing staff to walk a long way to obtain medication and created passing problems. Conflict was also caused between out-patients' and in-patients' staff movements. Access to the stores and offices by the rear of the dispensary gave rise to a through-flow of personnel, causing noise, interruption and obstacles.

Task analysis

All three types of prescriptions went through identical stages. The problems associated with each stage are listed below:-

Prescription In. This is different for all three types of prescription:

Out patients' prescriptions
- Confusion of knowing which serving window to go to.
- Difficulty of attracting the attention and communicating with Pharmacy staff.

In patients' prescriptions would be delivered by the pharmacist on returning from the ward rounds.
- If a nurse came to collect the prescription, they may have to wait to be served.

Tto. (to take out) prescriptions (for in-patients about to leave the hospital) were left in a letter box by porters or ward staff.
- Patients would arrive for medication before it was ready.
- Confusion of knowing which serving window to go to.
- Difficulty of attracting the attention and communicating with Pharmacy staff.

Pharmacist check.
- Locating pharmacist, pharmacist finding time to sign the prescription if busy.

Label Printing.
- Insufficient printer resources.

Dispensing.
- Walking a long way to obtain medication.

- Cramped spaces when dispensing, physically knocked by people passing by.
- Crossing of pedestrian paths and restricted space in aisles.

Check on outgoing medication.
- Locating senior member of staff, getting senior staff to check prescription if busy.

Medication out.
- Problems of patient loosing ticket.
- Patient does not notice when prescriptions are ready or patient unable to see the TV screen which lists completed prescriptions.
- Confusion of knowing which serving window to go to. Problems of communicating with Pharmacy staff and attracting attention if the patient arrives later.
- Problems of privacy.

Serving in shop.
- Problems with pharmacist's line of sight when witnessing transactions in the shop.

Work place measurements

Measurements of the dimensions of the waiting room and dispensary were taken. These included; current aisle widths, dimensions of equipment within the dispensary, shelf heights, door widths etc. Workstations were assessed using ergonomic checklists to record physical measurements and to note any problems or risks to staff.

The width of the aisles or walkways measured in the Pharmacy ranged from 1.19m to 1.47m. According to Pheasant (1986) a width of 1m allows for one person walking to pass another person pressed against a wall, and a width of 1.35m allows for two people walking abreast. In theory therefore the present aisles should present little problem to two members of staff wishing to pass each other. These widths however, make no allowances for people carrying medication from one location to another. The sides of the aisles are also used as locations to dispense medication and therefore there may be two people working on each side of the aisle as well as a further two people trying to pass each other in the middle. The aisles are therefore overcrowded and insufficiently designed to meet both the traffic flow and the needs of staff.

Shelf heights were located so that all the people using them had to bend, twist or reach. It would be very difficult to redesign the storage shelves to alleviate this problem as there is not enough space to store everything at a comfortable height. Bench height was 90cm and was reported to be too low by most of the staff. The height and depth of the serving hatches made it difficult to serve customers. No provision was made for wheelchair users whose eye height fell below the height of the serving hatches.

Problems noted from the check list are as follows:-
- Little or no leg room under the bench resulting in inappropriate working posture
- Glare from overhead lights on the VDU screen
- Chairs which were worn or broken, no lumber support provided
- Height of the counter slightly low for reading work while standing
- Depth of the counter was shallow for the amount of work placed upon it, making it cramped. Little space to write notes, to use the VDU mouse, or to rest hands.
- Little or inadequate storage space.

Measurements of lighting levels

Measurements of lighting levels were taken at various places throughout the Pharmacy. Emphasis was focused on workstations and shelf height where staff would be required to read written information and to operate VDUs. Lighting levels differed greatly throughout the Pharmacy which confirms the reports of the differing light levels mentioned by staff. Pharmiretrievers were so tall that they blocked overhead lighting and cast shadows. The recommended lighting level for VDU working is 300-500 lux, while for reading, writing and fine assembly work the recommended level is 500-700 lux, (Grandjean, 1988). Many of the recorded lighting levels were considerably lower than these recommended levels. The darkest readings were taken in the shelves and in areas where there was no lighting directly overhead. Glare was mainly a problem through inappropriate types of lighting and poor positioning of VDUs.

Discussion and Conclusion

In an attempt to elicit patients' views about the Pharmacy a questionnaire was distributed. Sixty completed replies were received, which while yielding interesting insights, is still a low response rate considering the number of patients who visit the Pharmacy in a day. There may have been some inadvertent self-selection of patients asked to complete the questionnaire which may have introduced bias into the sample. For example people with certain disabilities may not have been approached, as such patients may have had sufficient difficulty in obtaining their prescriptions. Such users may have had specific problems, relating to their disability, while using the Pharmacy which may not therefore have been highlighted here.

The staff questionnaire determined current problems which they perceived regarding interacting with patients and the environmental conditions. The latter were further confirmed and extended both by the task analysis and by the workflow study. Physical measurements of the current workplace and lighting levels again demonstrated inadequacies of the current layout.

Overall the inter-related approach taken both highlighted existing areas of difficulty and indicated areas of concern which need to be addressed in the future re-design of the department. This information has been fed back to all the pharmacy staff. Various recommendations have been provided to improve the ergonomic re-design of the Pharmacy. These are presently being discussed with hospital management and staff.

Acknowledgement

We are grateful to Dr D. Cousins and his staff for their co-operation with this project.

References

Buchanan, T.L., Barker, K.N., Gibson, J.T., Jiang, B.C., and Pearson, R.E., 1991, Illumination and errors in dispensing, American Journal of Hospital Pharmacy, Vol 48, Oct 1991.

Leach R.H., Pearse J.A. 1994, The new Pharmacy at North Manchester General Hospital. The Hospital Pharmacist. April, Vol 1

Leach R. H., Frosdick P. A., Friend R.R., 1995, The new dispensary at the Norfolk and Norwich Hospital. The Hospital Pharmacist, August, Vol 2

Pheasant S. 1986, *Bodyspace*, Taylor and Francis

AN EVALUATION OF THE PERFUSIONIST'S EQUIPMENT AND TASK IN NEONATAL OPEN HEART SURGERY

Joyce Lindsay and Chris Baber

Industrial Ergonomics Group,
School of Manufacturing & Mechanical Engineering,
University of Birmingham, Birmingham, B15 2TT, United Kingdom

This study aimed to assess if the configuration of perfusion apparatus matches the tasks undertaken in surgery and to determine the nature and consequences of potential failures. Perfusionists controls heart and lung bypass equipment in open heart surgery. Task analyses provided a coherent task description and reviewed the use of equipment by three perfusionists at Great Ormond Street Hospital for Sick Children (GOSH). An equipment configuration, more suited to the tasks, was proposed from this data. Perfusion involves high stakes and a multitude of potential failures; an FMEA highlighted potential errors and their consequences. Such errors are typically rare and perfusionists face the enormous difficulty of deciding appropriate corrective actions. A simulator could provide training to cope with such events; this research will contribute to it's development.

Introduction

In open heart surgery the function of the heart and lungs is taken over by the perfusion circuit, controlled by the perfusionist. A perfusionist is " *a skilled person, qualified by academic and clinical education who operates extracorporeal circulation equipment during any medical situation where it is necessary to support or temporarily replace the patient's circulatory and/or respiratory function*" (Kurusz, 1994; p. 212). The perfusion circuit is comprised of a large number of tubes, meters, dials and controls which are not manufactured by the same company and therefore don't always match each other. Perfusion equipment is not operated in isolation as the perfusionist (A) forms part of a multi-skilled team including: surgeons (B), nurse (C) and anaesthesiologists (D). Each subteam must achieve a balance between autonomy and teamwork to ensure that all the subgoals contribute towards the ultimate aim of surgery: successful surgery without patient injury. Action by one team member can impinge on the actions of others as heart surgery is dynamic and unpredictable. Constantly fluctuating physiological parameters, variation in operation types and patient response all contribute to the volatile status of the system which dictates the need to use different levels of expertise: i.e. from skill through to rule and knowledge-based behaviour (Rasmussen, 1987).

Initial interviews were conducted with a perfusionist to ascertain the basic procedure used for heart surgery.

• The circuit is assembled (A) and the patient is anaesthetised (D).

• The circuit is "primed" with blood, albumin and drugs (i.e. circulated round the circuit) (A) while the patient's thoracic cavity is opened in preparation for surgery (B).• The surgeon connects the patient to the circuit (B) and the patient is put onto bypass (A).

• The patient is taken to hypothermia (A) to allow heart and circulation are arrest (B).

• Once surgery is completed (B), the heart and circulation are restarted (B) so that the patient can be rewarmed (A).

• The patient is taken off of bypass (i.e. the circuit no longer circulates blood round the body) (A) and the patient is disconnected from the circuit (B).

• The circuit is dismantled (A) as the patient is prepared for and transferred to the intensive
care unit (B, C & D).

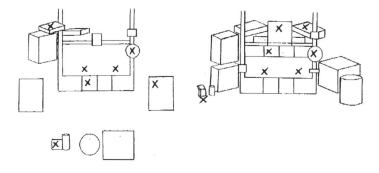

Figure 1a. Current circuit configuration.   **Figure 1b.** Proposed circuit

There are three main issues of concern regarding perfusion at GOSH. These issues serve to increase the workload of the perfusionist and require the skilled intervention of the perfusionist: the equipment has been designed to be used for adult patients, therefore excess fluid volume can cause oedema (swelling) of the patient's tissues; as the circuit is placed primarily for the convenience of the surgeon and the perfusionist, the circuit tubing is longer than required thus exacerbating the volume problem; a lack of standardisation between manufacturers typically means that component replacement involves problems such as discontinued models (due to limited markets and advancing technology) resulting in mis-matched equipment. The first element of the study investigated the circuit component configuration. Currently, the equipment is arranged in a circular fashion (figure 1a) but some tasks require that components positioned distally around the work space need to be monitored or controlled together or in close temporal succession. Although all of these problems potentially endanger the patient, this study only focused on the latter since circuit configuration could result in the loss of vital information or mistiming of a crucial action. A link analysis, based on observations of the perfusionists, resulted in proposals for redesign (figure 1b). It was felt that the initial redesign should be achievable using current equipment.

The study also focused on the potential for failure within the system; the large number of tasks and circuit components involved create potential for human and/or technical failure. The perfusionist's objectives are to ensure that the patient's body is sufficiently oxygenated and that none of the physiological parameters fall beyond safe limits. A variety of performance shaping factors (e.g. low temperature, changing cognitive demands and lengthy duration of surgery) may influence whether the perfusionist achieves the required goals by increasing the likelihood of error as suggested by literature on other vigilance tasks (Weinger and Englund, 1990). Besides concerns with equipment design the study was concerned with a second major PSF, the volatility of the workload level. During preparation to put the patient onto bypass the physical workload appears to be at a high level before falling to a low level during bypass which only requires vigilance and monitoring. Before the patient can be taken off of bypass the perfusionist experiences an increase in workload to a level similar as prebypass state. The conditions under which the perfusionist has to operate may lead to PSFs such as high visual demand from continuously changing displays; high risks for patient safety and time pressure. Periods of low workload give rise to other behaviour-influencing traits such as stress, boredom and fatigue. As the potential for error and failure is prolific at present, it is important that their consequences are clarified to highlight their severity and allow appreciation of error reduction technique benefits.

The objectives of the study are therefore threefold: to support the hypothesis that the configuration of equipment is not suitable for the nature of task sequences and frequency of component utilisation and to suggest a suitable configuration; to highlight errors and failures which be tackled as they are potentially life-threatening to the patient; to subsequently suggest effective error reduction techniques.

Methodology

As no similar previous studies were found in the literature the first step taken was to create a comprehensive task description which provided the basis for activity sampling carried out with three purposes: to determine if any quantitative interperfusionist differences existed (in terms of component use); secondly, to assess those differences in component use which seemed to occur between stages of the operation no matter who was operating the system and finally to provide data for a link analysis assessment of the equipment according to task suitability: i.e. are frequently used components located in peripheral parts of the circuit or are commonly frequented component sequences parts distally spaced? The second objective of the study (to draw attention to the need for error reduction techniques) was tackled by a FMEA (failure modes and effects analysis) to highlight possible failure modes, their effects and typical remedial actions which yielded potential preventative measures. Following the realisation of these objectives the results are to be channelled into the development of a simulator upon which future perfusionists could train, current perfusionists could receive refresher training and behaviour research could be undertaken. The methodology is briefly summarised below:-

• Four real operations were observed to familiarise the observer with the processes involved
    and to establish communication links with the medical personnel.
• Having ascertained a basic task description, a walkthrough of suitable stages (setting up the circuit and priming it) was added to this information to create a coherent task description and a subsequent HTA.
• With "MacShapa" (activity sampling) software, five real operations, from priming of the
    circuit through to taking the patient off bypass, were recorded on a lap-top computer to produce five spreadsheet of activity codes including a column for each stage of an operation and the frequency, sequence and duration of each activity recorded.
• By FMEA, the observer extracted the consequences of hypothetical failures and what their remedial actions would be from the perfusionists.

A number of practicalities limited the study mainly stemming from the fact that this was a field study and therefore there was no strict control of variables and conditions: e.g. operation times, operation types, patient responses, colleague actions and team structure. In addition, the methods had to be accepted by the whole medical team since any source of annoyance or distraction could compromise their performance and there are constraints on the equipment that can be taken into theatre therefore the lap-top computer was only just acceptable (or the safety of the patient). It was not practical to video record surgery for activity sampling for several reasons: theatre restrictions; no camera position provided a clear view of all the perfusionists activities; more than one camera would have created excessive analysing time.

Results

An HTA was compiled from the data produced by the walkthrough and initial observations. No interperfusionist differences were revealed by the activity sampling. As hypothesised, significant differences were found between each defined stage of surgery, probably because the nature of the task dictates the type, sequence and frequency of actions taking place. Findings from the link analysis were consistent with the hypothesis that the equipment configuration is suboptimal for perfusion tasks. This was established using frequency and sequence of use data which also allowed the proposal of a new arrangement (figure 2) to rectify the problem of distally spaced components which were commonly grouped temporally in use. For example, figure 1a & b depicts the components (marked by an 'X') used for a process called haemodilution; this highlights the relative superiority of the proposed circuit configuration. The suggested layout has been reviewed with enthusiasm by staff at the hospital.

The second stage of the results investigated the potential for human error, technical failure or unexpected patient responses. The results were presented in table format for coherence; Table 1 is one failure example of the whole FMEA table, looking at only one failure. This example has actually occurred in reality, although not to the perfusionists in this study. It was suggested that the surgeon decides on the administration of this drug because he/she has awareness of whether the sucker pumps are on or off.

Table 1 - FMEA

Task	Failure Mode	Aetiology	Result	Remedial action	Prevention
Controlling sucker pumps.	Coagulating substance administered when suckers were still on; perfusionist not informed.	Verbal indication of action not heard. It is the wrong drug to be added at that point.	If perfusionist has suckers switched on then the drug will cause the whole circuit blood to clot.	A new circuit would have to be set up and primed to replace the clotted one.	Timing of drug admin. should be decided by surgeon and more clearly indicated to the perfusionist

Discussion

Equipment

As outlined in the results section, the link analysis led to the suggestion of an alternative circuit configuration. This was a very basic equipment rearrangement as it was felt that initial suggestions should only involve current components to find out reactions to alterations and because time limitations prevented any analysis of the circuit interface, these details could not be included in any recommendation.

Failure Modes

Investigating hypothetical modes of failure highlights the importance of reducing the consequences of each failure by indicating their severity. Four properties of system failure become apparent from the results: the nature and severity of the consequence and the subsequent remedial action depends on how soon the fault is detected. Secondly, not all error induced failures can be rectified by the perfusionist just as some failures which the perfusionist deals with are not the result of his/her actions. Finally, the status of the patient can influence the type of remedial action employed: for instance, a clotted circuit cannot be replaced if the patient is not in a state of hypothermia. Thus, it is not only modification of the circuit configuration which can reduce the risk of failure; suboptimal procedures and poor communication can put the patient at risk by encouraging erroneous decision making and technical failure.

Summary

The study met the objectives set out: the first objective to formulate a coherent task description upon which further studies could be based was successful. As no significant interperfusionist differences were found by the methods employed we cannot say that there are any differences in how the perfusionists operate but at the same time (because of the limitations outlined for the methods employed) we cannot say that there are no differences; clearly this hypothesis needs to be tackled differently. Although the link analysis indicated that the equipment could be arranged more suitably for the task in hand, poor human factors considerations are not the only contributing factor to the currently suboptimal state of the circuit: practical (e.g. pumps and tubing need to be lower than the operation table as return of the blood from the patient to the system is gravity-assisted) and manufacturer (i.e. limited market, new technology and discontinued lines all lead to mismatched equipment) limitations contribute to this circuit status. Failures in the system seem to occur relatively infrequently considering the potential created by the complexity of the task and the number of tasks involved. This may be attributable to the possible occurrence of inconsequential problems: i.e. those failures which do not result in adverse consequences tend not to be noted or are dealt with before coming severe. A further possibility may be that as perfusionists gain experience they make fewer errors because they have developed skill and rule based knowledge and become increasingly competent in the application of knowledge-based behaviour with emergency scenario experience.

This study served as an introduction to a complex environment in which future studies will contribute towards the ultimate goal of simulator development. Currently, there is no such facility where perfusionists can be initially trained, receive refresher training or be monitored in different scenarios. To do this, techniques which might contribute in the foreseeable future include a full error identification and a study of perfusionists' skills (e.g. heuristics used; problem solving strategies and decision making processes).

References

Kurusz, M., 1994, Standards of practice in paediatric perfusion, *Perfusion*, **9**, 211-215.
Rasmussen, J., 1987, Cognitive control and human error mechanisms. In J. Rasmussen, K. Duncan and J. Leplat, Eds. *New Technology and Human Error*, 53-61, (New York: Wiley).
Weinger, M.B and Englund, C.E., 1990, Ergonomics and human factors affecting anaesthetic vigilance and monitoring performance in the operating room environment, *Anaesthesiology*, **73**, 95-102.

ATTENTIONAL DEMANDS OF DOCTORS' COMPUTERS

Derek Scott and Ian Purves

Sowerby Unit for Primary Care Informatics,
Faculty of Medicine,
University of Newcastle,
Newcastle-upon-Tyne NE4 2AA

A study is described which examined differences in doctors' (GP') attention to computers versus patients before and after a software upgrade was made. The upgrade, whilst designed as a decision support system for drug prescribing, no doubt necessitated increased use of the computer. It was found that, whilst there was no impairment in GPs' attention to their patients, the diversity in usage/performance between GPs far exceeded any before-after change.

Introduction

Computers have been known in General Practitioners'/physicians' (GPs') surgeries (examination rooms) for many years now and despite some pioneering work, largely in the early eighties, the time is now ripe for a further research launch into the effects of the ever-increasing cognitive and attentional demands imposed by the increasingly sophisticated software on the delicate doctor-patient relationship. In brief, this is because "computerisation" no longer refers to the likes of simply generating bills, but actually assists the GP in the *decision-making process* of choosing the most appropriate drug to prescribe.

Whilst many computerised formulary systems have been developed and provide doctors with details on suitable drugs, cost, and drug interactions, "the most ambitious project so far" (The Lancet, 1996, p. 1127) is the UK's "PRODIGY" (Prescribing RatiOnally with Decision-support In General practice studY) system. This system goes a step further by recommending preferred treatments. It is fully described by Purves (1995) and the methodological details of data collection by videoed consultations and patient questionnaires are described by Scott *et al.* (1996).

Previous research looking at the effects of computers in the surgery suggests that this technology has, with few exceptions, a minimally invasive effect on the doctor-patient relationship. However, the ever-increasingly sophisticated software being employed is likely to necessitate demands on the GP's attention and on their visual focus to the extent that thresholds of acceptability to the patient are exceeded. Such increasing demands on the doctor may negatively effect the patient's faith in the doctor, the perceived infallibility of medical advice, and hence compliance with prescription taking. On the other hand, there may be considerable scope for patient involvement within a three-way doctor-computer-patient (DCP) interaction, and attendant benefits of cost-effective patient education. This study aimed to examine, by before-and-after upgrade video analysis, differences within the parameters of non-verbal (e.g. eye contact) measures of interaction between doctor and patient and between doctor and computer. Essentially, data was to be collected from six practices throughout England before and after a software upgrade (more precisely, six slightly varying versions designed around the same basic requirements) had been introduced to the GP's desk-top computer. The various software systems ("pre-upgrade") *in situ* prior to the major software change to one of the six new generation ("post-upgrade") systems were all relatively simple; as described by one of the participating GPs as "little more than an electronic set of notes". This upgrade would no doubt necessitate increased visual and cognitive regard for the "intervening" computer; perhaps beyond an acceptable threshold to a point where the patient feels of secondary importance within the interaction.

Reviewing previous work relating to computers in the surgery, perhaps the most important message is from one patient's comment that: "Computers should complement but not replace the general practitioner" (Tooley, 1990, p. 167). This complementary role may well best be as an information-giver; i.e. a patient educational tool. Fitter and Cruickshank (1982) found that, according to patients' views, the prime quality is that doctors must listen and pay attention. Scott's (1996) review concluded by saying that: Previous research looking at the effects of computers in the surgery suggests that this technology generally has a minimal invasive effect on the doctor-patient relationship, yet there seems to be some instances which are a little too "close to the edge".

Methodology

The six GPs were chosen from those who had volunteered to be part of this sub-set of PRODIGY practices co-operating in videoing and patient questionnaire. All six sites were visited on two occasions; the second visit (post-upgrade data collection) taking place about five months after the first visit (pre-upgrade data collection). It was therefore felt that adequate time (four to five months) had been allowed for GPs to accustom themselves to the new PRODIGY software. Two surgery sessions (one morning, one afternoon/evening) were covered on each of the two visits. The patient sample is therefore felt to be a representative cross section of general practice consultations.

A total of 206 consultations were videoed and analysed.

Doctor-Patient Eye Contact

Total amount of time during which the doctor and patient had "eye contact" was tallied simply by the use of a stopwatch and video recording. Eye contact was interpreted quite loosely. For instance, timing continued during glances away for brief periods (a second or two), sometimes as if in thought. Conversely, a fleeting glance at the patient whilst the GP was clearly concentrating on the computer screen were not included in the timing. Time spent interacting in physical examinations were not included; e.g. although the GP may have been talking to the patient whilst taking a BP reading, the BP reading would be regarded as the primary concern. Thus, this may be regarded as a measure of time spent concentrating on the other person (whether speaking or listening), rather than "eye contact" within a strict operational definition. "Face to face" interaction may be more appropriate terminology. This measure was then calculated as a percentage of the total consultation time.

Doctor-Computer Interaction

This measure, a measure of attention given to the computer, as opposed to the GP's attention being primary towards the patient or towards medical notes, etc., was operationally taken as periods where the GP had eye contact with the screen or was using the keyboard. He may have been talking to the patient simultaneously, but his primary focus of attention was taken to be the computer. Brief, transient glances to the patient were ignored in timing, which was simply done cumulatively over the consultation with a stopwatch.

Results

Doctor-Patient Eye Contact

Table 1 summarises the length of time (mean percentage of consultation time) for the six GPs. Figure 1 depicts this data in the form of box plots.

Table 1. Mean (and standard deviations) of length of time (as a percentage of total consultation time) spent in "face to face" interaction between patient and GP.

GP	PRE-UPGRADE				POST-UPGRADE			
	Mean	sd	Min	Max	Mean	sd	Min	Max
1	32.321	21.645	4.000	75.342	27.321	15.400	9.870	67.847
2	35.639	15.775	11.481	65.124	33.205	13.841	9.568	62.626
3	22.115	12.179	10.244	50.000	23.640	16.112	2.210	58.333
4	55.775	15.370	26.596	94.681	55.441	18.385	22.910	89.545
5	32.389	20.920	15.937	70.741	29.659	7.925	19.091	46.970
6	40.630	24.128	0.782	75.895	30.821	19.609	3.125	67.674
Ave.	37.398	21.080	11.507	71.964	33.568	19.375	15.306	69.453

A two-way ANOVA showed there to be a highly significant difference between the GPs ($F = 12.633$; df = 5,1; $p < 0.001$), but no significant differences between pre- versus post-upgrade visits, and no significant interaction between these effects. No particular pattern emerged (see Figure 1) concerning changes between the two phases. In other words, the advent of the new software had no significant effects on the degree of eye contact between doctor and patient yet the GPs themselves showed considerable individual differences in the amount of overt attention which they gave their patients, as measured by this form of non-verbal communication.

A *post hoc* Tukey's HSD analysis revealed the differences to lie between GP4 (highest) and all other GPs ($p < 0.001$), and also a significant difference ($p < 0.05$) between GP3 (lowest) and GP6 versus GP1, GP4 versus GP2, and between GP4 versus GP3, with GP4 spending on average more than half the consultation time (56%) in eye contact with the patient whereas GP3 spent less than quarter of the time (22%) in this attentive mode.

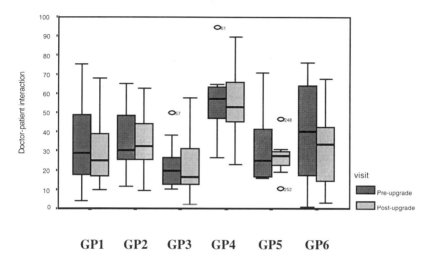

Figure 1. Box plot of length of time (as a percentage of total consultation time) spent in "face to face" interaction between patient and GP.

Two outliers emerged from the analysis. One, from the GP3 data, was a boy (Case 57) presenting with *mollosum contagiosum* (a skin condition). Exactly 50 percent of the consultation time was devoted to eye contact (between the GP and either patient or father) within an average percentage figure for this GP of 22 percent (*sd* = 12%). The doctor took time to explain the illness diagnosis and management (as measured by an independent objective measurement; Cox & Mulholland, 1993), was reassuring, and allowed time for the patient. The other, Case 61 from GP4, concerned a man in his forties presenting with impetigo (also a skin condition). Eye contact between patient and doctor was a total of 8 minutes 54 seconds of a consultation length of 9 minutes 24 seconds, accounting for 95 percent of the consultation, contrasting with a mean for this GP of 56 percent (*sd* = 15%). Again, by examination of the "GP's performance", it was seen that the doctor explained the condition to the patient, took time to reassure him, and gave individual attention in listening to the patient.

Doctor-Computer Interaction
Table 2 and Figure 2 summarise the length of time, expressed as a percentage of total consultation time, in which the GPs were primarily concentrating on the computer, whether inputting or outputting data.

Table 2: Mean (and standard deviations) of length of time (as a percentage of total consultation time) spent in interaction between GP and computer

GP	PRE-UPGRADE				POST-UPGRADE			
	Mean	sd	Min	Max	Mean	sd	Min	Max
1	9.075	7.439	1.923	25.278	12.264	15.496	0.000	64.872
2	6.423	6.938	0.282	21.29	14.400	10.253	0.377	37.037
3	16.644	11.800	0.500	41.538	24.386	16.559	0.435	57.500
4	5.043	4.086	0.000	15.637	3.272	4.608	0.000	17.404
5	19.566	12.972	0.741	38.372	18.226	11.763	2.488	38.430
6	22.980	15.788	3.333	65.556	21.721	13.468	1.163	56.983
Ave.	13.287	12.753	0.000	65.556	16.284	14.448	0.000	64.872

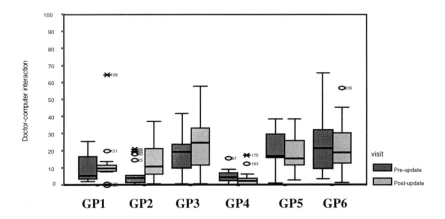

Figure 2. Box plot of length of time (as a percentage of total consultation time) the GPs spent interacting with the computer.

A two-way ANOVA again revealed a highly significant difference between the GPs in terms of their attention given to the computer ($F = 10.358$; df = 5,1; $p < 0.001$). Also, once again, no significant differences were found between pre- versus post-intervention visits, and there were no significant interaction effects between these variables. A *post hoc* Tukey's HSD analysis showed the differences to lie between GP3 (highest) versus GPs 1,2 and 4; between GP4 (lowest) versus GP5 ($p < 0.05$) and 6; and between GP6 (second highest) versus GPs1 and 2 (all bar one, $p < 0.01$). Thus, GPs once more vary considerably amongst themselves in their amount of time/attention devoted to the computer. What was striking, whilst also intuitively natural, was that from a comparison of Figures 1 and 2 it was clear that GPs who seemed to attend more to patients (e.g. GPs2 and 4) were those who least attended to their computer. In fact, there was a highly significant negative correlation ($r = -273$; $p < 0.01$) between these two measure of objects of attention. In other words, some GPs preferred to spend their time in face to face communication with their patients, whilst others seemed to optimise their time available for computer usage.

The most extreme case seems worthy of closer examination. Case 128 was a gout sufferer for whom 65% of the consultation time was recorded as devoted to doctor-computer interaction and only 25% of the time given to doctor-patient interaction. However, the important point is that the patient was actively drawn into the on-screen based information and actually read through, with the GP, two full text pages on the causes and treatment of gout. Therefore, whilst it is true that the GP's attention was primarily devoted to the computer, in that he was reading the screens and manipulating the keyboard, his behaviour provides an excellent example of a "triadic model" (Scott & Purves, 1996) of doctor-computer-patient (DCP) interaction; all three protagonists simultaneously interacting. As these figures show, and as Scott and Purves have argued, it can be misleading to simply view one dyadic aspect of a scenario such as that where the computer is playing an increasingly active role.

Conclusions

It is of note that with the exception of a handful of occasions (largely by the same GP) the patient was never allowed or encouraged to see the screen. Whilst to an extent this was largely due to inadequate positioning of the screen in relation to the patient, convention seemed to dictate who viewed what. This would seem a great waste of potential as the triadic DCP relationship offers considerable opportunity for education of the patient.

One issue which does occur, and in fact re-occurs frequently, is that of individual differences between the GPs; i.e. in terms of consultation style (content and duration), patterns of advice giving, interaction with computer and patient, etc. Clearly, *patients* are very different in terms of personalities and the situational demands dictated by their presenting illness (physical and psychiatric). Also, there were differences between the six various *computer* software devices. The third, equally important, component is the heterogenous GP *doctors* group, with their varying degrees of computer aptitude, general level of propensity to "educate" the patient, and degree of keenness to incorporate the computer within the learning situation. This simply reiterates our usual caveat of an awareness of the *human factors:* doctors, too, show *individual differences.*

References

Brownbridge, G., Evans, A. & Wall, T. 1985, Effect of computer use in the consultation on the delivery of care. British Medical Journal, **291**, 639-642.

Cox, J. & Mulholland, H., 1993, An instrument for assessment of videotapes of general practitioners' performance. British Medical Journal, **306**, 1043-1046.

Fitter, M.J. & Cruickshank, P.J., 1982, The computers in the consulting room: a psychological framework. Behaviour and Information Technology, **1**, 81-92.

Purves, I., 1995, *PRODIGY: Information for Participants.* Sowerby Unit for Primary Care Informatics. University of Newcastle-upon-Tyne, UK.

Scott, D., 1996, *Prodigy pre-intervention consultation videoes and patient questionnaire study.* Leeds: NHS Executive.

Scott, D., Purves, I. & Beaumont, 1996, Computers in the GP's surgery. In S.A. Robertson (ed.), *Contemporary Ergonomics 1996,* (Taylor & Francis, London) 397-402.

Scott, D. & Purves, I., 1996, The triadic relationship between doctor, computer and patient, Interacting with Computers, **8** (4).

The Lancet, 1996, The computer will see you now (editorial), **347**, 1127.

Tooley, P.J., 1990, Computers in general practice: Patients' views, British Journal of General Practice, **40**, 167.

COGNITIVE FACTORS IN THE USABILITY OF DECISION SUPPORT SYSTEMS FOR MEDICAL DIAGNOSIS

P J Simpson

Department of Psychology
University of Surrey
Guildford
Surrey GU2 5XH

Medical diagnosis is thought to involve a hypothesis testing process
reflecting the uncertain relationship between symptoms and underlying
causes. Computer decision systems have been designed which carry
out specific diagnostic functions with reliability and accuracy.
However the human user must provide information about symptoms
to the system. This task can involve context dependent judgements
reflecting case specific and general medical knowledge. Thus the
human user must have a significant degree of expertise to use the
system and this may limit the perceived usability of a decision support
system.

Introduction

For over 30 years work has been undertaken to create computer based
decision support systems to aid diagnosis, treatment and repair in wide range of
situations. These include the management of engineering systems, problem solving
in electronics and computer operations, power generation and information
transmission, and manufacturing processes. Applications can also be found in
financial management, welfare provision and the practice of medicine. Decision
support systems can provide the user with a rich information source which is open
to easy updating, extension and dissemination via computer networks. In the case
of medicine these systems can hold information about diagnosis and treatment in
general, as well as individual patients' records, Rennels and Shortcliffe (1987). An
ability to reason and plan on the basis of symptom information and the patient's
prior history is another facility which can be included. Decision support systems
offer reliability, consistency, and a medium for sharing expertise.

Studies of medical diagnosis suggest that diagnostic processes involve
complex recognition operations and hypothesis testing , Elstein, Schulman and

Sprafka (1978), and Lesgold (1989). The relationship between symptoms and disease is probabilistic and the diagnostic process is guided by knowledge of associations between disease types and age, sex, risk factors etc. The symptom information which is most readily available often involves effects which are indirectly related to the disease state. While symptoms generally have a clear causal relation with the underlying disorder, in some cases the symptoms can present which are consequences of the primary disease - for example, an anxiety state resulting from some chronic condition, Mevorach and Heyman (1995). The uncertain relationship existing between symptoms and causes, when combined with prior expectations about the likelihood of possible diseases, can lead to misinterpretation and errors in diagnosis, Davies (1994).

Decision support systems for medical diagnosis

The judgements involved in medical diagnosis are considered to be problematic. The task involved is complex. It requires access to, and utilisation of, extensive knowledge to reconcile symptom data with an interpretation and explanation, Evans (1989). Decision support systems offer an information and reasoning facility which can avoid errors of recall, reasoning, and inference shown by human diagnosticians. However, when using a decision support system, the human user has a crucial part to play in classifying and describing the patient's symptoms. This task requires both absolute and relative context sensitive judgements to specify the degree of change in key attributes linked to underlying disease states. In this role, the human diagnostician acts as a 'front end' to a system requiring information compatible with its internal representation scheme for symptoms and diseases.

Furukawa, Tanaka and Hara (1987) describe three sources of medical information. They are the patient's history, clinical manifestations and laboratory data. In the case of laboratory data, the information can involve a high level of measurement e.g. number of cells observed, blood pressure, temperature, presence of chemical marker. However, the significance of these measures relative to degree of abnormality and disease has to be categorised and coded, Thurmayr, Potthoff and Diehl (1987). Classification of clinical manifestations whether from direct sources (for example, the appearance of the patient's complexion, coordination, level of drowsiness), or from indirect sources (for example, medical images from X ray or ultrasonic sources) involves degraded information, Cuckle and Wald (1988). The perception and classification of clinical information depends on a 'reasoning' process, albeit an implicit process, which involves the integration of perceptual information and conceptual knowledge, Abercrombie (1969), Oatley (1978), Lesgold (1984).

Requirements for effective interface design

To be effective as a generator of information about symptoms, the human diagnostician would seem to require sufficient expertise to sustain a dialogue with the knowledge representation in the computer system. The scope of the dialogue will depend on the interface, for example, whether it supports menu lists as against natural language dialogue. It has been claimed that some of the early medical decision support systems , for example MYCIN, offered the user an awkward interface. As a results, interactions with the systems appeared to be overlong and offered little explanation of why the decision system required the information it requested , Furaka, Tanaka and Hara (1987).

One solution to these problems might be to provide an interface which supports a dialogue between the user and the decision support systems, Pollack, Hirschberg and Webber (1982). Pollack et al envisaged that the dialogue between user and system would take on the properties of a negotiation process. The authors developed their proposal following analysis of the dialogue between a human advice seeker and an expert. They suggest that the advice seeker must be allowed to participate in the definition and resolution of his/her problems.

Designing a decision support system which could sustain a natural language dialogue would involve solving many problems in computer language understanding, Gazdar (1993). Pollack et al's proposal requires that the decision support system embodies a sophisticated language understanding and conceptual ability which would 'comprehend' the user's perspective on the basis of the content of the dialogue. This facility is well beyond the brief of current systems.

Differences in conceptual frameworks

The scope of the dialogue will also depend on how far the user can articulate the basis for his/her reasoning. Expertise is said to depend on the integration of knowledge within a procedure based process which may or may not provide access to details of the rationale for the steps in the process, Oatley (1978). Studies of expertise have shown that experience and increasing competence within a problem domain, results in changes in the categories used to describe and organised reasoning with respect to a problem, Best (1995) . The rationale for this process has been described by Minsky (1968). Studies of experts in both the physical and medical sciences support Minsky's analysis, see for example Chi, Glazer and Rees (1982). One consequence of these findings is that the design of an effective user interface for a decision support systems should take account of differences in expertise, knowledge representation and descriptive schemes available to the intended user groups. In practice this requirement may be very hard to implement.

Given the difficulties providing a natural language interface able to adapted to the conceptual level of the user, it is not surprising that designers use simpler interface designs. Context dependent menus interfaces have been used. Potentially more acceptable is the use of a spreadsheet-like table entry format used for entering symptom information in the Georgia medical decision support system developed by Economou, Goumas and Spiropoulos (1996). This system provides advice on the diagnosis and treatment of 14 disease classes and 35 disease relating to lung disorders. The features or attributes which must be specified by the user in the table interface are based on a clinical differential diagnosis methodology which specifies an examining procedure. The system is built up by training artificial neural networks (ANNs) to recognise distinct disease classes and then additional ANNs to recognise the members of the classes. The system has been trained and optimised to achieve high levels of accuracy of between 88 and 95% on examples taken from documented hospital cases. Given this high level of performance, it is perhaps surprising to note that the system is used primarily for training purposes and by those who do not have easy access to specialist knowledge in medical centres, Economou et al (1996).

Integration of diagnosis and treatment

Medical decision support systems are usually developed within research labs and medical schools. But are they used by medical consultants/ experts in their everyday practice? The relatively simple style and level of communication supported by the interfaces of decision support systems may inhibit their use by medical specialists. Medical decision support systems generally have also been limited to small range of disease states. The doctor in a hospital must be prepared to interpret information and make a diagnosis which might involve a large number of underlying disease states. In that respect the Georgia system designed by Economou et al is a significant step.

Another major difference is that the medical practitioner is working in a context which requires that treatment of some kind be started even though the diagnosis of the disease state is uncertain. Treatment is started on an 'as if' basis which involves a working hypothesis. The validity of the hypothesis is tested by seeing if the patient recovers under the treatment plan indicated by the working hypothesis. Thus the distinction between the stage of diagnosis and treatment can be blurred. Of course by starting treatment, the state of the patient will change so that the information about initial symptoms is overlaid by new information. A related problem is that when the patient is first presented for diagnosis the stage in the underlying disease process, if one exists, may be difficult to determine. Thus the significance of the symptoms present, or absent, may be difficult to evaluate.

Case studies in diagnosis

These problems are illustrated by an example reported by Mevorach and Heyman (1995) which appeared initially to be a case of an elderly couple both of whom showed symptoms of heart problems. The coincidence of both the husband and wife showing distress prompted the physician to consider that the wife's symptoms reflected anxiety about her husband's illness. However tests suggested otherwise. Rest and simple treatment alleviated the symptoms in both patients and at that stage confirmed the initial hypothesis. However the couple returned after two days with effectively the same symptoms and this prompted a search for a common cause in relation to chemicals ingested or a hazard in their shared home environment. The matter was finally resolved by exploring the possibility of carbon monoxide poisoning which proved to be the source of their illness. This case had a happy outcome. But in other circumstances, the discovery of carbon monoxide poisoning could have been much more difficult because the individual symptoms are shared with many of other diseases.

The illness and subsequent death of 10 year-old boy described by Davies (1994) shows the problem of assessing the significance of symptoms which appear to be explicable in terms of a relatively minor complaint. The patient was admitted to hospital with vomiting and stomach pains. Fluids, nutrients and salts were replaced and he recovered. The diagnosis offered was gastroenteritis. However the patient became ill again after 4 months and shows a range of symptoms none of which pointed to a clear interpretation. Laboratory tests were planned but the patient died before the tests could be carried out. The post mortem revealed a failure of the adrenal cortex. Lacking hydrocortisone, the patient was vulnerable to a range of infections which most people can withstand. In this case, the initial recovery and period of good health, together with his age, may have set the context for the interpretation of subsequent pattern of illness leading to his death. The perilous state of his condition was not discovered. Both these cases illustrate problems arising from the uncertain relationship between symptoms and disease states, and the evaluation of a hypothesis on the basis of the response to treatment.

Conclusion

This paper set out to consider how cognitive factors can determine the usability of decision support systems in medical diagnosis. It has been reported that existing systems are used for training and to provide important advice for doctors who lack specialist knowledge. However constraints on the modes of communication available at the interface, the need for the user to make complex and context dependent category judgments, and differences in conceptual perspectives of the decision system and the human user may reduce the effective usability of decision support systems.

Diagnosis and treatment are often interleaved in medical practice. Changes in the patient's state following treatment can be a key source of information in diagnosis. Decision support systems based on pattern matching operations which are carried out on a patient's symptom profile can achieve high performance levels within specific disease classes. However, diagnosis under more uncertain conditions may require the evaluation of changes which occurs following treatment. This evaluation will occur in relation to a working hypothesis or hypotheses about the patient's disease states and a causal model linking symptoms and disease. This ability goes beyond pattern matching and requires an understanding of biomedical processes involved in a disease and treatment, Chandrasekaran and Mittal (1984). Consideration of causal and temporal relationships can be a key component underlying effective human diagnostic expertise. At this level, diagnostic skill depends on access to, and the integration of a range of conceptual elements within a reasoning process in order to establish the significance of the symptom and laboratory information available.

References

Abercrombie, M.L.J. 1969, *The Anatomy of Judgement* (Penguin Books).

Best, J.B. 1995, *Cognitive Psychology,* 4th edn, (West Publishing).

Chandrasekaran, B. and Mittal, S. 1984, Deep versus compiled knowledge approaches to diagnostic problem-solving. In M.J. Coombs (ed), *Developments in expert systems,* (Academic Press, London) 23-34.

Chi, M.T.H., Glaser, R. and Rees, E. 1982, Expertise in problem solving. In R.J. Sternberg (ed), *Advances in the psychology of human intelligence,* **1**, 7-76, (Hillsdale, NJ:Erlbaum).

Cuckle, H. and Wald, N. 1988, Britain's chance to get screening right, New Scientist, **120**, 48-51.

Davies, N. 1994, Inside Story: Medical Mystery, The Guardian Newspaper, December 24 1994, 20-24.

Economou, G.-P.K., Goumas, P.D. and Spiropoulos, K. 1996, A novel medical decision support system, Computing and Control Engineering Journal, **7**, 177-183.

Elstein, A.S., Shulman, L.S. and Sprafka, S. A. 1978, *Medical Problem Solving* (Harvard University Press, Cambridge, Mass).

Evans, D.A. 1989, Issues of Cognitive Science in Medicine. In D.A.Evans and V.L.Patel (eds), *Cognitive Science in Medicine:Biomedical Modeling,* (The MIT Press Cambridge, Massachusetts and London, England).

Furukawa, T., Tanaka, H. and Hara, S. 1987, FLUIDEX - A Microcomputer Expert System on Fluid Therapy Consultation. In M.K. Chytil and R. Englebrecht (eds), *Medical Expert Systems using personal computers,* (Sigma Press, UK) 59-82.

Gazdar, G. 1993, The Handling of Natural Language. In D Broadbent (ed), *The Simulation of Human Intelligence,* (Blackwell Publishers, Oxford).

Lesgold, A.M. 1984, Acquiring Expertise. In J.R. Anderson and S.M. Kosslyn (eds), *Tutorials in Learning and Memory,* (W.H.Freeman and Co) 31-60.

Lesgold, A.N. 1989, Context-Specific Requirements for Models of Expertise. In D.A.Evans and V.L.Patel (eds), *Cognitive Science in Medicine: Biomedical Modeling,* (The MIT Press Cambridge, Massachusetts).

Mevorach, D. and Heyman, S.N. 1995, Clinical Problem Solving: Pain in the Marriage, The New England Journal of Medicine, **332**, 48-50.

Minsky, M.L. 1968, Descriptive Languages and Problem Solving. In M. Minsky (ed) *Semantic Information Processing,* (The MIT Press) 419-424.

Oatley, K. 1978, *Perceptions and Representations,* (Methuen and Co).

Pollack, M.E., Hirschberg. J. and Webber. B. 1982, User participation in the reasoning processes of expert systems. In the *Proceedings of the AAAI-82 Conference,* (W.H.Freeman and Co).

Rennels, G.D. and Shortliffe, E.H. 1987, Advanced Computing in Medicine, Scientific American, **257**, 146-153.

Thurmayr, R., Potthoff, P., and Diehl, R. 1987, Algorithmic Classification of Chronic Diseases. In M.K. Chytil and R. Englebrecht (eds), *Medical Expert systems using personal computers,* (Sigma Press,UK) 165-174.

UNDERSTANDING MUSCULOSKELETAL DISCOMFORT IN MAMMOGRAPHY

J. May & A.G. Gale

AVRU, University of Derby, Derby, DE3 5GX, UK

This study aimed to determine the nature and extent of musculoskeletal discomfort experienced by radiographers working within the NHS Breast Screening Programme. The following interrelated approaches were used; task analyses, a national survey, observational studies, a body mapping study and workplace measurements. This paper details the results obtained from the task analysis and body mapping study. Potential causal or contributory factors while screening were identified and recommendations made to address these. A training video for radiographers has also been developed from this study.

Introduction

The most effective way of reducing the morbidity figures for breast cancer is by taking a mammogram (X-ray) of the breast. The national breast screening programme was established in 1987 to address the high mortality rate of breast cancer within the UK. For this programme to succeed it is important that the needs of both the radiographers performing screening and the women attending are met. Numerous reports of musculoskeletal discomfort have been received from screening radiographers but there are little data to substantiate these complaints. Although recent studies (Eckloff, 1993; Darnell, 1992) have highlighted problems of muscular discomfort in such occupational groups as radiotherapists and general radiographers, no such study has been conducted on radiographers in mammography and the nature and extent of any problems is unknown.

Methods

The following interrelated assessment methods were used :-
Task Analysis
Several task analyses were conducted by observing radiographers while they were breast screening. These are divided into the following sections:-

General Task Analysis. Hierarchical task analyses were conducted by observing radiographers' actions while they were screening. Interaction points in the screening cycle between the radiographer, the women attending for screening and the equipment used, were identified and problems that were encountered at each point noted. Nine radiographers of varying height (5'-6'6") were observed while performing screening on four different mammography units. Data were collected for a minimum of five medio-lateral view and five cranio-caudal view mammograms. From these data a hierarchical task analysis was constructed which detailed the radiographers' requirements. The first level of this analysis was used to identify the basic elements of the breast screening task and therefore to reflect what questions were asked in the questionnaire.

Task Analysis for each view taken. The radiographers' tasks and requirements were also analysed specifically in terms of which view of the breast she was taking. It was important to do this as the requirements for each view were slightly different.

Task analysis of radiographers who had different experience. Task analyses were also conducted to determine if there were differences in radiographers' screening technique. Three radiographers were observed while working on the same mammography unit. One radiographer had been screening since the beginning of the breast screening programme, the second for approximately two years while the third had been performing mammography for several weeks. They were observed while taking a minimum of six cranio-caudal and six medio-lateral views, three for each breast.

Radiographer Questionnaire

A questionnaire was administered to radiographers working within the breast screening programme. This is discussed in a previous paper (May et al., 1994).

Video Analysis

Videotaped observational studies were undertaken at two Breast Screening Centres. These have been detailed previously (May et al., 1995).

Mammographic Unit Measurements

Manufacturers of mammographic units were contacted for technical information concerning the present and future designs of all the mammography units in use within the breast screening programme. Additionally, direct measurements of various dimensions of mammographic units were recorded, along with measurements of room layouts and other equipment used by radiographers.

Body Mapping Study

This research aims to establish the nature and intensity of musculoskeletal discomfort experienced by radiographers throughout their working day. Reporting forms were constructed based on Corlett and Bishop's (1986) body mapping study. These contained a body map which divided the body into the same body part areas as contained in the diagram of the body on the questionnaire distributed in the earlier section. A seven point scale ranging from no pain/discomfort to extreme pain/discomfort was used to enable the radiographer to indicate the degree of pain/discomfort which they were experiencing. A scale was given for each area of the body.

Each breast screening centre was visited before the study and the purpose of the investigation explained to the radiographers. Once their consent to take part in the study had been obtained one radiographer was appointed to control the administration and the completion of the forms. The radiographers were asked to complete one form every hour throughout the working day. They completed the first form when they started work, before doing any screening. This was important as it then formed a baseline against which to monitor the subsequent build up of musculoskeletal discomfort while at work throughout the working day. Each radiographer was asked to use only one particular mammography unit throughout the day's screening. They were then asked to repeat the study on the same day the following week in a different environment or using different mammography equipment. The day selected for the study was when the most intensive screening throughout the whole week was taking place and also the day when radiographers could adjust their working schedules to ensure that they could work on one particular mammography unit at that centre on that day. Where radiographers could not adjust their schedules accordingly they were asked to perform the study on the closest day to the one identified. It was not possible to identify the same target day across all breast screening centres as screening sessions, assessment and symptomatic clinics were arranged on different days. Completed forms were returned for analysis.

Results

Preliminary results of the questionnaire and the results of the observational study are detailed elsewhere (May et al.; 1994, 1995) and are not discussed further here.

Task analysis

General Task Analysis. The task analysis revealed that the radiographers followed the same basic method of working on four different mammographic units, however there was a difference in working patterns between radiographers.

Task Analysis for each of the two views taken. The potential problems noted at each stage of the task analysis for each of the two views were listed. An example of these is shown in Table 1.

Task Analysis of radiographers with different periods of experience in mammography. The task analysis of each radiographer performing a cranio-caudal and medio-lateral view of each breast revealed that the radiographer with only a few weeks experience in breast screening on average took longer to screen the woman (79.1 seconds, cranio-caudal view; 92.7 seconds, medio-lateral view) compared to the other two radiographers (60.2 seconds and 54.5 seconds, cranio-caudal view; 87 seconds and 82 seconds, medio-lateral view). It was also noted that she often repeated some actions

Table 1. Examples of problems highlighted by the task analyses for radiographers performing the cranio-caudal view

---

1) **Places film cassette in bucky.** Problems noted:-
   a) Bending of spine to visually locate front of film in bucky.
   b) Hyper-extending thumbs when pushing cassette into bucky.
   c) Deviation of wrists while applying force.
2) **Selects appropriate markers.** Problems noted:-
   a) Repetitive force (every screening cycle) applied often with same finger/thumb.
   b) Deviation of wrists when attaching markers by applying pressure from thumb.

---

several times e.g. switching on the light, re-positioning and lifting the breast onto the bucky, or re-applying compression. This resulted in her using the controls more frequently. It also entailed them walking from one side of the unit to the other more often. To determine if inexperience was the causal factor the same radiographers were observed screening a woman on the same unit six months later. Comparison of the two task analyses revealed that the radiographer subsequently took less time to screen the woman on average (71 seconds, cranio-caudal view; 88 seconds, medio-lateral oblique view, - out of five screening cycles) and did not repeat actions so frequently.

Radiographer questionnaires

Further analysis of the questionnaire data revealed the following significant relationships:-

- Radiographers reporting working in small rooms or cramped conditions reported more discomfort
- Radiographers reporting difficulty in reaching some of the controls on the mammography units and having to stand on tip toe
- Taller radiographers experienced more neck, back and knee pain.
- Radiographers who had spent longer working in general radiography before specialising in mammography reported more discomfort
- Radiographers reporting having to twist or bend while using the mammography units reported more neck and back pain.
- Radiographers who reported that they never took their work breaks in the morning or afternoon reported more pain/discomfort
- Radiographers who reported performing lifting or handling tasks which they found difficult reported more back pain.
- Radiographers who reported greater levels of concern regarding applying too much compression to the breast reported more discomfort
- Radiographers who had smaller reach distances and arm spans reported more discomfort in their arms and wrists

Work place measurements

When compared to anthropometric guidelines (Pheasant, 1986), measurement of the available workspace and the equipment revealed the following problems:-

- When the C-arm is rotated for the medio-lateral view the radiographer may not be able to always maintain an upright posture without stooping or bending.
- Smaller radiographers may not be able to adequately reach the C-arm controls without stretching.
- The ratio between the height of the woman attending and the height of the radiographer may affect the screening technique used and hence their posture.
- On some units the rotation guideline cannot easily be seen by some radiographers.
- Some units have a handle running across the tube head. When the C-arm is rotated for the medio-lateral view, this handle reduces head clearance by some 5cm.
- Many buttons or controls are located near handles to facilitate use of the control and handle by the same hand. Some of these controls force the hand to adopt an awkward posture and reduce the effective grip provided by the handle.
- Markers on some mammography units are located out of reach.
- The exposure button on the control panel on many units is in a position which can only easily be operated by the right hand.

- Older units, requiring final compression applied by hand, place strain on the wrist.
- Cramped rooms, on mobiles and in static centres, do not allow adequate space for the radiographer to walk around the woman, and in some cases do not allow adequate space for the radiographer to stand comfortably when screening.
- Some pedals' leads are too short to enable them to be positioned appropriately.
- The height adjustment controls on some units are difficult to reach when the radiographer is standing towards the front of the bucky

Body Mapping

The data collected by this study indicates that the intensity of the pain reported generally increased throughout the working day. This is demonstrated for lower back pain in Figure 2. As the graph demonstrates on average the intensity of discomfort

Figure 2. The average intensity of discomfort experienced in the lower back throughout the working day.

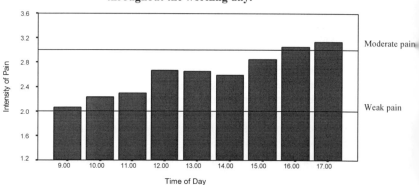

Time of Day

experienced at the beginning of the day is rated by radiographers to be 'weak pain' and this increases to 'moderate pain' as the day progresses. The intensity of pain decreases slightly at around 13.00 and 14.00hrs, at the time most of the radiographers recorded taking their lunch break and therefore had a rest from work. When they started their work again in the afternoon then by 15.00 hrs the intensity of the discomfort increased to levels above that recorded in the morning. The highest discomfort level experienced in the lower back was reported by the tallest radiographers and the lowest by the shorter ones.

The body areas where most musculoskeletal discomfort was reported were the lower back, neck and shoulders. Analysis of these results is still continuing in to the relationship between the intensity and frequency of pain reported in various areas of the body and the types of mammography unit used, working on static and mobile centres, age, height and length of time the radiographer has been performing mammography.

Discussion and Conclusion

The results from the body mapping study, the task analysis and the work place measurements complement the results obtained from the questionnaire and the observational study. The areas of the body where pain was most frequently reported in both the body mapping study and the questionnaire were the neck, shoulders and lower back. Radiographers who do not take their rest breaks reported more discomfort in the questionnaire results. The body mapping study revealed that if breaks are taken then the intensity of the discomfort does decrease for a while. However without sufficient rest the intensity still continues to rise.

It is difficult to link the development of specific problems with one type of mammography unit as many radiographers use several. There are, however, areas for improvement which should be addressed regarding the future design of mammography units. Radiographers are also noted to work using their own individual technique. There may be a standard technique that they always use or they may in fact vary it to suit the needs and body dimensions of the woman attending. Initially when starting breast screening their technique may vary from more experienced radiographers but as they become more efficient they become more competent at positioning the breast without compromising their posture.

Radiographers experience musculoskeletal discomfort which increases in intensity throughout the working day. While rest breaks may alleviate this discomfort it is shown to still increase during the working day. This may be aggravated by the design of certain units in relation to the radiographers' working technique and anthropometric characteristics. Some radiographers may have sustained an injury in previous years while performing general radiography and while this is not caused by breast screening, it is possible that performing certain actions may aggravate it. The workload for radiographers is presently increasing, with all centres recently being asked to perform two views instead of just one and the possibility of lowering the age limit to screen women under 50 years of age. Actions to eliminate the causal factors of discomfort are therefore of great importance and recommendations suggested by this study are presently being discussed by the Breast Screening Programme. The identification of different techniques and postures used by radiographers in this study has led to the production of a video to highlight the awareness of musculoskeletal discomfort and promote the use of good practice while screening.

Acknowledgement

We acknowledge the contribution Dr C. Haslegrave has made to this study.

References

Corlett, E.N. and Bishop R.P., 1976. A technique for assessing postural discomfort. *Ergonomics* **19** (2) pp175-182.

Darnell, C., 1992, The incidence, causes and effects of back pain among diagnostic radiographers. *Radiography Today*, **Vol 58,** No662, pp21-23.

Eckloff, K, 1993, Back problems among diagnostic radiographers, *Radiography Today*, **59,** 673, 17-20.

May J. L., Gale, A.G., Haslegrave C.M., Caseldine J. & Wilson A.R.M., 1994, Musculoskeletal problems in Breast Screening Radiographers. In *Contemporary Ergonomics*1994. (Ed). S.A. Robertson. (London), Taylor and Francis.

May J.L, Gale A.G. & Haslegrave C.M.: An investigation of the postures adopted by breast screening radiographers. In: *Contemporary Ergonomics*1995. (Ed). S.A. Robertson. (London), Taylor and Francis.

Pheasant, S. 1986 Bodyspace, Taylor and Francis.

ERGONOMIC ISSUES ARISING FROM ACCESS TO PATIENTS IN PAEDIATRIC SURGERY

I M Cowdery [**] and R J Graves[*]

[*]   Department of Environmental & Occupational Medicine,
        University Medical School, University of Aberdeen,
        Foresterhill, Aberdeen, AB25 2ZD

[**]   ACT, Oaktree Lane Centre, 91 Oaktree Lane, Sellyoak,
        Birmingham, B29 6JA

A study was carried out to assess the musculoskeletal risk to paediatric surgeons from their surgical activities. Hierarchical task and postural analysis with a workspace and biomechanical analysis were used to evaluate operating table systems and musculoskeletal risks to surgeons. A postal survey showed prevalence of neck and lower back pain predictable from the workspace analysis, and surgeons' experiences with equipment. Adult-sized mobile operating tables hindered patient access most. Recommendations focused on modification of operating tables and chairs for paediatric surgery to enable surgeons to vary their posture. Suggested changes to work practices included alternative patient positioning and minimising the use of foot controls.

Introduction

Operating staff in health care may be at risk from ergonomic factors in the working environment. For surgeons in particular, the necessity for accurate visual feedback and precise tissue manipulation are coupled with the stress of intense concentration and an awareness of the critical nature of the task. The requirement for manual dexterity and the constrained postures adopted for optimum visibility can themselves pose problems for the musculoskeletal system, especially since the postures are predominantly static. In addition, the high cognitive workload can also increase static muscle loading (Westgaard and Bjorklund, 1987).

The musculoskeletal risks facing surgeons and other operating room staff have been recognised and investigated in several studies. In a study of the subjective musculoskeletal complaints of both general and orthopaedic surgeons (Mirbod et al, 1995), it was found that both groups suffered from pain and stiffness in the shoulders, neck and lower back. Further, a survey of surgeons of all specialities by Irving (1992) revealed that 42% suffered from low back pain after operating (standing or sitting) for more than 2 hours. In a study of surgeons in four Scottish hospitals, Doherty (1993) found that 27% of surgeons had suffered at some time from health problems due to their work. Of these, 71% reported back pain, with neck pain also being a common complaint. The study cites awkward postures as being the main cause of musculoskeletal complaints, which may in turn be due to poor design of stools, surgical instruments and operating tables. In addition, approximately 20% of surgeons in four Scottish hospitals encountered problems in accessing the patient when operating. This was thought to be due to the design and function of the operating table and attachments. Complaints by 14% of the surgeons that the table top was of inadequate width or length demonstrated problems in accessing patients of above and below average height, or in operating on obese patients.

Paediatric surgery requires the surgeon to specialise in the operative care of children as well as being proficient in many different anatomical specialities. In addition to ergonomic difficulties encountered in general surgery, paediatric surgery is subject to two further sources of potential problems. Not only does the size of the patient mean that access to operative sites can be affected, but the lack of theatre furniture and surgical instruments designed specifically for work with children can exacerbate existing problems. There has been very little investment in the design or modification of equipment for work with children: certainly, none of the standards bodies have produced guidelines for specialist equipment design in this area, and there appears to be only one manufacturer of paediatric operating tables in the UK. The literature also lacks evidence that any studies have been conducted into the musculoskeletal problems of paediatric surgeons, which may well account for the lack of interest in specialist equipment design. In order to address some of these issues, a study was carried out to assess the musculoskeletal risk to paediatric surgeons as a result of their surgical activities (Cowdery, 1996).

Approach

Hierarchical task analysis was carried out during nine paediatric operations. Two levels of the hierarchy were examined, except in the case of the surgery task which was expanded to a further two levels. The latter task accounted for approximately 90% of the time in theatre, and therefore formed the focus for the rest of the study. During 12 operations, three of the main tasks involved in surgery were analysed in terms of the equipment used and how it is used. Whole body and hand/wrist postures were recorded, using Patkin's classification system (1981) to describe hand grips. A representative sample of these postures were analysed using the Rapid Upper Limb Assessment (RULA, McAtamney and Corlett, 1993) method. Since the postures were primarily static, photographs of the working postures were used to confirm the RULA classifications.

The dimensions of a combination of equipment most commonly used and most likely to cause problems to the surgeon were measured. The latter were a mobile operating table and a surgeon's chair with a large base and measures included height range, width, depth, base size and suitability and location of controls. Recommendations from the literature and standards bodies regarding the ergonomic design of work surfaces and chairs were compared with these measurements to provide some indication of the suitability of the equipment (see for example, Grandjean ,1988, and Sanders and McCormick, 1992). Where necessary, the dimensions were compared to anthropometric data (Pheasant, 1988) to analyse exactly where the problems lay in the interface between the surgeon and the operating room equipment. This data was then used to explain some of the observations made in the postural assessment.

The risk factor classification system of Keyserling *et al* (1991) was used as a framework for investigating sources of risk within the workplace. This examination drew upon the results of both previous stages of the study and of previous studies of the origins of musculoskeletal disorders at work. Each potential source of risk was examined with respect to the possible musculoskeletal disorders that could ensue. The loading of muscles and joints were calculated for areas of the body deemed to be particularly at risk.

This was complemented by a postal survey of a sample of Paediatric surgeons to determine the prevalence of musculoskeletal symptoms and investigate surgeons' experiences with theatre equipment. A pilot survey was administered to three paediatric surgeons at a local hospital. The pilot study utilised an examination of questionnaire responses coupled with interviews to investigate the suitability of the questionnaire. Several aspects of the questionnaire were amended and the new version of the questionnaire was administered by post. This was sent to all paediatric surgeons in Britain who are members of the British Association of Paediatric Surgeons (n = 81).

Analysis of the responses took the form of frequency counts of response categories, calculation of means and standard deviations and other simple descriptive statistics. Statistical tests such as t and F tests were used to explore the validity of the null hypotheses.

Findings and Discussion

The working postures most commonly observed in seated paediatric surgeons during 'normal' surgical activities (excluding endoscopy) were; neck flexed forwards and sometimes twisted; trunk flexed forwards; hips abducted, with the legs out to each side; feet rarely symmetrical; forearm and wrists unsupported (although the elbows sometimes rested on the mattress); wrists flexed and/or deviated and often twisted.

The most stressful aspect of these postures related to being sustained for long periods of time, and therefore were likely to result in postural fatigue. Postural fatigue in turn can lead to errors. According to Sanders and McCormick (op cit.), the accuracy of hand movements falls off rapidly as the hand moves away from or across the body. This strengthens the case for providing paediatric surgical equipment which promotes good access to the patient, keeping the surgeons' hands and arms close to his/her body.

An analysis of the working postures indicated that further investigation and changes were required soon, but not immediate. This is despite the fact that high levels of risk might be expected from the regularly constrained postures. This apparent inconsistency may be explained by RULA's rudimentary provision for recording static postures. The muscle use rating is extremely broad, with only two alternatives; 1 or 0, with a score of 1 indicating postures that are held for longer than one minute. For postures maintained for more than one hour, the actual static stresses are therefore substantially under-represented. It was also impossible to fully capture the extreme and asymmetrical postures of the legs with RULA, since there is no provision for recording the feet stretched backwards or in awkward positions in front of the body.

Observations of surgeons' seated postures during microsurgical operations by Congleton *et al* (1985) revealed very similar results such as; the surgeon sits on the front edge of the seat pan; the backrest is seldom used; the surgeon's abdomen is usually placed against table; the elbows are either rested on the table or totally unsupported; and the positions of the lower leg and feet are changed frequently.

There were many areas in which the equipment examined did not support the tasks of a paediatric surgeon. The chair did not have a suitable backrest, or means of tilting the seat pan, and the controls were difficult to operate. The table design led to inadequate feet and leg room. In addition, the bases of the table and chair prevent the surgeon from getting close enough to the patient without awkward postures. Improvements could be made to reduce musculoskeletal risk, however, constraints imposed by the surgical environment may hinder improvements. For example, the thickness of the mattress affects patient comfort and health (although the thickness of the table top itself could be reduced) and, the chair cannot be upholstered in a porous material for hygiene purposes.

The equipment analysis also highlights several conflicting equipment requirements. For example, positioning the operative site as close as possible to the surgeon to improve access increases the viewing angle and results in extreme flexion of the neck. Similarly, a need for a high work surface to accommodate critical visual requirements conflicts with a comfortable, relaxed posture for the upper limbs. In these situations, as is often the case in ergonomics, any final design must be a compromise between these opposing requirements.

The majority of respondents to the postal questionnaire believed their theatre equipment and furniture was adequate to fulfil the needs of paediatric surgery. Even the table attachments, with which they are the most unhappy, were rated as either excellent or adequate by 80% of surgeons. Only 5% were dissatisfied with surgical instruments and 10% judged the operating table to be unsuitable for paediatrics. Nevertheless, 45% suggested the design of the table could be improved. Table tops were found to be too wide and too thick, so surgeons could not get their knees under the table when sitting. It was also suggested that the height of tables was difficult to adjust and that table bases were too large, obstructing the feet, chair and foot pedals.

Only around 15 % of respondents use paediatric operating tables exclusively, with 30% using a paediatric table at some point in their work. Very little attention appears to have been directed to the design of operating tables for this speciality, perhaps because the size of the potential market does not justify the design effort and manufacturing costs. None of the respondents judged their operating table to be poorly suited to paediatric work, yet 23% of the surgeons stood while operating because the design of the operating table and seating facilities were so poor. Improvements to operating table attachments were advocated by 41% of respondents, highlighting their inflexible sizing (to fit both patient and operating table), and their awkwardness to position.

Adult-sized mobile operating tables affected access to patients most since feet and knee room was reduced, and surgical chairs could not be manoeuvred under the table. The table top was both too wide to position patients within easy reach and too deep to enable surgeons to adopt a comfortable seated position.

The work of a paediatric surgeon was associated with musculoskeletal discomfort with prevalence of neck and lower back pain among the survey respondents at 38% and 45% respectively. The survey results can be compared with the musculoskeletal complaints of groups of sedentary workers in other studies. Grandjean and Burandt (cited in Grandjean, 1988) examined bodily aches among 246 employees in traditional sedentary office jobs, and Linton (1990) investigated neck pain in 180 Swedish employees undergoing routine screening examinations using questionnaire methods.

Mirbod *et al* (1995) used a postal questionnaire to explore musculoskeletal discomfort in general and orthopaedic surgeons. The results of these studies are compared in Table 1. ⸜

Table 1   A comparison of the musculoskeletal problems of three groups of workers in sedentary jobs obtained using questionnaire methods

Part of body affected	Prevalence (%) of musculoskeletal problems			
	Paediatric surgeons Cowdery (1996)	**General surgeons** Mirbod (1995)	**Office workers** Grandjean & Burandt (1962)	**Office workers** Linton (1990)
Neck	38	42.9	-	18.5
Neck and/or shoulders	46	-	24	-
Lower back	45	36.5	-	-
Upper and/or lower back	55	-	57	-
Thighs and buttocks	15	-	35	-
Knees and feet	0	-	29	-

From these studies, the prevalence of neck pain in paediatric surgeons is significantly greater than in office workers. This may be due to the critical visual requirements and the increased viewing angle experienced by paediatric surgeons. It is also likely that paediatric surgeons have to maintain a posture uninterrupted for longer periods of time than an office worker who is generally freer to move around when a position becomes uncomfortable. Comparing paediatric with general surgeons, the prevalence of neck complaints are similar, whereas more paediatric surgeons have pain in the lower back. This may be attributable to a paediatric surgeon having to bend forward further across the table to reach the operative site, since the table is substantially wider than the patient.

Most of the postural problems experienced by paediatric surgeons may be attributed to insufficient access to the patient during surgical procedures. The relatively small operative fields encountered in paediatric surgery bring with them visual problems, not only in the short viewing distances necessary to achieve the level of detail required, but also in obtaining (and maintaining) an optimum viewing angle to enable the surgeon to see into deep cavities. Furthermore, the discrepancy between the size of the patient and that of an operating table designed to support adults makes matters worse, as the surgeon has to reach across the empty part of the table to reach the patient.

The equipment and working environment of paediatric surgeons examined in the study does not fully support the range of operative tasks. Adult tables appear to obstruct the feet and leg space of a surgeon, particularly when seated. The table tops were much wider than the patients so that seated surgeons could have severe difficulty in getting their knees under the table without raised shoulders and forearms at the operative site. Almost a quarter of surgeons would prefer to sit, or to be given the choice of sitting or standing to operate, but inadequate seating facilities are preventing them from doing so. Surgical seating does not appear to allow a variety of postures. Maintaining constrained postures needed to compensate for deficiencies in equipment design, often for a number of hours, produces high levels of static stress. Postural analysis using RULA revealed that changes to remedy poor working postures are required. From an analysis of the work methods and equipment, the neck and the lower back would seem to be most at risk due to having to lean forwards over the patient, and because of the large visual angles involved in looking down into the operative site for prolonged periods.

Recommendations were provided which focused on modification of operating tables and chairs for paediatric surgery to enable surgeons to vary their posture. Changes to work practices included consideration of alternative patient positioning and minimising the use of foot controls.

References

Congleton, J.J., Ayoub, M.M. and Smith, J.L. 1985, The design and evaluation of the neutral posture chair for surgeons, *Human Factors*, **27**, 589-600.

Cowdery, I.M. 1996, *The problem of patient access in paediatric surgery: Ergonomic evaluation of the effects of operating tables and surgical chairs on the musculoskeletal health of paediatric surgeons*, MSc Ergonomics Project Thesis, Department of Environmental and Occupational Medicine, University of Aberdeen: Aberdeen.

Doherty, S.T. 1993, *An evaluation of the user requirements of operating table systems and of the effects such systems have on the health of theatre personnel*, MSc Dissertation, Department of Environmental and Occupational Medicine, University of Aberdeen.

Grandjean, E. 1988, *Fitting the task to the man: a textbook of occupational ergonomics*, (Taylor and Francis, London).

Irving, G. 1992, A standing/sitting pelvic tilt chair - new hope for back-weary surgeons?, *South African Medical Journal*, **82**, 131-132.

Keyserling, W.M., Armstrong, T.J. and Punnett, L. 1991, Ergonomic job analysis: a structured approach for identifying risk factors associated with overexertion injuries and disorders, *Applied Occupational and Environmental Hygiene*, **6**, 253-363.

Kirwan, B. Ainsworth, L.K. 1992, A guide to task analysis, (Taylor and Francis, London).

Linton, S.J. 1990, Risk factors for neck and back pain in a working population in Sweden, *Work and Stress*, **4**, 41-49.

Mirbod, S.M., Yoshida, H., Miyamoto, K., Miyashita, K., Inaba, R. and Iwata, H. 1995, Subjective complaints in orthopedists and general surgeons, *International Archives of Occupational and Environmental Health*, **67**, 179-186.

McAtamney, L., Corlett, E.N. 1993, RULA: a survey method for the investigation of work related upper limb disorders, *Applied Ergonomics*, **24**, 91-99.

Patkin, M. 1981, Ergonomics and the surgeon, in W. Burnett (ed.), *Clinical science for surgeons*, (Butterworths, London).

Pheasant, S.T. 1988, *Bodyspace: anthropometry, ergonomics and design*, (Taylor and Francis, London).

Sanders, M.S. and McCormick, E.J. 1992, *Human factors in engineering and design*, (McGraw-Hill, Singapore).

Westgaard, R.H. and Bjorklund, R. 1987, Generation of muscle tension additional to postural load, *Ergonomics*, **30**, 911-923.

SCREENING FOR CERVICAL CANCER: THE ROLE OF ERGONOMICS

J. A. Hopper, J. May & A.G. Gale

Applied Vision Research Unit,
University of Derby, Derby,
DE3 5GX, UK

Regular cervical screening for women aims to detect early signs of cervical cancer. The National Cervical Screening Programme (NHSCSP) provides strict national quality assurance for such screening. As part of this QA the NHSCSP requested the establishment of minimum ergonomic guidelines for cytology laboratories. This paper details the initial exploration into the working conditions, defining the first widespread investigation of the cytology working environment. A human factors questionnaire was distributed to every cytology unit in the UK. The analysis of this illustrated poor ergonomic design in the current workplace. The development of minimum ergonomic recommendations and their implementation will aid screening performance and so reduce the likelihood of missed cancers.

Introduction

In the UK some 1600 women die annually from cancer of the cervix. Regular cervical screening is the first step in the process of combating this disease. Women aged 20 to 64 years are invited to have regular cervical smears taken every three to five years. Once a smear is taken by a nurse or GP it is sent to the cytology laboratory where each smear goes through a staining process, passing through a series of different chemicals, which stain the cells to make them more distinguishable. The smears are then mounted onto glass slides and individually examined under a microscope for the presence of abnormal cells. Approximately 32 slides are screened per day by each screener, also termed a cyto-screener, who typically carries out such examinations for fours hours a day.

To provide very early detection of any abnormal cells, including signs of cancer, it is vital that any cellular changes are discovered through screening. Individuals carrying out screening therefore use a microscope for much of their working day - if this task is not well designed then musculoskeletal and other problems may accrue which may diminish the screeners' detection capabilities. It is therefore important that due regard is given to all factors, including ergonomic principles, which may impinge upon this screening task. The main focus of such a study is on the prolonged use of the confocal microscope.

Historically, studies in microscopy have highlighted various problems. For instance, Emanuel and Glonek (1975) identified headaches, neck ache and visual strain as three main problem areas associated with microscopy. More recent studies have further confirmed that neck pain is still prevalent in microscope users (Krueger et al., 1989). Indeed there may well be a specific neck discomfort associated purely with microscopy (Robinowitz et al., 1981). Despite such studies showing the postural stress placed on the neck muscles, and recommended neck angles to minimise this (Lee et al., 1986), the design of the microscope has remained largely unaltered.

Much research has also focused on the visual strain experienced by microscope users. For instance, work by Baker (1966) indicated that the visual focal distance adopted for microscopy was inappropriate. Whilst the suggestions that microscopy leads to myopia and astigmatism are inconclusive there is much supporting evidence that the use of a microscope can lead to visual discomfort and strain.

The sedentary task of screening also invites more widespread problems in the design and layout of both equipment and workstations with regard to the postural requirements placed on the screeners. Static work postures, such as those required for screening, are known to create muscular stress and fatigue when adopted for long periods of time.

The NHS Cervical Screening Programme oversees cervical screening in the UK and implements strict quality assurance procedures. Following various informal reports about the working conditions of some screeners it was deemed important to investigate these fully. The first stage of this was to obtain an overview of the present workplace design of these laboratories and of the types of difficulties current users had. At the request of the NHSCSP we, as part of a Working Party, subsequently set about establishing minimum ergonomic recommendations for cytology laboratories.

Method

Initially several visits were made to a range of both new and older cytology laboratories to gain an insight into the actual task of screening smears and to obtain some working knowledge of existing conditions. Discussions took place with staff in these departments as well as with members of the NHSCSP team. Subsequently a human factors questionnaire was developed to elicit information from personnel involved in examining the smears concerning their working environment.

In order to obtain a widespread indication of current working conditions these questionnaires were distributed initially to all delegates at the annual NAC (National Association of Cytologists) conference in April 1996. Aimed at all staff working in cytology, the questionnaire was designed to gather information regarding the equipment, furniture, environmental conditions, muscular discomfort and the level of job satisfaction.

Subsequently further questionnaires were distributed to every Cytology unit in the UK, and a total of 523 questionnaires, representing some 134 cytology units, were completed and returned.

Results

Respondents

The questionnaires were completed by all cytology staff, from laboratory assistants to medical staff, as shown in Table 1.

Some 35.4% stained smears on a frequent or full-time basis, the majority of respondents (96.7%) screened smears either frequently or full-time, with 92.1% also reporting and signing them off.

Table 1. Summary of Questionnaire Respondents

Job Group Title	Percentage of Respondents
Admin. & Clerical Staff	1.7%
Trainee Cytoscreener	3.3%
Cytoscreener	39%
Trainee Biomedical Scientist	1.1%
Biomedical Scientist	27.7%
Senior Biomedical Scientist	13.6%
Chief Biomedical Scientist	9%
Senior Chief MLSO	1.3%
Medical Staff	2.3%
Other	1%

The questionnaire was very wide ranging and only some of the major findings are summarised here:-

Muscular Comfort

A total of 77.7% reported experiencing muscular discomfort, with 52.2% only ever experiencing pain at work. Table 2 shows the areas where such problems were reported.

Table 2. Reported areas of discomfort

Area of Discomfort	Percentage of Respondents
Neck	55.4
Shoulder	38.0
Upper Arm	6.8
Lower Arm	14.5
Elbow	15.9
Wrist	29.9
Chest	2.0
Abdomen	1.4
Upper Back	17.9
Lower Back	32.9
Hips/Thighs	7.8
Knee	12
Lower Leg	7.8
Feet/Ankle	8.8

Muscular discomfort was reported to be experienced every day by 28.6%, while a further 53.7% experienced muscular discomfort during each week. The majority of sufferers (52.7%) experienced such discomfort within the first two hours of work, while a further 37.2% experienced it within the four hour work period. Of those experiencing discomfort, 47.3% reported developing it within the last three years. Muscular discomfort was found to adversely affect the level of concentration ($r = -.02$, $p<.01$), as well as the level of job satisfaction ($r = -.18$, $p<.01$).

Microscope

The microscope is the key screening instrument. Whilst only 4.3% of respondents specifically reported their microscope to be unsuitable, some 54.5% reported problems with the functionality and control of the microscope. Principally, the height of the microscope (from the base to the top of the eyepiece) was reported to be unsuitable by 35.8% of staff, while the fixed eyepiece angle was reported to be unsuitable by 23%. Reported problems with microscope height correlated with neck ($r =.17$, $p<.01$), shoulder ($r =.18$, $p<.01$) and lower back ($r =.17$, $p<.01$) discomfort. Eye piece angle itself correlated with neck ($r =.24$, $p<.01$) and shoulder ($r =.2$, $p<.01$) discomfort. Some 20% of users also reported the position of the microscope stage and focus controls to be unsatisfactory.

No training on the setting-up and adjustment of their microscope had been received by 10.8% of respondents. Whilst the majority who had received no training claimed to know how to set-up the microscope anyway some 17.9% remained unable to do so.

A total of 73% reported that they sometimes experienced 'tired eyes' during or after screening, while 18.4% reported always having tired eyes.

Visual Display Unit

There has been a large increase in the number of cytology units using VDUs in the last few years, as reflected by the 82.4% of respondents currently using them for work. Problems reported with their use, however, centred around the overall lack of current workspace provided within which to operate the VDU equipment (70.7%), and the inability for users to sit at the workbench when operating the VDU (24.7%). Lower arm discomfort correlated with a lack of space to operate the VDU ($r =.19$, $p<.01$). Neck, shoulder, upper and lower back pain were all found to correlate with a reported lack of work space around the VDU (all $p<0.01$).

Work Benches

Many benches were reported to be unsuitable in terms of the bench height and the lack of leg room at the bench. Responses varied for the benches which were used for different tasks. For example the VDU bench was consistently reported as having insufficient work surface space whereas the microscope bench was often considered too high and the staining bench regarded as having restricted leg room.

Where both the staining and microscope bench heights were reported as unsuitable they were both found to correlate with neck, shoulder and elbow discomfort (all $p<0.01$).

Chairs

The age of chairs and their height adjustability were reported as problematic. With variable bench heights users need to employ the adjustability of the chair height to enable them to work comfortably at the bench. Some of the older chairs were reported to have lost their height adjustability which prevented the user from working at a suitable height for the equipment or the bench. Various problems with the chair support and the back rest were reported.

The absence of footrests was also reported. Many users had no footrest or found the footrest insufficient for their requirements.

Environmental Conditions

The room temperature was reported to be unsuitable by 87.4% who were unable to adjust the temperature within their own room. Ventilation was reported as unsuitable by 66%. While 95% had external windows in the room, some 27.5% could not open them to ventilate the room. Humidity levels were found to be inappropriate by 29%. The noise level in the screening room was reported as unsuitable by 52.9% of respondents.

The level of lighting was regarded as unsuitable by 20.2%. The light level between the microscope lighting and the general room lighting was reported to be incompatible by 18%, who experienced glare when looking down into the microscope or away into the room. Some 18%, who had blinds on the windows to reduce glare, could not open or close them, and therefore could not control the level of light from outside. Some 44.9% were unable to adjust the artificial lighting level within the room.

Concentration

The screening task requires full concentration from the user. While 17.9% found it difficult to concentrate, the majority of 47.9% found the conditions "OK" for concentration.

Environmental conditions (noise, temperature, lighting, ventilation, humidity) were cited by 96.1% as detracting them from the ability to concentrate on the screening task. The reporting of each environmental condition as a problem, was found to be inversely correlated variously (noise $r = -.39$, $p<.01$; temperature $r = -.12$, $p<.05$; lighting $r = -.23$, $p<.01$; ventilation $r = -.17$, $p<.01$; and humidity $r = -.18$, $p<.01$) to a deterioration in the level of reported concentration.

Job Satisfaction

A total of 63% felt a positive change in their working environment would increase their level of job satisfaction.

Discussion

Following some unsubstantiated complaints from cyto-screeners regarding aspects of their working environments a questionnaire study was carried out to ascertain the extent of any such problems.

The results show that difficulties do exist in many areas within cytology screening. The most fundamental finding is the high incidence of self-reported muscular discomfort experienced at work through adopting a static posture for some time. It is of interest to note that the distribution of muscular discomfort for these users differs from that found in work which we have carried out with other groups of health professionals (cf. May & Gale, this volume). The prevalence of discomfort, and the negative correlation between discomfort and the ability of the user to concentrate, make the design of a suitable and posturally efficient environment paramount to cervical cytology.

The appropriate ergonomic design of the microscope, in association with the work bench and chair, will optimise the working environment and working posture which are imperative to the quality of screening smears. Where other processes occur, such as staining or VDU work, then similarly appropriate ergonomic design of the workplace is vital. Furthermore ergonomic consideration needs to be given to the implementation of new and future technology.

Subsequently, further observational studies and various task analyses have been performed to supplement the work described here.

Conclusions

Ergonomically, the general working environments of personnel currently involved in cytology screening is less than ideal. The questionnaire served to highlight the common areas of concern across a large number of users. We are now addressing these issues. Additionally we are both developing and implementing minimum ergonomic guidelines for cytology screening. These should subsequently improve the ability of cyto-screeners to more ably accomplish their complex work task, by maintaining their concentration levels and minimising any physical discomfort.

Acknowledgements

We gratefully acknowledge the assistance of the many individuals who participated in this study. We are also grateful to Mrs J. Blaney, Mr C.W. Hancock and the other members of the NHSCSP Working Party.

References

Baker, J.R., 1966, Experiments on the function of the eye in light
   microscopy, *Journal of the Royal Microscopical Society*, 85(3), 231-254.
Emanuel, J.T., and Glonek, R.J. 1976, Ergonomic approach to productivity
   improvement for microscope work, Proceedings of the AIIE Systems Engineering
   Conference, Institute of Industrial Engineers, Norcross, G.A.
Krueger, H., Conrady, P., and Zulch, J., 1989, Work with magnifying glasses,
   Ergonomics, Vol 32, No7, 785-794.
Lee, K.S., Waikar, A.M., Aghazadeh, F., and Tandon, S., 1986, An
   electromyographic investigation of neck angles for microscopists,
   Proceedings of the Human Factors Society - 30th Annual Meeting, 548-551.
Robinowitz, M., Bahr, G.P., and Fox, C.H., 1981, Relieving muscle fatigue
   and eye-strain in microscopy, *Acta Cytologica*, 15, 585-586.

THE CULTURE OF
ERGONOMICS

DO ERGONOMISTS BELIEVE IN ERGONOMICS?

Elaine C. Mitchell

Scottish & Newcastle plc.
Abbey Brewery
111 Holyrood Road
Edinburgh. EH8 8YS

Even with the requirement to take an 'ergonomic approach' in several of the European Community Directives, industry has not rushed to recruit ergonomists. Yet the application of simple ergonomic principles can be enormously beneficial to the industry and its employees. Two case studies are presented to illustrate this point. A cost benefit analysis is given for implementing ergonomic changes in each case. The studies show how ergonomics can be developed within an organisation. Ergonomics can be used at strategic levels and is compatible with quality and other improvement initiatives. The need for the Ergonomics profession to develop with and within industry is discussed as is the importance of 'giving ergonomics away'.

Introduction

Simple ergonomic changes can have a major impact on the quality of working life. This can be achieved while providing the company with a competitive advantage. The two are not inconsistent. The following two case studies demonstrate this point.

Case Study 1 : Deshiving

Deshiving is a traditional process where the wooden discs (called shives or bungs) are removed from the side of aluminium containers (see figure 1). The shives are removed manually with a hammer and chisel. A container arrives to the operators right via a floor level conveyor belt (see figure 2, position 1). The operator pulls the container off the belt onto a metal platform (see figure 2, position 2). The operator uses the hammer and chisel to knock out the shive. The operator is in a fully stooped posture at this time and often the knee is hyperextended to place body weight on the container to prevent the container shifting. Once the shive is removed the container is tilted and rolled to a second conveyor behind the operator (See figure 2, position 3). The second conveyor leads to the washer which is the next step in the production process. There are three sizes of container 11 gallons, 22 gallons and 36 gallons.

Figure 1. Photograph of deshiving process before.

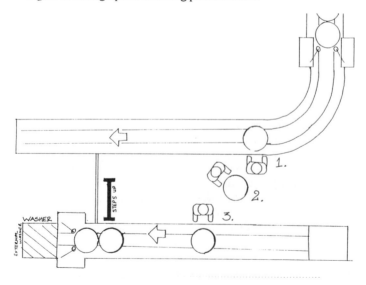

Figure 2. Plan drawing of deshiving process before.

Table 1. Benefits of ergonomic intervention with a Quality Team, Deshiving process.

---

* Production improved 71% : *from 210 shives per hour to 360 shives per hour.*
* More efficient shive removal : *average of 3-6 blows to remove shive reduced to average of 2-3 blows.*
* Container damage reduced: *due to improved posture and line of sight.*
* Shive damage to belt reduced: *shives getting caught in conveyor belts snapping them. The new design prevented this.*
  The cost of belt repairs 4 months before = £ 24 000
  The cost of belt repairs 4 months after changes = £ 500
* Slipping and tripping hazard: *shives on the metal platform not only caught in the belts they were often stepped or slipped on. This was reduced due to the new design and waste management system.*
* Injury related sickness absence reduced: *sickness absence as a direct result of an injury caused by the workplace for this team of operators.*
  12 months prior to any change = 54 days (estimated cost £5832).
  4 months after change = 0 days (possible Hawthorn Effect is accepted).
* Posture improved: *poor posture improved, stooping eliminated and static loading of back reduced, hyperextention of knee eliminated, line of sight greatly improved, double handling of containers eliminated (the gravity fed rollers require a light push only).*
* Tool tidy & sharpening procedures implemented.
* Indirect savings not costed: *e.g. improved productivity, reduced industrial claims management time to investigate accidents etc.*
* Cost of changes = £5000.
* Direct savings (prevention of belt breakdowns) = £72 000 per annum.

Case Study 2 : Retail Technical Services Vehicles

Technical services fitters are concerned with getting beer from the containers in the pub cellar to the customers glass. The pipework , cooling arrangements, control of nitrogen and carbon dioxide and the general quality of the beer are their responsibility. Beer is kept at the correct temperature by coolers. The coolers are fitted in pub cellars which are often underground. A common task for the fitters is to install or remove coolers. The cooler dry weight is 75kg although the wet or working weight is 140kg. The handles on the cooler prevent full access by an adult male hand and the positioning of the handles is poor. Figure 4 shows the removal of a line cooler from an estate car, the company vehicle asigned to this task. The size and weight of the cooler is a contributory factor as is the lip of the boot on the estate car and the vehicle boot height and width. The largest and heaviest type of cooler is shown. The task was potentially hazardous and unsatisfactory.

An ergonomic study concluded that increased space in the vehicle, mechanical assistance, and an improved cooler design would reduce the risks and increase the efficiency of the task. Several areas of the company trialled a new vehicle type for 12 months. Figure 5 shows the new vehicle type. The vehicle has increased headroom and space, the deck height is lower, and there is no lip to the edge of the boot. A Lucas swing lift was incorporated into the vehicle design (this can be seen on the left hand side of the picture).

When empty they weigh 10 kg, 20 kg, and 28 kg respectively. Occasionally the containers will contain waste beer (called ullage) so they may be heavier than their minimum weight. This was an unacceptable practice which resulted in musculoskeletal complaints (e.g. backache shoulder pain and hyperextention of the knee joints), damage to conveyors and containers, and poor productivity.

A quality team consisting of the operators, their team leader and the company ergonomist were required to reduce the number of musculoskeletal complaints being made and speed up the process to prevent holding up the rest of the production line. Using simple ergonomic principles an alternative layout and working procedure were explored (Figure 3). A mock up of the proposed workstation was evaluated by the team prior to construction of the new design. Two gravity fed rollers were introduced to link the two existing conveyors. This reduced manual handling and allowed the operators to work at a more optimum height. Space restricted the use of a sloping or height adjustable conveyor. A compromise of a fixed height conveyor (of 520mm high from the floor) resulted from anthropometric tables and trials on the mock up workstation. It was more cost effective to lower the platform the operator stood on than raise all the flat conveyors to the desired height. This made room for a waste management system (a skip for the shives!) and restricted an unofficial walkway. The benefits of the ergonomic intervention are listed in table one.

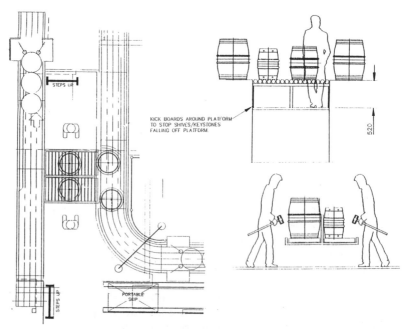

Figure 3. Plan, front and side elevations of deshiving process after redesign.

This was one of the first ergonomic projects undertaken at Scottish and Newcastle. It demonstrates the compatibility of participatory ergonomics and quality initiatives. The operators submitted the project and won the 1994 Scottish and Newcastle plc Quality Award.

Figure 4. Line cooler removal from an estate car.

Figure 5. Line cooler removal from a purpose designed van.

Table 2. Benefits of Ergonomic Intervention with Technical Services Vehicles.

---

* Almost eliminates manual handling.
* A safe systems of work was made possible.
* Reduced breakages to tools/stock. *New design incorporates storage shelving*
* Increased operator safety during transportation. *Safety grill fitted behind cab*
* Improved Quality of Service *due to increased room in the vehicle the operators can carry more consumables, stock and tools therefore:-* Store visits down 50%, mileage down, stock lines carried increased 12%, overtime down 300 hours saving £12,000.
* Reduced personal taxation on salaries for van.
* Indirect savings not costed *e.g. reduced absence, improved productivity, etc.*
* Total saving for pilot period for 7 vans for 12 months = £74, 000 per annum.
* When implemented estimated overall company saving of £2.4 million.

Discussion

The case studies show that the application of simple ergonomic principles can provide solutions that result in substantial and tangible savings. If ergonomics can be of such benefit why are there so few industrial ergonomists? The survival of ergonomics depends on the belief that ergonomics not only works but that it can be applied daily in the workplace by non ergonomists. Simpson (1993) summarises the position of ergonomics very clearly :-

"The potential contribution for ergonomics in industry remains considerable. In Europe in particular, this has been emphasised by the specific inclusion of the need to consider ergonomics in almost all of the recent EC Directives on health & safety at work.. If however ergonomists are to capitalise on this potential then they must re-consider their interaction with industry. Industry can manage to "limp along" without ergonomics as it has done for many years. However, as ergonomics is by its very nature an applied discipline, it is doubtful whether it can continue to exist in any meaningful way for much longer without a closer, more routine involvement in the day-to-day operations in industry."

One way to bridge the gap between ergonomic academia and industry is to do as Corlett (1991) suggests, "give ergonomics away". The success of the deshiving project was due to the team understanding the simple ergonomic principles being applied. By giving ergonomics away the team can then apply this knowledge to other situations. The teams propose the solutions so there is greater acceptance and less resistence to implementation. The team go on to sell their ideas to their colleagues and managers. Selling ergonomics is easy since they believe in ergonomics.

References

Corlett, E. N., 1991 Some future Directions for Ergonomics. In M. Kumashiro & E. D. Megaw (ed.), *Towards Human Work. Solutions to Problems in Occupational Health and Safety,* (Taylor & Francis, London) 417-421.

Simpson, G. C. 1993, Applying Ergonomics in Industry : Some Lessons From the Mining Industry. In E. J. Lovesay (ed.), *Proceedings of the Ergonomics Society's 1993 Annual Conference,* (Taylor & Francis, London) 490-503.

THE INTER-RELATIONSHIP OF PHYSIOTHERAPY AND ERGONOMICS : STANDARDS AND SCOPE OF PRACTICE.

Sue Hignett
Ergonomist, Nottingham City Hospital NHS Trust

Emma Crumpton
Lecturer, Faculty of Health and Social Care, University of the West of England

Lynn McAtamney
Consultant Ergonomist, COPE, Nottingham

This paper explores the relationship between the professions of ergonomics and physiotherapy in their complimentary roles of optimising human performance and well-being. The relationship within the area of physical ergonomics is explored using a model which has 'functional capacity' as the centre point leading through to 'ergonomic design and prevention', and 'rehabilitation and therapeutic care'. Concern is voiced over the scope of practice and the need for continued development of standards and recognition of core competencies to clearly define the boundary between the two professions. In conclusion it is believed that there is a considerable potential benefit from a closer relationship between ergonomics and physiotherapy, but in order for this to be achieved there must be mutual recognition and respect for the specialist skills of each profession.

Ergonomics

Ergonomics is both a science and a technology using research, knowledge and skills from a range of disciplines (Osborne, 1995). Scientific information about human beings (and scientific methods of acquiring such information) are applied to the problems of design, with the aim of defining the limits of human adaptability and diversity. Ergonomics is a relatively young profession, and as with many professions it is adapting to the changing world by modifying and extending it's scope of practice. Since it's initial inception ergonomics has grown from use in military engineering, through to space applications in the 1950's and onto a wider industrial usage in the following thirty years (pharmaceuticals, computers, cars and other consumer products). In the 1990's ergonomics is much more general with applications in commerce as well as industry. Health and safety legislation has made ergonomic input into most work environments a legal requirement ensuring that this particular role of the ergonomist is a rapidly growing one. This is where physiotherapy has scope for valuable input. One possible way of categorising the scope of practice of ergonomics is:

1. **Physical ergonomics**. This includes physiology, anatomy, and biomechanics providing information on physical capabilities and limitations, as well as physical dimensions (anthropometry) It also encompasses input from physics and engineering, looking at the design of tools, products, equipment and work environments (lighting, heat and noise).
2. **Cognitive ergonomics**. This category includes physiological psychology looking at the nervous system and behaviour, as well as experimental psychology providing information about cognitive functions e.g. perception, learning, memory etc.

3. **Organisational ergonomics**. The organisational structure, processes of work (including job design and analysis) and attitude and behaviour of people in the organisation will all effect the effectiveness and efficiency of an organisation.

Ergonomics has a huge scope of practice, and members of the profession come from a variety of background disciplines. In the USA most of the members of 'The Human Factors Society' have a background in psychology (45%) and engineering (19%) with only 3% coming from the medical field (Sanders and McCormick, 1993). This is believed to be a similar proportion to 'The Ergonomics Society' in the UK although the definitive figures are not known. This is in contrast to elsewhere in the world, for example, Norway, where most of the ergonomics practised in industrial settings is by Ergonomists with a background in physiotherapy (Bullock, 1990).

One of the drawbacks of having such a range of professionals, from many different backgrounds and with such a variety of skills and interests, is establishing an agreed level of competency. Bullock, (1994) advocated the definition of core ergonomic competencies (Table 1). She suggested that among professionals practising ergonomics a common basic foundation of knowledge was required to ensure an adequate level of communication which would support the need for Ergonomists to appreciate their own limitations and facilitate assistance being sought when required.

Table 1. Core Competencies for Ergonomists

---

1. Demonstrate an understanding of the theoretical basis for assessment of the workplace
2. Demonstrate an understanding of the systems approach to human integrated design
3. Demonstrate an understanding of the concepts and principles of computer modelling and simulation
4. Communicate effectively with clients and professional colleagues
5. Obtain information relevant to ergonomics from the client
6. Collect supplementary information relevant to ergonomics relating to the history of the problem and current management
7. Collect from the work place, in an appropriate manner, quantitative and qualitative data relevant to the perceived problem and ergonomics
8. Document ergonomic assessment findings
9. Recognise the scope of ergonomic assessments

The professional backgrounds and specialist skills may reflect in the approach or area of interest of the Ergonomist. A systems engineer may feel that the ability to carry through the design, development and implementation of a specific work station, tool or piece of equipment is of particular importance. Whereas, a physiotherapist may be more concerned with a specific musculoskeletal risk factor and modifications to the work station, task or tool to eliminate or reduce the risk of injury. The specialist skills of the Occupational Health Physiotherapist will be explored with a view to how they can be used to complement ergonomics practice.

Physiotherapy

Jacobs and Bettencourt (1995) suggest that the fundamental intent in the early days of physiotherapy was to *'assess, prevent and treat movement dysfunction and physical disability, with the overall goal of enhancing human movement and function'*. As with ergonomics the scope of physiotherapy has extended well beyond this initial intent to include, for example, health promotion, fitness training, incontinence treatments etc., and also into the field of occupational health.

The Chartered Society of Physiotherapy (UK) has produced a leaflet (Physiotherapy and Occupational Health) in which they outline the range of skills of a chartered Occupational Health Physiotherapist. (Table 2)

Table 2. Skills of an Occupational Health Physiotherapist

Expert in human movement
Advice to prevent further injury and helping to speed recovery
Evaluation of human task machine relationships and identification of problem areas which could cause pain
Specialised training in ergonomics and occupational health etc.

It is in this area of occupational health that the Physiotherapist has much to offer ergonomics. Physiotherapists are taught to be analytical about injury mechanism in order to be able to apply an appropriate treatment technique to alleviate symptoms. Therefore Physiotherapists have a specialist knowledge and the potential for insight into the scientific study of human work.

A Physiotherapists raison d'être is to restore health and well being. However there is little to be gained by successfully treating someone's back problem or Achilles tendonitis if they are then going back to the same faulty work station and gruelling lifting task, or the same pair of running shoes that led to the problem in the first place. They will continue to have recurrence of their problem until it becomes chronic. The successful resolution of work related musculoskeletal problems requires the elimination of those factors which lead to the continued over use of the bodily structures involved. Ergonomics is relevant to primary prevention and also to those branches of medicine which are concerned with the management of such conditions. Therefore ergonomics and physiotherapy can work together very successfully in order to give a complete approach to injury treatment and prevention resulting in the enhancement of performance.

Physical Ergonomics

Bullock (1994) proposed a model (outlined in Figure 1) which aimed to optimise human performance by maximising the expertise and input from both the physiotherapy and ergonomics professions in a combination of rehabilitation and therapeutic care together with design and prevention.

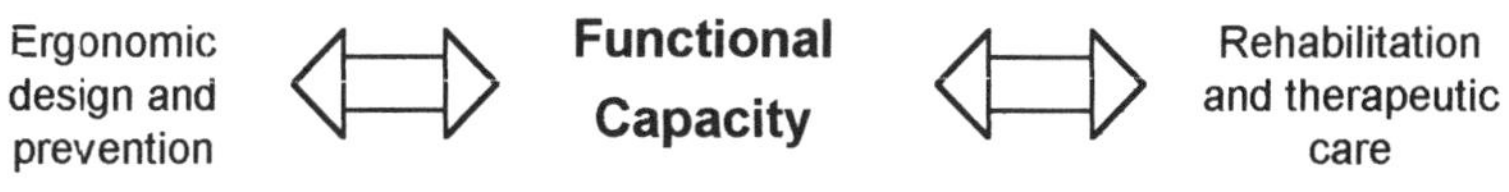

Ergonomic design and prevention ⟷ **Functional Capacity** ⟷ Rehabilitation and therapeutic care

Figure 1. Optimisation of human physical performance (modified from Bullock, 1994)

Functional Capacity

This is the ability of a client to carry out his / her job. A functional capacity evaluation can be carried out in order to assess and quantify this by relating the limitations of the individual to the demands of the job.

The scope of the Occupational Health Physiotherapist in the U K has tended to be limited to an advisory role with respect to general fitness criteria for return to work after a major sickness absence (Oldham, 1988) In the USA this advisory role has expanded into almost a new profession, with specialist skills, based in the application of work physiology and psychosocial research under the title Industrial Therapy. Industrial therapy is a system which encompasses a wide spectrum of treatment based upon the principles that a person working in industry, as anywhere else, has physical, emotional, vocational, educational, psychological and sociological needs which must be met to gain successful employment or, in the case of injury, rehabilitation and re-employment. This specialisation developed when it was noted that Therapists were moving away from the traditional role of treating injured workers in clinics and starting to provide services including injury prevention and rehabilitation as well as pursuing advanced work in related fields e.g. engineering or ergonomics providing functional capacity evaluations, work conditioning, work hardening within the practice of physical therapy, occupational therapy and certified vocational evaluators (as part of the physical or occupational therapy team).

Work Conditioning is the term used to describe a treatment programme which is specifically designed to restore an individual's systemic, neuromuscular and cardiopulmonary functions. It is used for clients who are more than six weeks post-injury and aims to restore physical capacity and function for return-to-work (Hart et al, 1994).

Work Hardening is the term used to describe a treatment programme designed to return the person to work. It uses real or simulated work activities and is suitable for clients who are more than twelve weeks post-injury and are physically deconditioned but medically stable (Hart et al, 1994).

Ergonomic Design and Prevention

Historically ergonomics has aimed at establishing working capacities of healthy individuals, and enhancing performance by using this data when designing tasks and work places to fit the users or workers. This contributes to functional capacity with a permanent engineering solution to work place problems.

Rehabilitation and therapeutic care

The role of physical ergonomics in physiotherapy practice allows a rounded and holistic view of patient care. Physiotherapists with skills and knowledge of ergonomics can not only address the direct therapeutic needs of their patients, but also contribute to the ergonomic design aspects of primary prevention. The following are examples of physiotherapists' involvement in physical ergonomics and illustrate the overlap between physiotherapy and ergonomics.

The first example is the advice that Physiotherapists give to their patients when they are returning to a job or leisure activity following treatment. Unfortunately patients symptoms are all too often addressed with little or no regard to their aetiology, despite the fact that epidemiological data suggests that a very high proportion of musculoskeletal problems are work-related. Often the patients, and therefore the physiotherapists' highest priority is to return to work as quickly as possible. This will frequently involve the alleviation of symptoms only to send the patient back to the very environment and activity that caused the problem in the first place. Primary prevention is obviously the key to breaking this cycle of recurrence, and this is where the physiotherapist may get involved in work place and work station assessments, using ergonomic techniques to recommend improvements to eliminate or minimise the risk of injury. In order to carry out ergonomic assessments there is an onus on the physiotherapist to ensure that they are adequately qualified, both in qualifications and experience, to be able to use and interpret the ergonomic tools. There is a need to recognise personal limitations and ask for expert support when required. Gaining access to a patients work place is in reality unlikely in most out-patient settings, but there is still much that a physiotherapist with the necessary knowledge and skills can do by questioning the patient, assessing the situation and contributory factors. A reasonable appraisal can lead to constructive, practical advice as to which of the factors are under the patients control and can be modified or eliminated.

The second example is the role of the Physiotherapist as an in-house manual handling trainer. There has been a tendency, particularly within the Health Care Industry, to use Physiotherapists as a pool of in-house experts but without always providing the appropriate ergonomics training to enable them to deliver the service required. Frost and McCay (1990) described a cascade system which was used in an NHS Hospital Trust to provide lifting and handling training, which was quoted as including ergonomics, biomechanics, teaching methods and attitudes. They used the hospital Physiotherapists as part of 'The Prevention of Injury Group' and found that one problem was the lack of formal ergonomics training within the group (although some members had limited knowledge). Buckle and Stubbs (1989) have also commented on the tendency within the Health Care Industry to use in-house Physiotherapists, they suggest that the ergonomic input is often left to professional groups that have not received formal training in this subject. In 1995 Crumpton carried out an extensive questionnaire survey of Back Care Advisors in the UK. She found that of the Physiotherapists employed in this role, 57% claimed to have had some training in ergonomics, but for 75% of these the training was limited to 1-3 day courses. She concluded that a greater awareness of the scope of ergonomics practice was needed in order to ensure that personal limitations were recognised.

This trend is of obvious concern to all involved. The vital contribution that ergonomics has to make to this important and growing area could be generally disregarded as ergonomic principles are misused by inexperienced staff and ergonomics is not seen to be doing what it claims to be able to. Also, Physiotherapists are being put into roles which are beyond their capabilities and they are expected to solve complex problems with no extra training by virtue of the fact that they have studied anatomy and biomechanics. This is not good for the physiotherapy profession as the inevitable result is that the problems are not solved efficiently.

The third example is the risk assessment of physiotherapy practice. There have been numerous articles published which have challenged the research basis for many of the techniques used by Physiotherapists. With health and safety legislation in the European Community now focusing on the risk of injury from manual handling activities, the justification for many of the physiotherapy techniques is again being called into question when the safety and well-being of Physiotherapists may be compromised. Ergonomics may have a role to play, using assessment techniques to evaluate the risk to the physiotherapy population (Hignett, 1995) and in the redesign of the work place and working environment (Fenety and Kumar, 1992).

Conclusion

Foster (1988) suggests that the professions of physiotherapy and ergonomics are divergent and convergent. Divergent in that the scope of ergonomics covers all aspects of peoples' interaction with their environment (physical, cognitive and organisational), whereas physiotherapy is concerned only with the physical well-being of an individual. Convergent in that both aim to optimise human performance and minimise any mis-match in the task requirements and physical ability. However physiotherapy achieves this aim by altering the person (e.g. work conditioning or musculoskeletal treatment), and in contrast ergonomics alters the task to bring it to an appropriate level for the person (including designing for disabled populations). Stubbs (private communication, 1996) concurs with the idea that physiotherapy has much to offer with respect to physical ergonomics, but comments that Physiotherapists tend to focus more on characteristics, capabilities and capacities of the individual rather than of populations. Hence they have what could be considered to be a more narrow perspective. Ergonomics knowledge could serve to broaden the outlook of the Physiotherapist, and ensure a more holistic approach to the patients problem by taking action on the environment as well as the individual

Physiotherapists are experts on the human musculoskeletal system, including work physiology, the effects of loads and forces and the prevention as well as the treatment of musculoskeletal disorders. McAtamney (1991) suggests that they are well qualified to provide advice on aspects of physical ergonomics. McPhee (1984) believes that many physiotherapy skills are wasted if only treatment services are provided, and that worker rehabilitation should form a large part of the treatment. She states that if Physiotherapists are to play a part in occupational health and ergonomics and develop the relevant skills they must first identify the areas in which their expertise can be used most appropriately. This type of role has already been seen in other countries, in particular in Scandinavia where ergonomics is a profession dominated by Physiotherapists (Bullock. 1990). The preventive role includes job analysis, work posture monitoring, task design, personnel selection and placement, education, supervision of work methods, influencing motivation and attitudes, provision of activity breaks and physical fitness programmes. In order to provide this range of occupational health and ergonomic services Physiotherapists need to obtain post graduate qualifications in occupational health and ergonomics and maximise their unique knowledge of the human musculoskeletal system.

There is a huge potential benefit from a closer relationship between Ergonomics and Physiotherapy, but in order for this to be achieved there must be mutual recognition and respect for the specialist skills of each profession.

References

Buckle, P. and Stubbs, D. 1989, The Contribution of Ergonomics to the Rehabilitation of Back Pain *Journal of the Society of Occupational Medicine* **39** 56-60

Bullock, M. I. 1990, *Ergonomics: The Physiotherapist in the workplace.* (Churchill Livingstone)

Bullock, M. 1994, Research to Optimise Human Performance *Australian Journal of Physiotherapy* 40th Jubilee issue 5-17

Crumpton, E. J. 1995, *An Investigation into the Role of the Back Care Advisor.* Unpublished M.Sc. Dissertation, University of London

Fenety, A.,and Kumar, S. 1992, An Ergonomic Survey of a Hospital Physical Therapy Department *International Journal of Industrial Ergonomics* **9** 161-170

Foster, M. 1988, Ergonomics and the Physiotherapist *Physiotherapy* **74** 9 484-489

Frost, H, and McCay, G. 1990, Prevention of Injury Group Initiative *Physiotherapy* **76** 12 796-98

Hart, D., Berlin, S., Brager, P., Caruso, M., Hejduk, J., Hoular, J., Snyder, K., Susi, J., Wah, M. D. 1994, Development of Industrial Standards in Industrial Rehabilitation *JOSPT* **19** 5 232-241

Hignett, S. 1995, Fitting the Work to the Physiotherapist *Physiotherapy* **81** 9 549-552

Jacobs, K, and Bettencourt, C. M. (eds) 1995, *Ergonomics for Therapists* (Butterworth-Heinemann)

McAtamney, L. 1991, Physiotherapy in Industry in Lovesey E J (ed) *Contemporary Ergonomics* (Taylor and Francis)

McPhee, B. 1984, Training and possible future trends in Occupational Physiotherapy *New Zealand Journal of Physiotherapy* Dec 12-14

Oborne, D. J. 1995, *Ergonomics at Work* 3rd Ed (John Wiley & Sons)

Oldham, G. 1988, The Occupational Health Physiotherapist's role in assessing fitness for work *Physiotherapy* **74** 9 422-425

Sanders, M. S. and McCormick, E.J. 1993, *Human Factors in Engineering and Industry* 7th Ed. (McGraw-Hill Inc.)

ERGONOMICS IN RENAISSANCE ART

Dr Kevin Tesh

Human Sciences Section, Institute of Occupational Medicine, 8 Roxburgh Place, Edinburgh EH8 9SU

In this paper, a story of working practices during the Renaissance art period will be told, demonstrating that artists were applying broad ergonomic principles during their work. These practices were employed not only in their art work (painting, sculpture and architecture) but also in their workshops and with their art tools and equipment.

Introduction

Although first coined in the 1950's, the systematic scientific study we now call ergonomics began, so they say, some 35 years earlier during the first world war. A study of working practices and equipment used during the Renaissance and art period reveals that artist and architects took ergonomic principles in their stride as custom dictated this approach. In face these practices long antedated the Renaissance period but only began to come to light during this period due to more frequent writing down of such information, which became customary after the general spread of literacy.

The science of ergonomics covers many topics and to provide a framework to ergonomic examples used during this period of art the paper will start with the person-machine interface, the main building block of ergonomics, and then move out onto other factors such as work equipment, environment and organisational conditions.

Person-machine

Anthropometric databases are invaluable tools which ergonomists use to ensure initially that the maximum numbers of users are provided with sufficient room to carry out their tasks. In a similar manner Leonardo de Vinci held a physiognomic database (facial features) that he used to fit to faces of his characters in order to emphasise the subject and meaning of the painting. The database consisted of heads, noses, mouths, lips, chins, throats, necks and shoulders. For example, with the nose sketches he had ten types in profile and twelve from the front. This database was first used in his Last Supper painting when depicting the twelve apostles. Sufficient data was available for eleven of the apostles but a special search was conducted for the facial features of Judas, the villainous betrayer.

This lack of data was one of the reasons why the fresco took longer to complete than expected.

One wonders whether Michelangelo had a similar database at his disposal when carrying out workplace fitting trials during his design of the reading room in the Laurentian Library at San Lorenzo in Florence. The sketch shows an average Florentine user (stature 1.52m or 5ft at the end of the 15th century) sitting at his workplace, with full backrest support, angled reading surface and suitable viewing distance.

Michelangelo was concerned with sitting and comfort levels as can be seen in his Sistine chapel vault fresco painting. The later painted nudes were provided with cushions for comfort but also to provide some protection from the cold marble podiums they were sitting on.

Work Equipment

Correct working heights were very much on Leonardo's mind when designing his workshop. A system of pulleys and weights were used to raise and to lower the working platform (cassa) on which the work was placed 'so that it would be the work, not the master, that would move up and down'. There were other advantages for this elaborate system, of which no models or sketches exist, including to hide the work away at night from the inquisitive eye; to prevent dust falling on the work; and to maintain a clear thoroughfare, reducing the chance of accidents.

A mahlstick, a long stick with a padded end still used in painting today was originally used to keep cuffs and frills out of the paint and off the panel. Illustrations of the tool in use also show that the tool was used as a wrist-rest that could be lent against the picture so that the painter could support and steady his hand when painting fine details.

Quill pens used for preparatory sketches were selected to suit the handedness of the user. Due to their curvature, goose feathers were selected from the right and left hand side of the bird to suit both right and left handed users respectively so that the arc of the feather fitted the arc of the hand.

Environmental conditions

Although the use of natural light rays to illuminate altars and spiritual figures pre-dates the Renaissance period Michelangelo used a method to angle the light down onto his sculptural work during his work on the San Lorenzo sacristy in Florence. In the design of the windows, the opening on the outside was located higher than that on the inside. This feature was built into the architectural plans at the drawing stage to ensure that the natural light was beamed in diagonally downwards onto his tombs of Giuliano and Lorenzo de'Medici.

Leonardo also recognised the adverse effects of too much light by providing large windows with adjustable blinds, which could be raised or lowered so as to throw the desired amount of light on the work or subject at the right height.

Organisational conditions

It is well recognised that an unreasonable pace of work can lead to health and safety problems. On painting the Sistine chapel Michelangelo used a fresco painting technique where the paint was applied to fresh wet plaster. The paint binds with the plaster to become part of the wall. This established technique required much vigour and promptness in decision making which was suited to Michelangelo's style. On the otherhand, Leonardo embarked upon a new and untried technique, using paint pigments mixed with molten wax on a double layer of dried plaster, first used on his Last Supper painting. Although Leonardo's primary reason for resorting to this technique was to recapture the Egyptian painting techniques used on the Fayum portraits in 4th century AD, another reason could be that this technique allowed a slower pace of work, with touch-ups and re-working and time to think, which better suited his pace on this work.

Another example of organisational consideration involves the criteria used by Michelangelo for the selection of plans drawn up by a variety of designers for the dome of St. Peter's in Rome. As well as considering the costs to build the style of the dome, and whether the building could be built to support the massive dome structure, he also had the forethought to consider the manning levels required for security. A rival design by Bramanté was dismissed as it had too many dark hiding places, was not easy to follow inside and would require up to twenty-five patrol staff to search out unwanted visitors.

Supervisory support and job satisfaction were two areas within a work environment that needed to be in sufficient measures to produce good work performance. Filipo Brunelleschi, the design of the Maria del Fieve dome in Florence was having these problems with his sculptors during its building. In order to improve staff morale and provide the workers with a place to rest and eat he built a canteen in the dome. There were other practical reasons for installing this eating place and this was to prevent 80 to 100 workmen scrambling up and down the scaffolding which was not only time consuming but also extremely dangerous.

Conclusion

This paper has drawn together, from a few artists of the Renaissance art period, examples of ergonomic principles being applied in their art work and working practices. Discussions with art historians seems to indicate that the artists were not aware of these ergonomic principles but custom at that time dictated this approach.

THE ROLE OF ERGONOMICS IN DEVELOPMENT AID PROGRAMMES

T Jafry and D H O'Neill

International Development Group,
Silsoe Research Institute,
Silsoe,
BEDFORD, MK45 4HS

In most development aid projects, economic and social issues are accorded paramount importance. Human-technical issues are often not addressed, or if they are, it is done implicitly rather than explicitly. Incorporating ergonomics into development aid projects can enhance project outcomes by providing greater benefits to poor people. These benefits can be achieved by improving working conditions and designing equipment to be better matched to human characteristics and capabilities. Through a process of creating awareness and providing training to field staff, project managers and other aid professionals, the human-technology gap can be bridged. This paper describes how we tackled the problem of incorporating ergonomics into development aid programmes and associated policy issues.

Introduction

There is enough evidence to prove that occupational health and ergonomics are two of the major determinants of the state of world health. Forty to fifty percent of the world's population are exposed to work-related health risks (Mikheev 1995), so these work-related health risks need to be properly addressed. However, a lack of awareness of the consequences of; work related diseases results in a total economic loss of up to 10-15% of GNP (Mikheev 1995) and what to do about this remains a problem. We have taken up the challenge of showing how ergonomics can provide benefits by improving the health of people in developing countries, in order to raise the priority of ergonomics activities within the British Overseas Development Administration (ODA) Aid Programme. At present there are few practical examples to illustrate the benefits that can come fromergonomics. In order to determine whether ergonomics should be a matter of policy within ODA, a three stage work plan was developed. The first stage was to establish what ODA Advisers understood by the term ergonomics. Providing practical examples to illustrate how ergonomics can provide benefits to ODA was the second stage. Thirdly, the views of other organisations with an interest in ergonomics, such as the International Labour Organisation, World Health Organisation and the Food and Agricultural Organisation were elicited.

Creating Awareness

Informal Interviews with ODA Advisers

Informal discussions were held to assess both the general level of awareness of ergonomics amongst relevant ODA staff and the contribution they perceived ergonomics could make to development aid. This was done using the semi-structured interview technique. The questions put to them followed a standard sequence in order to get a consistent approach with all Advisers. A template of questions was formulated for this but was not actually referred to during the interviews. A total of fifteen advisers was contacted of whom fourteen agreed to be interviewed and one agreed to be interviewed by telephone. A list of the questions asked, together with the responses, are presented in the results section. Prior to the interviews an ergonomics briefing note was circulated which gave a brief explanation of ergonomics and its application.

ODA Project Screening

In order to illustrate how ergonomics could benefit development aid projects, it was recommended that we work in close collaboration with three existing ODA projects. We therefore had to select three projects with which we could become involved, and, by adding an ergonomics component, provide "case study" information on the relevance and benefits of ergonomics. Three hundred projects from ODA's natural resources project data base were screened. During the initial screening, all projects completely unrelated to ergonomics were eliminated. During the second screening, all projects were scored for their likely ergonomics content using a checklist. Finally, during the third screening, the remaining eight highest scoring projects were scrutinised and discussed with the project managers; from this three projects were selected for ergonomics intervention.

The Views of Other Organisations with an Interest in Ergonomics

The views of other international organisations, International Labour Office (ILO), World Health Organisation (WHO) and the Food and Agricultural Organisation (FAO), who currently take an active interest in ergonomics for developing countries, were obtained by making short visits to these organisations. Key persons within these organisations were interviewed to determine their views on the role of ergonomics in development aid.

Results

Informal Interviews with ODA Advisers

The results of the interviews with the ODA Advisers (UK based) are summarised and given in Tables 1a and 1b.

Table 1a Summary of interviews with ODA Advisers (UK Based)

Question	Response	Specific comments
Have you heard of ergonomics?	All 15 Advisers had heard of ergonomics.	
What do you think ergonomics means?	There was a variety of responses to the question. Some gave more than one response: Very knowledgeable about ergonomics 1 Did not know much about it 1 Time & motion study 2 Workplace design 7 Tool & equipment design 3 Work on tasks with a repetitive nature 2 Physical workload 1 Improving efficiency/reducing drudgery 4 Manual handling 1	
Have you read the Ergonomics Briefing Note?	Yes 11 No 4	
Did you perception of ergonomics change after reading the Briefing Note? (this question put to the 11 that had read the Briefing Note)	Didn't really 2 No 5 Yes 3 No comment 1	did not realise ergonomics could deal with * tool and equipment design * accidents and injuries
Can you see a need for ergonomics in your Programme/Projects?	Yes 11 No 3 Other: Not in the business of looking out for projects.	
What contribution do you think ergonomics can make?	Tool and equipment design 12 Women and technology 9 Technology transfer 4 Work organisation & work Place design 5 Others: improve working conditions 1	

Question	Response	Specific comments
Do you envisage difficulties in trying to implement ergonomics in ODA Programmes/Projects?	No 7 Yes 8	**No problem provided that:** 1. The implementation of ergonomics is clear, using thought-out practical applications. 2. Ergonomics fits appropriately. 3. Ergonomics is integrated well into community problems and financial arrangements. 4. There is a need for ergonomics. 5. People are involved and the real problems are addressed. 6. Ergonomists work closely with the projects. 7. No specific comment. **Yes envisaged problems because:** 1. Ergonomics cannot so easily be implemented in the fisheries projects (2 Advisers commented on this). 2. Ergonomics should be considered for technical co-operation link not for research. 3. New ergonomically designed equipment costs money. People couldn't afford it. 4. Changing working practices would be quite difficult. 5. Ergonomics is not a priority in businesses. Credit more important. 6. People trying to survive. People need credit not ergonomics. 7. Questioned the importance of ergonomics relative to other subjects. 8. No comment.
Would you like to know more about ergonomics and its application?	Yes 12 No 1 No comment 2	Would like to attend seminar/workshop 11 Specifically to: * See well-argued case * For clarity * See benefits ergonomics can bring * See good demonstrations. Would prefer literature 1

Table 1b Other comments by Advisers (UK-based)

* Would like to see examples of ergonomics in future ergonomics concept notes.
* People lack awareness of ergonomics. Need to sell it better and increase awareness.
* Need to educate Non Government Organisations on ergonomics.
* ODA addresses ergonomics issues sub-consciously
* Change will become technology-driven. Need to consider the financial limits of change at the level of the individual.
* People would not do a task if it was not productive or efficient for them to do so.
* Would not like to see another checklist.
* Need to define ergonomics better so it stays in people's minds.
* Country specific exploratory study of ergonomics would be useful.
* Ergonomics deals with issues at ground level. ODA has only the capacity for bigger issues e.g economics and empowerment.
* Would welcome any science that reduces drudgery, improves efficiency, releases time to do other things.

Half of the Advisers stated that they did not envisage any difficulties in trying to implement ergonomics in ODA Programmes/Projects. The concerns raised by the other half of Advisers could be alleviated by providing examples on the practical application of ergonomics. It is interesting to note that approximately 75% of Advisers expressed an interest to attend a seminar/workshop to learn more about ergonomics and its application.

Project Selection

The three ODA projects selected, where ergonomics intervention could contribute to the project outcomes, were:

* East India Rainfed Farming Project (looking at farming systems in the Rainfed states of Bihar, Orissa and West Bengal).
* Community Participation in Forestry in the Caribbean; ergonomics and charcoal production in St Lucia.
* National Agricultural Research Project in Ghana; ergonomics and the processing of cassava into gari at a women's' cooperative.

For each of the projects, areas were identified where ergonomics could bridge the people technology gap. However, prior to any interventions, the project staff needed to be aware of how ergonomics could benefit their project. Several activities were undertaken to create this awareness.

India
* Ergonomics orientation workshop conducted with Indian ergonomics consultants.
* Training on how to conduct ergonomics survey.
* Recruitment campaign to identify ergonomist to join the project team.

Ghana
* Ergonomics seminar held for agricultural engineers at the University of Science and Technology in Kumasi.
* Discussions held with the women's food processing cooperative to identify problems and priorities.

West Indies
* Field visits to the charcoal producers to discuss problems associated with the production of charcoal.

- Conduct survey and collect work study data.
- Prioritisation of options for improvements.

Visits to ILO, WHO and FAO

All the visits had a positive outcome. Each of organisation stated that ergonomics could contribute to development aid. There were some common underlying issues aired by all three organisations as to why ergonomics is not given more recognition. These are:

* There is a lack of **awareness** of the benefits ergonomics can bring.
* There are not enough people trained in ergonomics so ergonomics problems may not be described as such.
* There are not enough data on the incidence of occupational health problems in developing countries.
* There is a need for ergonomics to bridge the human-technology gap but it must be considered as part of an overall strategy within an organisation.

Discussion

Informal discussions with the ODA Advisers revealed that they have some knowledge of the subject of ergonomics but the full extent of its application was not appreciated. In general, this lack of awareness and the request by Advisers to know more about ergonomics justifies the need to demonstrate how ergonomics can benefit development projects. Comments made by ODA Advisers have highlighted that there is a need and demand for ergonomics to be considered in development projects.

Involvement with the three ODA projects created an opportunity to provide examples of where ergonomics could benefit aid projects. The fact that all three selected project managers requested workshops and training on how to incorporate ergonomics into their projects endorses the view that *a lack of awareness* is the reason why ergonomics does not receive more recognition.

The recent experiences gained from this study show that it is important to provide practical examples of the benefits ergonomics can provide. An increase in the number of case studies is still needed to make a greater impact.

The ILO and the WHO both have strategies that recognise the importance of ergonomics. These are *Occupational Health and Safety* and *Occupational Health for All - a Global Strategy* respectively. Perhaps in the future the ODA will also have a strategy incorporating ergonomics into its development aid programme. One step towards this is the production and dissemination of *A Guide to Addressing Ergonomics in Development Aid.* This guide is being developed and tested as a practical tool for aid professionals to use to ensure that ergonomics components in future projects are addressed.

Acknowledgement

The authors would like to acknowledge the Overseas Development Administration of the Foreign and Commonwealth Office, for funding this work.

References

Mikheev, M I. 1995, Health at Work- Global Analysis. In E. Juengprasert (ed.), Proceedings of the International Symposium on Occupational Health Research and Practical Approaches in Small Scale Enterprises, 1995 (Division of Occupational Health, Ministry of Public Health, Thailand) 3-10.

CONSUMER
ERGONOMICS

USING SOUNDS TO IMPROVE THE USABILITY OF TELEPHONE TERMINALS

Denver Trouton

1/L 246 Bath Street
Glasgow
G2 4JW
☎ *+44 141 3531518*
Email : dtrouton@cs.strath.ac.uk

This paper describes an investigation, carried out at Telecom Sciences Corporation, into how sounds could be used to improve the usability of telephone terminals. An evaluation of the sounds used in current terminals was undertaken and recent Human-Computer Interaction work on using sounds was studied. The results of this showed that generally insufficient auditory feedback was supplied, and many of the advanced features were fairly complex to access. Some terminals use a small display with a soft key operated menu structure, while others use codes to select the more advanced features. It was thought that sounds could be used to improve this. A system of hierarchical earcons were developed to be used with the text menu system and a synthesised speech menu structure was created to replace the codes system.

Keywords : Auditory interfaces, earcons, menus, phone-based interaction

Introduction

This project was undertaken at Telecom Sciences Corporation, based in Scotland, in conjunction with the University of Strathclyde in an annual industrial placement scheme. Telecom Sciences are providers of PABX systems, CTI applications and mobile solutions. They also have interests in a security system for schools, internet applications and office and home office applications.

In our everyday life we use our eyes and ears to experience our environment. Anyone who has ever tried to cross a busy road while using a personal stereo will have noticed the extra effort in telling whether or not a car is coming. One of the earliest identifications of the use of sound in a computer interface was by Malone (1982) who showed that games players performed better with sound feedback. Recently however it has become increasingly common to expect the operator to use only the visual sense.

With the advent of more affordable displays and windowing systems on computers the little auditory feedback which was used now seems to have been abandoned.

Several people have attempted to find ways to rectify this, specifically for computer user interfaces. The simplest approach is to use synthesised speech, however there can be problems with this as it may take many words to impart basically simple information. Also as the average reading rate is 250 words per minute, as opposed to the optimal speaking rate which is 150 words per minute, synthesised speech can make the interface seem clumsy. Another approach is to use non-speech sounds. This has the advantage of allowing "auditory habituation". That is the ability of the auditory system to allow continuous sound of a restricted loudness range to "fade into the background" of consciousness. This is not possible with speech as vowels are often much more intense that consonants. The other advantage of non-speech sounds is international issues - sounds tend to be more global than any language. There have been two approaches to using non-speech sounds.

1.   Auditory icons. These are everyday sounds which have some natural association with the application. For example when clicking on an icon you might here a "tapping" sound or when closing a window a shutting sound might be heard. The problem with this approach is that there is often no "natural" sound which fits the application - what sound could you use for navigating through a menu structure? An alternative is
2.   Earcons. These are structured synthetic sounds which do not have a natural association with the application. Instead the association between the action and the sound is learnt by the user as they progress.

The reason for this experiment was to identify how or if any of this knowledge could be applied to the user interface of a more restricted system - a telephone terminal. In internal telephone systems more and more options are available to the user and it is becoming increasing complex to present all of these to the user in a fashion which is not confusing. It was suggested that auditory feedback might be an alternative to more expensive displays.

Approach

The first step was to examine which problems existed. The systems which Telecom Sciences produce have two types of interface to access the advanced options.

1.   No or limited display. In these systems the advanced options are accessed using codes. The problem here is that people often find this off putting and have difficulty remembering codes.
2.   Display terminals. These systems have a four line LCD display and a soft-key operated menu structure which allows the user to select the function they require. The soft-keys take the user down the structure and scroll buttons allow other groups of nodes on the same level to be accessed. The problem with this system is that people are often intimidated and confused by the menu system.

An examination of comparable systems produced by other companies showed no great difference in any of the systems employed. Certainly sounds were not used to any great extent in any of the systems examined.

Various solutions to these problems were considered taking into account the hardware restrictions. While in theory any sounds could be used by simply storing them on a chip and adding this to the terminal, the cost involved was thought to be too high. Instead the existing tone chip was to be used along with the speech synthesiser which is present on the system switch.

Menu System

Work by Brewster, Raty and Kartekangas (1996) has already shown that non speech sounds (carcons) could be added to a graphical menu structure with impressive results. It seemed logical therefore to see if we could use a similar system with our text based menu structure. The current menu system was examined carefully and a computer simulation created using Visual BASIC. Sounds were then created using a standard MIDI package and added to the simulation. The most useful properties of sound which can be varied are timbre, pitch, register, rhythm, and intensity. The most useful of these is timbre, however due to hardware restrictions that could not be varied here. The next best is rhythm.

The sounds used in the simulation were produced by adding a note to a rhythm each time the user went down the structure. In this way it should always be possible to know how far down the structure you are by the number of notes you hear. A different note was added depending on the button pressed and so the user should have some idea of where they were in the structure by the types of notes they hear. Also some parts of the structure used shorter notes to make the rhythm seem faster. This was used to distinguish between different node groups on the same level.

The idea was to test the simulation on two groups of people. One group would attempt to complete simple navigational tasks using first the silent version, and then the version with sounds. The second group would try the version with sounds first. Both groups were also asked to perform a recognition test to see how well the recorded sounds were recalled. This was for comparison with earlier work by Brewster (1996) where timbres were used, and also to show that the sounds used were suitable. This test consisted of the user being played a sound and being asked to mark on a map of the structure where they felt it belonged. Marks were awarded for identifying the correct level, the correct branch, and the correct node. Acceptance tests were also carried out by both groups on both simulations. The acceptance testing used NASA's raw TLX (Task Load Index) tests as in Brewster (1994) which cover such things as mental demand, annoyance and overall preference. Musical ability was also considered as due to the musical nature of earcons it has been suggested that musicians may fare better in tests.

Speech Menu System

A synthesised speech menu structure was also created to replace the codes system. The idea was to have a simple menu system which worked by using synthesised speech instead of a display which the user could scroll through. There were two problems encountered with this approach. Firstly due to the nature of the system switch only one user may access the synthesised speech at a time. A second user attempting to access the speech chip when it is busy will receive only an error beep. Clearly this is unacceptable. However it is possible within the system to allow one terminal to be given preference to the system. This would allow one terminal to be set up for a visually impaired operator. The second problem is that this is likely to suffer from all the problems of the existing menu structure with the added confusion which will result from the lack of visual feedback.

Unfortunately, due to time constraints, it has not yet been possible to test this system to see if it is a noticeable improvement on the codes system.

Results

Efficiency Tests
The efficiency of each user at completing each of the tasks was noted. The results can be seen in the graph below.

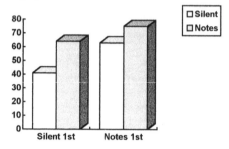

Figure 1 Results of Evaluation

As can be seen the testers on average performed much better when the version with sounds was used first. In fact their efficiency was almost 20% higher. Also interesting to note is that there was very little difference between the efficiencies of those who performed the sounds versions first, and those who used it after using the silent version. There is however a very large difference between the efficiencies of those who used the silent version first and those who used it after using the sounds version - in fact an increase of almost 40%. This can be seen on the graph below.

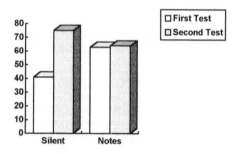

Figure 2 Comparison of results

This suggests that as well as sounds giving a better first time performance, they can aid learning. The difference of only 1% between the efficiencies of those who used the sounds version first, and those who used it after using the silent version, implies that little was learnt about the menu structure from using the silent version. However the fact that there is a very large difference between the two silent versions would suggest that a lot was learnt from the sounds version. Also comparison of results showed that musical ability was not a major factor.

Recognition Test

The overall recall rate was 66%. This compares favourably with the results of Brewster (1996) when you consider that we could not use many of the advanced features of the sounds due to the hardware restrictions, and also the training used in this case was not so formal. This result also tells us that the sounds used *were* of use.

Acceptance Tests

Overall people preferred the version with sounds to the silent version, however the sounds were considered more demanding mentally, and more annoying. It is interesting however to compare the results from the two groups. The first group, who used the silent version first, found the presence of the sounds *less* annoying than the confusion which they experienced in their absence. They also considered the tasks to be less demanding when the sounds were present. Those who used the sounds version first considered the sounds to be more annoying and demanding. This would seem to support the idea that people learnt very quickly from the sounds and so were not so confused, or annoyed, by the silent version.

Conclusions

The results from these experiments show that sounds can be very valuable in user interfaces, even if their construction is restricted. These results also suggest that sounds could be very valuable as an aid to learning and can remove some of the initial confusion users experience.

The major problem identified here was one of annoyance. Close examination of the acceptance tests showed that some of the results were in fact skewed by a very small sample who disliked the sounds intensely. Some of the annoyance is due to the fact that sounds were designed in a restricted fashion around an existing menu structure. A better approach would be to design the sounds and the menu structure together. In this way problem areas could be avoided. It is possible however that some people will always dislike the use of sounds.

In the future the speech menu structure described here will be evaluated. It would also be beneficial to examine the sounds used and attempt to improve them by varying the menu structure slightly to see if, as suggested, that could alleviate the annoyance some people felt. Also the learning differences between the two versions should be examined. It may well be that sound at the interface represents a very valuable learning tool.

D. Trouton

References

Brewster, S.A., 1994, Providing a Structured Method for Integrating Non-Speech Audio into Human Computer Interfaces

Brewster, Raty and Kartekangas, 1996, Earcons As A Method Of Providing Navigational Cues In A Menu Hierarchy

Handel, 1989, Listening - An introduction to the perception of Auditory Events,

Malone. 1982, Heuristics for Designing Enjoyable User Interfaces: Lessons From Computer Games, Human Factors In Computer Systems

PRODUCTS AS PERSONALITIES

Patrick W. Jordan

*Philips Corporate Design
Building W, Damsterdiep 267
P.O. Box 225, 9700 AE Groningen
Netherlands*

Traditionally, human factors approaches have tended to concentrate on usability issues. Such approaches encourage the perception of products as being tools which are used to complete tasks. A more holistic role for human factors includes looking at the wider relationship between people and products. Such an approach can involve looking at products as 'living-objects'.

Product Personality Assignment (PPA) is a new evaluation method whereby participants are asked to think about products as if they were people and to assign human personality characteristics to products.

A questionnaire and focus group based study in which this method was used is reported. Outcomes indicated that participants were able to make meaningful judgements as to product personality, that they were able to link product personality traits to particular product properties, and that products that reflected the participants' own personalities were preferred.

There are, however, a number of limitations on the study which leave unresolved questions about the method.

Introduction

Traditionally, human factors input into the design process has tended to emphasise usability issues. Human factors as a profession has become adept at evaluating products for usability as well as contributing pro-actively in the design process to ensure the usability of products.

Recently, however, a number of human factors professionals have advocated a role for human factors that goes beyond usability. They regard the analysis of more intangible issues — such as the emotions associated with product use and the values conveyed by a product — as falling within a human factors remit (e.g. Jordan and Servaes 1995, Rijken and Mulder 1996, Dandvante, Sanders and Stuart 1996, Jordan, in press).

Products as personalities

Usability based approaches to evaluation tend to see the product as a *tool* with which *users* complete *tasks* — hopefully as effectively and efficiently as possible and within reasonable levels of comfort and acceptability (based on the International Standards Organisation's definition of usability (ISO DIS 9241-11)). Such approaches are limiting. Products are not merely tools, they are *living-objects* with which *people* have *relationships* — usability issues affect only some part of these relationships. Products live in people's homes, their offices, their town. They are objects which can make people happy or angry, proud or ashamed, secure or anxious. Products can empower, infuriate, delight — they have personality.

OK, everybody knows that products don't really have personalities in the strict psychological sense, but try telling that to the guy who's computer has 'decided' to crash just before he was going to save that report he'd spent all morning working on. Tell it to the woman who's just given her cassette player a 'damn good thrashing' as a 'punishment' for chewing up one of her cassette tapes. Explain it to guy who's car has 'deliberately' got a flat tyre on the one day when it's absolutely essential that he gets to work on time.

In this paper a new evaluation method for investigating products' personalities — Product Personality Assignment (PPA) — is described and a small study to investigate its usefulness is reported.

Personality Types

Based on the theories of personality developed by Carl Jung in the 1920s (published in English in 1971 (Jung 1971)) Myers and Briggs created an instrument known as the Myers-Briggs Type Indicator (MBTI) (Briggs-Myers and Myers 1980). This instrument classifies people's personality type with respect to four characteristics. Approximate descriptions of these are given below:

Extrovert/Introvert (E/I)

Extrovert people show a preference for relating to the outside world. They are likely to be social and lively and have many friends. Introverts show a preference for internalising ideas, emotions and impressions. They are likely to be quiet, private people with a few close friends.

Sensible/Intuitive (S/N)

Sensible people prefer to rely on the evidence of their five senses. They are down to earth and practical. Intuitive people, on the other hand, often rely on a 'sixth sense'. They are often inspired and imaginative. They may appear to have their heads in the clouds.

Thinking/Feeling (T/F)

Thinking people have a preference for taking a logical approach to things. They are objective and analytical and try to look at things impersonally. Feeling people on the other hand take a value-based approach to decision making. They take a more subjective, sympathetic and intimate approach to things.

Judgmental/Perceptive (J/P)

Judgmental people like to live well organised lives, they plan ahead, are decisive and like working to firm deadlines. Perceptive people, meanwhile, prefer spontaneity. They are flexible, open-mined people who are comfortable with less definite deadlines.

A study to investigate Product Personality Assignment (PPA)

Participants

Thirteen people participated in the study. All were employees of Philips in Eindhoven, The Netherlands. Seven were designers, two were secretaries, one a social scientist, one a human factors specialist, one a marketing specialist and the other a software engineer. Six were men, seven were women and all were in their late twenties/early thirties.

PPA Questionnaire

The four personality dimensions were explained to participants. This was done through the investigator reading out the four descriptions of the dimensions as given above. Participants were shown photographs of products. They were then asked — on the basis of these photographs — to make judgements about the personality of each product.

They were asked to pretend that each of the products was a person and then asked what sort of a person it would be. Would it be extrovert or introvert, sensible or intuitive, thinking or feeling, judgmental or perceptive. They indicated their opinions via check boxes on a questionnaire.

There were four pairs of products — two vacuum cleaners, two alarm clocks, two kettles, and two toasters. Having indicated their impressions of the personality of each, participants checked another box to indicate which of the two they preferred.

The products evaluated in the study are illustrated in figure 1.

Figure 1. Products evaluated in the study. Participants judgements were based on colour photos of these products. (Top Row, left to right: Vacuum Cleaner A, Vacuum Cleaner B, Clock A, Clock B. Bottom Row, left to right: Kettle A, Kettle B, Toaster A, Toaster B).

After assigning personality characteristics to products, participants were then asked to indicate what they perceived their own personalities to be — again by marking check boxes on the questionnaire.

Focus Groups

Each participant completed the questionnaire independently, but after this the participants came together in two separate focus groups to discuss the reasons as to why they had assigned the various personality characteristics to each of the products. One focus group contained six people and the other seven.

Results

Assigned Personalities

Table 1 summarises the personality characteristics that were allocated to each of the products and the number of participants preferring each product to the other in its pair.

PRODUCT	ASSIGNED PERSONALITY	NO. PREFERRING
Vacuum Cleaner A	I (10) S (11) T (10) J (13)	2
Vacuum Cleaner B	E (7) S (10) T (7) J (10)	11
Clock A	E (7) S (11) T (9) J (10)	7
Clock B	I (8) S (8) T (7) J (9)	6
Kettle A	E (9) N (8) F (9) P (9)	8
Kettle B	I (8) S (9) T (9) J (12)	5
Toaster A	I (10) S (10) T (10) J (12)	2
Toaster B	E (11) N (11) F (12) P (12)	11

Table 1. Assigned personalities and preferences. Figures in brackets denote the number of participants assigning the characteristic to a product (n = 13). So, for example, in the case of Vacuum Cleaner A, ten participants assigned an introvert personality to the cleaner whilst three assigned an extrovert personality.

Product Properties and Perceived Personality

The discussions in the focus groups gave indications as to the properties of particular products that were seen as indicative of that product's personality type. A brief overview of comments is given below.

Extrovert/Introvert

Judgements about whether products were extrovert or introvert were usually made on the basis of the form of the product. Vacuum cleaner A, for example, was perceived as extrovert due to its "bold shape". Meanwhile, the design of Toaster B was seen as extrovert as the form and colours radiated warmth. In the case of Kettle A, the materials used were a strong influence, shiny chrome being seen as an extrovert material.

Similarly, when a product was regarded as being introvert, it was some aspect of the form that was most commonly cited. The design of Vacuum Cleaner A was seen as being unattractive and relying on strict shapes and straight lines, Clock A was seen as being made up of very basic shapes, and Kettle B as being something that would simply "blend into its surroundings".

Sensible/Intuitive

This personality dimension was partly associated with the product's form and partly with the functionality. Where the products form and functionality were seen as straightforward, limited and well defined, the product tended to be rated sensible (Vacuum Cleaner B, Clock B, Toaster A). "Dullness" was also associated with sensible products (e.g. Kettle B) as was a "lack of frivolity" and "lack of gadgets" (Vacuum Cleaner B). The formality of the digital display was cited as a reason why Clock A was sensible, whilst Vacuum Cleaner A was regarded as sensible due to the "sensible nature of cleaning as a task."

Kettle A was regarded as being intuitive because it was "unsophisticated" and because its traditional form had "traditional values" associated with it. Toaster B was rated as intuitive as it seemed to have a "friendly personality".

Thinking/Feeling

Products that were perceived as thinking were often perceived as being very "rational" and "practical" (e.g. Vacuum Cleaner B, Clock A, Toaster A). Having a "high-tech" image was also cited as a reason for perceiving a product as thinking (Kettle B). As with sensible products, thinking products were seen as being more "predicable" (Vacuum Cleaner A, Clock B).

Toaster B was classified as a feeling product because it was perceived as communicating "softness" in its design. Kettle A was seen as being "cute".

Judgmental/Perceptive

Participants tended to associate technical reliability with what they regarded as being judgmental products. Toaster A, for example, was cited as being "trustworthy" in this respect. A "limited and fixed functionality" was also associated with judgmental products (Vacuum Cleaner A, Vacuum Cleaner B, Kettle B). The reason given for regarding Clock B as judgmental was simply that it had a function (i.e. giving the time) that was linked to meeting deadlines.

Toaster B was regarded as being perceptive because its design was "unusual". Similarly, the design of Kettle A was seen as being "unexpected" considering that the jug style of kettle is now more common.

Participants' personality and preferred products

An analysis was carried out to see if participants self rating of their own personalities was related to their choices with respect to the preferences that they expressed with respect to the pairs of products that they rated.

Participants' self ratings of personality are given in table 2. The number given after each trait indicates the number of participants who assigned a particular personality trait to themselves . So, for example, six participants regarded themselves as being thinking whilst seven regarded themselves as being feeling.

PERSONALITY DIMENSION	E/I	S/N	T/F	J/P
USER SELF RATINGS	Extrovert 5 Introvert 8	Sensible 6 Intuitive 7	Thinking 6 Feeling 7	Judgmental 5 Perceptive 8

Table 2. Participants perceptions of their own personalities (n = 13).

A chi-square test, showed that with respect to all four traits no particular assignment occurred significantly more than its opposite. (e.g. there were not significantly more introverts than extroverts).

For each participant's assignments to each product a 'matching' score was given. This reflected the number of traits on which participant perceived his/her personality to match that of the product. This score varied between 0 and 4 — 0 if the product matched the participant with respect to no traits, 4 if the participant perceived a match on all four.

So, for example, if a user had rated themselves as being ENTP and had rated a product as being ESTJ, then his/her self rating would match his/her product personality assignment on two issues — extroversion and thoughtfulness — giving a matching rating of 2 out of 4.

For each participant, total matching scores on his/her preferred product were compared to total scores on non-preferred products. Because four pairs of products were compared there were, of course, a total of 4 preferred products and 4 non-preferred products. This gave a potential range of 0 to 16 for both preferred and non-preferred products. (So, a score of 16 would represent all four preferred or non-preferred products matching the participant with respect to all four traits, whilst a score of 0 would represent no matches on any traits for any products).

The mean matching score for preferred products was 9.0 whilst for non-preferred products the mean matching score was 6.5. A non-parametric Wilcoxen Signed Rank test was carried out to see if this difference was statistically significant. It was ($p < 0.1$). This indicates that the participants did indeed show a preference for products that they perceived as matching their own personality characteristics.

Discussion

The study reported represents a 'first pass' at investigating the usefulness of a methodology based on the assignment of personality characteristics to products.

Limitations of the Study

Of course this study has limitations. Photos were used instead of real products, the participants were a narrow sample (in terms of their professions and their ages) and the method of personality assignment was somewhat 'quick and dirty'.

Nevertheless, the outcomes of the study do give the opportunity to make some tentative comments about the potential usefulness of this method as an evaluation technique.

Product Personality Assignment as an Evaluation Technique

One indicator of how meaningful this technique is is the extent to which there was agreement as to the personality characteristics of each of the products. If there is a high degree of agreement between participants, then this might suggest that assigning personalities to products is meaningful. However, lack of agreement would suggest that such assignments may not be meaningful — that they may be little more than random assignments to meaningless categories.

The results were mixed in this respect. There was a high level of agreement about the personality characteristics of Vacuum Cleaner A, Toaster A and Toaster B, with at least 10 participants agreeing on the assignment of personality within each trait. For the other products, however, there was less agreement. Notably, there were no characteristics about which ten or more people agreed on either Clock B or Kettle A.

Another important indicator of how meaningful the assignments were is the degree to which participants assignments can be explained. Again, the results appear to give some cause for optimism and some for pessimism. During the focus group sessions, participants were able to give sensible and convincing reasons as to why they assigned particular personalities to a product. However, the responses given were not particularly rich in content, tending to focus on form and function (of course this may well be as a result of participants being asked to make judgements based purely on looking at photographs).

Responses from the second of the two focus groups also seemed to suggest that participants were using particular characteristics of the product to make classifications on all four personality dimensions. Indeed, a correlation matrix revealed that the assignment frequencies of extrovert, intuitive, feeling and perceptive correlated highly with each other (r being at least 0.6 for each correlation). This also meant, of course, that assignments of introvert, sensible, thinking and judgmental were highly correlated. In this study, then, the components of personality were not independent.

Again, these correlations give some cause for optimism and some for pessimism. That correlations between assignments were found is encouraging in that it suggests that the categories were meaningful to respondents. If they were not meaningful, then systematic relationships between the categories would not be expected. However, when the correlations are *this* high it raises questions as to whether it is necessary to ask about each of the four characteristics individually.

The data provided by the focus groups were useful in gaining an understanding as to why participants preferred some products to others and the personality assignments gave a useful label to 'hang' some of the differences on. However, this alone does not necessarily mean that the data provided by this method are any richer than that that could be gained from simply asking participants why they prefer one product over another.

An interesting result was that the match between participant and product characteristics was a predictor of product preference. This suggests that people have a preference for products that they perceive as being 'made in their image'. By understanding both the personality traits of a product's target group and the links between product properties and the perceived personality of that product, it may be possible to tailor a product's design to give advantages in terms of the extent to which the product will be accepted by that group.

Conclusions

This novel method of evaluation seems to offer a potentially useful framework within which to analyse and quantify what might, hitherto, have been seen as 'intangible' characteristics of a product. It is, then, a potentially useful tool for those who recognise a wider role for human factors than just usability engineering.

This study is inconclusive. There is some evidence that the method can provide a useful insight into the properties of a product that give it a particular image. There is evidence that people could make meaningful assignments and that there is a preference for products that match people's perceptions about their own personalities. It also appeared that people could make meaningful associations between properties of a product and particular personality characteristics.

Perhaps the main questions left unanswered are whether or not the method would work with real products rather than photos and whether or not the method can provide rich enough information upon which to base practical design solutions. These are questions worthy of future research in the search for methods that can take human factors beyond usability.

Acknowledgements

Thanks very much to all the participants: Dominic Butler, Wim Dejonghe, Stephanie van Dijck, Sandra Hekklelman, Mark van Huystee, Roeland Lundahl, Joanna Kabala, Joyce Lamerians, Paul Moore, Els Quist, Ellen van der Ros, Casper Ulijn, Inge Verheijden.

References

Briggs-Myers, I. and Myers, P. 1980, *Gifts Differing*, (Consulting Psychologists Press)

Dandavante, U., Sanders, E.B-N., and Stuart, S. 1996, Emotions matter: user empathy in the product development process. In *Proceedings of the Human Factors and Ergonomics Society 40th Annual Meeting — 1996.* (Santa Monica: Human Factors and Ergonomics Society) pp 415-418.

ISO DIS 9241-11, *Ergonomic requirements for office work with visual display terminals (VDTs):- Part 11: Guidance on Usability.*

Jordan, P.W. 1996, A vision for the future of human factors, *Keynote Address to the Annual Conference of the Europe Chapter of the Human Factors and Ergonomics Society, November 7-8,1996, Haren, The Netherlands.* (Copy of paper available from the author on request)

Jordan, P.W. and Servaes M. 1995, Pleasure in product use: beyond usability. In S. Robertson (ed), *Contemporary Ergonomics 1995*, (Taylor and Francis, London) pp 341-346.

Jung, C.G. 1971, *Psychological Types* (translator: H.G. Baynes), (Princeton University Press).

Rijken, D. and Mulder ,B. 1996, Information ecologies, experience and ergonomics. In P.W. Jordan et al (eds), *Usability Evaluation in Industry.* (London: Taylor and Francis).

HOW DO WARNING LABELS AND PRODUCT FAMILIARITY INFLUENCE HAZARD PERCEPTIONS?

Clare Hyde & Elizabeth Hellier

*Department of Psychology,
City University,
London, EC1V 0HB.*

This study investigated how hazard perceptions of consumer products may be influenced by the presence of a simple warning, and by the familiarity of the product itself. Twelve product labels were categorised as familiar or unfamiliar, and had either a warning added to them, or their labels left unchanged. 24 university undergraduates rated the product labels on a variety of variables including perceived hazard. The results revealed that the products were perceived as more hazardous when a warning was present, and when the product was unfamiliar, and that the presence of a warning per se, has more of an effect on the hazard perceptions of unfamiliar than familiar products. The implications of these findings for product manufacturers are discussed.

Introduction

The decision to behave safely depends to a certain extent on an individual's perceptions of the risk or hazard involved in a potentially hazardous situation (Wogalter, Desaulniers and Brelsford, 1987; Young, Brelsford and Wogalter, 1990). Consequently, warning research is often aimed at measuring the hazard perceptions generated by the warning, and calibrating them to the level of hazard present in the situation (Braun and Silver, 1995). This paper proposes that the considerable time and effort spent on warning calibration research to support the design of every warning may not always be necessary. Other factors less frequently tested may also have an impact on people's hazard perceptions, and therefore may shed light on the likelihood of safe behaviours being performed. It is suggested that in some circumstances, hazard perceptions may be sufficiently raised by the presence of a simple and relatively neutral warning. This study focuses on the effects of two particular factors, which are assumed to influence safe behaviour; the presence of a warning per se, and the familiarity of a situation or product.

Several groups of researchers have concluded that on-product warnings do not increase the safety of products (McCarthy, Ayres and Wood, 1995; Horst, McCarthy, Robinson, McCarthy and Krumm-Scott, 1986), whereas others dispute this view believing that warnings can produce safer behaviour in product users (Friedmann, 1988). The most recent analysis of warnings literature employing a no-warning control group has applied a meta-analysis technique to examine the data from 15 empirical studies. Cox, Wogalter, Stokes and Tipton Murff (in press) performing the first analysis of this kind on this category of studies, concluded principally that warnings do increase safe behaviour.

The second factor which may influence perceptions of hazard, and therefore may also influence compliance, is the familiarity of the situation. Potentially dangerous situations may be perceived as less hazardous and therefore produce less cautious behaviour if the situation is familiar and has been experienced safely many times, than if the situation is unfamiliar and rarely experienced at all. This "familiarity effect" has been frequently demonstrated in warning labels research. Studies have reported that when a product is familiar, ratings of perceived hazardousness are lower than when the product is reported to be unfamiliar (Godfrey, Allender, Laughery and Smith, 1983). Research investigating familiarity effects on behavioural compliance have also shown that people more familiar with the products and scenarios presented, displayed lower compliance rates with the warnings used (Otsubo, 1988; Wogalter, Barlow and Murphy, 1995).

The present study builds on this previous research by investigating the additive value of a warning per se, over and above the hazard cues present in the situation. It also establishes whether the magnitude of the added value depends on cues perceived in a situation which is familiar or unfamiliar. As Edworthy and Adams (1996) note, the technique of introducing control conditions in a warnings study will distinguish between the situation itself with all it's inherent cues, and the warning which represents it, which has not explicitly been shown before. For the purposes of this experiment, a pilot study was conducted to assess product familiarity before the products were displayed with a warning present.

Method

Participants

Pilot study: Ten male and ten female volunteers studying at City University, London, with a mean age of 24.45.
Main experiment: An opportunity sample of 12 male and 12 female undergraduates with a mean age of 19.67 years. All participants spoke English as their first language.

Pilot study stimuli

Twenty-six consumer products found in chemist's shops and DIY stores were selected. The rear labels on each of the products included either a written or pictorial warning, often indicating that protective clothing should be used, or safety measures taken. These product features were taken to indicate that the product was sufficiently hazardous to require the provision of a warning by law. The front labels on the products were then photographed, and presented in a pilot study to 20 independent judges who rated each label for familiarity. Of the 26 products rated for familiarity, 6 were chosen to represent the familiar products, and 6 were chosen to represent the unfamiliar products (mean ratings were 1.00 and 5.75 respectively).

Stimuli Booklets

Following the procedure employed by Wogalter, Jarrard and Simpson (1992), the twelve product labels were digitised and stored on computer. Using paint and draw software, the message "Warning- Read back label carefully" was produced using a black or white font in a style which closely matched the print on the original labels. Colour versions of the product images were produced before and after the addition of the warning message using a laser printer. Twelve images were used in the experimental conditions; bathroom cleaner, bleach, cold relief powder, creosote, liquid paper, outdoor disinfectant, paint, plant-food, predye, rooting powder, sugar soap and Persil washing liquid.

Six "filler" photographs were also scanned into the computer, and two sets were reproduced in colour. The filler products were relatively safe products which contained no warning of any kind. The purpose of including the fillers was to help maintain belief in the market study by reducing the likelihood that participants would notice that the study was investigating the effects of warning labels. The experimental products were manipulated to represent one of four conditions; warning present and product familiar; warning present and product unfamiliar; warning absent and product familiar; warning absent and product unfamiliar.

The stimuli were presented to participants in a booklet. All the participants saw the 6 familiar and 6 unfamiliar products together with 6 filler items. Each experimental item was presented both with and without a warning, but the warning presence or absence was balanced between participants so that no participant saw the same product with or without a warning present. A balanced Latin Square design was used to determine the order of the experimental items, and the fillers were randomly inserted between them.

Rating scales

Participants were given a questionnaire (based on that used by Wogalter et al., 1992) requesting responses based on nine-point Likert-type scales. The three questions critical to the study, together with their numerical and verbal anchors are shown below. The participants also rated the products on likelihood of purchase and attention-grabbing scales.

1) *Familiarity*: "How familiar are you with this product (or a product of the same type)?" The anchors were: (0) not at all familiar, (2) slightly familiar, (4) familiar, (6) very familiar, (8) extremely familiar.
2) *Hazard*: "How hazardous is this product?" The anchors were: (0) not at all hazardous, (2) slightly hazardous, (4) hazardous, (6) very hazardous, (8) extremely hazardous.

3)   *Caution in use*: "How cautious would you be in using this product?"
      The anchors were: (0) not at all cautious, (2) slightly cautious,
      (4)cautious, (6) very cautious, (8) extremely cautious.

Procedure

Participants were told that the study was a market research survey dealing
with people's perceptions of consumer products. They were given the questionnaire
which asked them to rate the products on the five dimensions described earlier.

Results

The data was analysed using three different criteria of familiarity:-
i) the familiarity ratings generated from the pilot study
ii) the participant's own familiarity categorisations with familiarity anchored
at the "3" point on the anchor scale (so that judgements of 3 and above were
categorised as "familiar")
iii) the participant's own familiarity categorisations with familiarity anchored
at the "5" point on the anchor scale
Two (presence vs. absence of warning) x two (familiar products vs. unfamiliar
products) analyses of variance were performed on the hazard perception and the
intent to comply ratings.

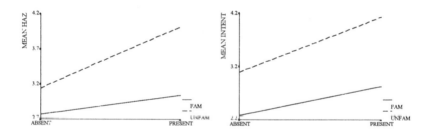

Figure 1. Interaction between warning presence and product familiarity for hazard
perception and intent to comply data.

There were main effects of warning presence on hazard perceptions for all 3
familiarity categorisations (p≤0.05). There were main effects of warning presence on
intent to comply when familiarity was anchored at 3 and at 5 (p≤0.05). With the pilot
study classifications, the trend of the mean intent to comply ratings were in the
predicted direction. Overall, these results suggested that the presence of a warning
increases perceptions of hazard and intent to act cautiously
There were main effects of familiarity on hazard perceptions with the pilot
study classifications and when familiarity was anchored at 3 (p≤0.05).). When
familiarity was anchored at 5, the trend of the mean hazard perception ratings were in
the predicted direction. There were main effects of familiarity on intent to comply for

all 3 familiarity categorisations (p≤0.05), overall suggesting that a familiar product reduces hazard perceptions and the intent to act cautiously.

There was an interaction between warning presence and product familiarity for the hazard perception data for all categorisations of familiarity (p≤0.05). There was also an interaction for the ratings of intent to comply when familiarity was anchored at 3 and at 5 (p≤0.05). With the pilot study classifications, the trend of the mean intent to comply ratings were in the predicted direction. As displayed in Figure 1, the presence of a warning increases hazard perceptions and the likelihood of cautious behaviour more with unfamiliar products , than with familiar products.

Discussion

This experiment found that when a warning is present people do perceive products as being more hazardous, and that people show an associated intent to act more cautiously, compared to when a warning is not present. Therefore it seems that the warning does provide an added effect on both hazard perceptions and intent to comply, suggesting that people do take hazard cues from the warning, as opposed to focusing on the cues inherent in the situation alone. This finding not only lends support to the likes of Cox et al. (in press) by confirming that warnings do increase the likelihood of safe behaviour, but further suggests that the presence of a simply constructed warning, similar to the one designed for this experiment, may be an adequate safety measure to take for many product warnings. This has implications for consumer-protection litigation involving allegations of negligence, since it supports manufacturers in their use of even simple warnings as a safety mechanism.

This study also found that unfamiliar products were consistently perceived as more hazardous, and produced a greater intent to behave cautiously than familiar products, providing evidence that the situation also has an effect on the likelihood of safe behaviour. This finding supports the research on warning compliance conducted by researchers such as Otsubo (1988), and the research measuring changes in hazard perceptions conducted by Godfrey et al. (1983).

The interaction displayed how the provision of a warning has more of an effect on hazard perceptions of unfamiliar products than of familiar products. A behavioural study conducted with subjects using actual products would more fully establish the influence of these factors on safe behaviour. The use of more salient warnings on familiar products appear desirable to produce raised hazard perceptions, since situational variables such as past experiences with the product, seem to override the influence of a simple warning. However, a simple warning may be an adequate safety device to use on an unfamiliar product. With unfamiliar products from which such situational cues are unavailable, even a simple warning provides additive hazard information over and above that provided by the product itself, so will be used to perceive hazard cues. The recommendations for manufacturers here are therefore to provide a more prominent and salient warning on products with which the user is likely to be familiar. This is particularly important for products which are packaged identically, but have had their active ingredients modified, thereby possibly introducing a new potential hazard to some users. It may also be important to provide more salient warnings when manufacturers launch a new product which although in itself is unfamiliar, includes a familiar manufacturer's name and label design belonging to a familiar range of product.

In summary, this research has shown that warnings do provide additive hazard information over and above that implied by the situation. This additive data is increased for unfamiliar situations and when a warning is present.

References

Braun,C.C.and Silver,N.C.1995, The interaction of signal word and colour on warning labels: differences in perceived hazard and behavioural compliance. *Ergonomics,* **38(6).**

Cox,E.P., Wogalter,M.S., Stokes,S.L. and Tipton Murff,E.J. (in press) Do product warnings increase safe behaviour?: A meta-analysis. To appear in the *Journal of Public Policy and Marketing* in 1997.

Edworthy,J. and Adams,A. 1996, *Warning Design : A research prospective.* (Taylor and Francis, London.)

Friedmann,K. 1988, Effect of adding symbols to written warning labels on user recall and behaviour. *Human Factors,* **30(4),** 507-515.

Godfrey,S.S., Allender,L., Laughery,K.R. and Smith,V.L. 1983, Warning messages - Will the consumer bother to look? *Proceedings of the 27th Annual Meeting of the Human Factors Society,* 950-54. (Santa Monica:Human Factors Society.)

Horst,D.P., McCarthy,G.E.,Robinson,J.N., McCarthy,R. L. and Krumm-Scott,S. 1986, Safety information presentation : Factors influencing the potential for changing behaviour. *Proceedings of the 30th Annual Meeting of the Human Factors Society,* 111-5. (Santa Monica:Human Factors Society.)

McCarthy,R.L., Ayres,T.J. and Wood,C.T. 1995, Risk and effectiveness criteria for use of on-product warnings. *Ergonomics,* **38(11),**2164-75.

Otsubo,S.M.1988, A behavioural study of warning labels for consumer products: perceived danger and use of pictographs. *Proceedings of the 32nd Annual Meeting of the Human Factors Society,* 536-540. (Santa Monica:Human Factors Society.)

Wogalter,M.S.,Barlow,T.and Murphy,S.A. 1995, Compliance to owner's manuals warnings: influence of familiarity and the placement of a supplemental directive. *Ergomics* **38,** 1081-1091.

Wogalter,M.S.,Jarrard,S.W.and Simpson,S.N.(1992) Effects of warning signal words on consumer-product hazard perception. *Proceedings of the 36th Annual Meeting of the Human Factors Society,* 935-939. (Santa Monica:Human Factors Society.)

Wogalter,M.S.,Desaulniers,D.R. and Brelsford,J.W. 1987, Consumer products : How are the hazards perceived?. *Proceedings of the 31st Annual Meeting of the Human Factors Society,* 615-19. (Santa Monica:Human Factors Society.)

Young,S.L., Brelsford,J.W. and Wogalter,M.S. 1990, Judgments of hazard, risk and danger : Do they differ? *Proceedings of the 39th Annual Meeting of the Human Factors Society,* 503-7. (Santa Monica:Human Factors Society.)

DRIVERS AND DRIVING

ON-CAMPUS PEDESTRIAN CROSSINGS: OPPORTUNITY FOR LOCALE BASED DRIVER IMPROVEMENT

Tay Wilson and Christine Chisel

Psychology Department
Laurentian University
Ramsey Lake Road
Sudbury, Ontario, Canada
P3E 2C6
tel (705) 675-1151
fax (705) 675-4823

Assessment was made, for some 15000 vehicles and 800 pedestrian groups, of driver assistance or obstruction rates at three pedestrian crossings of a local university during autumn and winter sessions. One third of drivers interacting with pedestrians were driving 50% or more above the speed limit and they obstructed 20 times as often as they assisted pedestrians. Those driving at speeds ranging from 5 km/h to 50% above the speed limit obstructed 7 times as often as they assisted. Finally those driving at or below the speed limit obstructed twice as often as they assisted. Surprisingly, public service vehicles (100%), university security/maintenance vehicles (80%) and general drivers (90%) obstructed at similar rates which, in turn, were much than those found in earlier studies in England and in Canada. A prototype locale based driver improvement programme is presented.

Wilson (1991) noted that a neglected but promising opportunity for effective driver improvement programmes lay in the field of specific locale based driving style assessment and subsequent modification. In pursuance of this concept, Wilson and McArdle (1992) developed and applied a direct observation technique for assessing driver - pedestrian incidents on a set of corners and pedestrian crossings in Greater London. In this technique, audio recording of all events observed to transpire when a group of pedestrians were at or on a crossing and one or more vehicles were within a given distance of the crossing. Out of 551 pedestrian groups, 110 incidents of infringements of pedestrian road rights according to the highway code were noted and only 10 instances of courteous behaviour on the part of drivers. Wilson and Godin

(1994) applied this observational technique to a comparative study of driver pedestrian incidents on a set of crossings near shopping centres in Sudbury, Canada. When like situations were compared in the two studies, British figures yielded about 1/3 of driver - pedestrian interactions as resulting in obstructions by the driver and only 2% resulting in driver assistance, while in Canada, the driver obstruction/assistance rates were about 20% and 30% respectively (Wilson and Godin, 1994).

In Laurentian University, Sudbury, Canada most of the 4000 students and 700 faculty and staff drive to university and use one or more pedestrian crossings on a 3 km private road network on campus. Because of the circumscribed nature of the traffic, an ideal opportunity for is presented for locale based driver assessment and modification if warranted. The question asked in this study is "Does the nature of driver-pedestrian interactions at crosswalks at Laurentian University indicate the need for modification by measures such as a locale based driver improvement programme?"

Method

Three pedestrian crossing areas were chosen on the 3 km of private roads of Laurentian university in Sudbury, Ontario: Willet, a two metre wide crosshatched crosswalk, located on a wide curve in the road, connecting the Willet-Green mining/research centre with the physical science buildings; Education, an unmarked crossing site at the major intersection dividing physical education buildings from other academic faculties; and Affiliated Colleges, a two metre wide cross hatched crosswalk at the bottom of a moderate hill connecting the main campus with the affiliated colleges. Vehicles were classified as either university security and maintenance (painted Laurentian university vehicles with logos), service (off-campus buses, delivery trucks and other service vehicles) and general (including students, staff and all other visiting drivers). Time measured velocity was assessed for all vehicles interacting with pedestrians over the 50 metres immediately preceding the crossing site. Traffic volume was assessed by counting traffic for five minutes before and after observation sessions.

Using the Wilson and McArdle (1992) pedestrian vehicle interaction assessment method, transpiring events when pedestrians were within a metre of the crosswalk and simultaneously vehicles were within 50 metres were tape-recorded, later transcribed and analysed for 27 hours during two weeks in October and 1 week in November, 1994 and 27 hours during four weeks in January, 1995. Days, times (one hour intervals 8 a.m to 5 p.m.) and crossing sites were randomly chosen - within limits of the observer's lecture times - for each observation. Six hours of pilot studies were carried out to develop techniques and reliability of observations was assessed by using a second observer for 11% of the sessions. In recording vehicle- pedestrian interactions in these latter sessions, 100% agreement between raters was obtained.

Results

About 15,000 vehicles and 816 groups of pedestrians traversed all three crossing sites during the 54 hours of observations. Of these events, 612 pedestrian group crossings were uneventful, 166 groups interacted with one or more drivers from the general population, 14 groups interacted with university security or maintenance vehicles and 12 groups interacted with off-campus service vehicles. No significant differences were noted in the relative frequency of assistance/obstruction by time of day or by crossing location; thus the data was pooled. In table 1 can be found the assistance/obstruction frequencies of interacting pedestrian groups by type and speed of vehicle. Interactions classified as obstruction (167), in which pedestrians were prevented from or interfered with in using the crosswalk by a driver, were about 7 times as frequent as interactions classified as assistance (25) in which interacting vehicles slowed down or stopped to allow pedestrians to use the crossing (cumulative binomial probability p < 0.001).

The speed limit on all campus roads was 40 km/h. The speed of oncoming vehicles when interacting with crossing pedestrians was classified as very fast (60 - 90 km/h), fast (45 - 59 km/h) and moderate (15 - 44 km/h). There was a significant difference (χ^2 = 22.5, df = 2, p < 0.001) in the ratio of assistance to obstruction of pedestrians for very fast vehicles pedestrians (3/63) compared with the corresponding values for fast (11/79) and moderate speed (11/25) vehicles. When pedestrians were present at crosswalks, 80% of drivers were travelling more than 5 km/h above the speed limit and 1/3 of the drivers were travelling more than 50% above the speed limit (carrying two times as much energy as they would if driving at the speed limit). In sum, when interacting with pedestrians, vehicles going from 5 km/h above the speed limit up to 50% more than the speed limit obstructed pedestrians about 7 times more than they assisted, those travelling 50% more than the speed limit to more than twice the speed limit obstructed about 20 times as much as they assisted. Finally, while somewhat more sociable than speeding drivers, those operating within the speed limit still obstructed about twice as often as they assisted pedestrians.

Table 1. Assistance/obstruction of crossing pedestrian groups by type (general public, university security and maintenance or off - campus bus or truck service vehicle and speed of driver in the last 50 metres before crossing.

Speed	60 - 90 km/h		45 - 59 km/h		15 - 44 km/h		
Driver	Ass	Obs	Ass	Obs	Ass	Obs	sum
General	3	60	11	64	8	20	166
University	0	2	0	6	3	3	14
Service	0	1	0	9	0	2	12
sum	3	63	11	79	11	25	192

For drivers interacting with pedestrians at crossings, the breakdown of type of driver by the above three speed classifications (very fast, fast and moderate) was general drivers (63/ 75/ 28), university security and maintenance drivers (2/ 6/ 6) and off campus service vehicles (1/ 9/2). When interacting with pedestrians, a significantly higher proportion (χ^2 = 67, df = 2, p < 0.001) of general drivers (40%) travelled at speeds greater than fifty percent above the speed limit compared with universitysecurity and maintenance drivers (15%) and off-campus service vehicles (10%). Only about17%

of both general drivers and off-campus service drivers travelled within the speed limit when interacting with pedestrians at crossings while 43% of campus security and maintenance drivers did so. Finally, 70% of off -campus service vehicles travelled 5 to 20 km/h above the speed limit. In sum, although there was overlap in the middle speed category, a tendency was noted for different types of drivers to drive at different speeds. That is, about 40% of general drivers but few other types of drivers travelled above 60 km/h, three quarters of off campus service vehicles travelled between 45 - 59 km/h and half of university security and maintenance vehicles travelled at or below the speed limit.

No significant difference (χ^2 = 3.24, df = 2, p < 0.2) was noted in the relative ratios of assistance/obstruction among the three types of drivers: general (22/144), university security and maintenance (3/11) and public service vehicles (0/12). In sum, 100% of all off-campus service vehicles (city buses etc), 90% of general traffic and 80% of university security and maintenance vehicles obstructed pedestrians with whom they interacted.

Discussion

A sample of recorded pedestrian - driver interactions protocols might elucidate the nature of the data collected in this study. "Two female pedestrians are in middle of crosswalk, bus driver continues to go, does not slow down; female driver approaching from opposite direction stops since the two pedestrians are in middle of crosswalk, meanwhile a male pedestrian walks across crosswalk to middle of road, driver does not remain stopped for him but continues on her way." "One female pedestrian forced to wait at curb until university security - maintenance vehicle drives by". "Two female pedestrians begin crossing with no approaching vehicle, halfway across a male driver appears travelling 60 km/h and comes to an abrupt stop to let pedestrians across". "University - security driver travelling within speed limit sees female pedestrian, slows down sufficiently to let her cross but does not stop". "Female pedestrian is in middle of crosswalk, female driver coming from University of Sudbury has stopped at three way intersection one metre into roadway, male driver drives through crosswalk at 60 km/h; second female driver drives across crosswalk and then turns corner; the stopped vehicle turns across crosswalk. No-one allows pedestrian to continue crossing." "One male driver comes to rolling stop when he sees pedestrian walking on crosswalk, pedestrian runs the rest of way across the road." "Three female pedestrians at crosswalk wait until procession of four vehicles drive by, one female then runs across crosswalk while other two females wait for another two vehicles to drive by. The three females are heard commenting to the effect that "don't people realize this is a crosswalk" and about finger gesturing crossing practices used in Toronto." "Two female pedestrians in middle of crossing when male driver approaches at 45 km/h; he does not stop but slows down and swerves around them." "One male pedestrian is forced to wait at Willet crossing while 10 vehicles go by." "Two female pedestrians standing on crosswalk wait for 5 vehicles, the last of which is a city bus, to pass then run across since another vehicle is in sight." "Two male pedestrians more than half way across the road have two vehicles approach and traverse the crosswalk; the first, does not slow; it comes very close to the pedestrian; the second vehicle also comes close to the pedestrian who then leaps to the curb." "Two female adults with seven children stand waiting at crosswalk, university security - maintenance vehicle and then two further vehicles go by without stopping." Finally, on the good side, "one female pedestrian enters crosswalk, male driver driving at 45 km/h comes to a complete stop, waits until pedestrian completes crossing and then continues to drive."

It is appropriate to note that, overall, the 87% rate of obstruction of pedestrians by drivers at crossings at Laurentian University is extremely high and does not compare favourably with the 23%, 19% and 17% rate of obstruction note in Sudbury signalled crossings, private road marked shopping mall and unmarked shopping mall crossings or even the 35% rate of obstruction noted in Greater London, England (Wilson and Godin, 1994).

What contributing factors might help to account for the high level of obstruction (80 - 100%)of pedestrians on Laurentian campus? First, explanation in terms of Laurentian being a private road, per se, fails because Wilson and Godin (1994) report obstruction levels below 20% on a different set of Sudbury private roads. Second, a truth seeker would be remiss is failing to point out that Laurentian University operates under an cultural/special interest group based agenda which, in many instances, *de facto*, gives preference to certain cultural groups, distorts hiring practices, administration appointments and university policy decisions and creates culturally based ghettos. It would be natural enough for resulting latent hostility and alienation of considerable numbers of community and university members to be expressed on the road. Third, explanation in terms of Laurentian administration action is and apparent attitude towards pedestrians is possible. Rather than act in such manner as to encourage more sociable behaviour on the part of drivers towards pedestrians, administration has within the last couple of years carried out the following actions. First, signs have been placed at some pedestrian crossings saying in two languages "uncontrolled walkway proceed with caution". Second, a crosshatched pedestrian crosswalk has been removed from a major pedestrian crossing at the Physical Education Building. Third, a second crosshatched pedestrian crosswalk was relocated from a location likely to be used by pedestrians to a more dangerous location and moreover away from where pedestrians were travelling. Fourth, large building identification signs have been placed in positions restricting turning vehicles vision of oncoming traffic and hence attending time for crossing pedestrians. The placement of the "uncontrolled walkway" signs, in particular, contrast greatly with signs placed on the outskirts of Calgary to the effect that pedestrians were to be given the right of way at all intersections and crossings. Calgary is a major city of 700,000 in Alberta, which orentian, the individual, as person, rather than the special interest group as unit of analysis and on the principal of equality of treatment of individuals. Two interpretations of the signs have emerged from conversations with individuals on campus regarding the "uncontrolled walkway"signs, neither of them desirable: first, that vehicles have the right of way over pedestrians and should not yield to them and second that the administration is trying to abrogate its responsibility for the safety of people on campus and so reveals a basic non-caring attitude towards persons on campus. In postscript, it might be noted that the senior author has been at some pains to get at least one of the building identification signs and one cross-walk re-located and that pressure from other quarters has resulted in the return of the Physical Education crosswalk.

We can answer affirmatively the question posed in our study and conclude that the nature of driver pedestrian interactions at crossings warrants improvement. A locale based driver improvement programme might well be warranted on campus. In outline, such a programme might proceed as follows. Signs would be placed at the two entrances to campus to the effect that pedestrians should be given the right of way at all intersections and crosswalks. Building upon the already sociable travelling speed of half of university security and maintenance vehicles, such drivers would be encouraged to lead the way as models of appropriate driving style by driving within the speed limits and giving pedestrians the right of way at all times at the said locations. City bus administrators might be approached to modify bus driving practices. About fifty metres from crossings signing of the crosswalks with instructions such as "stop for pedestrians at crossings" as well as possible widening of some or all of the pedestrian painted crosswalks might be entertained. A concentrated attempt to point out concretely (see Wilson, 1991) desirable driving style on campus might be made during student orientation. Similar targeted concrete descriptors of desirable on-campus driving style could be given to faculty and staff. Specific driving style suggestions to students, faculty and staff using the road up to the affiliated colleges might well prove profitable. Periodic surveys of campus driving style might be taken and published in the Gazette. Finally, administration might be encouraged to change there apparent stance on driving on campus, make appropriate noises on the topic and ensure that their traffic and road initiatives are consistent with encouraging more sociable driving on campus.

References

Wilson, Tay, 1991. Locale Driving Assessment - A Neglected Base of Driver Improvement Interventions. In *Contemporary Ergonomics*, Praeger, London, pp 388-393.

Wilson, Tay, and McArdle, G., 1992. Driving Style Caused Pedestrian Incidents at Corner and zebra Crossings. In <u>Contemporary Ergonomics</u>, Praeger, London, pp 388-393.

Wilson, Tay and Godin, Marie., 1994. Pedestrian/Vehicle crossing incidents near shopping centres in Sudbury, Canada. In <u>Contemporary Ergonomics</u>, Praeger, London, pp 186 -192.

ASSESSING THE DETERMINANTS OF DRIVER RISK: UTILITY VS. ENGINEERING

Di Haigney[1], Charlotte Kennett[2] and Ray.G. Taylor[2]

[1] *Road Safety Dept.*
RoSPA
353 Bristol Road
Birmingham B5 7ST

[2] *Applied Psychology Division*
Aston University
Aston Triangle
Birmingham B4 7ET

The main premise of Risk Homeostasis Theory (RHT) - that compensatory behaviour will mainly be realised through the utility mechanism - is tested. Utility and engineered safety were operationalised as experimental conditions in a pilot simulated driving task on the Aston Driving Simulators (ADS), through an adaptation of the time/distance methodology developed by Hoyes (1992) and via a damage vulnerability/invulnerability mechanism respectively. Analysis of driver performance data revealed that statistically significant differences were only recorded as a main effect of engineered safety. This result brings into question the adequacy of the'utility driven' compensation models of driver behaviour and risk assessment against a 'passive' engineering formula. The implications of the analysis are discussed and suggestions for future work are made.

Introduction

A utility-based theory of behavioural compensation in drivers has been developed in the last decade which has been the subject of much controversy (Trankle and Gelau, 1992). The model is 'Risk Homeostasis Theory' (RHT) which was formerly referred to as the 'Theory of Risk Compensation'.

RHT was initially proposed by Wilde to account for the rate of accident loss observed per time unit of exposure, per capita and per kilometre driven (Wilde, 1989) and as a framework from which interrelations between these measures can be explained. RHT has also been posited as a potential 'general theory of behaviour' (Wilde 1985, 1989), although this paper is concerned with the theory as it relates to driver behaviour.

In RHT, the population temporal accident rate (the summed cross-products of the frequency and severity of accidents per time unit of road user exposure) is regarded as arising not from a general attempt to minimise perceived risk levels as is postulated in some theories (e.g. Naatanen and Summala, 1976), but rather to match the subjectively perceived risk against a 'target' or 'desired' degree of risk.

The 'target' level of risk is determined by the interaction of four utilities associated with those behavioural options which are regarded as either riskier or safer alternatives to the road-users current activity - and so each utility varies relative to the situation the user currently deems him/herself to be in. According to this model, road-users engaged in a driving task will, through assessing the four utilities simultaneously, experience a 'net' subjective level of risk which is deemed as being the 'acceptable' or the target level of risk. This is frequently compared with the 'perceived' level of risk at a 'pre-attentive' level of cognition (Wilde, 1982a, p.210).

Should a difference between these two variables become great enough for the discrepancy to be perceived in the road-users central consciousness, the road-user will then behave in a manner aimed at reducing this difference to a 'subjective zero', i.e. they will engage in certain 'compensatory behaviours', until the difference between perceived risk and target risk lies just below the 'just-noticeable-difference' or 'JND' level. Such compensation is not hypothesised in RHT as occurring through specific pathways (Wilde, 1989, p.277), but may be realised spontaneously through any behavioural option available to the individual, depending upon the 'net' utility associated with each alternative course of action.

Wilde regards the target level of risk as the fundamental determinant of accident rates, as all compensatory behaviour serves to subjectively match perceived risk against it. (Haigney, 1995). As a result, Wilde maintains that 'conventional engineering' road safety solutions are not effective in the long term as they do not directly address utility issues and therefore not target risk.

Hoyes (1992), argued that certain methodologies, such as the Aston Driving Simulators (ADS) provide the opportunity for 'collapsed experience',which can effectively account for the 'long term' negation of compensation in engineering interventions in the 'short term' of an experimental run. The ADS are therefore regarded as a suitable means for testing for compensation via relatively short term 'engineered' environmental interventions.

This study attempts to induce differences in both utility and 'the degree of engineered safety', so that the relative weighting of either in the compensatory process may be isolated and identified.

The operationalisation of utility was achieved through an adaptation of the methodology described in Hoyes (1992), in which participants undertaking some experimental run on the ADS were informed that they either had to 'drive' for a specified period of time or for a specific 'distance' in the simulated driving environment (all subjects in actual fact drove for the same amount of time). Hoyes (1992), argued that subjects in the 'distance' condition, would attach a greater positive utility to risky behaviours which would allow them to complete the task more quickly than otherwise, and that subjects in the 'time' condition would have no reason to attach utility to such behaviours. As a result, Hoyes (1992), asserted that any nonsignificance in ADS performance across time/distance utility conditions could indicate that utility is not necessarily the single factor governing accident rates - as claimed in RHT.

Unfortunately, Hoyes (1992) was not able to develop this means of operationalising utility further and as a result, the methodology as it stands is open to considerable criticism through the major assumption (of participant motivation and consequent action) held within it (Wilde, personal communication, 1992).

The adaptation of this methodology described in this study seeks to address this criticism through the use of penalty and reward systems as described below. Furthermore, the efficacy of these systems in inducing differences in utility across the 'time' and'distance' conditions was evaluated through post-run group discussions with subjects.

Engineering was operationalised through the provision, or non provision of information on the engineered 'robustness' of the simulated vehicle if it were to be involved in a collision reinforced through direct 'feedback' (through sound, visual stimuli and resulting performance of the simulated vehicle), after a collision was experienced on the ADS.

In this study 'compensation' has been operationally defined as those significant differences within the ADS variables logged between the experimental conditions of utility ('time' and 'distance') and engineering ('feedback' and 'non feedback').

Method

The ADS consists of a car seat, a steering wheel, accelerator and brake pedals placed so as to mimic operating conditions within an automatic car. The 'windscreen view' is represented by a computer graphic output of a road image, displayed on a terminal placed in front of the participant. When the simulation program is initiated, the 'view' also incorporate images of 'other traffic' on a single carriageway road, travelling on both carriageways. The 'other traffic' was capable of 'intelligent action', such as overtaking the user.

Participant responses, through the the use of pedals and steering wheel, are not only used to update the screen output (so that moving the steering wheel caused the 'car' to move towards one edge of the 'road', for example), but also stored as 10 driving performance variables (mean speed, mean acceleration, mean braking, number of [successful] overtakes, steering variance, collisions with others on the righthand side of the carriageway, collisions with others on the lefthand side of the carriageway, right headway, time spent on the verge on the righthand side of the carriageway, time spent on the verge on the lefthand side of the carriageway).

After a 'practice session', participants were randomised between the two utility conditions on the ADS. Half the participants 'drove' the simulator for ten minutes under a 'time' utility condition and half under the 'distance' utility condition, with both sets of participants experiencing both engineering conditions of 'feedback' and 'non feedback' conditions for five minutes each in a random order.

All participants were presented with standard instructions on the simulator screen throughout the run which informed them of the utility conditions which they would be experiencing, any penalties/rewards associated with the run overall and the vulnerability/invulnerability of the simulated vehicle as appropriate.

All subjects were informed that a monetary penalty would be levied for each collision experienced on the ADS. Subjects in the 'distance' utility condition were also informed that this penalty could be reduced by a substantial amount if they completed the driving task quickly, 'freeing up' simulator time.

When participants had completed both feedback conditions on the ADS, they were required to complete a short questionnaire and took part in a group discussion on the effectiveness of the utility operationalisation methodology.

Results

Repeated measures ANOVAs were performed for all the dependent variables provided by the ADS for both feedback conditions, between subjects ANOVAs for ADS dependent variables resulting from the utility conditions, two-way Chi squares aswell as Pearson correlations between ADS variables and questionnaire responses were calculated. Group discussions were subjected to qualitative content analysis.

Data are discussed by 'Risk' and by 'Utility'. The term 'Risk' refers to the mean values of the ADS dependent variable for the 'feedback' and 'non feedback' conditions and the term 'Utility' refers to the mean ADS variable values for the 'time' and 'distance' conditions.

Mean Speed

Mean speed was found to be significant across Risk ($F[1,30]=6.749$; $p=.0144$) although it was nonsignificant across Utility ($F[1, 30]=.205$; NS) and no interactive effect was determined for UtilityxRisk ($F[1, 30]=1.53$; NS).

Mean Acceleration

A main effect of Risk was established for mean acceleration ($F[1, 30]=14.126$; $p=.0007$), though no significance was recorded for Utility ($F[1, 30]=2.358$; NS), or for UtilityxRisk (F[1, 30]=1.918; NS).

Mean Braking
Both the main effects of Risk (F[1, 30]=1.02; NS) and Utility (F[1, 30]=.012; NS) did not reach significance. The interactive effect of UtilityxRisk (F[1, 30]=.369; NS) was also nonsignificant.

Number of Overtakes
The 'overtakes' variable logs the number of times a subject has successfully managed to pass the vehicle ahead of them in the lefthand carriageway (i.e. through driving into the righthand carriageway and returning completely into the lefthand carriageway).
Risk was marginally significant as a main effect for mean braking (F[1, 30]=4.226; p=.0476), whilst both Utility (F[1, 30]=2.252; NS) and UtilityxRisk (F[1, 30]=.587; NS) were nonsignificant.

Steering
As the 'overtakes' variable logged only those overtaking manoeuvres which were successful, aborted attempts at overtaking may be deduced through reference to the 'steering' variable, which recorded the variance in simulated vehicle tracking, relative to the centre-line of the road.
No significant effects for Risk (F[1, 30]=.22; NS), or for Utility (F[1, 30]=.398; NS) were gained. A nonsignificant result for UtilityxRisk (F[1, 30]=.392; NS) was also recorded.

Collisions with others on the righthand side of the carriageway
The mean number of collisions with others on the righthand side of the carriageway was found to be significant with a main effect of Risk (F[1, 30]=5.351; p=.0278). Utility (F[1, 30]=.0057; NS) was nonsignificant as a main effect and UtilityxRisk (F[1, 30]=1.224; NS) was also nonsignificant.

Collisions with others on the lefthand side of the carriageway
The mean number of collisions with others on the lefthand side of the carriageway was nonsignificant for Risk , Utility and UtilityxRisk at: (F[1, 29]=1.044; NS), (F[1, 29]=.016; NS) and (F[1, 29]=.51; NS) respectively.

Right headway
'Mean right headway' is the value in 'simulated metres' between the subjects vehicle when in righthand side of the carriageway (such as during an overtaking manoeuvre) and of any oncoming vehicle in the righthand carriageway. It therefore allows the degree of risk taking during either successful and/or aborted overtaking manoeuvres to be established.
Utility (F[1, 28]=.242; NS), Risk (F[1, 28]=.003; NS) and UtilityxRisk (F[1, 28]=.017; NS) all failed to produce any significant results.

Time spent on the verge on the righthand side of the carriageway
During the driving task, it was possible for subjects to 'run their cars off the road' and onto the 'grass verges' on either side, generally as a result of oversteering during some collision avoidance measure.
Neither Risk (F[1, 30]=.99; NS) or Utility (F[1, 30]=1.18; NS) gave significant results for a main effect. UtilityxRisk (F[1, 30]=.017; NS) was also nonsignificant for this variable.

Time spent on the verge on the lefthand side of the carriageway
Risk was found to be a main effect (F[1, 30]=7.58; p=.0099), although Utility (F[1, 30]=2.808, NS) and UtilityxRisk (F[1, 30]=.627; NS).

Questionnaire analysis and Group Discussions

Of all the questionnaire items, only two produced any significant relationship with the ADS variables, namely question B3 ('How do you rate your overall driving performance against the average driver?') and question D5 ('It is a waste of a fast/sporty car if it is not driven fast'), both of which correlated positively with mean steering at (r=.445 and r=.423) respectively.

As noted above, the steering variable allows risk taking through successful aswell as aborted overtakes to be examined. As the questionnaire items tested for self rating of task skill and risk acceptance, it would appear that these items are useful predictors of ADS performance and risk taking propensity. The lack of any relationship with other questionnaire items is a cause for concern however, and may indicate a need for some considerable revision of the questionnaire items and format.

As subjects were run in groups of four, group discussions were held directly after each group had completed their run on the ADS and centred on evaluating the operationalisation of utility adopted in the study. As the utility conditions were balanced (i.e. two participants per run would have experienced either the time or distance condition), either condition was represented equally in each group.

The majority of subjects stated that they felt the reward/punishment systems had noticeably (in their view) affected the way in which they drove the ADS, with those under the 'time/feedback' condition, tending to emphasise the care with which they had taken to avoid collisions. Those experiencing the 'distance/non feedback' conditions reported more risk taking than other subjects.

Discussion

Whilst a number of ADS variables recorded a main effect by Risk (i.e. mean speed, mean acceleration, overtakes, collisions on the righthand side of the carriageway and time spent on the righthand verge), none of the ADS variables recorded a main effect of Utility, or an interactive effect of Utility×Risk.

Although this result could have arisen through the utility operationalisation technique failing to create a great enough distinction in utility between the 'time' and 'distance' conditions, the group discussions would appear to confirm that the utility methodology did allow for subjects to distinguish between the conditions successfully in terms of the values attached to driving styles.

These differences are not reflected in those ADS variables which can be taken as indicating either safety acceptance (Wagenaar and Reason, 1990), or risk acceptance (Matthews, Dorn and Glendon, 1991) however, even though indications of possibly compensatory behaviours have been determined via a main effect of Risk in the ADS variables.

As the ADS have been evaluated as sensitive to changes in driving style (Dorn, 1992), non significance across Utility conditions *could* suggest that a change in a utility exceeding the JND was not realised through task relevant behaviour directly (Adams, 1988), or it may be that the subjective 'size' of the difference perceived between utility conditions was sufficiently great to warrant recognition by the subjects, but not great enough to result in direct behavioural change.

In conclusion, it would seem that the infamously 'woolly' (Haigney, 1995), concept of utility is more elusive, fluid and dynamic than has been appreciated to date. It would appear that there is scope for a 'grey area' in the utility evaluation mechanism in RHT, where although the net evaluation of the four utilities is shifting beyond the JND, subjects may not be responding with compensatory behaviour, even though they recognise a 'need' for it. On the other hand, some compensation appears to occur with little reference to the direct manipulation of utilities but is rather more 'reactive' than 'pro-active', although RHT would tend to suggest the latter. Further work is required in the area of utility in risk - especially in terms of its measurement and operationalisation, as until more progress has been made, the empirical status of utility driven theories of behaviour cannot be established.

References

Adams, J.G. 1988, Risk homeostasis and the purpose of safety regulation,
    Ergonomics **31**(**4**), 407 - 428.

Dorn, L. 1992, *Individual and group differences in driving behaviour*. Unpublished
    Doctorate Thesis, Applied Psychology Division, Aston University.

Haigney, D. E. 1995, Compensation - implications for road safety, *InRoads* **17** (1),
    21-33.

Hoyes, T. W. 1992, *Risk Homeostasis Theory in simulated environments*,
    Unpublished Doctorate Thesis, Applied Psychology Division, Aston
    University.

Matthews , G., Dorn, L. and Glendon A.I. 1991, Personality correlates of driver stress,
    Personal and Individual Differences **12**, 535 - 549.

Naatanen, R. and Summala , H. 1975, A simple method for simulating danger-related
    aspects of behaviour in hazardous activities, *Accident Analysis and Prevention*
    7, 63 - 70.

Trankle, U. and Gelau, C. 1992, Maximisation of participative expected utility or risk
    control? Experimental tests of risk homeostasis theory, *Ergonomics* **35** (1), 7 -
    23.

Wagenaar, A.C. and Reason, J.T. 1990, Types and tokens in road accident causation,
    Ergonomics **33** (**10 - 11**), 1365 - 1375.

Wilde, G. J.S. 1982a, The theory of Risk Homeostasis: Implications for safety and
    health, *Risk Analysis* **2**(**4**), 209-225

Wilde, G.J.S. 1985, Assumptions necessary and unnecessary to risk homeostasis,
    Ergonomics **28**, 1531 - 1538.

Wilde, G.J.S. 1989, Accident countermeasures and behavioural compensation: The
    position of Risk Homeostasis Theory, *Journal of Occupational Accidents*
    10(**4**), 267-292.

TAKING THE LOAD OFF: INVESTIGATING THE EFFECTS OF VEHICLE AUTOMATION ON MENTAL WORKLOAD

Mark Young and Neville Stanton

Department of Psychology
University of Southampton
Highfield
Southampton SO17 1BJ

In the quest to relieve operators of workload - physical or mental - an engineering solution is often to implement automation in some form. Recent technological advances have meant that the driver of the modern automobile will soon be faced with such automation. However, devices such as Adaptive Cruise Control and Active Steering pose new concerns for the vehicle ergonomist. Some of the classic problems of automation, such as vigilance and automaticity, and in particular those concerns surrounding mental workload, are addressed in this paper. It is suggested that automated vehicle systems reduce driver workload to such an extent that it may be detrimental to performance. Future research needs to determine this; the current paper draws on evidence from workload and automaticity literature to make the case.

Introduction

In the quest to relieve operators of workload - physical or mental - designers of complex systems often use the technology at their disposal to implement automation in some form. The automotive industry is no exception to this, with systems such as automatic transmission and power assisted steering being familiar on today's roads. More recently, increasingly novel technologies are being developed for the vehicle of tomorrow. Devices such as Adaptive Cruise Control (ACC), which controls both speed and headway of the user's vehicle; and Active Steering (AS), which keeps the car in its current lane until interrupted, are likely to be on the road within the next decade.

Although these innovations will undoubtedly benefit the driver to some extent, it is anticipated that there will also be some disadvantages. Stanton & Marsden (1996) foresee potential problems in automobile automation, including equipment reliability, training and skills maintenance, error-inducing equipment designs, and overdependence on the automated system. They argue that driver workload is only excessive in exceptional circumstances, thus appropriate allocation of function may not involve complete automation - this could result in increased workload (cf. Reinartz & Gruppe, 1993). Conversely, with increased automation comes a transition in the driver's role from operational to supervisory control (Parasuraman, 1987), raising problems of inattention and vigilance.

It is these latter problems which are of primary concern in this paper, for it is recognised that mental underload is at least as serious an issue as overload (Brookhuis, 1993; Leplat, 1978; Schlegel, 1993). In a series of studies, we hope to demonstrate that automation will lead to driver underload to such an extent that it may actually impede performance. Furthermore, the issue of automaticity will be explored by comparing the performance of novice and expert users with ACC and AS. By discovering adverse effects of automation in the largely uncharted territory of mental underload, one may contribute to designing for the user and optimise the performance of the system as a whole.

Automation and Mental Workload

Both psychologists and ergonomists have been concerned with automation for the past two decades. Seminal papers in the field (e.g. Bainbridge, 1983; Norman, 1990; Reason, 1990) have argued that automation poses problems to the human operator in terms of skill degradation and either inappropriate or insufficient feedback. These issues are generally symptomatic of the transition in the role of the human from operational to supervisory control (Parasuraman, 1987). Such a situation has the potential for simultaneously imposing overload and underload: reduced attention during normal operations, resulting in the classic vigilance decrement (Parasuraman, 1987; Singleton, 1989); however difficulties increase when faced with a crisis or system failure (Norman, 1990). In the latter scenario, the human is forced to immediately return to the operator role, gather information about the system state, make a diagnosis and attempt a resolution. Indeed, Gopher & Kimchi (1989) argue that monitoring and decision-making skills in automation already stretch capacity limitations, and there is a need for knowledge and expertise before entering this kind of work. This possibly represents *the* irony of automation - that it is of least use when it could be the most helpful (i.e. in a failure scenario).

Thus we see that automation can have bidirectional effects on mental workload (MWL). Concern with what constitutes underload or overload is rife, and has stemmed primarily from the aviation industry (Sanders & McCormick, 1993). However, there are investigations of MWL in car driving. Certain driving tasks have been found to increase MWL and consequently present potentially dangerous situations (see e.g. Dingus, Antin, Hulse & Wierwille, 1989; Hancock, Wulf, Thom & Fassnacht, 1990). Some researchers (e.g. Brookhuis, 1993; Fairclough, 1993; Wildervanck, Mulder & Michon, 1978) have therefore explored the use of monitoring systems to detect situations of driver underload or overload, and to intervene either directly or indirectly if the situation becomes critical. This type of monitoring popularly utilises physiological channels. Other researchers (e.g. Schlegel, 1993; Verwey, 1993) are interested in the determinants of driver workload with a view to developing adaptive interfaces which may reduce such workload in future. These studies are generally in recognition of the fact that modern drivers are presented with increasingly complex information, with the advent of Intelligent Vehicle Highway Systems (IVHS).

As far as we are aware, there is only one piece of research so far which has attempted a structured evaluation of vehicle automation. Nilsson (1995) investigated the effects of ACC in critical situations. It was found that ACC did influence behaviour, such that for the situation in which collisions occurred (when the car approached a stationary queue), 80 per cent of the collisions occurred when ACC was engaged. Nilsson attributed this to the expectations that drivers have about ACC, rather than to increased workload or decreased alertness.

Questions about workload become more difficult as technology changes work, resulting in mental load being predominant (Rumar, 1993; Singleton, 1989). Automation is usually intended to reduce workload, although this is not necessarily a good thing - the goal should be to optimise workload, with implementations such as the electronic copilot (Parasuraman, 1987) and human-centred automation (Reichart, 1993; Rumar, 1993). Such optimization will inevitably involve a balancing act between demands and resources of both task and operator. Perhaps this view, which essentially captures the spirit of ergonomics itself, could offer more satisfactory solutions.

Automaticity

Automatic processing is defined as being fast, attention-free, unconscious and unavoidable, and is the converse of controlled processing (see e.g. Anderson, 1995; Eysenck & Keane, 1990 for reviews). It is thought (Baddeley, 1990) that these elements lie on a continuum, rather than being discrete from each other. At first appraisal, there seems to be a lot in favour of automatic processing, however there are drawbacks associated with the unconscious and unavoidable criteria. The classic example is the Stroop effect, however the cost of automaticity is manifested in any procedure which is highly routinised yet requires close attention, such as verbal checklists (Barshi & Healy, 1993). Errors occur due to reduced workload and vigilance, similar to the ætiology of automation errors outlined above. Barshi & Healy (1993) see a paradox in the checklist procedure - automatic performance is equated with expert performance; however routine tasks are complex and susceptible to error, thus the operator must execute controlled processing in order to avoid errors. Yet, by definition, it is impossible to perform at once in an automatic *and* a controlled manner.

Gopher & Kimchi (1989) believe that real-life tasks actually employ a balance of automatic and controlled processes, and that mental load is determined by the proportions of each. Furthermore, it has been demonstrated that whilst subjective workload is influenced by the presence of automation, a secondary (time estimation) task can discriminate automatic from nonautomatic processing Liu & Wickens, 1994).

Bainbridge (1978) provides some inspirational thoughts on automaticity. Skill is viewed as a change in knowledge and decisions, such that the expert performs by implicit anticipation (i.e. "open-loop" behaviour) rather than feedback. Learning increases the knowledge base and leads to these expectations. Automaticity, then, is a situation of low uncertainty and high predictability, thus drawing little from attentional resources. In a high demand situation, predictability breaks down and a feedback (i.e. novice) strategy is returned to.

Driving is a skilled behaviour which is a classic example of automaticity (Stanton & Marsden, 1996). Indeed, the cost of automaticity is again all too evident in certain accidents (Hale, Quist & Stoop, 1988). However, some automatic aspects of driving are advantageous. In the Nilsson (1995) study described above, drivers avoided an accident when a car pulled out in front of them because braking was considered to be an overlearned and automatic response (a view shared by Baddeley, 1990).

Automation can be related to automaticity in a number of ways. It is posited here that drivers essentially satisfy the criteria for automaticity when faced with automation (that is, fast, attention-free, unconscious). However, with automation, this is true for both experts and novices. Consider Bainbridge's (1978) point that increased demand essentially transforms an expert into a novice. It is surely plausible to assume that the reverse would be true in a situation of unusually low demand. However, whereas the expert has an enhanced knowledge base and can anticipate events, the novice is deprived of this ability. Thus they will not react as experts in critical situations, such as the overlearned braking response in Nilsson's (1995) example.

These issues make the link between automation and automaticity worth exploring. Thus the major problem is how to cope with the further reduction in attentional demand which automation presents, and how to react in a failure scenario. This could be the real 'problem' with automation - that all situations will be subject to autonomous control, rather than associative or cognitive control (cf. Anderson, 1995). This would apply whether the antecedent was environmental (e.g. road junctions) or internal (a novice driver). The Institute of Advanced Motorists circumvents this problem by teaching drivers to always concentrate on their driving (this is facilitated by a commentary). It is likely that the technological solution will involve improved feedback, however this is a subject for future research to resolve.

Research in the Southampton Driving Simulator

It is our intention to assess specifically the effects of ACC and Active Steering on driver mental workload, and to explore how this may affect driver performance. Although many authors expect automation to increase workload, we anticipate that under normal circumstances, the effects of underload will be more substantial, particularly if a failure situation arises. Imagine a driver who has been travelling for at least 30 minutes with both automated systems engaged, thus has been "out of the loop" for long enough that their vigilance has degraded. Suddenly, one of the automated systems fails. Is it realistic to expect that driver to be able to reclaim control in a safe and timely manner?

Initial studies suggest not. Stanton, Young & McCaulder (1996) used the Southampton Driving Simulator to explore the effects of ACC failure on driver performance. It was found that one-third of all participants collided with the lead vehicle when ACC failed. Although not a majority, this is a substantial proportion of drivers. In addition, the use of a secondary task demonstrated that under normal circumstances, workload is significantly reduced when ACC is engaged.

The present study (Young & Stanton, 1996) intended to reinforce the findings of reduced workload under automation conditions by comparing performance under different combinations of manual and automated driving. Participants drove for 10 minutes in each of four trials: manual driving; using ACC; using Active Steering; using both ACC and Active Steering (i.e. fully automated vehicle control). The simulator was set up for automatic transmission in all conditions. Instructions to participants were such that a relatively constant speed (and consequently distance travelled) were maintained across all trials. This was hieved by using a "follow-that-car" paradigm - a lead car travelling at 70mph was to be followed for the duration of the trial. This procedure was relatively successful, as evidenced by the primary task data (see below).

Workload measures for this study included primary task variables, a secondary task measure, and a subjective mental workload scale - the NASA-TLX (Hart & Staveland, 1988). All primary and secondary task data was recorded automatically by the simulator software. The secondary task arguably draws upon the same attentional resource pools as driving (i.e. visual input, spatial processing and manual response). This was to ensure the secondary task was indeed measuring spare attentional capacity, and not the capacity of a separate attentional pool.

30 participants were used in this study, all of whom held a full British driving licence for at least 1 year (mean 6.9 years), and who drove for an average 5650 miles per year. 17 of the participants were male, and the average age was 25.3.

Results

Primary Task Data: As stated above, the primary task data suggests that the instructions to participants to maintain constant speed and headway were heeded. For headway, significant differences only arose when comparing ACC to non-ACC conditions $\chi^2_{(3)}$= 15.6; p<0.005), implying that although participants chose different headways than the automated system, these headways were consistent within participants. Similarly, the speed data only exhibit significant differences between ACC and non-ACC conditions. Indeed, it is apparent that significance was only achieved due to the limited dispersion of these data (the mea s were 70.0 for manual; 70.2 for ACC; 70.3 for Active Steering; and 70.5 for fully automated control). Thus it can be reasonably assumed that participants adhered to the instructions of following the leading car at a steady headway and speed of 70mph. Furthermore, we can then be confident that any apparent differences in workload from the secondary task or subjective measures are due to actual differences and not artefactual ones (e.g. different driving styles).

Secondary Task Data: Comparing number of correct responses on the secondary task across the four automation conditions, a Friedman Two-Way ANOVA was highly significant ($\chi^2_{(3)}$= 68.87; p<0.0001). A series of Wilcoxon Matched-Pairs Signed Ranks tests found all comparisons to be highly significant *except* when comparing manual driving to ACC supported driving - this was nonsignificant. In all cases, the direction of the difference was as expected - more correct responses (i.e. lower workload) when automation was engaged (means: 106 for manual; 114 for ACC; 179 for Active Steering; 215 for fully automated control).

Subjective Workload Data: The Overall Workload (OWL) scores presented a very similar picture to the secondary task data. Again, a Friedman Two-Way ANOVA was highly signific t ($\chi^2_{(3)}$= 63.25; p<0.0001), and the Wilcoxon comparisons were all highly significant ex t for manual versus ACC, which was again nonsignificant. Once more, the means suggested lower workload (i.e. lower OWL score) correlating with increased automation (63.3 for manual; 61.4 for ACC; 35.4 for Active Steering; 19.6 for fully automated control).

Discussion

Thus we see that automation does indeed have a significant effect on driver mental workload, although the specifics of this effect are not quite as expected. Nilsson (1995) found no significant differences in workload between manual and ACC-supported driving using the NASA-TLX. Yet on the basis of Stanton et al. (1996), it seemed that these results would be contradicted - ACC appeared to reduce workload as evidenced by the secondary task results. The present results, however, support Nilsson, with no effect of ACC on workload, and only significant reduction in workload when Active Steering is engaged (neither of the previous studies explored Active Steering).

The possibility that these contradictions are due to different measuring techniques (i.e. NASA-TLX in Nilsson's study; secondary task by Stanton et al.) are effectively ruled out when one examines the correlation between the two. As both techniques were used in the present study, it was possible to explore this relationship. A simple visual inspection of the results outlined above suggests that there is an association between the two variables, and a significant correlation confirms this (r=0.691; p<0.001). Thus almost half the variance in either variable is accounted for by the other. It is also notable that subjective workload results for ACC and manual conditions were equivalent across the present study and Nilsson (1995), despite the latter being in an apparently high fidelity moving-base simulator. It is therefore reasonable to conclude that the fidelity of our fixed-base simulator is satisfactory.

Returning to the issue of workload, then, it is apparent that in this situation, Active Steering has a far greater influence on workload than ACC. It has not yet been determined as to why this may be the case, however the practical implications are more obvious. More work needs to be conducted on whether this reduction in workload is too extreme; that is, resulting in a dangerous vigilance decrement, and potentially critical situations should the system fail. All of this is speculative, of course, nonetheless we can now be confident that automation at some level reduces driver workload significantly. It remains to be seen whether or not this is desirable.

Further speculation arising from these results introduces the concept of *malleable resource pools*. Evidence is accumulating that reducing demand per se is not necessarily the answer. Possibly, resources also shrink to accomodate this, in a converse of the 'work expands to fill the time available' tenet. This would explain why there is no apparent change in workload when ACC is engaged, yet accidents occur when it fails, implying excessive demands. Hopefully, research will determine the cause of such accidents, whether it be that ACC distracts the driver from operational aspects of driving, or that there is indeed some quantitative and/or qualitative change in the resource pool.

An ongoing programme of research plans to investigate further the effects of automation failure on workload and performance, by replicating and extending the studies of Nilsson (1995) and Stanton et al. (1996) to examine failures of ACC and Lane Support. Following from the automaticity arguments above, it is likely that if the demands of the situation are outweighed by the resources available, performance will deteriorate in the same way as if the converse were true. This returns us to the conclusion that workload should be optimised rather than reduced.

Conclusions and Future Research

It may be demonstrated that the introduction of novel technologies into the automobile has the potential for deteriorating driver performance rather than facilitating it. Devices such as ACC and particularly Active Steering can affect drivers' mental workload, lowering it to the extent that they suffer from the vigilance decrement, and keeping them out of the feedback loop such that - should a device fail - the driver faces an explosion of activity to undertake in order to avoid an accident. Research so far suggests that in certain situations, many drivers cannot cope with this eventuality, and a collision is the inevitable result. It is possible that this is related to the degree of expertise and automaticity the driver possesses, and future research will compare the performance of drivers of differing levels of experience to test this.

Ultimately, the series of studies in the Southampton Driving Simulator will use all of these results to propose suggestions for the design of future systems. It is anticipated that such designs will follow the dynamic allocation of function precedent, and provide enhanced feedback to keep the driver in the control loop. This would reflect a trend to assist drivers rather than to replace them. Some authors have already advocated the use of adaptive interfaces (Verwey, 1993) or human-centred automation (Reichart, 1993) as a path towards the goal of optimal driver workload. If advanced driving involves constant concentration, perhaps designers of automated systems should follow this precedent and provide constant feedback for drivers. Future research will determine whether this is the answer.

References

Anderson, J. R. 1995, *Cognitive Psychology and its Implications*, 4th ed. (W. H. Freeman & Co., New York)

Baddeley, A. 1990, *Human Memory: Theory and Practice*, (Lawrence Erlbaum, Hove)

Bainbridge, L. 1978, Forgotten Alternatives in Skill and Work-load, Ergonomics, **21**, 169-185

Bainbridge, L. 1983, Ironies of Automation, Automatica, **19**, 775-779

Barshi, I. & Healy, A. F. 1978, Checklist Procedures and the Cost of Automaticity, Memory and Cognition, **21**, 496-505

Brookhuis, K. A. 1993, The Use of Physiological Measures to Validate Driver Monitoring. In A. M. Parkes & S. Franzen (Eds.), *Driving Future Vehicles*, (Taylor & Francis, London) 365-376

Dingus, T. A., Antin, J. F., Hulse, M. C. & Wierwille, W. W. 1989, Attentional Demand of an Automobile Moving-Map Navigation System, Transportation Research-A, **23**, 301-315

Eysenck, M. W. & Keane, M. T. 1990, *Cognitive Psychology: A Student's Handbook*, (Lawrence Erlbaum, Hove)

Fairclough, S. 1993. Psychophysiological Measures of Workload and Stress. In A. M. Parkes & S. Franzen (Eds.), *Driving Future Vehicles*, (Taylor & Francis, London) 377-390

Gopher, D. & Kimchi, R. 1989, Engineering Psychology, Annual Review of Psychology, **40**, 431-455

Hale, A. R., Quist, B. W. & Stoop, J. 1988, Errors in Routine Driving Tasks: A Model and Proposed Analysis Technique, Ergonomics, **31**, 631-641

Hancock, P. A., Wulf, G., Thom, D. & Fassnacht, P. 1990, Driver Workload During Differing Driving Maneuvers, Accident Analysis and Prevention, **22**, 281-290

Hart, S. G. & Staveland, L. E. (1988). Development of NASA-TLX (Task Load Index): Results of empirical and theoretical research. In P. A. Hancock & N. Meshkati (Eds.), *Human Mental Workload*, (Elsevier Science, North-Holland) 139-183

Leplat, J. 1978, Factors Determining Work-load, Ergonomics, **21**, 143-149

Liu, Y. & Wickens, C. D. 1994, Mental Workload and Cognitive Task Automaticity: An Evaluation of Subjective and Time Estimation Metrics, Ergonomics, **37**, 1843-1854

Nilsson, L. 1995, Safety Effects of Adaptive Cruise Controls in Critical Traffic Situations, Proceedings of the Second World Congress on Intelligent Transport Systems, **3**, 1254-1259

Norman, D. A. 1990, The 'Problem' with Automation: Inappropriate Feedback and Interaction, not 'Over-Automation', Phil. Trans. R. Soc. Lond. B, **327**, 585-593

Parasuraman, R. 1987, Human-Computer Monitoring, Human Factors, **29**, 695-706

Reason, J. T. 1990, *Human Error* (Cambridge University Press, Cambridge)

Reichart, G. 1993, Human and Technical Reliability. In A. M. Parkes & S. Franzen (Eds.), *Driving Future Vehicles*, (Taylor & Francis, London) 409-418

Reinartz, S. J. & Gruppe, T. R. 1993, April, Information requirements to support operator-automatic cooperation. Paper presented at Human Factors in Nuclear Safety Conference, London.

Rumar, K. 1993, Road User Needs. In A. M. Parkes & S. Franzen (Eds.), Driving Future Vehicles, (Taylor & Francis, London) 41-48

Sanders, M. S. & McCormick, E. J. 1993, *Human Factors in Engineering and Design*, 7th ed. (McGraw-Hill, New York)

Schlegel, R. E. 1993, Driver Mental Workload. In B. Peacock & W. Karwowski (Eds.), *Automotive Ergonomics*, (Taylor & Francis, London) 359-382

Singleton, W. T. 1989, *The Mind at Work: Psychological Ergonomics*, (Cambridge University Press, Cambridge)

Stanton, N. A. & Marsden, P. 1996, Drive-By-Wire Systems: Some Reflections on the Trend to Automate the Driver Role, Safety Science, Manuscript in press

Stanton, N. A., Young, M. S. & McCaulder, B. 1996, Drive-By-Wire: The Case of Driver Workload and Reclaiming Control with Adaptive Cruise Control, Manuscript submitted for publication

Verwey, W. B. 1993, How can we Prevent Overload of the Driver? In A. M. Parkes & S. Franzen (Eds.), *Driving Future Vehicles*, (Taylor & Francis, London) 235-244

Wildervanck, C., Mulder, G. & Michon, J. A. 1978, Mapping Mental Load in Car Driving, Ergonomics, **21**, 225-229

Young, M. S. & Stanton, N. A. 1996, Automotive Automation: Investigating the Impact on Driver Mental Workload, Manuscript submitted for publication

OVERTAKING ON THE TRANS-CANADA HIGHWAY: CONVENTIONAL WISDOM REVISED

Tay Wilson

Psychology Department
Laurentian University
Ramsey Lake Road
Sudbury, Ontario, Canada
P3E 2C6
tel (705) 675-1151

An on-road assessment of overtaking while travelling at the posted speed limits on primarily two-lane roads over a 9,000 km route across Canada from Ontario to British Columbia and return aimed at encouraging "data based" rather than "folk wisdom" based driving style yielded several major findings. First, "clear road" driving was experienced (only about 50 cars and 30 trucks were overtaken and about 200 cars and 20 trucks overtook the experimenter, i. e., two to three interactions per driving hour). Second, overtaking drivers showed a clear preference for purpose built passing lanes. Third, only three dangerous or "double yellow" overtakings were observed. Use of findings for driver improvement programmes is examined.

Funds for highway upgrading are limited. Many major highways, including the Trans-Canada highway, cover large distances and yet have relatively low traffic volumes. Consequently, considerable sections of these highways consist of two-lane roads with occasional opportunities for overtaking on purpose built passing lanes on the upside of long or steep hills or where visibility for the driver is curtailed by geography. Unfortunately, political pressure groups frequently lobby for up grading of two-lane highways with much heat and relatively little data. There exists a relative dearth of on-road driving data collected from the point of view of a driver travelling along a route. Wilson and Neff (1995) drove at the posted speed limit immediately after a highway was re-opened after closure due to traffic mishap. Few traffic holdups were noted. Purpose built overtaking lanes relative to their length were used significantly more frequently than the ordinary two-lane roads. Wilson (1996) studied on-road overtaking on the same route during normal non-mishap conditions. Again a significantly greater relative use of purpose built passing was observed. Evidence did not support increased relative frequency of dangerous overtakings after highway closure due to mishap but did support increased risk taking as estimated by traffic speed in such conditions. The question is "Would these results be obtained on other highways?" To answer this question, on-road overtaking was assessed along a 4000+ km stretch of highway comprising for the most part the Trans-Canada highway from Sudbury, Ontario to Victoria, British Colombia.

Method

The experimenter drove a 1986, Oldsmobile Custom Cruiser Station Wagon at the posted speed limit on a 4000+ km route to Victoria, B. C. and back. In an attempt to replicate normal tourist travel, the Trans-Canada highway was followed from Sudbury to Medicine Hat whereupon was followed Highway 3 to Lethbridge, Highway 5 to Waterton National Park, Highway 6 to Pincher Creek, Highway 22 to Black Diamond, Highway 2 to Calgary, Highway 1 and 56 to Drumheller, Highway 9, 72, 567 and 1a to Banff and hence back to the Trans-Canada to Victoria. The route back from Victoria was the same except that the diversion from the Trans-Canada from Medicine Hat through Lethbridge, Waterton and Pincher Creek-Calgary was deleted and a short section was added from Drumheller to Medicine Hat by way of Highway 56 and Highway 1. The sections through Ontario from Sudbury to the Manitoba border are largely 90 km/h two-lane, winding roads with many steep hills which are interspersed with purpose built passing lanes occupying much less than 10 percent of the total distance. The sections through Manitoba and Saskatchewan from the Manitoba border to the Alberta border are largely 100 km/h rather straight flat prairie roads with good visibility, one third of which are two-lane and two thirds of which are four-lane roads. The section from the Alberta border to Medicine Hat is four-lane with a speed limit of 110 km/h, from there to Cardston, Waterton, Pincher Creek, Black Diamond and hence to Highway 2 leading to Calgary is two-lane 100 km/h prairie or rolling foothills. Highway 2 into Calgary and Highway 1 out of Calgary are four-lane 110 km/h roads. The 100 km/h roads to and from Drumheller are rolling two-lane, while highway 1a to Banff is foothills and high country with a speed limit of 100 in some sections and 80 km/h on other sections. While travelling at the posted speed limit the experimenter recorded the overtakings by and of the experimenter according to whether the road was two-lane with a purpose built passing lane, two-lane without such a passing lane or four-lane and other salient traffic events. The experimenter drove generally morning to night with some rest days. Nine driving days were spent on the outward journey and six driving days were spent on the return journey.

Results

In tables 1 and 2 can be found, for each leg of the journey, distances, the number of overtakings of and by the experimenters vehicle on normal two-lane roads, two-lane roads with purpose built passing lanes and on four-lane roads. Four separate chi-square comparative analyses of relative distribution of overtakings were carried out: four-lane versus two-lane overtaking, two-lane versus purpose built passing laneovertaking, outwards versus return journey overtaking on two-lane versus purpose built passing lane and outward versus return journey four-lane overtaking versus being overtaken.

Table 1. Overtaking on special passing lanes (PL), two-lane (2L) and four-lane roads (4L): Sudbury To Victoria

Start Date		OVERTAKE ME						I OVERTAKE						
06/21/96	Dist	Cars			Trucks			Cars			Trucks			4L
Destination	km	pL	2L	4L	pL	2L	4L	pL	2L	4L	pL	2L	4L	Dist
Sault S Marie	299	5	4	5	2			2	2		1			32
Pancake Bay	51	4	3					1	1			1		
Terrace Bay	543	6	2		2									
Nipigon	105		5											
Thunder Bay	105	2	1		1				1		1			
Ignace	224	2	4		1							1		
Man Border	342	5	5		1									
Port La Prairie	249			5						8				249
Regina	476		9	6		5	4		1	5				255
Moose Jaw	71			3						1				71
Swift Current	172			6			2			3			1	172
Medicine Hat	223		2	1					2			2	1	128
Lethbridge	167													
Cardston	77		3						2					
Waterton	54		1											
Calgary	245		3						3					
Drumheller	140		2	2		1						1		50
Banff (old Rd)	276		2									1		26
Golden	135	5		2				7				1		2
Revelstoke	148		2						1					
Kamloops	210	6	2	6				1	2		1	2		20
Victoria (4lane)	386													386
SUM	4698	35	50	36	7	6	6	11	15	17	3	9	2	1391

Table 2. Overtaking special passing lanes (PL), two-lane (2L) and four-lane roads (4L): Victoria To Sudbury

Start date/time 07/24/96 Destination	Dist km	OVERTAKE ME Cars pL	2L	4L	Trucks pL	2L	4L	I OVERTAKE Cars pL	2L	4L	Trucks pL	2L	4L	4L dist
Kamloops (4lane)	386													386
Revelstoke	210	2				1		1			1	1		20
Golden	148	4						1						2
Banff	135	2		10				1			1			
Drumheller	276	1	4									1		26
Medicine Hat	247			3		2	2					1	2	175
Sask Border	60													60
Swift Current	163		2									2		68
MooseJaw	172			2								1		172
Moosamin	294		4									1		139
Portage LP	253			2										187
Vermillion	448	4			1			1			3			249
Nipigon	473	6	13		1				1		3	1		
White River	276	3	4					1			2	4		
Wawa	91	4							1		1			
Sault S Marie	225	4	3					2						
Sudbury	299	3	2	1				2						32
Sum	4156	33	32	18	2	3	2	9	2	0	11	12	2	1516

In the first analysis comparing four-lane to two-lane overtaking, overtakings were pro-rated by relative total distance of four and two-lane roads (e.g. 2640 km of two-lane roads compared with 1516 km of four-lane roads on the return journey). On the return (but not the outward) journey significantly more vehicles overtook the experimenter per mile of two-lane roads compared with four-lane roads ($\chi^2 = 7.75$, df = 1, p < 0.01) and more cars (but not trucks) overtook the experimenter per mile of two-lane roads compared with four-lane roads ($\chi^2 = 7.71$, df = 1, p < 0.01). On the return (but not the outward) journey the experimenter overtook significantly more vehicles on two-lane compared with four-lane roads ($\chi^2 = 14.8$, df = 1, p < 0.001); moreover the experimenter overtook significantly both more cars ($\chi^2 = 6.31$, df = 1, p < 0.02) and trucks ($\chi^2 = 8.7$, df = 1, p < 0.01). By province, the following significant results obtained. In British Columbia, significantly more overtakings of the experimenter occurred in four-lane relative to two-lane roads for both outwards ($\chi^2 = 39.0$, df = 1, p < 0.001) and return ($\chi^2 = 70$, df = 1, p < 0.001) journeys. In Alberta, no significant four-lane versus two-lane differences were noted. Taking Saskatchewan and Manitoba together (similar road configuration), on the return (but not the outward) journey) significantly more vehicles in two-lane relative to four-lane roads overtook the experimenter ($\chi^2 = 4.7$, df = 1, p < 0.05) and were overtaken by the experimenter ($\chi^2 = 8.8$, df = 1, p < 0.01). In Ontario, significantly more vehicles overtook the experimenter on two-lane roads relative to four-lane roads (χ^2 = 11.4, df = 1, p < 0.001).

In the second analysis comparing two-lane and purpose built passing lane overtaking frequency of overtaking was pro-rated on the conservative basis of assuming one km of purpose built passing lane to 10 km of ordinary passing lane (weighted against hypothesis of greater passing lane overtaking). Despite this conservative criterion, overall on the outward journey significantly more overtakings were observed in purpose built passing lanes compared with other two-lane roads of the experimenter (χ^2 = 4.7, df = 1, p < 0.05) and by the experimenter (χ^2 = 3.1, df = 1, p < 0.08, borderline). Overall, on the return journey, significantly more overtakings were observed in purpose built passing lanes compared with other two-lane roads of the experimenter (χ^2 = 6.4, df = 1, p < 0.01) and by the experimenter (χ^2 = 8.3, df = 1, p < 0.01). Considered by province, the following significant results were noted. In Ontario, on the outward journey significantly more overtakings were observed in purpose built passing lanes compared with ordinary two-lanes of the experimenter (χ^2 = 8.5, df = 1, p < 0.01) and by the experimenter (χ^2 = 3.7, df = 1, p < 0.05). In Ontario, on the return journey, significantly more overtakings were observed in purpose built passing lanes compared with ordinary two-lanes of the experimenter (χ^2 = 7.3. df = 1, p < 0.01) and by the experimenter (χ^2 = 12.3, df = 1, p < 0.001). In Manitoba, Saskatchewan and Alberta insufficient purpose built passing lanes existed to analyse data. In British Columbia, on the outward journey, significantly more overtakings in purpose built passing lanes compared with ordinary two-lanes were observed of the experimenter (χ^2 = 8.4, df = 1, p < 0.01) and by the experimenter (χ^2 = 7.8, df = 1, p < 0.01). In British Columbia, on the return journey, significantly more overtakings in purpose built passing lanes compared with ordinary two-lanes were observed of the experimenter (χ^2 = 12.7, df = 1, p < 0.001) and by the experimenter (χ^2 = 9.4, df = 1, p < 0.002).

In the third analysis, overtaking on purpose built passing lanes compared with other two-lane roads showed no overall significant differences in the pattern of such overtakings on the outward journey compared with the return journey. However, for cars considered alone, the experimenter overtook significantly more cars in purpose built passing lanes versus other two-lane roads on the return journey compared with his overtaking pattern on the outward journey (χ^2 = 4.9 df = 1, p < 0.05).

In the fourth analysis, a comparison of outward versus return journey pattern of four-lane overtaking of the experimenter versus by the experimenter, the experimenter overtook significantly fewer vehicles than he was overtaken on the return journey compared with the outward journey overall (χ^2 = 4.2, df = 1, p < 0.05) and also for cars alone (χ^2 = 7.6, df = 1, p < 0.01).

The following other salient events were recorded. On the outward journey only two vehicles were observed to overtake on "double yellow" (passing forbidden) lines, both in Ontario, one coming towards the experimenter and one passing him. No traffic mishaps, but a single abandoned car in a B.C. ditch was observed. Some suggestion emerged of relatively increased overtaking just before and after cities even when discounted for local traffic. Finally, when driving rolling country during rain showers some being overtaken and then overtaking the same vehicle was observed (perhaps due to different rain to hill speed selections). On the return trip some of the overtaking in the high mountain country around Golden was due to my selection of a motor saving slower speed on passing lanes. Two small four-lane sections near

Banff just before major highway construction areas was associated with increased traffic speed and consequent overtaking. Near Swift Current one car attempted to overtake the experimenter's vehicle and a preceding truck with insufficient room and was compelled to squeeze in between the experimenter and the truck. One truck was seen to be in the ditch in Ontario. Travelling at the speed limit, a short traffic delay was experienced but three times. On the outwards journey six vehicles followed a truck moving 10 km/h below the speed limit for 10 km just before the Ontario border where four-lane road begins. On the return journey, the experimenter and one car followed a truck moving at 10 km/h below the speed limit for 8 km on the way to Swift Current and then the experimenter and three vehicles followed two trucks going 10 km/h below the speed limit for 15 kilometres in the Ontario border to Kenora section. There was also a moderate delay of about thirty minutes due to construction east of Banff, Alberta. Overall, total traffic caused trip delay over eight thousand km would be measured in minutes.

Discussion

As far as these findings are concerned, the following conclusions might be drawn which contradict some popular conventional wisdom. The 8000+ km of roads for someone driving the speed limit offer clear sailing; moreover, such drivers are not causing traffic pile-ups. There is a great tendency among drivers to wait for purpose built passing lanes to overtake; moreover such waiting does not cause great delays in trip time. Those drivers overtaking on dangerous double yellow sections should see themselves to be among a very small high risk taking minority. It is suggested that these findings might be utilized in driver instruction packages which aim at developing sociable driving style particularly among new drivers who might be more susceptible to inaccurate conventional wisdoms about driving.

References

Wilson, Tay and Neff, Charlotte. 1995. Vehicle Overtaking in the Clear Out Phase After Overturned Lorry has Closed a Highway. In *Contemporary Ergonomics*, Praeger, London, pp 299-303.

Wilson, Tay 1996. Normal traffic flow usage of Purpose built overtaking lanes: a technique for assessing need for highway four-laning. In *Contemporary Ergonomics*, Praeger, London, pp 329-333.

MIND THE BRIDGE ! DRIVERS' VISUAL BEHAVIOUR WHEN APPROACHING AN OVERHEAD OBSTRUCTION

T. J. Horberry, K. J. Purdy and A. G. Gale

Applied Vision Research Unit
University of Derby
Mickleover,
Derby DE3 5GX

Poor traffic signing is often cited as a major reason for vehicles hitting low bridges. To evaluate the performance of bridge warning signs subjects performed a simplified driving task in a simulated lorry environment and their eye movements with respect to the road image were recorded. Visual behaviour with respect to these warning and other traffic signs were analysed in terms of whether the signs were detected and for how long each sign was fixated. The results demonstrated that the bridge warning signs were detected by less than half the subjects. In addition, these signs were generally looked at for less time than other, matched, traffic warning signs. The effectiveness of the current design of the bridge warning sign in attracting the visual attention of drivers is therefore called into question.

Introduction

In the UK almost 1000 rail over road bridges are struck each year by road vehicles (Department of Transport, 1993). While the exact cause of a bridge strike is extremely difficult to quantify, recent figures suggest that almost 15% of strikes are attributed to poor traffic signs (Department of Transport, 1993). This large percentage questions whether the current design of such signs is optimum. The majority of such traffic signs which warn drivers of a low bridge ahead are standardised in the UK as blue and white 'information' category signs (see Horberry, Halliday, Gale and Miles, 1995, for a more detailed description).

Following Young (1991) it is argued that for a traffic warning sign to be effective it is necessary for it to be detected, to be understood (comprehended) and to promote appropriate driver behaviour (i.e. in the case of a low bridge warning sign for drivers not to attempt to pass under the bridge if their vehicle is too high). Thus the first stage in examining the effectiveness of currently used signs is to determine whether they are actually detected by drivers. Although a great deal of research has previously been performed in the general area of visual behaviour and driving (e.g. Zwahlen, 1981; Hughes and Cole, 1986), no work has previously specifically examined drivers' visual behaviour when driving towards low bridges.

The study reported here is part of an ongoing research project that aims to improve the visual warnings which a driver of a high sided vehicle receives when approaching a low bridge. Specifically, this study focuses on whether drivers of large vehicles, performing a simplified driving task while watching a video of driving toward low bridges, actually look at (fixate) the signs warning them of the low bridge ahead. The aim was to establish how many subjects actually detected the bridge warning signs. A related question which this study examined was for how long were the signs actually looked at. Although it may not be necessary for a driver to fixate a road sign for too long (and in some cases this may even be dangerous if they neglect other aspects of the traffic environment), a certain minimum fixation time period to read the information contained in the sign is essential. An additional aim was to compare the number of subjects who detected the bridge warning signs to other data collected during the trial which considered the detection of different types of traffic signs which were matched to the bridge signs for position, sign size, traffic density and amount of sign information, based on the recommendations of Hall, McDonald and Rutley (1991). If the bridge warning signs were detected by significantly fewer subjects as compared to the other carefully matched traffic signs (or were looked at for too short a time period) then the effectiveness of the current design of the bridge sign as a visual warning must be questioned. The hypothesis of this study was that the bridge warning signs would compare badly on measures of experimental visual performance when compared to the other traffic signs.

Method

Subjects

Twenty seven people participated, all held current vehicle licences. Almost half the group were male (44.4 %) and half female (55.6 %). The age range was 20-49 years.

Equipment

The study was performed in the Unit's vehicle laboratory. The specialist equipment involved was:
1.   A Ford Cargo lorry cab. This was mounted to a realistic height on a metal frame. The steering wheel was locked to prevent eye movement data being lost when subjects turned the wheel (as eye position was being recorded through a hole in the vehicle's dashboard in the middle of the steering wheel). The brake and accelerator pedals were fitted with small springs to give them the 'feel' of being real pedals controlling the vehicle.
2.   An ASL 4000 remote eye movement measuring system.
3.   A large, high resolution monitor - positioned in front of the driver at a distance of 86 cm from the eye position of the driver (i.e. just outside the vehicle's windscreen). The video displayed on the monitor subtended 25.55 degrees horizontally and 19.14 degrees vertically from the eye position of the subjects.

Stimuli

A video camera was positioned in a vehicle close to the eye height of a driver in a light truck, after Cobb (1990) who found the mean driver eye height in a light truck was 1.63 m. Driving footage was then recorded from the vehicle travelling at 25 mph (thus being similar to the speed a lorry would possibly to be travelling at when approaching a low bridge) and was filmed in urban environments that all had generally light traffic densities. The time of day when the recording was made was between 10 a.m. and 1 p.m. and the weather was cloudy but fairly bright. The raw video footage was then edited to include the approaches to two low bridges together with other footage of general urban driving. This final stimulus tape consisted of a few minutes of static calibration points (used to set up the eye movement recording system) followed by eight minutes of driving footage.

Procedure

Each subject was tested individually. Subjects were first seated in the lorry cab and the eye movement measuring system was calibrated. The experimental procedure was then read to the subjects:

"I want you to try to imagine that you are driving this large lorry. On the screen in front of you I'm going to show you a tape of general driving. Please try to watch the video as you would normally when driving. Please try to sit as still as possible when watching the footage. As well as watching the video I also want you to use the accelerator and brake pedals as you would normally when driving. However, the video will not change due to your responses. The footage will last about 8 minutes, but I will not be recording the first minute to give you chance to get used to things. Any questions before I start the video?"

Results

The recorded eye movement data consisted of the video scene viewed by each driver together with a superimposed marker indicating their eye fixation position.

Eight traffic signs on the stimulus video tape were identified first and then the drivers' eye movements with respect to these signs were coded and analysed. The signs were:

1. Routing sign (No. 1).
2. Bend ahead warning sign.
3. Bridge warning sign (No. 1).
4. Road junction ahead warning sign.
5. Height limit sign on the bridge.
6. Routing sign (No. 2).
7. Speed limit 40 mph.
8. Bridge warning sign (No. 2).

The video data were analysed by an experienced researcher and later independently checked to confirm accuracy. Initially it was analysed in terms of whether subjects fixated on the various road signs. A fixation was empirically defined as being in excess of 200ms, this figure is normally regarded as the minimum time needed for a fixation to occur (Widdel, 1984). Following Theeuwes (1996), it was considered that once the eye fixation falls within a predefined sign target zone, the sign can be considered as being detected. This produced the following for the eight different signs (in ascending order of detection).

Table 1: Percentages of subjects who detected the different traffic signs

Name of sign	Percentages of subjects who detected the sign
Bridge warning 1	4 %
Bend ahead warning	19 %
Road junction ahead warning	22 %
Routing 1	37 %
Speed limit 40 mph	41 %
Bridge warning 2	48 %
Height limit on the bridge	52 %
Routing 2	63 %

A Wilcoxon Matched-Pairs Signed-Ranks Test was applied to the above data between pairs of signs matched to each other according to their location and their environmental complexity - at least one of which was one of the bridge warning signs. The following significant differences were found between pairs - with the signs in column A being detected by significantly more subjects than those in column B (at $p < 0.01$).

Table 2: Significant differences obtained between pairs of signs

Column A	Column B
Routing 1	Bridge warning 1
Bridge warning 2	Bridge warning 1
Height limit on bridge	Bridge warning 1

There were no significant differences between the following matched pairs involving the bridge warning signs (again in terms of whether one sign is detected by significantly more subjects than the other sign): 'Road junction ahead' and 'Bridge warning 1', 'Bend ahead warning' and 'Bridge warning 1', 'Routing 2' and 'Bridge warning 2', 'Speed limit 40 mph' and 'Bridge warning 2', and 'Height limit on bridge' and 'Bridge warning 2' (all 2-tailed $p > 0.05$).

Further analysis examined dwell times for selected signs. The routing signs were chosen to be paired to the bridge warning signs as they were in matching traffic environments, were similar sizes and were judged to be like the bridge warning signs in terms of the amount of information contained on the sign (i.e. both types of signs contained a similar number of words of text). Total dwell time, defined here as the total amount of time that subjects fixated on a sign, was converted into total mean dwell time for the whole group of subjects. This allows all the data to be considered - as subjects who did not look at the sign would be included in the analysis (with a dwell time on that sign of 0 seconds). Total mean dwell times were calculated for the two bridge warning signs and the two routing signs, the table over shows the results produced.

Table 3: Total mean dwell time for all subjects

Name of sign	*Total mean dwell time*
Bridge warning 1	0.03 seconds
Bridge warning 2	1.08 seconds
Routing 1	0.70 seconds
Routing 2	0.67 seconds

Applying a Friedman Two-Way Anova to the above total mean dwell time for all subjects produces a highly statistically significant difference between the four figures. To find where the difference occurs a t-test for paired samples was applied, this found that bridge warning 1 has significantly lower (at $p < 0.01$) amount of mean dwell time compared to the other bridge sign (No. 2) and to the two routing signs. No significant differences were found between the mean dwell times of the other three signs.

Averaging the above data for the two bridge warning signs produces a total mean dwell time of 0.56 seconds. The average total mean dwell time for the two routing signs was 0.69 seconds. A t-test for paired samples found no significant differences between these figures ($p > 0.05$), thus the routing signs do not have a significantly higher mean dwell time compared to the bridge warning signs.

Discussion

When driving a large vehicle it would be hoped that all drivers would detect signs warning them of a reduced clearance ahead (plus, of course, read the sign, process the information and use this information to behave accordingly). In the scenario here this was found clearly to be not occurring. In the worst case (bridge warning sign 1) only 4 % of drivers fixated and therefore detected the sign. In the best case (bridge warning sign 2) 48% of drivers fixated the sign - thus over half of the drivers tested did not detect the sign warning them of a potentially very dangerous situation ahead.

The bridge warning sign 1 had a lower number of subjects detect it than two other signs matched to it - the 'bend ahead' and the 'road junction ahead' (the difference was not, however, statistically significant). There were statistically significant differences when this bridge warning sign was compared to the routing sign 1 or to the height limit sign on the bridge (both at $p < 0.01$). Thus overall this bridge warning sign compared badly to the four other traffic signs in terms of the number of subjects detecting it.

The other bridge warning sign (No. 2) performed better. There were no significant differences (in terms of the number of subjects detecting the sign) between this sign and the other signs used to compare to it, i.e. the speed limit 40 mph, the routing 2 and the height limit sign on the bridge. Indeed, this bridge warning sign was fixated by significantly more subjects than the first bridge warning sign (at $p < 0.01$).

Analyses of the total mean dwell times for the bridge warning and the routing signs found similar results. When the total mean dwell time for all subjects was calculated it was found that bridge warning sign 1 performed significantly worse (at $p < 0.01$) than the other three signs. The average total mean dwell times for the bridge warning signs was lower than for the routing signs (but not significantly so). While it may not be necessary (or even desirable) for traffic signs to be looked at for long periods, previous work in the area (see Agg, 1994) generally suggests that at least 0.3 seconds is needed for signs of this complexity to be read- thus the 0.03 seconds all subjects spent, on average, fixating the bridge warning sign 1 was far too short to read the information it contains.

Thus the hypothesis forwarded earlier that the bridge warning signs would compare badly to other traffic signs on measures of visual attention has to be accepted for the bridge warning sign 1 compared to several of its controls - both in the number of subjects who detected it and on total mean dwell times.

It must be noted, of course, that this was an observational not an experimental laboratory study. No variables (in this case, road signs) were manipulated, thus caution must be applied when comparing one type of sign in a specific location with another type of sign in a different location. The bridge warning signs were, however, carefully matched as closely as possible to the control signs in terms of traffic density and location - and followed the recommendations of the research by Hall et al., 1991. Thus the generally poor performance of the bridge warning signs are valid results within the context of this study.

References

Agg H. J. 1994, *Direction Sign Overload* (TRRL, Crowthorne)

Cobb J. 1990, *Roadside Survey of Vehicle Lighting 1989* (TRRL, Crowthorne)

Department of Transport 1993, *Progress to Reduce Bridge Bashing* (HMSO, London)

Hall R.D., McDonald M. and Rutley K.S. 1991, An experiment to assess the reading time of direction signs. In A.G. Gale, I.D. Brown, C.M. Haslegrave, I. Moorhead and S.P. Taylor (Ed.) *Vision in Vehicles III* (Elsevier, Amsterdam)

Horberry, T. , Halliday M., Gale A.G. and Miles, J. 1995, Road signs and markings for railway bridges: An ergonomic design intervention, *Proceedings of the IEA World Conference*, Rio 1995.

Hughes P. K. and Cole B. L. 1986, What attracts attention when driving?, *Ergonomics*, **29** (3), 377-391

Theeuwes J. 1996, Visual search at intersections: An eye-movement analysis. In A.G. Gale, I.D. Brown, C.M. Haslegrave and S.P. Taylor (Ed.) *Vision in Vehicles V* (Elsevier, Amsterdam)

Widdel H. 1984, Operational problems in analysing eye movements. In A.G. Gale and F. Johnson (Ed.) *Theoretical and Applied Aspects of Eye Movement Research* (Elsevier, Amsterdam)

Young S. 1991, Increasing the noticeability of warnings: Effects of pictorial, color, signal icon and border. *Proceedings of the Human Factors Society 35th Annual Meeting* (Human Factors Society, Santa Monica)

Zwahlen H. 1981, Driver eye scanning of warning signs on rural highways, *Proceedings of the Human Factors Society 25th Annual Meeting* (Human Factors Society, Santa Monica)

GUIDELINES

GAMBLING WITH DESIGN

Claire Raistrick

*Principal Consultant
Hu-Tech Associates Ltd
Saxon Court
29 Marefair
Northampton NN1 1SR*

Ergonomists know that ergonomics can contribute to design but is this message reaching those responsible for introducing new workstations? This paper illustrates how the ergonomics process contributes to workstation design. The early application of ergonomics guidelines and the removal of as many constraints as possible have been found to produce the best results. Where the application of ergonomics guidelines has been arbitrary, or where the organisation was unable to remove constraining factors, the resulting workstation has been inferior. If organisations wish to be serious about fulfilling their legal obligations to minimise employees exposure to risk, they should ensure that their workstations are designed to match the capabilities and take account of the limitations of those who use them.

Introduction

The ergonomics process enables ergonomists to take an eminently practical look at how people interact with their environment. This has tremendous benefits when one comes to measure the acceptability of a design to those who should matter most - the people who use it.

Ergonomics incorporates knowledge of the behaviour, physical capacity and understanding of people in a particular context to ensure that their needs and limitations are complemented when they interact with 'things' in their environment.

This paper considers how the ergonomics process has been applied to the design of workstations used by cashiers in Banks and Building Societies. In some cases the extent of the ergonomics contribution was defined, unknowingly, by decisions that each client made, before the ergonomist was even involved.

Each client recognised that there was a need for ergonomics criteria to be incorporated into the design of cashier workstations. Their awareness of this need had stemmed from the concerns of health and safety personnel who:

1. Had knowledge of health and safety legislation with respect to risk assessment, particularly in connection with those personnel who relied on display screen equipment to carry out their duties.
2. Had information regarding the cases of upper limb symptoms which had already been reported to the organisation.
3. Were keen to minimise the potential for harm which might lead to claims for personal injury compensation.

The ergonomics input was sought at different stages in the three case studies discussed in this paper. Table 1 shows the stage at which the ergonomist was first involved in each case.

Table 1. Stage at which ergonomist was first involved

Case Study 1	Case Study 2	Case Study 3
Before design work was started.	1½ years after design work started and three weeks prior to completion of the design phase.	After an alternative security system was installed using an off-the-shelf counter.

In each instance the ergonomists were required to produce the same outcome - ergonomics guidelines for input into the design.

Constraints on the ergonomics process

Once any constraints have been identified and minimised it is possible to establish the key design criteria. These are based upon the potential users physical and mental capabilities and lead to the development or selection of appropriate ergonomics guidelines.

The constraints in each of these three case studies affected the extent to which the ergonomists and designers were able to incorporate the ergonomics guidelines. The influence of ergonomics on the final design is shown in Table 2.

Table 2. Extent of ergonomics influence on design

Case Study 1	Case Study 2	Case Study 3
High	Moderate	Low

Case Study 1

In Case Study 1 the ergonomics input began before any work on the design of the cashiers' workstation. The ergonomics process commenced with an audit of the current situation. The audit included:

1. Identification of the nature of the work carried out and the organisational arrangements.
2. Identification and evaluation of the current problems.
3. Identification of wider issues that may have influenced the new design.

This stage provided information on the organisations' business objectives and its chosen method of achieving these. A clear picture of the existing ergonomics problems and history of occupational illness was obtained. It was also possible to identify those, other than the client department, who could, or should, have an interest in and influence upon the project. These potential 'key players' included: premises; computer services; security; purchasing; health and safety; and cashiers - the users.

This counter design project represented the first use of ergonomics by this organisation. Their previous experience had concentrated more on the image that the designers wanted to create rather than focusing on the implications of using the design.

The initial ergonomics audit identified that the long term success of the new design would depend upon the extent of any constraints. There were two key constraints which compromised an ergonomically sound design. The security system had to stay, as staff safety remained the over-riding priority. There were alternatives, but the organisation was committed to the system they had already selected. However, greater flexibility could be exercised with the display screen equipment. The current hardware had given good service and the ergonomics case to make changes persuaded the computer services department to commit themselves to make changes in line with the overall aim of the redesign project. For the first time the two key departments, premises and computer services, were convinced (by the ergonomics case) to work together to achieve the best result for the users.

Case Study 2

The contribution of the ergonomist was restricted in Case Study 2 because of the late stage at which they joined the project team, only 3 weeks before the final drawings were required. The designers had been working on the project for over 1½ years. They had established and developed their designs to such an extent that the prospect of any significant alteration was extremely threatening and inconvenient.

Following a preliminary evaluation of existing branch operations the ergonomist identified essential information such as the operational requirements of the cashiers' role, the sequence and frequency of use of various items and the postures that were used to carry out these activities. The ergonomists' preliminary observations revealed aspects of the cashiers' role that had not been observed by the designers. Consequently they had not been taken account of in the design. For example, the cashiers needed to use specific forms but no storage space had been allowed.

The ergonomists were particularly concerned about the potentially harmful postures which the cashiers were using. The new design had not taken account of these. Although it was possible to make some recommendations to be incorporated at this stage, there were basic constraints which had been built in because they had not been recognised as constraints. As in Case Study 1 the security system and the display screen equipment presented constraints. The organisation was also committed to using the existing cash storage solution The shape and layout of the worksurface had been determined and limited the options for positioning equipment where it could be accessed using a safe posture. It was clear that neither the client or the designers had expected that the ergonomists' involvement would require more than a few adjustments to the dimensions.

Case Study 3

In Case Study 3 there was an urgent security need and the priority was to protect the lives of cashiers by introducing an alternative security system which was installed with an off-the-shelf counter. This combination had been installed in one branch and the same security system was due to be installed at other branches during the following weeks. It was necessary for the ergonomics recommendations to be limited to modifications to the counter design rather than a redesign solution. Information on the ergonomics problems and occupational ill-health attributed to this counter was collected while developing recommendations for modifications to the design.

As well as the time constraint it was necessary to continue to utilise the existing display screen equipment, printer and cash storage provision. However, it was still possible to make improvements to the counter design.

Exploration of solutions

The ergonomics guidelines that were developed were based upon anthropometric criteria for the user populations. The main dimensions for which guidelines were presented encompassed reach, legroom and visual requirements. Dimensions were relevant to the workstation itself; other furniture, such as chairs and footrests; and the layout of equipment and other work items on the workstation.

Case Study 1

The project team began the process of exploring solutions using the design criteria they had established. Several designs were created and reviewed to establish which had the best potential. No one design seemed best at this stage and the project team progressed with a number of variations.

Case Study 2

The design was reviewed to identify how it compromised the guideline dimensions. Similarly the workstation layout was considered to anticipate the effect that this would be likely to have on the cashiers' posture. Once the potential problems had been identified the ergonomists and the designers had to attempt to make adjustments. It was possible to alter the lateral leg room to allow sufficient space for the chair to be pulled in. Similarly the thickness of the worksurface could be altered to allow cashiers to achieve a desirable height for keyboard entry tasks. However, the poor location of the cash storage could not be improved in any significant way. Although it was raised slightly the cashiers would still need to twist and bend to the side when using it.

Case Study 3

The ergonomics guidelines were established and outline recommendations given to the client for consideration prior to further details being provided by the ergonomist. A modification to the counter profile allowed the printer to be positioned in a more ergonomically sound location. Also a more suitable chair and foot support were recommended. The client went ahead and started to implement the recommendations without any further discussion.

Once the 'improvements' had been made the ergonomist was asked to review the workstation and effectively 'sign it off' as ergonomically satisfactory. However, although the client had provided foot support by building a false floor under the worksurface they had not been able to perceive the wider implications of this alteration. The users had become painfully aware that the false floor that had been added prevented them pulling their chairs close enough to the worksurface. However, because the client did not wait for the necessary detail from the ergonomist, once they had decided that they wished to install a false floor, it was built to the wrong dimensions. The false floor had to be replaced when the ergonomist provided dimensions which matched the requirements of the users and continued to allow the safe and effective positioning of each cashiers' chair.

Testing and evaluation of designs

Case Study 1

The design variations were tested. Firstly as 'quick and dirty' cardboard or MDF mock-ups and later through more sophisticated constructions. Each mock-up was evaluated by the 'key players', each of whom would present their own viewpoint for consideration. A series of user trials was run to evaluate a simulated working counter and eventually a fully installed working counter in a High Street branch.

This whole process was continually iterative and adaptive. Users who were not involved themselves have been informed of progress at all stages via internal publications. A video has been made to show the design process, explain how 'key players' and representative users were involved and to reinforce safe ways of working. This will be shown at all new branches to promote understanding and ownership of the change process.

Case Study 2

The design was not tested until it was installed in an operational branch. The ergonomics review confirmed that the initial recommendations were necessary and further changes had to be made to the design, where possible. However, these changes were limited by the constraints that have already been discussed.

Case Study 3

There was no opportunity for testing the modifications before they were made to existing branches. The counter design was altered to comply with the ergonomics recommendations on an on-going basis.

Realisation of benefits

Case Study 1

The new counter design removed many of the risks that had been inadvertently incorporated into previous counter designs by design teams which did not include ergonomists. The potential for harm has been reduced significantly.

The 'key players' whose actions and decisions affect the potential for risk have been informed and have an understanding of the ergonomics process. They can continue to effect change from an informed viewpoint.

The client is committed to monitoring future changes to counters to ensure that they do not stray from the ergonomics guidelines which have been established. There is documentation and an internal system to enable this process. This ensures that the client will continue to benefit from their investment in ergonomics.

Case Study 2

The ergonomics guidelines were incorporated to some extent. However, it was unfortunate that the opportunity for ergonomics input was delayed until the final stage of the design. Had the ergonomists been involved at the outset the client could have made more informed decisions and perhaps been able to avoid some of the compromises. It was nevertheless worthwhile to alter the design in line with the ergonomists recommendations as otherwise the client would have installed a counter design which, although aesthetically pleasing, presented significant problems to the users.

Case Study 3

The incorporation of ergonomics guidelines, albeit in a limited way was beneficial. The modifications made to the counter reduced the risks to users and increased the clients' awareness of the ergonomics issues affecting workstation design. The client realised the potential benefits of an ergonomically satisfactory design and asked the supplier of the 'off-the-shelf' counter to work with the ergonomists on future projects.

Conclusions

Organisations can be reluctant to use the ergonomics process to manage and take control of the risk presented by hazards within their business. This is often because they do not have personal experience of its effectiveness and are not familiar with the techniques used. The use of ergonomics techniques does require a different focus but the approach becomes easier to apply once the basic concept has been accepted. It can be helpful to see good examples and to speak with others who have successfully introduced change.

The ergonomics process is more effective if used from the start of a project. This helps to ensure that the ergonomics constraints are identified whilst there is sufficient time to make changes. Equally ergonomics guidelines can be developed and conveyed to the designer before they or the client develop any fixed ideas. It is essential that the health and safety implications of a design are considered and that the design is tested to ensure that it is satisfactory before the client is forced to install it regardless of any remaining problems. By incorporating ergonomics guidelines into a workstation design an organisation will be taking positive action to minimise their employees exposure to risk.

ADECT - AUTOMOTIVE DESIGNERS ERGONOMICS CLARIFICATION TOOLSET

Andrée Woodcock and Margaret Galer Flyte

Department of Human Sciences,
Loughborough University,
Ashby Road,
Loughborough, LE11 3TU

Ergonomics is playing an increasingly important role in product specification and design. In competitive markets, such as automotive design, companies need ways to provide their products with a leading edge. An ability to identify factors which will satisfy and delight customers and translate these into design features will distinguish successful and unsuccessful products. Ergonomics is one of the key factors in this. Taking automotive and automotive product design as an example, a survey has shown that although ergonomics is considered throughout design, it is not applied as rigorously as it could be. ADECT (Automotive Designers Ergonomics Clarification Toolset) has been developed to support design teams in their development of an ergonomics specification during the concept stages of design.

Introduction

Incorporating ergonomics early into the product design life cycle reduces problems with usage, increases the acceptance and functionality of the finished product, Lim, Long and Silcock (1992). This is especially true with regard to products requiring heavy investment during design and development, such as automotive design. Increasing the use of ergonomics during initial design phases is "the most efficient and effective method of ensuring that the occupant-vehicle interface is satisfying the needs of the customers", Thompson (1995). When design changes are made later in the development cycle it has three consequences: firstly, the changes are reactive and preventative, Simpson and Mason (1995), rather than proactive i.e. they are attempts to correct inadequate and inappropriate features; secondly, they are more likely to be superficial; and thirdly, as interdependencies increase (such as retooling), they become more costly, Grudin, Erhlich, and Shriner (1987). No amount of manufacturing excellence can compensate for poor design. If the design fails fully to meet user requirements it will fail to realise its full market potential, for example the Sinclair C5 vehicle, Pugh (1990).

As part of their training, most designers now receive modules or lectures in ergonomics. However until very recently, the extent of the ergonomics has been limited to static and occasionally dynamic anthropometrics. There has been no inclusion of any systematic approaches to understanding the range of users, tasks and environments in which the product will be used. Even so, once they become practitioners, ergonomics and user requirements might not be considered with the same degree of rigour and importance as they had during training. Ergonomics again becomes just one of a host of factors to be considered. Unless very fortunate, a designer cannot allow her/himself the luxury of conducting detailed user surveys, experimentation, literature searches etc. due to limitations on knowledge, time, cost, resources and often management will. Instead, practising designers rely on their own experience, Pheasant (1986), the ability of users to adapt themselves or the artefact to alleviate problems brought about by poor design, and the fact that appearance and styling influence purchase to a greater extent than ergonomics. Whilst one would expect customers to consider ergonomics important when choosing products, good (or bad) ergonomics only becomes manifest after purchase, Ward (1990).

A recognition that products might be failing in the market place due to poor ergonomics has provided an incentive to increase the use of good quality ergonomics in design. Fortunately, this has resulted in increased employment for ergonomists where companies have established in-house ergonomics expertise or sought input from consultancies. In cases where direct involvement of an ergonomist is not possible other means of increasing the use of ergonomics in current and future design practice need to be developed. Development of support tools to do this require a knowledge of current and future design practices, present problems in the current provision of information and use of ergonomics. This paper reports a study of the use of ergonomics in automotive design practice. These issues were studied by a postal survey of automotive designers and engineers. The paper presents and discusses the results of this study and then the development of ADECT (Automotive Designer's Ergonomics Clarification Toolset). ADECT is system to enable a design team or designer to follow a more user centred (ergonomic) approach to design.

Ergonomics in automotive design

Automotive design was chosen because ergonomics is used by people from a number of different disciplines (not necessarily trained ergonomists) and is of critical importance to the success of the final product. However, it is believed that the results of the study as well as the ADECT system are generalisable to other design domains.

A postal survey was undertaken of European automotive designers and engineers. The questionnaire contained questions relating to current and future use of ergonomics, the way design was carried out at present, and the use of computers during design.

The 40 respondents came from 6 automotive companies (Ford, Jaguar, Toyota, Rover, Fiat, Daimler Benz) where they worked as managers, stylists, designers, engineers. 10% had a first degree in design, with nearly 20% having qualifications in both design and engineering. 90% had been in the automotive industry for over 5 years. All had experience of and used ergonomics in a wide range of contexts including exterior and interior packaging design, the design of instruments and controls, facias, seats and door trims. In defining ergonomics 58% responded with 'user oriented' issues such as acceptability, comfort, usability. The remaining 42% of the answers focused on the design of specific objects such as door controls, spare wheel access.

From the depth and variety of answers it was clear that the respondents had between them a wealth of experience of and familiarisation with applying ergonomics information to real automotive design problems. Ergonomics was used in some form at all stages during the design and development of new cars, from pre-concept, through concept and evaluation and in making last minute changes.

The main sources of information on which design decisions were made were the designers' own knowledge and experience and that of her/his colleagues, design guidelines and sales brochures. 40% never used company guidelines, and 30 % had never referred to the standard ergonomics journals (e.g. Ergonomics, Applied Ergonomics, Human Factors). Guidelines and standards were used patchily (depending on the activity). Many of the standards had been internalised as company documents, and were referred to in these documents. Respondents did report conducting their own investigations (e.g. buck trials, clinics).

In terms of the use of computers during design the industry is still characterised by islands of technology. Pre- concept and concept design have proved most difficult to support by computers. However, computers are becoming more widely used at all levels e.g. administrative tasks, design logs, teleconferencing and decision support systems as well as CAD/CAM systems.

In conclusion, ergonomics was reported to be practised widely by designers and engineers in the automotive industry throughout the product lifecycle, although it is not formally accessed or provided. However, important new research findings, which would be of relevance to designers might not be available unless the company has commissioned the research. Likewise vital information about the user populations is not necessarily fed into the design process.

Design decisions relating to users' characteristics and needs were informed by individual designers beliefs and experience, reference to guidelines and, sometimes, national and international standards. Respondents expressed dissatisfaction with their somewhat 'cavalier' approach, and reported that they were aware they could do better in terms of ergonomics, that vital ergonomics issues were overlooked until production. However, as was pointed out, 'ergonomics has to exist amongst other needs - package, cost, styling wants, downstream engineering, manufacturing and the eventual customer. The vehicle is a compromise - nothing is absolute'.

From the results of the survey it would seem that designers and engineers would welcome a system which supported their (group) consideration and appreciation of ergonomics aspects of the product, early on in the design process. A system, similar in principal and nature to HUFIT, Allison, Catterall, Dowd, Galer, Maguire and Taylor (1992) was envisaged.

The HUFIT (Human Factors in Information Technology) Toolset was developed to support software engineers/designers when designing the user interaction of IT products. It enables the designers to systematically take account of the users of the product, the task the product is to support and the environment in which it is to be used.

Development of ADECT - Automotive Designers' Ergonomics Clarification Toolset

Two systems have been developed to support the use of ergonomics by automotive designers, taking into account present and future design practices. It is believed that the principles can be applied to any design domain. One system is an information resource environment providing ergonomics guidelines. The development of these on-line guidelines has been the topic of previous papers, Woodcock and Galer Flyte (1995 and 1997). However, merely providing ergonomics information to designers does not mean that it will be used. This is especially true when other factors have to be considered, as in automotive design. The other system is ADECT an ergonomics decision support tool.

In order to be used effectively any information relating to the task, user or environmental conditions should be fed into an ergonomics specification produced before the concept design stage. This should, ideally, have the same 'weight' as engineering specifications, and be used to guide and evaluate design solutions. The output of ADECT is an ergonomics specification, generated by those engaged in the early design stages, similar in construction to engineering specifications.

The ADECT system aids designers in approaching the design brief, by highlighting the ergonomics issues they should consider during concept design. It systematically addresses and supports decisions relating to the users, the tasks and the environment in which the vehicle or component is to be used. Designers without access to such a system may continue to design products which suit themselves and their native environment. In today's markets this is no longer an appropriate strategy.

ADECT is being developed as both a paper and computer based system. The full system will support the free exchange of ergonomics information amongst small, multi-disciplinary teams, of temporary membership working synchronously or asynchronously. As a brainstorming tool, it emphasises the recording of diverse opinions and collective group learning. It can also be used by an individual designer wishing to consider ergonomics issues relating to a current project.

The three main stages of ADECT are:

1. Clarification of the design brief.
It is still common practice in automotive design for projects to proceed without all team members having a clear understanding of, or seeing the design brief. This is especially true as new members are introduced to the project during its progress. This first stage ensures that all stakeholders (e.g. designers, managers, engineers) understand the functions of the new product, its target market and the reason it is being developed e.g. facelift, move into new market sector.

2. User definition
This stage is concerned with identification of the users of the product, and their high level requirements and goals. For example products are sometimes designed only with the main user in mind. Cars are often designed from the driver's perspective - however attention should also be given to other occupants and road users e.g. passengers and safety systems. People also have different needs from their product (e.g. driving quickly, efficiently, economically, carrying loads). Once the principal user groups have been

established a user cost benefits assessment is undertaken to determine whether it is worth proceeding with the design.

3. Characteristics which should shape the product requirements. This stage supports a systematic consideration of:

- Users e.g. anthropometry, literacy, age of the main user groups.
- Tasks the user wishes to perform with the product. Although driving is the primary activity, it can be broken down into sub tasks such as navigation, maintenance of correct position, monitoring the state of the vehicle. Car's are also used more frequently as offices, bedrooms and entertainment centres. These secondary tasks also contribute to the shaping of product requirements.
- Usage of the product e.g. frequency and discretion of use, the extent of task and product knowledge, pacing of the task, distraction, privacy.
- Environmental factors, both the internal and external environment. If products for global markets are being designed, account must be taken of environmental characteristics such as terrain, temperature, road and traffic conditions and legislation.

4. Product Requirements (or Functionality) Matrix.

This is a visual representation which is generated automatically from the characteristics outlined above and product requirements derived from group discussion among the design and development team. These are based on an understanding of the proposed users of the product (not on the designer). From here, the matrix can be used to shape the design solutions which are proposed during concept development. It can also be used to evaluate competitor products.

The prototype ADECT

The present prototype system has been developed for a single user - either the lone designer or a co-ordinator/facilitator, recording information from group discussions, or designers working asynchronously (file sharing). The system is PC based, running under Delphi, and consists of a series of edit fields which enable characteristics to be entered, prioritised and linked to users and relevant product requirements. Memo fields are available allowing further information or discussions to be recorded. Toggling one of the cells of the functionality matrix enables potential 'solutions' to be entered and prioritised. It is possible to generate a full ergonomics specification before the design commences.

The system has been developed bearing in mind current and future design practices. It is simple and straightforward to use, requiring no extensive training, yet powerful enough to keep track of all design decisions. Design decisions and outcomes are presented, and can be updated immediately. Future work will focus on usability studies and iterative design, and extension of the current system to provide support for group working.

Conclusions

Automotive design in the future will be characterised by an increasing enthusiasm for ergonomics amongst design teams. Although companies are now establishing their own ergonomics units many design teams will still operate without direct involvement from an ergonomist. In addition the use of computers in design is becoming much more common. The development of a design decision support system that is computer based, which enables designers to systematically take account of ergonomics at the concept stage of design, meets a currently growing need. ADECT and the ergonomics information resource environment provides such a design support environment.

Acknowledgements

This work was sponsored by Loughborough University, and the Daphne Jackson Research Memorial Fellowship Trust, 1994-1996

References

Allison, G., Catterall, B.J., Dowd, M., Galer, M.D., Maguire, M. and Taylor, B.C., 1992, Human Factors Tools for Designers of Information Technology Products, in M.D. Galer, S.D.P. Harker and J. Zeigler (eds.) *Methods and Tools in User Centred Design for Information Technology*, North Holland, Elsevier Science Publishers, Amsterdam, The Netherlands, 13- 42.

Grudin, J., Erhlich, S. F., Shriner, R., 1987, Positioning human factors in the user interface development chain, *CHI and GI,* ACM Press, New York, 125 - 131.

Lim, K. Y., Long, J. B. and N. Silcock, 1992, Integrating human factors with the Jackson System Development method: an illustrated overview, Ergonomics, **35**, 10, 1135-1161.

Pheasant, S., 1986, *Bodyspace: Anthropometry, Ergonomics and Design*, Taylor & Francis, London, Washington DC.

Pugh, S., 1990, *Total Design*, Addison - Wesley Publishing Company, England.

Simpson, G. C. and Mason, S., 1983, Design aids for designers; an effective role for ergonomics, Applied Ergonomics, **14**, 3, 177 - 183,.

Thompson, D. D., 1995, An ergonomic process to assess the vehicle design to satisfy customer needs, International Journal of Vehicle Design, **16**, 2/3, 150 -157.

Ward, S. J., 1990, The designer as ergonomist: Ergonomic Design - Products for the consumer, *Proceedings of 26th Annual Conference of the Ergonomics Society of Australia*, 101 - 6.

Woodcock, A. and Galer Flyte, M., 1995, The opaque interface - the development of an on-line database of ergonomics information for automotive designers, *HCI'95, Adjunct Proceedings*, 96-104.

Woodcock, A. and Galer Flyte, M., 1997, Development of computer based tools to support the use of ergonomics in design practice, CADE'97 (in press)

GENERAL ERGONOMICS

ON POSTURES, PERCENTILES AND
3D SURFACE ANTHROPOMETRY

Pyter N. Hoekstra

Delft University of Technology
Faculty of Industrial Design Engineering
Department of Product and Systems Ergonomics
Jaffalaan 9, 2628 BX Delft, the Netherlands

In 3D Surface Anthropometry one of the issues is what postures should be defined as standards for surface scanning. It is argumented that next to the already accepted classical postures like standing erect and sitting up straight, at least one functional posture (like e.g. sitting relaxed) should be incorporated in order to assess automatic 3D anthropometric model generation. Next, since 3D surface datasets also allow the calculation of values for the classical (one-dimensional) anthropometric variables, the concept of their percentiles can still be useful. Since a 3D surface scan implicitly contains the spatial relationships between anthropometric variables the concept of percentiles could possibly be extended and transformed into some kind of density function.

Introduction

At this moment the research project CAESAR (Civilian American and European Surface Anthropometry Resource), see e.g. Rioux and Bruckart (1995), is well under way: a combined effort between a.o. the American CARD Laboratories, CARD (1996), and the Human Factors Research Institute of the Netherlands Organization for Applied Scientific Research (TNO), TNO (1996), aiming to 3D Whole Body surface scan a total of 15.000 civilian subjects from the USA (8.000), the Netherlands (3.500) and Italy (also 3.500). Their goal is to have the combined database available on CD in the year 2000. An even larger research was undertaken a few years ago by the Japanese Research Institute of Human Engineering for Quality Life, HQL (1995), resulting in a compilation of anthropometric data (standard as well as 3D surface) from 50.000 Japanese civilians. Also at this moment the University of Surrey in Guildford, UK, is distributing a 3D Network Questionaire, Douthwaite (1996), in this way taking the lead in trying to set up a new network for people interested in using, researching or developing 3D design, manufacturing and other soft and hardware packages applied to anthropometry. The above may illustrate the growing importance that the industrial and scientific community attaches to 3D Surface Anthropometry.

3D Surface Anthropometry

A major advantage of 3D surface anthropometry over traditional ("classical" or one-dimensional) anthropometry is the fact that it provides quantized information (depending on the scanning resolution) about the surface or outside geometry of the human subject, see e.g. Brooke-Wavell, Jones and West (1994) or Rioux and Bruckart (1995) for scanning techniques and data collection. While traditional anthropometry measures distances between points, 3D surface data represents a "copy" of a human subject's surface that can be visualized, stored, manipulated and evaluated. A 3D dataset still allows us to obtain the values for the standard anthropometric variables but we can now also calculate e.g. surface areas and volumes. While there is hardly any information about where traditional measures are in relation to each other, in 3D surface data this is implicitly available, see e.g. Robinette (1993), Robinette and Daanen (1996) or Daanen (1996). This weekness of classical anthropometry is also its strength: if the measurement of e.g. one out of seventy anthropometric variables is faulty, this will have less influence on the other sixty-nine (and the fault quite often can be traced and rectified). In contrast, a faulty 3D scan would influence all its constituent variables. We will come back to this in the next paragraph.

Postures

As in classical anthropometry, some well defined postures have to be chosen in 3D surface scanning to serve as standards in order to make exchanging and comparing data possible. As was decided by both the CEASAR group, Daanen (1996) and the HQL institute, HQL(1995), two postures were "borrowed" from classical anthropometry: standing erect and sitting up straight (see Figure 1.a and b). This obviously makes sense: since next to 3D scanning of the subject's surface (supplied with a host of relevant markers) also standard anthropometric measurements were and will be taken, this will establish the relationships between the two technologies and will allow a comparison between these surveys and data from earlier "sourcebooks".

The next posture to be adopted as standard (see Figure 1.c) is the implicit result of one aspect that is inherent with non-invasive scanning techniques: shadowing. When the arms aren't raised, parts of the armpits will remain in the scan-shadow of the arms and will not be visible. The same holds for the legs. In this posture the hands are supinated into the front plane to enable complete scanning of the hands and lower arms. The last posture depicted in Figure 1.d, looking somewhat like being in a "hands-up" situation, is derived from posture 1.b, again with the objective to completely scan the lower and also the upper arms.

A note should be made here that for the HQL survey the arms of the subject in 1.c were not supinated but held in a neutral position. In this posture the hands were used to grab some handles to minimize body-sway - even though scanning technology is rapidly advancing, a complete high resolution whole body scan can take up to twenty seconds. Bearing this in mind it will be understood that respiratory movement and body sway can result in scandata that has to be corrected for these kind of movement artifacts. We like to end this part by stating that the postures described above seem a rational mix of efficiency and an anthropometric minimum. However, we would like to argue that at least one more posture - preferably relaxed sitting - should be added as a standard. The ergonomic or human factors community hardly ever encounters people

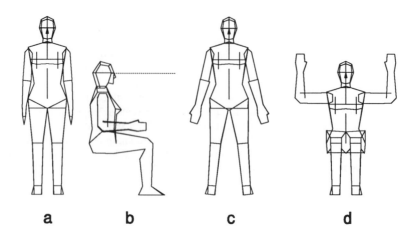

a   **b**   **c**   **d**

Figure 1. Possible postures for 3D surface scanning: a. and b. standard anthropometric
postures: standing erect and sitting up straight, c. standing with arms and
legs spread and lower arms supinated in the front plane, d. sitting up
straight with "hands-up".

in real life situations, be it working or resting, that can be described by the static
postures depicted above. Designers of cars, buses, coaches, airplanes or seats in
general, would still have to transform the original 3D scan dataset into necessary
functional anthropometric design information. Software for automatic generation of
3D anthropometric models is commercially available and already incorporated in some
computer aided anthropometric assessment packages, see e.g. Carrier (1996). If we
would have scan data for both postures - sitting up straight and sitting relaxed - we
would probably be able to better evaluate the relationships between these two postures
in regard of automatic model generation, body segmentation and animation.

Percentiles
         As mentioned above, we can calculate values for the standard (one-dimensional)
anthropometric variables from 3D surface scans of human subjects (when preferably
supplied with the relevant anthropometric markers). As such the concept of percentiles
can still be useful. Percentiles however are by definition statements about strictly one-
dimensional variables. To combine them into an entity like a 3D anthropometric model
is difficult, quite often speculative, misleading and sometimes plain nonsense (when
speaking e.g. of a "P_5-Head"), see e.g. Robinette (1993). In contrast, 3D surface scans
inherently represent information about three-dimensional space. Since these scans
implicitly contain the spatial relationships between anthropometric variables they can
possibly be used to expand the notion of percentiles and transform it into something
that has some aspects of a density function. To illustrate this we have depicted in
Figure 2. some simplified ADAPS models as stickfigures.

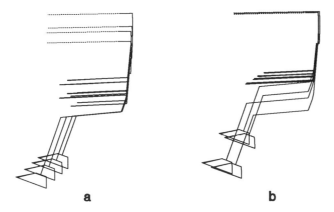

Figure 2. Eight simplified ADAPS models as stick figures: two models (one short and one tall) in two versions (short legs and tall trunk and vice versa), each with either short or long arms; a. centred at the seat reference point, b. centred at the eye reference point.

Starting with two models representing statures of 1,58 m and 1,93 m, we created two more versions: the first with short legs and a tall trunk, a second with long legs and a short trunk; each model then equiped with either short or long arms. In Figure 2a the models are centred at the seat reference point, in Figure 2b at the eye reference point. The same kind of pictures would emerge when combining the 3D surface scans of eight corresponding subjects and grouping them in a like manner. Although the relevant percentiles in Figure 2a are, by definition, *exactly* the same as in Figure 2b, the spatial densities for e.g. the feet or the arms are completely different. It would be hard or even impossible to extract this kind of information from classical Anthropometric Sourcebooks. This centre-dependent aspect of spatial density should of course be better defined and expressed mathematically. The concept could well be used e.g. when aligning 3D head scans at the back of the skull to determine the cloud of eye-reference points for helmet mounted display research. The same holds for quantifying the place of hand-operated controls. Since this again concerns a functional posture incorporating this as a standard scan posture would be very valuable especially for the automobile industry, see e.g. Porter (1995).

Discussion

This rather incomplete and abridged review of only some aspects of 3D surface anthropometry may nevertheless illustrate the enormous potential of this technology. The advantages of real spatial information about the human body over a list of one-dimensional statistics are obvious. Other advantages of surface anthropometry include observer independency (as measurers we do not contact the human body anymore), the possibility to integrate the results from our non-invasive measuring with the results from invasive measuring techniques like CT and MRI. Also measurement speed, the possibilities for automatic 3D model generation and body segmenting (relevant for the clothing industry) have been mentioned. We must bear in mind that the scans, once they are stored as digital copies of the objects surfaces, can again be used long after the scans were made to re-measure a variable or define new ones (e.g. the contours of seat-belts). The strengths of standard anthropometry are well known: it is simple, cheap and can be performed almost everywhere. This is as yet hardly the case for surface scans. The advantages of this new technology however far outweigh the current disadvantages. A host of data extraction tools and software packages are readily available (quite often developed for 3D Surface scanning's mirrored counterparts: computer aided sketching, drafting, designing, visualization, surfacing, rapid prototyping, testing and manufacturing).

Regarding the choice of standard postures: they form an adequate minimum to realise the anthropometric goals though we would like to stress again here the importance of adding at least one more posture, one that could really be described as ergonomically functional. It would present us with more realistic information about the human body's shape, especially in cases where the body is supported and its shape as a consequence will be deformed.

Since 3D surface scans also contain the information about the traditional type of measures the concept of percentiles is still useful (if properly expressed). An extra advantage is that we have a direct link with scanning and visualization. Hopefully expressions like a "P_5- Head" can now become obsolete. We think that the concept of percentiles can be expanded into some kind of centre-dependent density function that could prove fruitful.

References

Brooke-Wavell, K., Jones, P.R.M. and West, G.M. 1994, Reliability and repeatability of 3-D body scanner (LASS) measurements compared to anthropometry. *Annals of Human Biology*, **21**-6, 571-577.

CARD, 1996, Computerized Anthropometric Research and Design Lab, Fitts Human Engineering Division of Armstrong Laboratories at Wright Patterson Air Force Base, Ohio, USA. http://www.al.wpafb.af.mil/~cardlab/

Carrier, R. 1996, Consultants Genicom inc, Montreal, Canada; http://www.safework.com/

Daanen, H.A.M. 1996, 3D Surface Anthropometry, TNO-TM-Colloquium, Soesterberg, the Netherlands. E-mail: Daanen@tm.tno.nl

Douthwaite, C. 1996, Robens Institute and BioMedical Group, University of Surrey, Guildford, UK. Personal communication. E-mail: C.Douthwaite@surrey.ac.uk

Hoekstra, P.N. 1995, Levels of Validity in Computer Aided Anthropometric Assessment. In A. de Moraes and S. Mariño (eds.), *Proceedings of IEA World Conference 1995, 3rd Latin American Congress, 7th Brazilian Ergonomic Congress*, (ABERGO, Rio de Janeiro) 77-80.

HQL, 1995, Research Institute of Human Engineering for Quality Life, Osaka, Japan. http://www.hql.or.jp/eng/index.html

Kohn, L.A.P., Cheverud, J.M., Bhatia, G., Commean, P., Smith, K. and Vannier, M.W. 1995, Anthropometric Optical Surface Imaging System Repeatability, Precision, and Validation. *Annals of Plastic Surgery*, **34**-4, 362-371.

Phillips, S. and Stevenson, M. 1993, An anthropometric data base for combined populations, and the importance of its use on manual handling. In W.S.Marras, W.Karwowski, J.L.Smith and L.Pacholski (eds.) *The Ergonomics of Manual Work*, (Taylor & Francis,London), 211-214.

Porter, J.M. 1995, The Ergonomics Development of the Fiat Punto - European 'Car of the Year 1995'. In A. de Moraes and S. Mariño (eds.), *Proceedings of IEA World Conference 1995, 3rd Latin American Congress, 7th Brazilian Ergonomic Congress*, (ABERGO, Rio de Janeiro) 73-76.

Rioux, M. and Bruckart, J. 1995, Data Collection. In K.M.Robinette, M.W.Vannier, M.Rioux and P.R.M.Jones (eds.) *3-D Surface Anthropometry: Review of Technology*, Draft Report, AGARD, personal communication.

Robinette, K.M. 1993, Fit testing as a helmet development tool. *Proceedings of the 37th Annual Meeting of the Human Factors and Ergonomics Society*, (The Human Factors and Ergonomics Society, Santa Monica CA), **1**, 69-73.

Robinette, K.M. and Daanen, H.A.M. 1996, CEASAR *Proposed Business Plan*, Armstrong Lab.(AFMC), Wright-Patterson AFB, Ohio, USA, personal communication.

TNO, 1996, TNO Human Factors Research Institute, Soesterberg, the Netherlands. http://www.tno.nl/instit/tm/tm.html

THE ACCLIMATISATION OF INTERNATIONAL ATHLETES

T. Reilly[1], R.J. Maughan[2], R. Budgett[3] and B. Davies[3]

[1]Liverpool John Moores University
[2]Aberdeen University
[3]British Olympic Medical Centre, Northwick Park Hospital

Introduction

Many international sports competitions are held in environmental conditions that impose stresses on performers additional to those associated with extreme exercise. These have included the World Cup Championship finals for association football in Italy (1990) and USA (1994) and all of the summer Olympic Games since Moscow (1980) where heat stress has been a persistent feature. The Olympic Games in Atlanta in summer 1996 provided such a situation where high ambient temperatures and humidity were envisaged. For British competitors, travelling to the Games also entailed a transition of five time-zones and the transient desynchronisation of circadian rhythms that gives rise to the syndrome of jet-lag.

The combination of jet-lag and heat stress presents a novel cocktail of stresses on the unacclimatised athlete. On arrival in the new time-zone, there is need to avoid heat stress and sunburn, adopt a sound hydration regimen for training in the heat and ensure that fluids, electrolytes and energy lost during exercise are restored afterwards. This must be accomplished whilst at the same time stimulating physiological adaptations to the heat and developing behavioural mechanisms to cope with it.

As part of the preparation for the 1996 Olympic Games in Atlanta, an acclimatisation strategy was devised to enable British competitors to deal with these challenges. The strategy incorporated education, sports science support, experimentation and heat acclimatisation programmes.

Methods

Education Programme

The education component of the acclimatisation strategy consisted of developing detailed written guidelines on travel and preparation for heat and dehydration. These were provided to all national governing bodies affiliated to the British Olympic

Association. The guidelines were the first tier of a three-tiered system and operated at a generic level. These were further tailored to the needs of particular sports, for example weight-controlled events, by dedicated sports-specific personnel. The support programme was enhanced by a series of lectures or work-shops and counselling work with individual sports squads. The third tier encompassed individual - specific recommendations.

A training camp in Tallahassee, Florida was used during the summers of 1994 and 1995 to provide an opportunity for individual athletes to experience environmental stress comparable to that of Atlanta in mid-summer. It also afforded athletes the opportunity to develop their own coping strategies and allowed the support staff to monitor adaptations to the environmental conditions and time-zone changes. The education programme was enhanced by formal and informal contacts with the athletes who attended the Florida camps.

Experimental Work

Research work at the 1995 camp in Tallahassee was conducted to finalise the protocols for monitoring heat-related responses and travel regimens. The former included monitoring of body mass each morning and analysis of osmolality in urine samples in selected groups of athletes.

Research focused on jet-lag was aimed at:-
i) monitoring subjective, physiological and performance variables in elite athletes and sedentary subjects following a westerly flight across five time zones;
ii) examining whether promotion of good sleep by means of a sleeping draught (10 mg of temazepam) influences these responses to transmeridian travel.

Subjects comprised eight members of the British men's gymnastics squad and nine members of the British Olympic Association's support staff (4 females, 5 males). Each subject was pair-matched (except for one subject) for age, sex and athleticism and assigned to either the treatment group (n=9) or placebo group (n=8).

All subjects travelled from U.K. to Atlanta, Georgia (eight hour flight) and onwards to Tallahassee, Florida (one hour flight) using the same flight schedule, arriving in Tallahassee at approximately 22:00 hours. All subjects resided in the British Olympic training camp in Tallahassee, eating and socialising at similar times of day.

A test battery was administered to the subjects at 06:30, 12:00, 17:00 and 21:00 hours on the first full day of residence (this was designated day one) and then on every other day (day 3, day 5 and day 7). Immediately before retiring to bed on days 1, 2 and 3, subjects ingested either 10 mg of temazepam or a placebo. This meant that day 1 was the control day on which the test battery was administered without any treatment or placebo intervention.

Measures in the test battery included (in order of measurement) sleep quality (0 to 10 simple analogue scale), sleep length, number and timing of nocturnal awakenings; subjective jet lag (0 to 10 simple analogue scale); oral temperature, tympanic temperature; one, two, four and eight choice reaction time; grip strength (left and right) and leg and back strength (using a portable dynamometer, Takei, Tokyo). For the 24 hours of each test day, subjects also recorded the volume of each urine void.

Data pertaining to each variable were analysed using a repeated-measures ANOVA model with two within-subjects factors (test day and time of day) and two between-subjects factors (treatment/placebo and athlete/nonathlete. The assumption of "sphericity" in the ANOVA model was tested using the Huyn-Feldt epsilon, the degrees of freedom being corrected if the epsilon was found to be <0.75. A priori determined multiple comparisons were performed using Tukey tests.

Heat Acclimatisation

A network of facilities was established throughout the United Kingdom for Olympic Games competitors to experience heat stress and acclimatise prior to travelling to Atlanta in Summer 1996. The acclimatisation laboratories had facilities to simulate the conditions predicted for the Atlanta Games (32°C; 70% relative humidity). Exposures were generally at an exercise intensity of 70% maximal oxygen uptake (VO$_2$max) and for durations of 45-60 minutes, once athletes were familiarised with the procedures. During these sessions observations were made on heart rate (short-range radio telemetry), body temperature (infra-red thermometry) and sweat losses (change in body mass, allowing for volume of fluid ingested whilst exercising). Data are presented for illustrative purposes from the women's hockey team.

Results and Discussion

Education Programme

It is difficult to determine precisely the contribution of the education programme on its own towards helping the team managements to have their athletes at a peak at the time of the Olympics. The programme did play a role in translating the acclimatisation strategy guidelines for the competitors. Feedback from the squads indicated that the camp exposures provided purposeful experiences for their preparation, so that the acclimatisation programme is likely to have worked in conjunction with its education component. The use of the facilities in Tallahassee as a staging post prior to moving into Atlanta and the Olympic Village seems to have been generally beneficial. This period during summer 1996 is likely to have been effective in completing the acclimatisation programme begun in the United Kingdom and in overcoming jet lag.

The jet lag symptoms were monitored day by day in the synchronised swimming squad during summer 1995. Results were compared with responses of all athletes monitored a year earlier. Attention to all aspects of the 'guidelines' (particularly life-style) contributed to an earlier adaptation, hastening the disappearance of jet-lag symptoms from 6 days in 1994 to 5 days in 1995.

The background educational work had attempted to get advice to all athletes, directly or through their mentors, competing at the Atlanta Games. Nevertheless, some athletes did arrive in Tallahassee prior to the Games with an inadequate appreciation of the time course of adjusting to the heat and the changes in time zones. This suggests that the communication chain to the athlete was not always completed, although the Olympic representatives did include some athletes who were selected near the deadline date.

Experimental work

The research on the male gymnasts and support staff yielded data on subjective responses, performance measures and physiological variables following travel from the United Kingdom to Florida, a 5-hour time zone transition. The experiment also helped to establish whether use of a sleeping draught had any influence in resynchronising the body clock following such a time change (Reilly et al., 1997).

Subjective jet lag reduced from 4.6 units to zero, and sleep quality improved by 2.0 units from day 1 to day 5 (P<0.001), after which no further alterations were noted. Subjective jet lag, tympanic and oral temperature, left and right grip strength and choice reaction time all showed post-flight day x time of day interactions (P<0.05). On day 1, these variables deteriorated as the day progressed to reach the worst recorded values. On days 3, 5 and 7, circadian variations with the conventional peaks in the early evening and peak-trough differences of about 10% were evident. It appears that the sequence of adjustment is:- sleep > subjective jet-lag > physiological measures > performance.

The ingestion of temazapam did not influence any of these findings (P>0.10). There was an interaction between athlete/nonathlete and time of day in body temperature, grip strength and simple reaction time, with the difference between morning and early evening values being more marked in the athletic subjects (P<0.05).

These results suggest that the administration of temazapam has no influence on subjective, physiological and performance measures following a westward flight across five time zones. In agreement with studies carried out under entrained conditions (Atkinson et al., 1994), circadian rhythms of athletes following transmeridian travel differed from those of sedentary subjects. Subjective jet lag and performance were worst in the evening of the first full day after arrival. This general effect of jet lag is currently being examined in more detail by comparing the data of the pooled sample to a control group which did not travel across time zones. Additionally the differences between the sedentary subjects and the athletes are being further explored in order to ascertain the individual circadian characteristics associated with a fast adjustment of the body clock.

Heat Acclimatisation

The exposures to elevated laboratory temperatures induce adaptations referred to as 'acclimation', since specific environmental variables were manipulated. This entailed three separate sequences of exposures in the final 6 months prior to the Olympics. The adaptations were complemented by the acclimatisation process (responses to the naturally hot environment) following arrival in the U.S.A.

Observations during the acclimation sessions of the women's hockey players typically included an increased sweating rate, a fall in exercise heart rate and a reduction in the rise in body temperature, that is over a period of 60 min. At the end of the period of simulations, typical sweating rates approached 1.5 l/h. The average fall in exercise heart rate in mid-session approached 12 beats/min. The players were able to drink approximately 750 ml fluid during the laboratory simulations. This proved to be useful in preparation for fluid intakes of 1 litre in the hour pre-match in Atlanta.

Overview

The experience of the camps in 1994 and 1995 demonstrated the need to modify the normal training routines in the early days of residence in the USA. The camp also served as a familiar and friendly 'home away from home' during the days adjusting to the USA and its climate.

There were no major heat casualties among the British athletes in Atlanta and it seems that the majority had developed strategies to cope with the climatic conditions. The environmental conditions experienced were less severe than had been anticipated. The circadian phase adjustment to be experienced in Sydney in 2000 will be more extreme - a larger number of time zones to cross and in an adverse easterly direction. There is also the inevitability that more countries will be better resourced for the purpose of optimising their performances than there were in 1996.

The overall outcome of these measures was the avoidance of any serious heat-related failures at the Olympic Games. The use of this strategy for the Olympics has implications for other occupations, notably for military and industrial workers deployed overseas for short periods in hot countries.

Acknowledgement

Gratitude is expressed to the British Olympic Association (Olympic Preparation Camp), Neal Pollock (Florida State University, U.S.A.), Greg Atkinson (Liverpool John Moores University), Jackie Davis (Lilleshall Human Performance Centre) and Alison Purvis (University of Sunderland, U.K.) for assistance with this project.

References

Atkinson, G., Coldwells, A., Reilly, T. and Waterhouse, J. 1994, A comparison of circadian rhythms in work performance between physically active and inactive subjects, *Ergonomics,* **36**, 273-281.
Reilly, T., Atkinson, G., and Budgett, R. 1997, Effects of Temazapam on physiological and performance variables following a westerly flight across five time zones, *Journal of Sports Sciences*, **15** (in press).

THE ERGONOMIC DESIGN OF PASSENGER SAFETY INFORMATION ON TRAINS

Simon Layton and Jayne Elder

Human Engineering Limited
Shore House
68 Westbury Hill
Westbury-On-Trym
BRISTOL
BS9 3AA

This paper describes the design and development of on-train safety information for trains with slam doors (or 'central door locking' (CDL)). The levels of passenger awareness of door operation and emergency procedures were established. Solutions for improvement were proposed and validated. Results showed that the new signs and safety information gave significant improvement in passengers' understanding of emergency procedures and door operation. The results support the use of combined text and icons to present information and a phased delivery of safety information.

Identifying the Problem

The first stage of this programme of work involved conducting surveys and interviews on in-service trains to collect data on passenger knowledge of door operation and emergency procedures and to identify the media, both on-board trains and in the station environment, to which they pay attention. Analysis of the data revealed that there was a lack of awareness and confidence concerning the proper operation of the doors, particularly amongst less frequent travellers. This was also the case for knowledge of emergency procedures. However, on matters of general safety, even *daily* passengers were unaware of some important issues. The main issues to be addressed were identified as:

- Information was required to raise passenger awareness of the operation of the doors and, in particular, whether or not they are locked

- Information was required to raise passenger awareness of general safety issues, but also specific items such as the location and use of emergency equipment

- The signage at the point of need required clarification

- Passengers expressed a preference for safety information to be available when seated

Design Rationale

New signs, posters and leaflets were designed for trials on an in-service train. The following principles were adopted:

- Clarify messages using pictograms with text

- Add an exterior sign with central door locking information
- Provide general information at seats and carriage ends
- Phased delivery of information

The phased delivery of information aims to minimise the amount of information that the passenger must absorb at the point of need. (The phased delivery should begin before passengers even board the train, i.e. at ticket offices, on platforms, etc.) It is particularly important in cases where the passenger will have to perform a sequence of behaviours or non-intuitive actions. The door opening procedure is the most obvious example. The safety leaflets, read while passengers were still in their seats, were designed to present information which would be recalled at the door, reducing the amount of time the passenger would need to comprehend the sign on the door itself.

A number of the existing signs were re-designed with the emphasis on clarity, use of icons combined with text and appropriate sign location. (Symbols were combined with text as there is evidence that using them in combination is more effective than using either on its own (Edworthy & Adams, 1996).) The situations addressed were:

- Exterior door locking
- Door opening
- Emergency door release
- Danger sign (leaning on doors, etc.)
- Fire extinguishers
- Emergency stop
- Emergency window hammers
- Emergency procedures

The New Signs

Individual signs were designed for use at the point of need, i.e., at the door, by the fire extinguisher, etc. Due to limitations on space in this paper, only the safety poster/leaflet will be presented here (although the results refer to all new signage). (Detailed reports of the study are contained in unpublished documents; Elder & Gosling.) The poster/leaflet contains general safety information (some of which was also used at the point of need) and was displayed on A3 sized posters and A4 sized leaflets. (Although the results of the initial data collection suggested that passengers would prefer information on labels fixed to the seat backs and window area this was rejected by the train operating companies due to the likely aesthetic impact and cost of maintenance. Therefore, end of carriage notices and leaflets were used instead.) Leaflets were placed in all seat back pockets and posters were located in the vestibule area (between the luggage racks and the external doors.

There are three main areas of information presented on the poster/leaflet as shown in Figure 1.

- Central door locking (CDL)
- Location of emergency equipment
- Emergency procedures

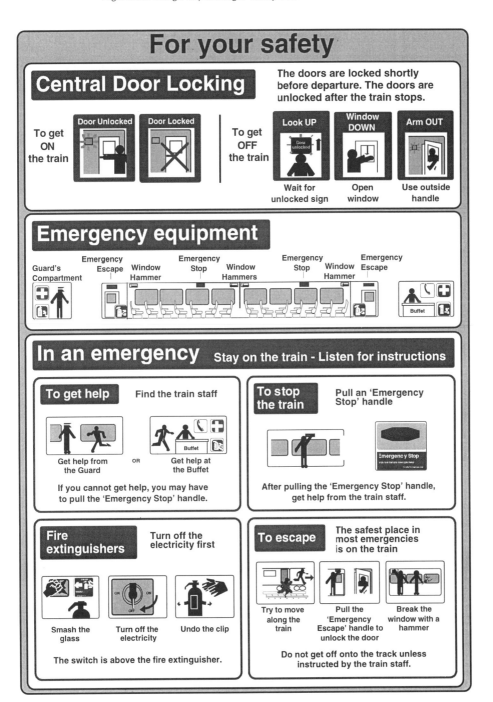

Figure 1. Trial safety poster / leaflet

Central Door Locking

The first two pictograms illustrating getting on the train are also repeated on a sign placed on the exterior of the train next to the door. This information aims to raise passenger awareness of CDL and its indicators on the outside of the train; that is:

- When the orange light is ON, the doors are unlocked and passengers can board the train

- When the orange light is extinguished the doors are locked and passengers cannot enter the train

(N.B. The use of the orange indicator light is counter-intuitive for passengers as the intended purpose of the lights was to act as a warning for station staff if doors are unlocked.)

The second set of pictograms is also repeated on labels on the inside of the door. The original sign contained only text. The new sign includes pictograms and intentionally concise text.

Location of emergency equipment

To inform passengers of the location of emergency equipment a representative diagram of a train carriage is provided. This uses a combination of standard icons (first aid, fire extinguishers, telephones) and text to illustrate the locations of equipment.

Emergency procedures

The main message for emergency situations is that passengers should stay on the train. Again the information is presented using a combination of pictograms and concise text. Fire extinguisher information is located at the point of need as well as on the leaflet.

The information on emergency procedures emphasises the fact that passengers should try to get help from train staff. A point to note is that, previously, the Emergency Stop handle was referred to as the 'Alarm' handle. This gave passengers the false impression that it would alert train staff as to where the alarm handle had been pulled and that help would arrive. In fact it simply stops the train and train staff then have to locate the cause of the incident. Therefore, the new sign refers more intuitively to the Emergency Stop handle.

Validation trial results

Did passengers notice and read the new information?

- 12% of the passengers surveyed had read a safety poster (located in the vestibule)

- 21.5% of the passengers surveyed had read a leaflet

- 44.9% of the passengers with a leaflet in the seat pocket in front of them read it

For a passenger to read a leaflet they had to have a leaflet in front of them, notice the leaflet and finally choose to read it. The percentages completing each of these steps were:

- 48.5 % of surveyed passengers had a <u>seatback</u> in front of them

- 66% of <u>seatback</u> passengers <u>noticed</u> a leaflet

- 67 % of <u>seatback</u> passengers who <u>noticed</u> the leaflet <u>read</u> it

Although the overall proportions of passengers who read leaflets or posters were small, the numbers should be seen in relation to the accessibility of the information. When information was made available to seated passengers, nearly half of them (45%) spontaneously read it in the short space of time (5-10 minutes) before they were given a questionnaire.

Did more passengers read the leaflet or the poster?

23% of passengers noticed the poster, whereas 66% of passengers with a leaflet in the seat pocket noticed it.

Of those who noticed leaflets or posters, a slightly greater proportion read the leaflet than read the poster (67% for the leaflet compared to 54% for the poster).

61% of passengers suggested the information should be displayed in locations visible from the seat.

Did the new signs increase door understanding?

Passengers were asked a number of questions about door operation and central locking indicators while facing the doors, i.e. able to see the signs. Results for passengers viewing the exterior signs showed an improvement in understanding of 30% although this was not statistically significant due to the low numbers of subjects available for this part of the trial.

The results for the new interior door sign show improvements resulting in 80% or 90% correct responses (depending on the specific question asked). Results are statistically significant.

The trial door signs did increase door understanding. The percentage improvements must be considered in relation to the original baselines. The response to the question "Where is the 'Doors Unlocked' sign?" is a good example, with an improvement of 20% from an original score of 72% correct giving a threefold reduction in incorrect answers (from 28% incorrect to just 8% incorrect).

Did the leaflets and posters increase door understanding?

A large number of passengers (205) completed questionnaires while seated. Passengers with a seat back in front of them read a safety leaflet, while those at tables did not. (All 'seatback passengers' were requested to read the safety leaflet before completing the questionnaire whether they had spontaneously done so or not.) There were statistically significant differences between the two groups when asked about door operation and central locking indicators. Improvements in correct answers were of the order of 15% to 20% resulting in percentages of correct answers between 60 to 70% for poster and leaflet readers.

Some of the passengers who completed questionnaires had read the trial safety posters displayed in the vestibule. These passengers showed percentage improvements in correct answers of 28% for the two main questions about door operation and central locking.

The greater improvement for poster readers compared to leaflet readers may result from the fact that all passengers who read posters did so spontaneously whereas roughly half of the leaflet readers did so only when asked to in the questionnaire.

Did the new signs increase emergency knowledge?

There were improvements in the percentage of correct answers of the order of 30% in questions relating to the emergency door release when comparing the in-service sign against the trial sign.

Questions relating to the new signs in the carriage (Emergency Stop and window hammer) showed statistically significant improvements of the order of 10 to 14% taking the percentage correct to more than 80% . These improvements are for passengers who did not read the poster or leaflet, so are assumed to be a result of the new 'Emergency Stop' and hammer signs.

For the fire extinguisher, there were no statistically significant improvements. The problem highlighted in the original survey still exists - passengers do not expect to have to turn the electrical supply off before using the fire extinguishers.

Did the leaflets and posters increase emergency knowledge?

Passengers were asked three general safety questions (safest place in emergencies, fire in carriage, which side to evacuate the train). The percentage improvements in correct answers, when comparing passengers who had read leaflets with those who had not, were from 10% to 22% with 2 of the 3 questions reaching levels over 85% correct for poster and leaflet readers. Results were statistically significant for non-commuters.

Questions relating to the location of emergency equipment showed improvements in percentage correct of the order of 20% for poster and leaflet readers.

Conclusion

This paper illustrates the successful application of the ergonomic principle that combining images and text is generally more effective than either used on its own. It also highlights the point that safety information cannot be clear unless the procedures themselves are clear and logical. For example, it is unreasonable to ask a passenger to "seek help at the buffet - unless it is closed - in which case find the guard - who may not actually be in the guard's compartment...".

The work also supports the notion that passengers are more likely to read safety information made available at their seats rather than general notices and specific signs at the point of need. If the phased delivery of information is used, signs at the point of need are made more relevant and serve as reminders rather than new instructions.

References

Edworthy, J. & Adams, A. 1996, *Warning Design: A Research Prospective* (Taylor & Francis, London)

Elder, J. & Gosling, P. 1996, *Central Door Locking Safety Review: Review of Passenger Awareness of Safety and Emergency Regime.* (unpublished)

Sanders, M. S. and McCormick, E. J. 1993, *Human Factors in Engineering and Design,* 7th Edition (McGraw-Hill International)

EFFECT OF SITTING POSTURE ON PELVIS ROTATION DURING WHOLE-BODY VERTICAL VIBRATION

Neil J. Mansfield and Michael J. Griffin

Human Factors Research Unit
Institute of Sound and Vibration Research
University of Southampton
Southampton, SO17 1BJ
England

Pelvis motion was measured during random vertical motion in the frequency range 1.0 to 20 Hz at 1.0 ms^{-2} r.m.s. Subjects sat in nine body postures: 'upright', 'anterior lean', 'posterior lean', 'kyphotic', 'back-on', 'pelvis support', 'inverted SIT-BAR', 'bead cushion' and 'belt'. Individual data showed differences between subjects, although all showed a peak in the pelvis rotation between 10 and 18 Hz. Differences in the seat vertical to pelvis rotation transmissibilities occurred between 'upright' and 'posterior lean', between 'upright' and 'inverted SIT-BAR', between 'upright' and 'bead cushion', and between 'upright' and 'belt' postures. As no significant differences occurred in the 4 to 7 Hz frequency range, it is concluded that the pelvis rotation does not contribute to the previously observed change in the 5 Hz apparent mass resonance frequency with posture change.

Introduction

Fatigue, discomfort and pain of the lower back have often been associated with exposure to whole-body vibration (e.g. Seidel and Heide, 1986). Reduced potential for vibration injury might be achieved by minimising the magnitude of vibration at source but might also be possible by optimising the posture of the exposed person.

Previous studies of the effects of posture on the biodynamic responses of the body have studied the driving point impedance and the transmission of vibration to the head. Frolov (1970) showed that for one subject, the seat-to-head transmissibility had peaks at 3.9, 11.9 and 35.8 Hz in a normal seated posture. When the subject held his knees to his chest and forced kyphosis of the spine, only one peak in the transmissibility was observed, at 8.0 Hz. Holding arms above the head also changed the response to give two resonances in the transmissibility, at 12.7 and 25.0 Hz. His study demonstrated that changes in posture can affect the transmissibility of the seated human body.

Sandover (1978) measured the effect on the apparent mass of a subject of sitting on two 25 mm wooden cubes beneath the ischial tuberosities of a subject, and the effect of visceral support provided by many turns of a 40 mm wide webbing belt. He found that the apparent mass resonance frequency increased from 4 to 6 Hz when sitting on the wooden blocks but that the visceral support did not change the resonance frequency.

If it is possible to reduce the rotation of the pelvis caused by whole-body vibration, then the inter-vertebral discs may experience less compression and a beneficial health effect may result. There are no known studies of pelvis rotation during whole-body vibration, although some studies (such as Kitazaki, 1992) have presented data at the front and rear of the pelvis from which the rotation could be calculated. This paper reports on a study investigating the effect of sitting posture on the rotation of the pelvis during whole-body vertical vibration.

Apparatus and Procedure

The experiment was performed using a 1-metre stroke electro-hydraulic vertical shaker. Subjects sat on a flat rigid seat with the seat surface 470 mm above their feet which were supported by the shaker table and moved with the seat. A loose lap strap was fastened around the subjects. A stop button was held by the subjects and another was within reach of the experimenter at all times during the experiment.

Motion of the seat was measured using an Entran EGCSY-240*-10 accelerometer mounted on the shaker table directly beneath the seat. Motion of the pelvis was measured by fixing small lightweight (1 g) Entran EGA-125-10D accelerometers to the skin above the iliac crest and posterior superior iliac spine.

In each posture, subjects were exposed to 60 seconds of Gaussian random vertical acceleration with a flat constant bandwidth spectrum in the range 1.0 to 20 Hz. The vibration had a magnitude of 1.0 ms^{-2} r.m.s. The excitation stimulus was generated using an *HVLab* data acquisition and analysis system. Signals from the accelerometers were conditioned and acquired into the *HVLab* system at 100 samples per second with anti-aliasing filters set at 25 Hz.

Subjects were instructed to sit in nine postures during the experiment. Postures are described in Table 1. The inverted SIT-BAR (Whitham and Griffin, 1977) is a rigid indenter which is flat on one side and contoured on the other. Sitting on an inverted SIT-BAR increases the pressure at the ischial tuberosities relative to that on a flat rigid seat.

Table 1. Description of postures.

Posture	Description
upright	comfortable upright posture, no backrest contact
anterior lean	leaning forward 10 degrees, bending at the pelvis
posterior lean	leaning back 10 degrees, bending at the pelvis
kyphotic	as 'upright' with slouched upper spine
back-on	back in contact with the backrest
pelvis support	rear of pelvis supported in rigid frame
inverted SIT-BAR	increased pressure at ischial tuberosities: sitting on inverted SIT-BAR
bead cushion	sitting on cushion of polystyrene beads (rigid up to 20 Hz)
belt	subjects wearing an elasticated lifting belt

Eleven male subjects participated in the experiment with mean height 1.81 m (s.d. 0.03 m) and mean weight 73.5 kg (s.d. 6.7 kg).

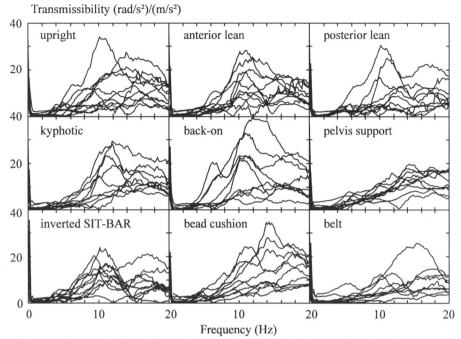

Figure 1. Seat vertical to pelvis rotation transmissibilities for eleven subjects in nine postures.

Analysis

Rotational acceleration of the pelvis was calculated using the vertical accelerations measured at the iliac crest and posterior superior iliac spine. The acceleration time histories were corrected to eliminate local skin-accelerometer motion (after Kitazaki and Griffin, 1995) and subtracted to find the difference in motion. Dividing by the accelerometer separation gave the rotational acceleration in rad.s^{-2}.

Transfer functions between the vertical seat acceleration and rotational pelvis acceleration were calculated using the cross spectral density (CSD) method. The transfer function, $H_{io}(f)$, was determined as the ratio of the CSD of acceleration at the seat and pelvis, $G_{io}(f)$, to the power spectral density (PSD) of the acceleration at the seat, $G_{ii}(f)$:

$$H_{io}(f) = \frac{G_{io}(f)}{G_{ii}(f)}$$

Transfer functions were calculated with a resolution of 0.195 Hz corresponding to 48 degrees-of-freedom.

Results and Discussion

The seat vertical to pelvis rotation transmissibilities for all subjects in all postures are shown in Figure 1. Most subjects show a greater transmissibility magnitude in the frequency range of 10 to 18 Hz. Wearing a lifting belt and sitting with a pelvis support reduced the pitch motion of the pelvis. The largest transmissibilities occurred for the 'back-on' and 'bead

Transmissibility (rad/s²)/(m/s²)

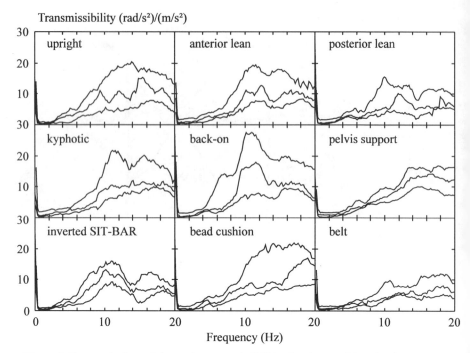

Figure 2. Seat vertical to pelvis rotation transmissibilities for eleven subjects in nine postures -
median and inter-quartile ranges.

cushion' conditions with transmissibility magnitudes reaching 40 rad.s^{-2}/ms^{-2}. These data
show that there was a large inter-subject variability in the transmissibilities, particularly for
those conditions displaying a larger magnitude at resonance.

Median and inter-quartile ranges of the data are shown in Figure 2. These data show
clear resonances at about 10 Hz in the seat to pelvis transmissibility for the back-on and
inverted SIT-BAR conditions. All conditions show more pelvis rotation in the frequency
range of 10 to 18 Hz than at lower frequencies. The transmissibility measured with the
subjects wearing the belt was lower than that measured in the normal upright posture in the
frequency range 6 to 20 Hz. Relative to the normal (i.e. upright) posture, the median
transmissibility was lower for the 'posterior lean' condition at around 16 Hz and for the
'inverted SIT-BAR' condition at frequencies greater than 14 Hz.

The posture with the least inter-subject variability was the 'belt' condition with the
transmissibility at most frequencies showing an approximate 2:1 difference between the 25th
and 75th percentiles. Data were more variable for the 'back-on' and 'bead cushion' postures
than for the other measurement conditions.

The Wilcoxon matched-pairs signed-ranks test was used to compare the
transmissibilities measured in the upright posture with the transmissibilities in other postures.
The results are summarised in Table 2. Significant differences ($p<0.1$) were observed
between the transmissibilities measured in the upright posture and all other postures at certain
frequencies. In particular, there were differences between measurements made in the
'upright' and 'posterior lean', 'upright' and 'inverted SIT-BAR', 'upright' and 'bead
cushion' and the 'upright' and 'belt' conditions.

Over the 4 to 6 Hz frequency range there were no significant differences in the seat vertical to pelvis rotation transmissibilities between any postures. It has been widely reported (e.g. Fairley and Griffin 1989, Mansfield 1994) that the seated body has a peak in apparent mass at around 5 Hz. Fairley (1986) and Kitazaki (1992) showed that the apparent mass resonance frequency of the seated body changes with posture. This study showed no significant differences in the pelvis rotation at the frequencies normally associated with the resonance in the apparent mass (4 to 6 Hz). Therefore it is suggested that rotation of the pelvis does not contribute to the variation in the apparent mass at resonance caused by postural changes.

Table 2. Wilcoxon matched-pairs signed-ranks test results: comparison of transmissibility with upright data. (* $p<0.1$, ** $p<0.05$, *** $p<0.01$).

Posture	Frequency (Hz)																			
	1	2	3	4	5	6	7	8	9	10	11	12	13	14	15	16	17	18	19	20
anterior lean	**	**						*			*									
posterior lean	**													**	**	**	*			
kyphotic		**	**					*	*	*		*								*
back-on									**											
pelvis support								*												
inverted SIT-BAR													*	**	***	**	*	*		
bead cushion	*																*	**	**	
belt	*	*					**	***	***	**	**	**	**	**	**	**	**	**	**	*

The median resonance frequencies and magnitudes at the principal resonance of the pitch motion of the pelvis are listed in Table 3. Median resonance frequencies occurred at around 11 Hz for most of the postures. The 'anterior lean' and 'inverted SIT-BAR' conditions had the lowest median resonance frequencies of 10.9 Hz. Median resonance frequencies of 14.1 Hz and 13.7 Hz were found for the 'pelvis support' and 'bead cushion' conditions respectively. The 'upright' posture showed the largest difference (4.5 Hz) between the 25th and 75th percentiles of resonance frequencies. The differences between resonance frequencies measured in the upright and other postures were only significant for the 'pelvis support' condition ($p<0.1$, Wilcoxon).

Table 3. Transmissibility resonance frequency and magnitude at resonance; inter-quartile ranges using 11 subjects in 9 postures.

Posture	Frequency (Hz)			Magnitude (rad.s^{-2}/ms^{-2})		
	25%	50%	75%	25%	50%	75%
upright	9.4	11.9	13.9	7.3	16.4	21.2
anterior lean	10.2	10.9	12.9	10.5	15.8	20.3
posterior lean	10.2	11.5	13.1	6.1	12.6	16.7
kyphotic	10.9	11.9	13.9	9.3	14.6	23.3
back-on	10.2	11.5	13.1	9.6	19.0	29.3
pelvis support	10.9	14.1	14.8	8.6	11.2	17.8
inverted SIT-BAR	10.0	10.9	14.3	11.0	14.8	20.5
bead cushion	10.0	13.7	14.1	7.4	9.7	28.2
belt	11.9	12.5	15.2	5.3	8.7	12.5

There was a wide range in the median resonance magnitudes measured using the nine postures. A median magnitude of 8.7 rad.s^{-2}/ms^{-2} was measured for the 'belt' condition in contrast to a magnitude of 19.0 rad.s^{-2}/ms^{-2} measured for the 'back-on' condition. The magnitudes of the transmissibility at resonance compared to those measured for the normal upright posture were significantly less for the 'posterior lean' ($p<0.1$) and 'belt' ($p<0.05$) conditions.

Conclusions

Measurement of the rotation of the pelvis for 11 seated subjects in 9 postures while exposed to whole-body vertical vibration showed that posture affects the pelvis rotation. Measurements of the pelvis motion showed the greatest pitch accelerations in the 10 to 18 Hz frequency range. Significant differences in the seat vertical to pelvis rotation transmissibility between the upright posture and all other postures were found. Differences between measurements made in the 'upright' and 'posterior lean' postures occurred in the 14 to 17 Hz range, 'upright' and 'inverted SIT-BAR' in the 14 to 19 Hz range, 'upright' and 'bead cushion' at frequencies above 18 Hz and the 'upright' and 'belt' condition at frequencies above 7 Hz. In comparison with the 'upright' posture, no condition showed a significant difference in the 4 to 6 Hz frequency range, implying that changes in pelvis rotation do not contribute greatly to the variation in the apparent mass at resonance caused by postural changes. A significant difference was observed between the resonance frequencies measured in the 'upright' and 'pelvis support' postures. The magnitudes of the transmissibility at resonance were significantly different from the 'upright' posture for the 'posterior lean' and 'belt' conditions.

References

Fairley, T.E. 1986, Predicting the dynamic performance of seats. Ph.D. Thesis, Institute of Sound and Vibration Research, University of Southampton.

Fairley, T.E. and Griffin, M.J. 1989, The apparent mass of the seated human body: vertical vibration. *Journal of Biomechanics*, **22**, 81-94.

Frolov, K.V. 1970, Dependence on position of the dynamic characteristics of a human operator subjected to vibration. In: Dynamic Response of Biomechanical Systems. ASME, 1970.

Kitazaki, S. 1992, Application of experimental modal analysis to human whole-body vibration. Proceedings of the United Kingdom Informal Group Meeting on Human Response to Vibration, held at the Institute of Sound and Vibration Research, University of Southampton, 28-30 September 1992.

Kitazaki, S. and Griffin, M.J. 1995, A data correction method for surface measurement of vibration on the human body. *Journal of Biomechanics*, **28**, 885-890.

Mansfield, N.J. 1994, The apparent mass of the human body in the vertical direction - the effect of vibration magnitude. Proceedings of the United Kingdom Informal Group Meeting on Human Response to Vibration, held at the Institute of Naval Medicine, Alverstoke, Gosport, Hampshire, 19-21 September 1994.

Sandover, J. 1978, Modelling human response to vibration. *Aviation, Space and Environmental Medicine*, **49**, 335-339.

Seidel, H. And Heide, R. 1986, Long term effects of whole-body vibration: a critical survey of the literature. *International Archives of Occupational and Environmental Health*, **58**,1-26.

Whitham, E.M. and Griffin, M.J. 1977, Measuring vibration on soft seats. International Automotive Engineering Congress, Detroit. Society of Automotive Engineers. SAE 770253.

PICTURES AS SURROGATES: THE PERCEPTION OF SLOPE IN THE REAL WORLD AND IN PHOTOGRAPHS

Jörg Huber[1] and Ian Davies[2]

[1] *School of Life Sciences, Roehampton Institute, Westhill, London SW15 3SN and*
[2] *Department of Psychology, University of Surrey, Guildford, Surrey GU5 5XH, UK.*

We investigated how judgements of slopes in photographs correspond with judgements of real slopes, and the importance of viewing the photographs from the 'correct' distance. Subjects judged slopes in the real world, inclined planes in the laboratory and photographs of these inclined planes taken with either a 35 mm lens or a 50 mm lens. The photographs were judged under free-viewing conditions or from the correct station-point. Judgements were overestimates, but photographs and real slopes produced similar results. In free-viewing, subjects viewed the images from too far, and their estimates were larger than for the correct station-point condition. Judgements with the 35 mm focal length were lower for the 50 mm lens. Implications for using photographs to inform us about the real world are discussed.

Introduction

The first virtual reality displays were probably produced in paeolithic times on cave walls. These early techniques were refined during the renascence when the rules of linear perspective were made explicit by for instance Alberti and Leonardo da Vinci. These principles were brought to their purist form in photography this century. Photographs or paintings which follow the principles of linear perspective within certain limits, reproduce the pattern of light to the eye that would have come from the real depicted scene, provided they are viewed from the centre of projection (the station-point). The latter pairing of image and viewing point can result in trompe l'oeil images: the observer perceives the image as though it were a real 3-D scene rather than a 2-D projection (e. g. Pozzo's ceiling in St. Ignathio, Rome). More prosaically perhaps, photographs of real scenes, viewed though reduction screens from the appropriate station-point, can be confused with views of the real scene (e.g. Smith & Gruber, 1958).

This projective isomorphism account of how photographs work requires elaboration when it is realised that most photographs are not viewed from the station-point (see Gibson 1970 for an analysis of this account). Viewing them from other than the station-point results in a pattern of light to the eye that corresponds to a different, but related scene (Rosinski & Farber, 1980). If the function of the photograph is to afford the observer 'generic' identification of objects – it is a tree – or to provide a canonical form (a good likeness of Auntie), then it may be that deviations from the station-point are unimportant. However, if the function of the photograph is to inform the observer of the precise spatial layout of the depicted scene, then deviations from the station-point may have more serious consequences. Alternatively, the observer may be able to 'compensate' for viewing from the wrong position by calculating what the optic array would be like viewed from the station-point – the Einstein-Pirenne hypothesis (Pirenne, 1971). This conjecture is plausible in the case of viewing from the wrong angle – photographs are normally viewed with the surface at right angles to the line of regard –, but less so for deviations in viewing distance – how would the observer know what the correct distance was? Viewing photographs from other than the station-point produces the predicted 'errors' in perception. Judgements are in accord with the optical projection to the eye, rather than being in accord with the depicted scene (e. g. Smith & Gruber, 1958).

In this paper we investigate how observers view photographs under 'natural' conditions, and we compare spatial judgements under these free-viewing conditions with judgements of observers placed at the station-point. To the extent that spatial judgements follow the optic array to the eye, then observation from further than the station-point should lead to compression of visual space – perceived slopes should be overestimates (Lumsden, 1980). We also assess the accuracy of slope judgements in natural scenes and for experimentally controlled slopes in the laboratory to provide a baseline for comparison with judgements of photographs.

Experiment 1: baseline judgements

Method

Baseline judgements were obtained under two conditions: 'natural scenes' and 'laboratory slopes'. For natural scenes ten subjects were taken one at a time for a walk in the countryside and asked to judge the angle of the slope they were standing on. There were seven preselected slopes varying in angle from about 2 to 20 degrees from the horizontal. The slopes were footpaths in the country with varying surrounding terrain. Subjects provided numerical judgements of the slope with the aid of an 'angle guide' depicting angles from 0 to 90 degrees in 5° steps viewed from the side. The sequence of slopes was varied across subjects as far as practically

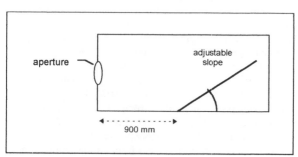

Figure 1: viewing box for slopes using laboratory conditions (not to scale)

possible. The laboratory slopes were made from hardboard and were either 600 mm by 900 mm or 200 mm by 300 mm. These slopes were placed inside a viewing box with a small aperture in the front 900 mm from the pivot point of the slopes (see Figure 1). The slopes were pivoted about the edge in contact with the floor of the box. The surface of the slopes varied in two ways combined factorially to give four surface conditions. First, half of them had a rectangle drawn on the surface to provide subjects with perspective information, and half had no such figure. Second, half the surfaces had a random texture on them while the other half had no such texture. The angle of the slopes varied from 5 to 45 in 10 steps, which in combination with the size, perspective and texture variables produced a total of 40 different slopes.

Twenty subjects judged each of the 40 slopes twice in a counterbalanced sequence. They provided numerical estimates of the angle of slope in the same way as for natural scenes.

Results and discussion

Figure 2 shows the mean angular judgement across subjects for each of the natural slopes. There are two main features of the results. First, for all angles of slope, subjects overestimate the angle by about 10°. Second, judgements of slope covary with the real slope reasonably well ($r = 0.88$). ANOVA showed that the effect of angle of slope was highly significant: $F = 45.72$; $df = 7,56$; $p < 0.001$.

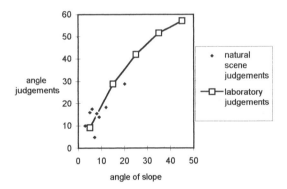

Figure 2: angle judgements for natural and laboratory slopes (in degrees)

Figure 2 also shows the mean across subjects as a function of angle of slope, collapsed across size, perspective and texture for the laboratory slopes. As with natural scenes, judgements of angle of slope overestimate actual angle of slope. In this case the overestimation is about 15°. Again, estimates of angle covary well with real angle ($r = 0.97$) and ANOVA showed that there is a highly significant effect of angle: $F = 527.5$; $df = 4,72$; $p < .001$.

In summary, the results of the baseline experiment have established that subjects' judgements covary with the actual angle of slope, but in both cases there is a general overestimation of the slope of between 10° and 15°. However, the regularities shown by the means in Figure 2 conceal quite large variations across subjects in the degree of overestimation. This may reflect the need for an anchoring point. Their relative judgements of slope accord well with the relative position of the real slopes, but this is accompanied by a constant error which could probably be eliminated by providing an anchor.

Experiment 2: judgements of slope in photographs

The main aim of Experiment 2 was to compare judgements of slope in photographs under free-viewing conditions with equivalent judgements made with the observer at the station-point. If the observer is further away than the station-point, the optic array presented is equivalent to that coming from a steeper slope than that photographed, and conversely, if the observer is closer than the station-point, then the optic array is equivalent to that coming from a shallower slope (see Lumsden, 1980). In addition, in order to make it more likely that subjects would place themselves at the wrong station-point, we used two different focal length lenses: a standard 50 mm lens and a wider angle 35 mm lens. The distance of the station-point from the photograph is given by the focal length of the lens multiplied by the magnification used during film processing. (Standard commercial prints for a 50 mm lens tend to use a magnification of between five and seven, resulting in station-points of between 250 mm and 350 mm).

Within the general design of the experiment, we address four questions. First, under free-viewing conditions, how close to the correct station-point do observers view the photographs? Second, if they view them from the 'wrong' viewing point, do their judgements accord with the optic array obtaining at that distance, or is some kind of compensation possible such that their judgements accord with the real slope? Third, do the different lenses lead to different viewing distances; if not, are there differences in slope judgements in accord with the discrepancies in viewing distances? Fourth, to what extent do judgements of slope in photographs correspond to judgements of real slopes?

Method

Each of the 40 laboratory slopes used in the first experiment was photographed from the aperture shown in Figure 1. This was done once with the 50 mm lens and once with the 35 mm lens. The films were developed using a magnification of 5.7. resulting in station-points of 285 mm and 199.5 mm respectively for the 50 mm and 35 m lenses. The photographs were laminated to protect them from soiling, and the final prints measured 200 mm by 150 mm.

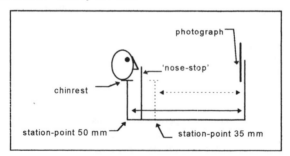

Figure 3: apparatus for viewing photographs at correct station-points for 35 mm and 50 mm focal lengths; chinrest and 'nose-stop' are adjustable (not to scale)

Twenty subjects judged the angle of slope of the 80 photographs under free-viewing conditions. The sequence of photographs was randomised for each subject by shuffling the sets. Subjects indicated their estimates using the same procedure as in the first experiment.

The experimenter observed the subjects informally and estimated the viewing distance chosen by each subject.

A second group of eight subjects made equivalent judgements to the first group, but they viewed the photographs from the correct station-point. This was achieved by means of a chin rest and 'nose-stop' which positioned the subject's eye at the correct distance from the photograph (see Figure 3). The photographs were placed in a frame at the correct distance, such that their surface was at right angles to the line of regard. The two focal length conditions were blocked, and half the subjects did the 35 mm condition first and the other half did the 50 mm condition first. Within blocks, the sequences were random for each subject.

Results

Observation of the subjects in the free-viewing condition showed that subjects tended to view the photographs from further away than the station-point. Figure 4 shows the mean scores across subjects for each combination of focal length and viewing condition, as a function of the real angle. It can be seen that for all four conditions estimates of angle of slope covary closely with the real angle of slope (r ranges from 0.991 to 0.997). Further, in all cases judgements of angle of slope are overestimates of actual slope, but the overestimation reduces with real angle (gradients less than 1 in all cases). There is a consistent ordering of the size of judged angle across the four conditions. Estimates made at the correct station-point are smaller than those made under free-viewing; and estimates of the 35 mm focal length photographs are lower than those from the 50 mm focal length photographs. These impressions are supported by ANOVA of focal length by angle by viewing condition with repeated measures on the first two factors. All three main effects were significant. First, there was a significant difference between the two viewing conditions : F = 6.94; df = 1,26; p < 0.2. Second, the effect of focal length was highly significant: F = 69.43; df = 1,26; p < .0009. Third, the effect of angle of slope was also highly significant: F = 125.47; df = 4, 104; p < .009. None of the interactions were significant. Comparing Figures 2 and 4, it can be seen that the judgements of the laboratory slopes and of the photographs of these slopes produce estimates of about the same order.

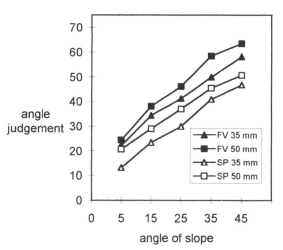

Figure 4: angle judgements (in degrees) for free viewing (FV) and correct stationpoint (SP) for 35 and 50 mm focal length

General discussion

There are a number clear patterns in the results. First, for real and photographed scenes, subjects overestimate the slopes. Second, there is reasonable correspondence between judgements of real slopes and photographs of slopes. However, this latter generalisation needs qualifying in terms of the viewing conditions of the photographs. Under free-viewing conditions, observers place themselves further away than the station-point and estimates of slope follow the optic information – judged slope is greater than real slope. Further, use of inappropriate station-points is exacerbated by variations in the focal length of the lens. Judgements of angle of slope in 35 mm focal length photographs were consistently less than equivalent judgements of 50 mm photographs. However, even when placed at the station-point there remain consistent differences between photographs taken with different focal lengths.

The consistent overestimation of slopes may be a relatively trivial effect of the method of measurement. Judgements are accurate to an ordinal scale of measurement but without an anchoring point, they may be indeterminate. Certainly, photographs can only specify spatial layout to a scale factor rather than specifying absolute values, unless there is an anchoring value available. It may be that if spatial judgements were assessed in terms of how people act rather than how they describe space, actions may be more in accord with veridical spatial layout than we found here. For instance, Smith and Smith (1958) found that asking subjects to throw a ball at a target in a scene depicted in a photograph lead to as accurate throws as the real scene did.

The most important implication of these results is that if photographs are to be used to inform observers about the spatial properties of the real world where accurate judgements are required, then viewing the photographs from the correct station-point will lead to more veridical transfer than under free-viewing.

References

Gibson, J. J. (1971). The information available in pictures. *Leonardo*, **4**, 27-35.
Pirenne, M. (1970). *Optics, painting and photography*, (Cambridge University Press, London).
Lumsden, E. A. (1980). Problems of minification and magnification: an explanation of the distortion of distance, slant, shape and velocity. In M. Hagen (Ed.) *The Perception of pictures* Vol. I, (Academic, New York).
Rosinski, R. R & Farber, J. (1980). Compensation for viewing point in the perception of pictured space. In M. Hagen (Ed.) *The perception of pictures* Vol. I, (Academic, New York).
Smith, O. W. & Gruber, H. (1958). Perception of depth in photographs, *Perceptual and Motor Skills*, **8**, 79-81.
Smith, P. C. & Smith, O. W. (1961). Ball throwing responses to photographically portrayed targets, *Journal of Experimental Psychology*, **62**, 223-233.

FURTHER EXTRACTS FROM ERGONOMIX FILES: TIME, THE THIRD FUNDAMENTAL UNIT

Mic L. Porter

Department of Design,
University of Northumbria at Newcastle,
Ellison Place, Newcastle-upon-Tyne.
NE1 8ST

The ergonomist is constantly concerned with time, as an absolute point of reference, as measure of duration or a divisor in the measurement of work rate and when photography is used notions of time will be incorporated into the images. The photographs may be made to go beyond the pure objective and to take on an artistic or aesthetic dimension. How do those who view this poster display react to the images, originally made in connection with various ergonomic projects but now presented out of context?

Introduction and context

When an ergonomist takes a photograph a piece of scientific equipment is used as part of professional scientific activity either as a data capture device or with the intention of recording the task or experimental context. Rarely will the ergonomist seek to manipulate the image beyond ensuring that the frame contains all the information required and that the focus, illumination, exposure, etc. are acceptable.

Using chemical processes and technical expertise images may be cropped or manipulated further but it is the recent development of digital systems that permits, indeed encourages, much more. It will be interesting to see how such facilities are exploited by the ergonomist and whether, or not, an artistic dimension to their presentations and publications will be sought.

A photograph is taken for a purpose, journalistic, scientific, commercial, social record, artistic, etc. The ergonomists purposes are usually to accurately record, the layout of experiments or to capture transient information for later analysis. It will be the intention of the ergonomist to make these records neutral and without intent to mislead or deceive. At some significant moment a team photograph or other social record may also be made. Ergonomists do not, however, appear to routinely attempt to create more aesthetic or artistic photographic images. Arguably, it is the very absence of any such artistic intent in the taking of the photograph that makes it an acceptable and repeatable tool for the scientist and ergonomist.

The chemistry of a traditional photographic film or the physics of the latest electronic camera ensures that a record is made of an instant in time. Today, with simple manipulation, the instant captured by readily available equipment might easily vary from one ten-thousandth of a second (using electronic flash) up to any number of minutes from a camera held on the "B" setting. Thus to record a behaviour or action the ergonomist usually makes a choice of one of the first two strategies; a third is, however, available:

1. Open shutter, "long" exposures. A subject, fitted with LEDs (possibly flashing), undertakes an action in a darkened space in front of a camera with an open shutter and the resultant single image records dotted lines produced by the motion of the point illumination. Alternatively, "flash" illumination might be fired several times, resulting in multiple images on the same frame.

2. Conventional, but "series/multiple" exposures. Using motorised film winders many cameras can expose film at rates of many frames per minute; much faster rates can be obtained from motion picture film, video and digital electronic systems. These images are usually presented as a linear sequence in time order.

3. Multiple exposures "joined". In which the many frames are not presented in a linear timeline format but rather jointed together for forming a single, two dimensional, image in which time meanders in a space. David Hockney (1993) calls such pictures "joiners". *"The Scrabble Game, Jan 1 1983"* is one of the first published which shows multiple movement unlike, for example, the earlier, purely scenic, *"Grand Canyon Looking North"*.

Photographs - The X factors

Any photograph will contain an aesthetic dimension, albeit unintentionally, yet the use of the image only in a scientific role suppresses this and inhibits consideration of it. However, a photograph presented out of context encourages the artistic response. This, for example, commonly occurs when pictures made to describe scientific apparatuses are reprinted later to be viewed in a historical context. For example, Sutcliffe's photograph of a Jet Workshop in Whitby (1890), the photograph of Lord Rayleigh (1842-1919) undertaking a Chemical Experiment (undated) (Sansom, 1974) or those of Amundsen's expeditions (Huntford, 1987) now take on aesthetic qualities and promote discussion that goes far beyond that conceived when the original records were made.

A few, generally journalistic or record images, are of events so disturbing that they resist any artistic response from most people and remain images that horrify, inform, enlighten and promote emotive responses only. The content of photo-journalistic images such as Edward Adam's (Langford, 1986) of the summary street execution of a Viet Cong suspect or Lee Miller's images of post liberation Dachau and Buchenwald force most viewers to respond to the content alone and to give little consideration to the composition or the manipulation of light, contrast, viewpoint point, depth of field, etc. In such images it is the content that promotes the response and this is little dulled with time. It is, however, the viewer that makes the response and they may be encouraged, by the context of publication to view the aesthetic aspects of the image and not dwell upon the content. For example the recent Victoria and Albert Museum's presentation of the Museum of Modern Art, New York's touring exhibition "American Photography 1890-1965" contains Arthur Felling's ("Weegee the famous") *"Harry Maxwell Shot in car"* (Galassi, 1995).

Few ergonomists, except perhaps those that concern themselves with health and safety or accidental injury issues, make images with such "strong" content.

The ergonomist will routinely consider how they will present the results of their work. However, we rarely consider the aesthetic possibilities of our work and images yet these can be the gateway to productive responses from the client and even, the general public. Time series of people at work, photographs of laboratory equipment, etc. can all be used to bring ergonomics to a wider audience and to encourage appropriate debate as to the benefits that may be obtained from its application.

The technology is reliable and easily controllable and thus the "snaps" that ergonomists use will be suitable for showing in exhibitions. The images, once taken away from the original scientific context can entice a response from the viewer. To make future aesthetic use of the photographic possibilities of our projects requires a little creative planning before the shutter is pressed or the experiment dismantled. At it simplest the camera must be moved to cover the whole field of view so that Hockney type "joiners" can be made later. These multiple images can trigger open responses from the client that results in creative, less formal discussion. Thus they can promote the type of environment from which higher level systemic solutions may emerge. In this role such a "joiner" can be used to establish the context in which a Checkland Soft Systems Approach can flourish (Scholes and Checkland, 1990)

Hockney's photographic "joiners" can be seen as just recent examples of the multiple image produced, in paint for many years. Pieter Bruegel (c1525-1569), for example, produced a wide range of such images now much used for Christmas Cards, Jigsaws, etc. His pictures are often full of the detail that interests ergonomists, for example, in the *"Peasant Wedding Feast"* (c1566) the use of an old door and poles by two members of the catering staff to distribute food raises many questions, not least whether or not it will shortly overbalance when the next plate is removed! However, unlike Hockney most of these "narrative" or "illustrative" artists control perspective, although not always rigorously.

Stanley Spencer has painted work that can be said to fall between the scenes painted by Bruegel and the photographic "joiners" of Hockney. Newton (1947) describes Stanley Spencer as an "illustrator" who unlike a true "narrative" painter requires text to support and guide the viewer. Newton suggests that the nearest counterpart to Stanley Spencer is Bruegel but then goes on to dismiss the formers "photographic painting".

> "Spencer's purely photographic painting, then can be left out of account in this essay. It needs no explanation.... I have included just as much of it... as is necessary to show its quality. To those who admire keenness of eye coupled with skill of hand it is admirable. To me it has a dry unlovable harshness which begins to set my teeth on edge once the admiration has worn off. And yet, without that skill and keenness of eye not one of his imaginative illustrations could have been painted...." (Newton, 1947)

Stanley Spencer produced paintings which celebrate people at work throughout his career; from the early *"Apple Gathers" (1912-13)* and *"Swan Upping at Cookham" (1915-19)*, to *"Workmen in the House"* and *"The Builders" (both 1935)* and *"The Sausage Shop" (1951)*.

However, it is Stanley Spencer's "Port Glasgow"/"Shipbuilding on the Clyde" series and the associated pictures (1940-1950) (Patrizio and Little, 1994) that most clearly demonstrate the joining of many small images to form the whole with multipoint perspective. These illustrations were produced from a vast number of individual sketches, made on rolls of toilet paper, from direct observations of those working on the building of Merchant ships on the Clyde during the Second World War. To the ergonomist the wealth of detail is of immense interest and shows the power of debate possible from viewing composite images. (It is also interesting to consider the response to these "illustrations" from the Art Historian. Newton dismisses them as photographic yet they contain technical errors, and in the definitive catalogue (Bell, 1992) several are incorrectly titled, eg *"Riverters"* as *"Riggers"*, *"Plumbers"* becomes *"Bending the Keel Plate"*. Illustrating, perhaps, that the focus is on the artistic elements of the painting and not the content. It is; probably, the reverse for the ergonomist who views the task then the aesthetic!)

Now that many of the technical details of capturing images have been solved and the ability to manipulate images is becoming much easier we can make increasing use of the images in our presentations. Typically, a photograph made as part of a project will contain all the appropriate information and, perhaps with additional text, clearly show some aspect of our work. It is also increasingly possible for use to build upon the work of artists and produce "joiners". These multiple images encourage an aesthetic response from the viewer that can yield interpretations and responses that might not be obtainable in other ways. With a little planning they may be made in most situations using easily available equipment and materials.

Conclusion

Few ergonomists will be able to produce the most powerful artistic and aesthetic images but we can all give the same attention to the visual presentation of our work as we would do to the rigorous writing of the report or journal article. Should we show "time-line" series of images, multi exposure photographs or disregard the time and assemble composite "joiners" from our "snaps"? All can be produced, with relative ease, from equipment available to us all. The presentation of photographs encourages a much wider range of responses than commonly found among our clients and can promote a wide range of discussion of the circumstances of the project. This can be emergent property of our work and should be exploited as an important step towards the development of systemic and Meta-ergonomic solutions.

References

Bell, K. 1992, Stanley Spencer: *A Complete Catalogue of the Paintings.* (Phaidon Press in association with Christie's and the Henry Moore Foundation, London).

Checkland, P. and Scholes, J. 1990, *Soft Systems Methodology in Action.* (John Wiley & Sons, Chichester).

Galassi, P. 1995, *American Photography 1890-1965.* (The Museum of Modern Art, New York).

Huntford, R. 1987, *The Amundsen Photographs.* (Hodder & Stoughton, London).

Hockney, D. 1993, *That's the way I see it.* (Thames and Hudson, London).

Langford, M. 1986, *Basic Photography,* 5th Edition. (Focal Press, London).

Newton, E. 1947, *Stanley Spencer.* (Penguin Books, Harmondsworth).

Patrizio, A. and Little, F. 1994, *Canvassing the Clyde: Stanley Spencer and the Shipyards.* (Glasgow Museums, Glasgow).

Sansom, W. 1974, *Victorian Life in Photographs.* (Thames and Hudson, London).

USER-TRIALLING IN THE DESIGN OF A BICYCLE PUMP

L.W. van Hees, H. Kanis and A.H. Marinissen

Faculty of Industrial Design Engineering
Jaffalaan 9
2628 BX Delft, the Netherlands

In this project, users' trialling features as an explorative technique in generating design requirements, as a means to check anticipated usage, and also as a source of inspiration triggering innovative solutions. Thus, it is shown that user trialling by no means necessarily ties designers' hands to solutions which happen to be negotiated. The significance of users' trialling in a design context seems to go hand in hand with a flexible adoption of this type of empirical observation, rather than with forcing it into an academic straightjacket of quantification.

Introduction

User activities, i.e. use-actions as external, physical activities aimed at the activation of product functions, and also the perception/cognition of featural and functional cues, are allways addressed in user trialling of products. User trialling can be carried out for various purposes at different stages of a design process (see Green et al. (1997)):
- in making an inventory of the current usage of an existing product, which may be of help in redressing observed difficulties in a redesign on the basis of similarity (in the design analysis phase),
- in anticipating possible future usage on the basis of models which feature negotiated design solutions (in the design simulation phase), and
- in checking user activities anticipated in the new design on the basis of a working prototype (in the design evaluation phase).
In addition to these three types, user trialling has yet another role, which may emerge in any design stage, i.e.
- as a source of inspiration, triggering innovative solutions which may break away from current practice or from developments in a design already underway.

To some extent the present study of the design of a new bicycle pump features all these aspects. The operation of existing pumps was observed, as well as the simulated use of models and the actual use of a new prototype. The trial with the models, in particular, gave rise to the choice of a new direction for the design of the prototype. The circumstances surrounding this development will be discussed.

Objective

The aim of the project was to design an alternative to the existing portable bicycle pump. Publications in cycling periodicals show the difficulties arising from the operation of these pumps, e.g. the uncomfortable posture which must be adopted when connecting the pump to the valve, whilst exerting considerable force with the arms which may result in overstress of the elbow joints, see Figure 1. Such difficulties occur to a much lesser extent with stirrup pumps, designed for use at home, but these are far too unwieldy to be carried around. Carbon-dioxide cartridges, which are occasionally used, are criticised for being ineffective and expensive - at least two cartridges seem to be needed each time.

The initial approach in the project was to look at the possibility of dispensing with the need for a pump
- by considering whether Dunlop's tyres could be 're-invented', or
- by evaluating various kinds of patented mechanisms which could be incorporated into a bicycle, perhaps driven by the chain or the saddle-pin while cycling.

Unfortunately, such proposals seem as unrealistic as they are inefficient. It was therefore necessary to focus on the development of an alternative device for inflating bicycle tyres, which would be convenient to use, and sufficiently compact and light-weight to be easily carried around.

Principles for a design solution

Given the lack of success of the alternatives, it seemed inevitable that any technical solution would involve the compression of a contained volume of air by muscle force. Achievement of a sufficiently pressured tyre in one or two strokes appears to be unrealistic. The aim would then be to tune a type of repetitive compression to a convenient and efficient transfer of muscle force. It soon became clear that the force would have to be produced by the upper extremities. The idea of using leg-power was abandoned because of the difficulty of maintaining a repetitive movement with the leg. Also the necessary robustness of any equipment in that case might easily prejudice the requirements of being compact and light-weight.

Figure 1. Postures during tire inflation, and different ways of clamping a pump with the hand at the valve of the wheel.

Compression can be built up by two types of movements - rotational and lateral. Human powered rotational systems were found to be out of the question due to the high number of revolutions required for an acceptable level of effectiveness (for example 3000 rev./min. for 50 % efficiency in case of a screwcompressor). Enforced lateral movements can be produced in various ways, e.g. by pinching, pushing and pulling. Pinching within the hand was ruled out due to the relatively limited amount of achievable force exertion. Thus, to improve on the current practice, the basis of the new design had to be a more favourable combination of a technical principle motivated by arm movements.

Two principles have been chosen as possible design solutions to be incorporated into models, which explore the user-product interaction by simulation, see Figure 2. As to repetitive compression, a favourable principle would be to begin inflating an empty tyre with relatively large volumes of air at low pressure, and ending up with small volumes, in order to accomodate the rising tyre pressure, thus flattening the required effort involved during inflation. This is the principle on which model 1 is based, using two cylinders which telescope into each other. The idea is to begin inflating by moving the whole grip up and down, by which, effectively, only the cylinder with the large diameter is used, see picture at the left. As the tyre pressure rises, higher pressures can be reached by operating only the upper half of the grip, at the cost of smaller volumes of air. Model 2 in Figure 2 aims at quick and easy movements of a grip up and down a thin cylinder. The piston is moved to and fro by wires running across two fixed pulleys at both cylinder ends, and across two pulleys in the grip which function as loose pulleys. In this way, the piston travels twice the displacement of the grip. Sealing problems, although considerable in this concept, might be solved by the application of techniques developed in other areas. A disadvantage of both models is that, like current pumps, they feature a dead stroke.

The first trial including users' own pump and two cardboard models

The aim of this trial is to explore user product interaction by simulation, given the observed ways users operate their own pump.

Method
Figure 3 shows the models which were used in the simulation. Both the telescope model (1) and the wired model (2) were varied, in length as well as in the provision of a flexible connection. The telescope model was also equipped with a

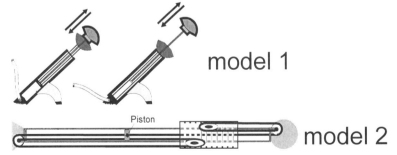

Figure 2. Working principles incorporated in two models.

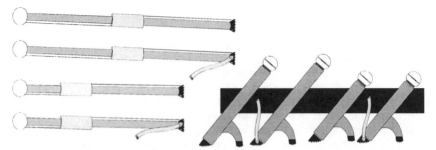

Figure 3. Cardboard models in the simulation.

claw in order to facilitate the grip at the rim. This provision was omitted from the other model as this concept is not intended to be positioned at the rim.

Eight subjects were involved, all of whom had current experience in dealing with punctures. Each subject was videod whilst demonstrating their normal method of inflating a tyre on their bike, using their own pump. After a brief explanation of the intended working principle, the subjects were asked to 'operate' both versions of the two models. The models were presented in cardboard in their actual dimensions, with the sequence balanced between the two types. Subjects were invited to think aloud about their experiences during the trials, and afterwards the subjects evaluated the usability of the models in a debriefing.

Results

As well as confirming the operational difficulties mentioned above (e.g. the amount of force exertion needed in an uncomfortable posture), the simulation revealed firmly engrained use habits, for example in the adoption of body posture, in positioning the valve (if it is not at the lower point of the rim then the claw of the telescope model is useless), and in clamping the pump with a hand on the valve. It seems that a different working principle, or product feature such as the length, do not divert users from long held habits. The overall similarity of the models to pumps in current use may even prompt subjects to assume force exertion, and also to adopt habitual ways of operating the pump while balancing the bicycle. Only in the case of the flexible connection does there appear to be any variation in posture and operation from those demonstrated when using their own pumps. The unfamiliar working principle of the telescope model did not seem to be fully understood by the subjects. Its possible benefits were not recognised in the trial. The length of the pump is seen as an important characteristic, as subjects realise that a bicycle-mounted pump is unused for most of the time. Regarding this, the wires of the one model are considered to be too vulnerable.

More important than these average findings and observations was the growing awareness during the trial that, in order to avoid simply a re-shuffle of current inconveniences, a different approach was needed. In addition, the dead stroke in both models showed signs of becoming a design embarrassment. This triggered the idea of developing a new prototype combining the two-container principle from model 1 with the to and fro movement of model 2 in such a way that no dead stroke would occur, see Figure 4 for the technical principle. The to and fro movement is eased by a lever, which should be foldable in order to facilitate transportation. Further technical details concerning valves and sealings are not discussed here.

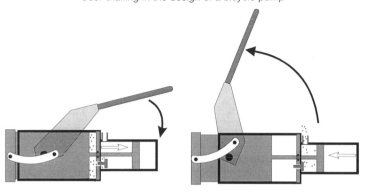

Figure 4. Working principle of the new prototype.

Users' trial with the prototype

The aim of the second trial was to check the operation of the new principle for difficulties in use, such as emerged in the first trial.

Method
With the same subjects, the trial was carried out in our Ergonomics Usability Lab. The prototype had to be used with a flexible connection, and without, see Figure 5. No explanation was given about the working principle. For the rest, the procedure largely resembled the first trial.

Results
Only one of the eight subjects needed some help in getting started with the pump operation. For the rest, the way in which to use the pump seems self-explanatory, even though some subjects indicated that, initially, they were uncertain whether the pump was working at all. Most of the subjects felt that to inflate a tyre sufficiently with the new pump, took them less effort than they were used to (actually, with the new pump, roughly 3 'to and fro's' are needed in order to equal the effect of 2 'up and downs' with existing pumps). As in the previous trial, the flexible connection allowed for more posture variation. Actually, most of the subjects stood upright, a posture which was adopted by only one subject with his own pump. With an absent flexible connection, the direct positioning of the pump on the valve between the spokes turned out to be troublesome. Another difficulty was caused by the possibility of folding the lever in its transport position on the grip. The last part of the stroke towards the grip is empty, which is cued by some extra force needed to stretch both braces alongside (see Figure 4). But all subjects tended to 'finish' the stroke, thus trapping their fingers. Finally most subjects appreciated the shortness of the pump for carrying it in a bag, but a few subjects feared that in case of a bicycle mounted pump, their calves could rub against it (the maximum width of the pump amounts to 7,0 cm).

Adjustments of the prototype
The user trial prompted reconsideration of the construction of the lever (to avoid trapping of fingers), to apply the flexible connection with the one-way valve (to eliminate pressure loss), and to minimise the diameter of the pump. In addition,

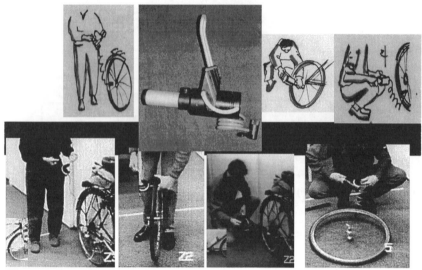

Figure 5. From the users' trial with the prototype.

the interior can be further optimised, e.g. by enlarging the reactivity of one-sided permeable sealings, in order to improve on the maximum achievable pressure, which for the prototype amounts to 3,5 bar.

Discussion

Aside from all kinds of useful information which is difficult or impossible to generate in other ways, this project shows that user trialling by no means ties necessarily designer's hands to solutions which happen to be negotiated. In addition to being a means to correct or to anticipate, user trialling may equally well serve to inspire (cf. van der Steen et al., 1996). It is true that it is uncertain what the result would have been, had it been decided to elaborate one of the two models. The same holds, had the prototype featured as a model in the simulation. By the course it takes design has a memory which cannot be erased. In this respect, no single perfect solution can ever be claimed. Besides, this project shows the significance of user trialling in a design context by a flexible application of this type of empirical observation, rather than by adhering to quantitative imperatives which tend to dominate academic endeavours in empirical research.

References

Green, W.S., Kanis, H. and Vermeeren, A.P.O.S. 1997, Tuning design of everyday products to cognitive and physical activities of users. In S. Robertson (ed.), *Contemporary Ergonomics*, (Taylor & Francis, London) this issue.

Steen, van der V.B.D., Kanis, H. and Marinissen, A.H. 1996, User involved design of a parking facility for bicycles. In S. Robertson (ed.), *Contemporary Ergonomics*, (Taylor & Francis, London) 50-55.

SUBJECTIVE LOUDNESS COMPARISON BETWEEN A HEAD PHONE AND A BONE VIBRATOR

Jack K.W. Chung and Richard H.Y. So

*Department of Industrial Engineering and Engineering Management,
Hong Kong University of Science and Technology,
Clear Water Bay, Kowloon
Hong Kong*

A headphone will cover a listener's outer ear and reduce his / her auditory situation awareness of the surrounding. Unlike headphone, bone vibrator can transmit sound without covering the outer ear. In order to study the benefits of replacing headphone with bone vibrator, factors affecting the subjective loudness comparison between a headphone and a bone vibrator will need to be investigated. Three experiments have been conducted to study the effects of practice, static contact force, and the order of presentation on loudness comparison ratings between a 1 kHz tone transmitted via a headphone and a bone vibrator. Results showed that neither practice nor order of presentation affect the loudness comparison data significantly. As the static contact force between a bone vibrator and a mastoid increased from 0.1 to 3.6N, the perceived loudness peaked at 0.7N and then decreased.

Introduction

The most common way to transmit a personal message is through the use of a head phone. However, a head phone will cover the outer ear and reduce the ability to localize surrounding sound sources. This can reduce the auditory situation awareness of one's surroundings. Unlike a head phone, a bone vibrator, readily available in the market as a hearing aid, can transmit sound messages to a person without covering the outer ear. It is hypothesized that a person listening to a bone vibrator has a better auditory situation awareness than a person listening to a head phone. In order to test this hypothesis, it is necessary to compare his / her ability to localize secondary sound sources while listening to a primary sound message transmitted through a bone vibrator and a head phone. However, before such a comparison can be made, the primary sounds transmitted via the bone vibrator and headphone have to be set to the same loudness. The most direct method to calibrate the loudness difference between a bone vibrator and a head phone is by subjective assessment. As there are many factors which may affect the sound transmission through a bone vibrator, studies have been conducted to study the effects of these factors.

The aims of these studies are to investigate the effects of practice, presentation order, and static contact force on the subjective comparison of loudness between signals presented via a headphone and a bone vibrator. Three experiments have been conducted and it was hypothesized that: (i) repeated practices would increase the consistency of loudness assessment of sound signals; (ii) the order of presenting a pair of sound signals will affect the loudness assessment between them; and (iii) increasing static contact force between a bone vibrator and the mastoid would first increase the loudness and then reduce it.

Details of experiments

Task and apparatus

In each of the three experiments, subjects were instructed to listen to a pair of consecutive one kHz pure tones of 20 seconds duration. They were then asked to assess the relative loudness between the two presentations using a continuous scale from -50 to +50 (see Figure 1). A positive rating indicates that the second presentation is louder than the first presentation and a zero rating indicates equal loudness. The two pure tones were presented via a pair of bone vibrator (Viennatone 90AN, model 145-158-2/2L) and a pair head phone (Vega Mono Wire) respectively. The sound levels used were set according to the experimental conditions. Contact force between the bone vibrator and the mastoid was measured using a UPEX2 system (Ergonomic Concepts Ltd.) force sensor module. The experimental set up is shown in Figure 2. Only the left arm of the bone vibrator was used during the study investigating the effects of contact force.

1st presentation is **Much louder**	Equal loudness	2nd presentation is **Much louder**
!------------------------------------	0 ------------------------------------!	
(-50)	(0)	(+50)

Figure 1. Subjective rating scale for loudness comparison between two consecutive signals (subjects were asked to mark on the scale).

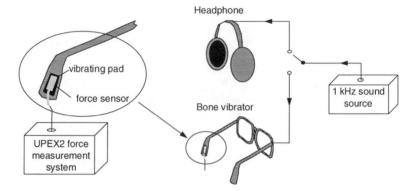

Figure 2. Experimental set up

Method and design

The same six subjects participated the three experiments. They were university students with normal hearing. Their ages ranged from 21 to 24. All subjects were restricted to male Chinese because Roysters *et. al* (1980) reported that race and gender could influence hearing levels.

The independent variables investigated included: eight repeated runs, two orders of presenting the pair of pure tones (bone vibrator first and head phone first), and six static contact forces between the bone vibrator and the mastoid (0.1N to 3.6N).

For the 'effects of practice' experiment, subjects first listened to a pure tone presented from the bone vibrator and then a second pure tone presented from the headphone. They were asked to rate the relative loudness between the two presentations. This was repeated eight times. The levels of tones were arbitrarily chosen by the experimenter.

For the 'effects of presentation order' experiment, subjects listened to three pairs of consecutive pure tones, in two different orders: bone vibrator first and headphone first. This formed six conditions and the sequences of presenting these conditions to the six subjects were assigned according to a 6x6 Latin square design. The volume levels of the three pairs of tone presentations were chosen arbitrarily by the experimenters. Each condition was repeated three times.

For the 'effects of contact force' experiment, subjects listened to six pairs of pure tones presented only by bone vibrators. The levels of all twelve presentations were the same although different static contact forces were applied between the bone vibrator and the mastoid. The first pure tone of each pair was presented with a static contact force of 0.1N while the second pure tones were presented with six different static contact forces. The sequences of presenting the six pairs of pure tones among the six subjects were balanced with a 6x6 Latin square design. ISO 389-3 (1994) recommends a 5.4N static contact force to be used with bone vibrators. This level was also used in some published studies with bone conduction (e.g. Haughton and Pardoe, 1981; Richter and Brinkmann, 1981). However, a 5.4N force was thought to be quite high and might cause dis-comfort if the bone vibrator was to be worn for more than one hour. A preliminary test had shown that the contact force of wearing the Viennatone 90AN bone conductor spectacle ranged from 0.05N to 0.5N depending on the size of the head (Jack and So, 1996). Therefore, the ranges of forces used in this experiment were chosen to be 0.1, 0.7, 1.8, 2.4, 3, and 3.6N.

Results and discussion

Effects of practice

The loudness comparison ratings are shown as functions of practices (Figure 3). Friedmann two-way analysis of variance has indicated that the practices had no significant effect on the loudness comparison ratings ($p > 0.5$). Inspections of Figure 3 suggest that the inter-quartile ranges were approximately constant during the first four practices and then increased with further practices. This dis-agrees with the hypothesis that practices would reduce the inter-quartile range. One possible reason for the increase in inter-quartile range was that the subjects lost their concentrations after four practices.

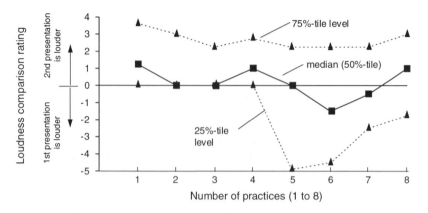

Figure 3. Loudness comparison ratings between two consecutive 1k Hz tones as functions of practices (a -ve rating indicates that the 1st tone was louder).

Effects of presentation order

Three pairs of 1kHz pure tones with different sound levels were used in this experiment (pair A, B, and C, Figure 4). Each pair consisted of a pure tone presented from a bone vibrator and a pure tone presented from a headphone. The effects of presenting the tone from the bone vibrator first or the tone from the headphone first were investigated. Loudness comparison ratings of the three pairs of pure tones with two presentation orders are shown in Figure 4. Each condition was repeated three times, and the results shown were the median across of the 6 subjects' data with the three repetitions (see last section for the effects of practices). Inspection of the figure shows that the loudness comparison ratings changed sign when the order of presentation was changed. This seems logical. Wilcoxon matched-pair signed ranked tests were performed to compare the absolute magnitude of the ratings with different presentation orders. Results of the tests indicated that the absolute values of the ratings were not significantly affected by the switching of presentation order ($p>0.1$).

Effects of static contact force

The loudness comparison ratings between a pair of 1k Hz pure tones presented via the same bone vibrator with different static contact force are shown in Figure 5. A 0.1N force was used in the first presentation while the contact forces used in the second presentation ranged from 0.1N to 3.6N. Inspection of Figure 5 shows that as the contact force of the 2nd presentation increased, the perceived loudness of the 2nd presentation decreased. Results of Friedmann two-way analysis of variance indicated that the ratings were significantly affected by the contact forces ($p<0.002$). It can be observed from the figure that the rating peaked at 0.7N although the differences between the ratings at 0.1N and 0.7N were not significant ($p>0.5$, Wilcoxon matched-pair signed ranked test). As the force increased to 1.8N and beyond, the sound transmitted via a bone vibrator reduced and the reduction became significant at 3.6N ($p<0.01$, Wilcoxon matched pair signed ranked test performed to compare the ratings at 0.1N and 3.6N). It is worth noting that the 0.7N level is much lower than the 5.4N static force recommended by the ISO 389-3

(1994). Further research is needed to confirm the relationship between the static contact force and the loudness of a bone vibrator.

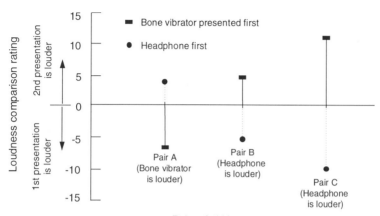

Figure 4. Loudness comparison ratings of three pairs of 1kHz tones presented consecutively via a bone vibrator and a headphone. The comparisons were repeated with 2 orders of presentation: bone vibrator first and headphone first. (A -ve rating indicates that the 1st presentation was louder). Data shown are median of 6 subjects with 3 repetitions.

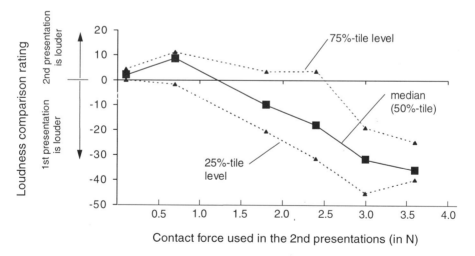

Figure 5. Loudness comparison ratings between two 1kHz tones of equal volume applied with the same bone vibrator but with different static contact force. The force used in the 1st presentation was 0.1N and the forces used in the 2nd presentation are shown. Median and inter-quartile range of 6 subjects' data are shown.

Conclusions

With eight consecutive practices, no significant trend was found in the subjective loudness comparison ratings between a pair of 1k Hz tones. However, there was a slight observable trend for the inter-quartile variability to increase after four repetitions. When comparing the loudness between two consecutive presentations of twenty seconds duration, switching the order of presentation would not introduce any bias in the comparison results.

When the static contact forces increased from 0.1N to 3.6N, the loudness rating of sound transmitted from a bone vibrator peaked at 0.7N and then decreased with increasing force. As these results were obtained with only six subjects, further research to confirm the findings are desirable.

Acknowledgment

This research was supported by the Hong Kong Research Grant Council through Direct Allocation Grant, DAG 95/96.EG.06.

References

Chung, J.K.W. and So, R.H.Y. 1996, Ability to localize a secondary sound source while listening to a bone conducting device: a comparison with the use of ear phone, Research report No. I.E. 96.02. Department of Industrial Engineering and Engineering Management, Hong Kong University of Science and Technology, Hong Kong.

Haughton, P.M. and Pardoe, K. 1981, Normal pure tone threshold for hearing by bone conduction, *British Journal of Audiology*, **15**, 113-121.

International Standard Organisation 1994, Acoustics - reference zero for the calibration of audiometric equipment. ISO 389 - Part 3, 4-5.

Richter, U. and Brinkmann, K. 1981, Threshold of hearing by bone conduction, *Scandinavian Audiology*, **10**, 235-237.

Royster, L.H., Royster, J.D., Thomas, W.G. 1980, Representative hearing levels by race and sex in North Carolina industry, *J. Acoust. Soc. Am.* **68**, pp. 551-566.

TUNING THE DESIGN OF EVERYDAY PRODUCTS TO COGNITIVE AND PHYSICAL ACTIVITIES OF USERS

W.S. Green, H. Kanis and A.P.O.S. Vermeeren

Faculty of Industrial Design Engineering
Jaffalaan 9
2628 BX Delft, the Netherlands

The relationship between the phases of a user-centred design process and the possibilities for intervention by established techniques of usability evaluation is discussed. The role of the various experts and expert techniques is assessed, and it is concluded that, because of the inevitable interpretative role of the designer, the present practises of evaluation are not likely to be very effective unless very closely linked to design, or actually conducted by designers.

Introduction

It is becoming increasingly inappropriate to view the user product interaction interface as a dichotomy between physical interfaces on the one hand and computers on the other. Evidence of this is the continued and accelerating computerisation of common domestic products, often associated with extension of functions which frequently contribute to confusion and operational difficulties. The extent to which this may be seen as a transitional phase, whilst we wait for a generation brought up with computer technology to become adult product users, is debatable. There is no clear evidence that the rate of change will decrease or be less demanding for future generations. There is also the phenomenon of what may be termed retro-development: for example the introduction on some new designs of the familiar rotary control for the analogue adjustment of intensities, or the incorporation of analogue presentations of digitally derived data. Given this mixture of interface modalities, user activities rely ever more evidently on the perception/cognition of sometimes complex featural and functional cues linked to the external, physical activities of users aimed at activating product functions.

This paper discusses the identification of these activities in user centred design, which involves both the anticipation of these activities at the design stage and the subsequent confirmation or denial of them by means of research of some kind . In particular, it focuses on the key role of designers. This role may range from ill or well-informed guessing, to being in charge of an empirical study.

User Centred Design

In the so-called standard design process, there are several identified stages which are common to most if not all of the variations. For the purpose of this paper we have used three steps: Information and Analysis; Synthesis and Simulation; Evaluation.

Information & Analysis

This stage is the time when the designer(s) define or accept the problem, decide on the weighting of the various imperatives, decide on the information they need and begin to look for information sources. The information & analysis phase is a divergent process of gathering information which concludes with a list of design requirements. (In real terms it is often difficult or impossible to determine the division between this and the following phase. The actual point where the divergence of information gathering becomes the convergence of solution identification is rarely well defined).

Synthesis and Simulation

In the creation phase designers apply the gathered information, together with their knowledge and experience of the problem to the generation of possible solution alternatives, that are captured in the form of sketches, models, dummies etc. The creative process of designers is an explorative, iterative activity in which a pre-determined set of requirements serves as a submerged set of boundaries which may be raised from time to time to the level of output-to-date and a judgement of fit made. Therefore, the creation phase has a highly iterative character, consisting of a recurring cycle of generating ideas, evaluating ideas, finding and analysing additional information, generating new ideas and even re-defining the problem. Evaluations during this phase are aimed at generating new ideas, modifying solutions and selecting ideas for further development. It finally converges into a small number of design proposals to which the first more formal evaluations may be applied, aimed at identifying a design that is ready for engineering documentation leading to a prototype. Within this process there are many variations, especially in terms of models and simulations, some of which are primitive and which never appear beyond the designer's drawing board. Subject to evaluation by the designer only, these are often extremely important in that they may determine which possible directions are chosen or discarded at an early stage.

Evaluation

In this phase evaluations are conducted in order to asses to what extent the proposed design meets the design requirements. Then, a formal decision follows on whether the design can be taken into production, whether it needs some extra design attention, or even whether or not it will be produced at all.

Information on user activities in design

In this paper we start from the simple assumption that a designer needs information. The broad range of data required in an engineering, economic or marketing sense is assumed, and the focus is on data required for user-centred design. Implicit in the statement that the designer needs information is that the designer is central to the process. Finding out the size of a person, or collecting data from a user trial, or analysing an

heuristic evaluation, is not designing. The data must ultimately be interpreted by the person or team who forms the product, and who always has a great variety of imperatives. With respect to anticipation or identification of user activities four types of information may be mentioned which designers can or should resort to:
- designers own experience,
- data from handbooks,
- experts' inspection,
- empirical data (user trials; field observations).

Designers own knowledge/experience

The designer(s) has a number of resources:
- The design brief, which will certainly impose boundaries, some of which may be helpful if drawn from expert assessment of a previous design.
- Known information derived from recent exposure to similar situations or problems.
- Experience. A loose term for the amalgam of fact and fiction which make up the normal human repertoire, and in the case of an experienced designer, a more focused and relevant compendium of ideas and memories.
- Reference to actual examples of the product or device - perhaps a first generation of the product from the same manufacturer, or a similar product from a different manufacturer.
- Possible feedback from marketing/consumer services/complaints departments etc. concerning these previous versions.

Data from handbooks

These data concern human characteristics like anthropometrics, capacities, reaction times etc. They are usually derived, with varying degrees of difficulty, from the standard reference sources and have varying degrees of applicability. Significantly, there is very little in the standard reference texts/data sources, which seem to be more generally used by designers, which helps with user activities. Human characteristics data is useful for setting general boundary conditions. It tells us little about what people actually perceive, how perceived cues are understood, and what users actually do. Authors such as Donald Norman have taken some important steps towards identifying common problems and raising general awareness.

Experts inspections

Experts of some kind could be brought in to comment on a design or design proposal in some form. The way in which this can be done ranges from the informal unstructured manner in which an expert comments on a design using his 'expert' intuition, to more structured ways of assessing a design such as Heuristic Evaluations (Nielsen and Molich, 1990) and Cognitive Walkthroughs (Polson et al., 1992). These are referred to further as inspection methods. Both with respect to inspection methods as well as to the more informal ways of expert assessment of a design it should be stated that the type and quality of such assessments largely depend upon the specific level and type of knowledge and experience of the expert. One could be an expert on the specific type of product that is designed, on general human factors/ergonomics or human computer interaction, or in the specific domain of application of the product

Inspection methods provide a way for evaluating the usability of a software design with respect to some criteria (Mack & Nielsen, 1993) . These criteria may be derived from guidelines or from some more or less formal theory (e.g., of how novices learn, as in Cognitive Walkthrough, see Polson et al., 1992) . There are differences in how inspection methods are conducted. Some inspection methods focus on a specific set of features or usability issues which can be expressed in the choice of guidelines or heuristics or questions. Other methods focus on a specific set of functions or set of user tasks. These functions or tasks can be expressed in task scenarios, and inspectors apply the inspection criteria to the steps of the scenarios (e.g. in Cognitive Walkthroughs). Heuristic evaluation is a relatively new term used to describe a long-standing, commonly-used evaluation method for identifying difficulties with user interfaces (Bailey et al., 1992). It is done by looking at an interface and trying to form an opinion about what is good and bad about it.

Cognitive Walkthroughs help in assessing the usability of a system, and in assigning causes to usability problems. They thereby focus on ease of learning. The reviewers step through the required user actions, considering the behaviour of the interface and its possible effects on the user, and attempting to identify those actions that would be difficult for the proposed user to choose or execute. Claims that a given step would not cause any difficulty (as well as claims that a given step is problematic) must be supported by theoretical arguments, empirical data, or relevant experience and common sense of the team members (Polson et al, 1992).

Inspection methods can be applied to primitive design models as well as to functioning products. It even seems to be possible to conduct a (partial) inspection if a design is only specified in the form of a flow-chart (Desurvire in Mack and Nielsen, 1992). The applicability of these methods to physical interfaces has not been clearly demonstrated.

User trialling

In a user trial, activities of people are observed in dealing with artefacts. The artefact may be a primitive design model such as a drawing or a storyboard presentation of an interface, but may also be an existing product which has to be redesigned. The ideal is to observe user activities as these actually occur, i.e. uncontaminated. In fact, this ideal can only be approached for use-actions with a functioning product. As distinct from expert inspections, user trials offer the opportunity to single out particular causes of use problems on charting perceptions and cognitions, rather that negotiating several possible causes. Perceptions/cognitions have always to be registered via the intermediary of language, be it for a functioning product or any model. Subjects may be stimulated to think aloud or have their mental activities examined by retrospective questioning. Neither of these processes are beyond dispute, for example in the case of disturbance of normal usage by thinking aloud, and in the production of desired or allegedly desired reasoning in retrospect, see Ericsson and Simon (1993). Sometimes, for partly working design models or prototypes it is still possible to limit the intrusiveness of use-action simulation by concealing the actual point of interest, e.g. by suggesting a distracting focus to subjects. Since user trialling in a design context is about identification of different types of possible usage, rather than averaged frequencies, it's efficiency may be enhanced by the selection of subjects to reflect constraints which may be thought to be the source of various use patterns (see Kanis and Vermeeren, 1996). This enhances the credibility of small sampling techniques.

User trials may also function as a source of inspiration both in the Information and Analysis phase (see van der Steen et al, 1996) and in the Synthesis phase (see van Hees et al, 1997) In particular, the latter study illustrates the significance of user trialling in a design context by a flexible application of this type of empirical observation to design goals, rather than the adherence to quantitative imperatives which tend to dominate academic endeavours in empirical research.

Tuning design to user activities

In the information and analysis phase

The predominant information source in this phase is the designer's own knowledge and experience. Data from handbooks serve to provide human characteristic information - see previous section. Some information from expert evaluation and user trials is possible if derived from previous designs. It is unfortunate that the feedback from previous designs rarely occurs with the prime motivation of improving usability (see Green and Barnett, 1995). The fact that a design may be 'friendly' or 'unfriendly' is often masked in sales data, or exists in the shade of some other powerful design feature which subsumes criticism of the interface. In particular, users often seem willing to endure inconvenience or worse for the sake of e.g. beauty or status, or blame themselves for things which go wrong (see Weegels, 1996)

In the synthesis and simulation phase

The designer is all powerful in the first stages of this process: even the design brief may be changed or modified in response to the iterative idea generation process. Evaluation is not possible by any external means because there is nothing to evaluate. Only as the first, possibly crude, documentation begins to appear is there anything on which to apply expert judgement. It seems that software is a good deal more susceptible to evaluation at this stage than the physical aspects of designs, and for that it is possible to have systematic evaluation procedures operating from this point on. When a three dimensional design is involved, there is clearly no possibility of physical trialling until the first model appears. Virtual reality offers some prospect of usability trialling in advance of the physical reality, but there is some fear that virtual reality may lead to virtual solutions (see Bruckmann and Gottlieb, 1995.) In the simulation stage some aspects of the interface may be modelled in sufficient detail to allow expert assessment. It is virtually impossible to assign generalised rules, since the dimensionality of the design, it's purpose, it's environment, it's scale, and it's degree of development all have a profound effect on the suitability of the various evaluation techniques. However, for the purposes of this discussion, it is accepted that evaluation techniques are applicable, and become more so as the design is developed. It is at this point that user trialling may be first applied.

In the evaluation phase

It is in this phase that experts other than the designer can make a dominant contribution. Broad physical parameters can be checked against available data, expert evaluation techniques are applicable, and user trialling is possible with a greater degree of certainty that the results will reflect actual in-service behaviour. The conduction of inspections or user trials by an expert in the technique (as opposed to a designer) is attractive for two reasons - one, that the evaluations/trials are likely to be better structured and two, that it guarantees independency. The other side of this coin is the point made previously - that the designer will very likely act as a filter/interpreter in any case, and furthermore, that this filtering will only apply to the information s/he has, and not to the knowledge implicit in the usability expert.

Discussion

Frequently the design process is seen as a sort of reverse delta model, in which a great many tributaries converge to create the great stream - the finished design. In this paper, a user centred design approach turns this around, and the designer is placed in the single generative position from which the design flows out through a number of channels to a multiplicity of users Ultimately, design considerations are directive for any charting of user activities. Even when substantial information is derived from all or any of the techniques mentioned, it must metamorphose into designed forms, and in this the role of the designer may be analogous to the role of language in the consideration of perception/cognition. That is, capable of acting as a perfect conduit, or of perverting or contaminating the process. In order to gain insights into why problems occur, designers need to see users struggling, and this can be achieved by conducting user trials or attending user trials conducted by usability specialists. There is a strong case for usability specialists to conduct the trials, but then they must find a way of transmitting more than just a list of problems to designers. Ideally, then, designers should be regularly involved in user trials. At least initially, this should occur under the guidance of usability specialists, but the primary purpose must be to allow designers to better anticipate user activities and to apply usability inspection methods to designs in progress, thus moving the established techniques upstream to the point of maximum effect.

References:

Bailey R.W., Allan R.W., Raiello P. 1992, Usability Testing vs. Heuristic evaluation: a head to head comparison. *Proc. of the Human Factors Society 36th Annual Meeting*, 409-413.

Bruckmann, R. and Gottlieb, W. Daimler-Benz AG. 1995, Spatial perception of vehicle interior. In *Virtual Reality World*, 459-461.

Ericsson, K.A. and Simon, H.A. 1993, *Protocol Analysis: verbal reports as data*. MIT Press, Ma, USA.

Green, W.S., and Barnett, C.S. 1995, Ergonomics of heavy plant and equipment. Report to the Dept. for Ind. Affairs., Gov. of South Australia.

Hees, L.W. van, Kanis, H. and Marinissen, A. H. 1997, User trialling in the design of a bicycle pump. In S. Robertson (ed.) *Contemporary Ergonomics*, (Taylor & Francis, London).

Kanis, H and Vermeeren, A.P.O.S. 1996, Teaching user involved design in the Delft curriculum. In S. Robertson (ed.) *Contemporary Ergonomics*, (Taylor & Francis, London) 98-103.

Mack, R. and Niesen, J. 1993, Usability inspection methods: report on a workshop held at CHI'92. In SIGCHI Bulletin, 25, I, 28-33.

Nielsen, J. and Molich, R. 1990, Heuristic Evaluation of user interfaces. In *Proceedings of CHI*, ACM, New York, 249-256.

Polson, P.G., Lewis, C., Rieman J. and Wharton, C. 1992, Cognitive walkthroughs: a method for theory based evaluation of user interfaces. Int. Jounl. of Man-Machine Stud., **36**, 741-773.

Steen, V.B.D. van der, Kanis, H. and Marinissen, A.H. 1996, User involved design of a parking facility for bicycles. In S. Robertson (ed.) *Contemporary Ergonomics*, (Taylor & Francis, London) 50-55.

Weegels, M.F 1996, *Accidents Involving Consumer Products*. Delft University Press.

MILITARY ERGONOMICS

HUMAN FACTORS REQUIREMENTS IN C³I SYSTEM PROCUREMENT

Ronald W. McLeod Ph.D.* and Dianna Bishop B. Eng (Hons)†

* *Nickleby & Co. (Scotland) Ltd., 239 St. Vincent Street, GLASGOW G2 5QY*
†Ash 1, MoD(PE), Abbey Wood #93, PO Box 702, BRISTOL BS12 7DU

Keywords: Command and Control; Requirements Analysis HFI

Naval command systems are critically dependent on the ability of their human operators to interact with them effectively and efficiently. In a fixed price procurement environment, it is essential that operator requirements are adequately specified before a contract is placed for the development of a command system. The requirements of a systems human operators are of two sorts: Functional and Non-Functional.
Functional Human Factors requirements are concerned with the interactive facilities the system is required to provide to its operators, and how the system is required to behave in response to operator actions. Non-Functional Human Factors requirements (NF HFRs) are concerned with quality related issues, such as how well operators are required to be able to perform, how easily, and at what cost. These Non-Functional requirements can be critical in determining the systems overall fitness for purpose, even if the Functional requirements are satisfied.
This report contains the results of a short study into methods for capturing and specifying NF HFRs to support the procurement of naval command systems. The study developed a method to demonstrate the process by which NF HFRs can be captured and specified.

Introduction

Many human-related aspects of a system design impact on the extent to which the system is fit for use in an operational environment. In some cases, these aspects can be the primary factors limiting a system's effectiveness and efficiency. They are also important in determining the cost of developing, operating and supporting a system. Examples include:

- The time and effort spent by users in accessing and manipulating information

- Delays or inconveniences arising from the way in which the system requires the operator to interact with it

- The ability of operators to communicate meaningfully and effectively

- The ease with which operators can learn to use the system

- The ease with which maintenance personnel can locate faults or error conditions, diagnose the source of the problem and carry out effective maintenance.

Ensuring that operators will be able to use equipment effectively and efficiently depends on i) identifying critical Human Factors issues early in the procurement process and ii) specifying these issues as requirements in contractual procurement documentation: i.e. in Requirements Specifications.

Most definitions of usability (e.g. Chapanis, 1991, Shackel, 1991, ISO, 1990) emphasise that for a specification of a usability requirement to be sufficient, it needs to define five features: i) the tasks to be performed, ii) the usability objectives (e.g. "easily and effective"); iii) the target user group iv) the support expected to be needed (including user training and on-line HELP) and v) the environment in which the system is to be used. Shackel (1991) illustrates how information derived from Human Factors analyses activities, principally Task and User Analysis, can be used to support such a specification.

ISO 9241, Part 11, (1990) contains guidance on how the usability aspects of products can be specified. The Part includes examples of methods for specifying the context in which computer equipment is to be used, and identifies a range of measures which are often appropriate. ISO 9241 does not however explicitly define a means of specifying usability requirements.

System Requirements

Whatever the particular operational purpose, at the highest level, those procuring systems are always interested in ensuring efficient and effective operation. "Effectiveness" relates to the extent to which the system satisfies its operational objectives. The Captain Naval Operational Command Systems (CNOCS; BR 8710, 1993), define effectiveness as "The extent to which a task, in a defined context, is achieved". This does not make allowance for the effort, resources or costs of achieving the tasks: it is solely concerned with the ability to satisfy operational demands. 'Effectiveness' therefore includes consideration of: i) the time to perform each of the system operations, including preventive and corrective maintenance (speed of response); ii) the extent to which the results of each system operation has the desired effect (accuracy of the response), and iii) the appropriateness of the response.

Speed and accuracy are both amenable to measurement. Appropriateness cannot however be measured directly. It includes dimensions of whether the response is commensurate with the threat and whether the response increases risks in some other way. Further analysis would be necessary to identify ways in which the requirements addressed by 'Appropriateness' could be adequately assessed objectively.

"Efficiency" is concerned with the amount of effort and resources expended in achieving the operational objectives. From the point of view of Human Factors, efficiency includes consideration of the number of people needed to operate the system, the effort required of those operators, and the resulting short and long-term effects on those operators.

The effort required of the system operators includes both the time for which operators are allocated to an activity and the amount of effort operators are required to exert in performing the activity The resulting effects on operators arising from system operations includes consideration of possible impairment to operators' health, safety or well-being, the levels of fatigue arising, and the degree to which operation of the system leads to operator dissatisfaction. The notion of 'Efficiency' also includes a range of dimensions typically addressed under other high level objectives, such as procurement costs and Integrated Logistics Support (ILS) aspects.

Functional and Non-Functional Requirements

Achieving Efficiency and Effectiveness involves a combination of two sorts of requirements: Functional and Non-Functional. Functional requirements identify the set of operations which a system is required to be able to perform. Functional requirements define what features the system will have, and what it will do. Functional requirements are satisfied through the implementation of specific lines of computer software, sets of mechanical operations or electrical events.

Assessing whether functional requirements have been satisfied depends only on observation of the presence or absence of a feature, or the success or failure in achieving a pre-defined result in response to a specified event. The only criteria which needs to be applied in making the assessment is one of whether or not the feature exists, or whether the expected result was achieved.

Non-functional requirements, in contrast, refer to the set of quality related issues which are associated with system functions, but which are not of themselves the operations or transformations performed by the system. They specify the standards to which the functionality has to be provided, such as performance, effort, safety, health, and learnability. Non-functional requirements are not specifically related to the execution of particular pieces of software or the occurrence of mechanical or electrical events. Many of them arise as a consequence of the way in which the total system or sub-system is designed. (Note: Non-functional requirements are sometimes considered as Constraints).

The assessment of whether Non-functional requirements have been satisfied depends on the collection of data using an appropriate device to measure the characteristic of interest.

Functional Human Factors requirements can be illustrated with the following example:

"the colour of objects shown on a tactical display shall represent the systems current belief of the standard identity of the real-world object represented"

This requirement is expressed from the point of view of the system only. Software needs to be developed which, in the event that the identity of the object changes, will ensure that the colour of the object on the display changes. Nothing in the statement requires that the users of the system are able to detect or discriminate the colours used on the display, or that they are able to relate the displayed colour to standard identities.

The same requirement, expressed from the point of the system operator, might be:

"system operators shall be able to determine the standard identity of objects represented on the tactical display through its displayed colour"

The functional requirement is the same as previously. However, because this requirement includes the condition that operators are to be able to determine the object's standard identity from the displayed colour, the requirement is now Non-Functional. It is only by paying attention to the visual capabilities of the operators, and the viewing and task conditions in which the display will be used, that it will be possible to determine whether this requirement has been satisfied.

To provide an adequate specification of the requirement at a level which is testable, the (non-functional) requirement needs to be expressed in a manner such as: "Representative system operators shall be able to identify the standard identity of an object represented on the tactical display through its visual appearance within 2 seconds of its appearance and with less than 2% confusions following a 1 hour period of training and after a 6 hour period of continuous usage in a realistic environment, without experiencing undue visual fatigue or experiencing mental strain due to the effort required to be exerted".

Requirements Specification and Acceptance

An appropriate method for expressing NF HFRs needs to satisfy a number of criteria:

- The specification should only address those Human Factors considerations which directly impact on the operational effectiveness, efficiency, safety or costs of setting-to-work or supporting the system. It should, however, address all Human Factors considerations which directly impact on these objectives

- Requirements should be mutually exclusive

- Requirements expressed at a high level should be decomposable such that design features can be derived in a manner which is directly traceable to the high level requirement. Implementing the resulting set of design features correctly should therefore, in principle, ensure that the high level NF HFRs are met

- The requirement should be clear and unambiguous. The method of specification should ensure that the assessment of whether a design satisfies the requirement is straightforward and subject to minimal interpretation.

- The means of demonstrating compliance should involve minimal dependence on subjective opinion.

- Specifications should identify both the criteria against which a design will be assessed, as well as standards required to be achieved for the requirement to be accepted as having been satisfied.

Conducting valid Human Factors evaluations within a formal system acceptance process can involve considerable expense in terms of time, effort and other resources. It can also impose significant demands on projects; for example in obtaining access to a sufficient number of representative users for the duration necessary to conduct a valid assessment. A realistic balance must be struck between demands for equipment suppliers to demonstrate that NF HFRs have been satisfied, and, on the other hand, requiring only that level of demonstration necessary to provide confidence that the design will satisfy the needs of the operators.

The Human Factors Requirements Process (HFRP)

Recent MoD initiatives (such as HFI/MANPRINT and the Sea System Controllerate Publications (such as SSCP11, 1994)) aimed at integrating Human Factors activities within equipment procurement programmes do not explicitly provide procedures for capturing and specifying NF HFRs. A brief review of current systems analysis methods and existing Human Factors standards and guidelines was conducted. None of the methods, standards or procedures reviewed provide explicit support to the process of capturing NF HFRs. A method was therefore developed to demonstrate the process by which NF HFRs can be captured and specified. For convenience, the method has been termed the "Human Factors Requirements Process" (HFRP). The steps involved in the HFRP are as follows:

1. State the highest level objective(s) (assumed to be an input to the process) as requirement(s)

2. Identify the HF issues necessary to satisfy each requirement

3. Prioritise the issues arising from each requirement

4. Validate the issues to ensure they are necessary and sufficient to satisfy the requirements at that level

5. For all issues of sufficiently high priority, re-state them as requirements at the next lower level

6. Iterate steps 2, 3, 4, and 5 until requirements are adequately specified

7. Validate the hierarchy.

Identifying the set of Human Factors issues that are necessary and sufficient to satisfy the requirement, could involve consulting experienced operators (e.g. CNOCS), HF practitioners, results of structured HF analysis activities (Task Analysis, etc.), considering predecessor of similar systems as well as information contained within relevant standards or guidance (such as SSCP 11, Def Stan 00-25). For example, the process could involve experienced Human Factors analysts, using appropriate analysis activities to mediate between experienced users and system designers to capture the relevant Human Factors issues.

Prioritising Human Factors issues could involve consideration of factors such as; the extent to which satisfying overall system objectives is dependent on that issue; experience gained from previous systems; environmental and other conditions in which operators will use the system; the scope of the procurement (e.g. in terms of areas in which new design is anticipated); the budget and development timescales.

A validation exercise must be carried out to ensure that all the issues identified at each level are necessary and sufficient to ensure the relevant requirement will be satisfied. This review could, for example, involve current system operators and other representative users; HF practitioners and appropriate use of rapid prototyping (particularly part-task prototyping).

The following rules can be used in analysing Human Factors requirements to the required level of detail:

Rule 1: All requirements must be uniquely numbered. Issues at any level shall have the same numbering as the requirement at the level below. This provides traceability through the hierarchy

Rule 2: All issues must be necessary to satisfy the associated requirement at the same level

Rule 3: The total set of issues arising from a requirement must be sufficient to satisfy that requirement

Rule 4: Issues of particular importance shall be restated as a requirement at the next lowest level in the hierarchy

Rule 5: Issues become requirements at the next lower level when they are necessary to meet the higher level requirement and they are not yet expressed at the level of detail needed for acceptance purposes

Rule 6: Analysis continues until the requirement is specified at a level which is appropriate for system acceptance purposes.

The number of levels of analysis needed to produce an adequate specification will vary depending on the complexity of the issues arising from a requirement. Requirements at the lowest level should contain the level of detail necessary to form part of the contractual baseline for a system development contract.

Summary

Current MoD practice in preparing User Requirements Specifications for naval command systems do not define non functional Human Factors requirements to the level of detail or completeness necessary for use in system acceptance. These requirements can however be captured and specified at an appropriate level of detail. A relatively straightforward process (HFRP) can be used to achieve this, although it is dependent on two things: i) the appropriate and timely application of Human Factors analysis activities within a development programme, and ii) effective collaboration between experienced user representatives and experienced Human Factors practitioners. Successful integration of analysis and specification activities is likely to involve experienced Human Factors analysts, using appropriate analysis activities to mediate between experienced users and system designers to capture the relevant Human Factors issues.

References

BR8710 (1993) *Design of Combat Management Systems User Guide* (CSUG) User Interface Guidelines and Interface Glossary D/DOR(Sea) D/24/1/62 Chapter 14

Chapanis, A. (1991) *Evaluating Usability In: Human Factors for Informatics Usability* Shackel, B. & Richardson, S. (Eds.) Cambridge University Press

Defence Standard 00-25 *Human Factors for Designers of Equipment*, MoD

ISO 9241 (1990) *Ergonomic Requirements for Office Work with Visual Display Terminals*

Sea Systems Controllerate Publication No. 11 (1994) *Human Factors Guide for Management and Design in Royal Navy Combat Systems* HMSO

Shackel, B. (1991) *Usability - Context, Framework, Definition, Design and Evaluation In: Human Factors for Informatics Usability*, Shackel, B. & Richardson, S. (Eds.) Cambridge University Press

FRIEND OR FOE?: AN EVALUATION OF SYMBOLOGY EFFECTS ON TARGET RECOGNITION

Susan C. Driscoll[1], Ian R. L. Davies[1] & Stephen J. Selcon[2]

[1]Department of Psychology,
University of Surrey,
Guildford,
Surrey.

[2]Systems Integration Department
Air Systems Sector
DRA Franborough
Hants.

Despite new sensor and electro-optical technology, the final decision to fire upon a ground target depends upon the pilot making a positive visual recognition. Two display response experiments have been carried out in order to examine the possible disruptive perceptual effects caused by the use of current Target Designator (TD) symbology. The first experiment was concerned with possible environmental reference framing effects, and the second was evaluating the proximity effects between the TD and the target. Significant results were found supporting the environmental reference framing hypothesis but not for the proximity hypothesis. Therefore, an alternative circular TD symbol has been proposed as a more suitable design to aid ground target recognition.

General

The importance of accurate and rapid object recognition is rarely greater than when it is made during an air-to-ground military attack. Within less than a minute it is the pilot's responsibility to detect a potential target, determine its status as hostile or friendly, acquire the target and destroy it. Often the pressure of making such a decision is increased by factors such as obscure terrain, poor display resolution and merged boundaries between the opposing forces. Since the final confirmation of a ground target is made with the human eye, ensuring the successful elimination of hostile vehicles and, equally important, the non-engagement of friendly vehicles, becomes an extremely important and difficult process.

A great deal of military research has been carried in order to investigate the dynamics of armoured vehicle recognition, however, little has attended to the possible visual effects resulting from the integration of visual reality (seeing the vehicle from the cockpit) and display symbology (that which is superimposed over the 'real world' by the Head Up Display, HUD, or Helmet Mounted Display, HMD). The aim of this investigation is move the emphasis from that of recognition errors resulting from lack of operator training or experience, to an evaluation of the possible visual effects on recognition resulting from the display symbology.

For the majority of flight time such displays will be showing basic flight information, however, during military engagement the pilot can select a weapons system which will change the display format to include further symbology. One such piece of vital symbol is the Target Designator box. Currently the TD box is a square of 1° visual arc.

Once the designator button has been pressed the TD box remains fixed over the target in reference to particular ground co-ordinates. If the target moves then it will have to be re-designated, but as the aircraft moves the TD will remain over the target as the co-ordinates are recalculated. If the aircraft banks to the left or right the TD box will rotate around the target, staying fixed as a square to the display and to the pilot's eyes. Therefore, the pilot will perceive the ground target as rotating within the TD box.

Experiment One

Introduction

Pilots already face a recognition problem from being in a situation where the target will often be seen from unusual angles. However, in addition to this the target is superimposed with a square which is constantly rotating to various degrees, around it. It is the possible disruptive effects of the rotating TD box which are investigated. Theorists such as Rock (1973) and Butcher and Palmer (1980) propose a theory of referencing which provides the theoretical foundation for this potential problem. Rock refers to the importance of an object's environmental frame of reference remaining constant. That is, if a shape or objects remains in constant reference to its environment it will relatively easy to recognise, however, if the environment is changed in reference to the shape then recognition will be disrupted. Rock claims this is due to the fact that the environment which surrounds a shape or an object imposes its own frame of reference, once this is changed the reference frame is lost and, thus, the shape becomes more ambiguous. Similarly, the TD box can be regarded as such an environmental frame of reference and, therefore, has potential to influence the perception of the target within it.

In order to test this theory the current TD square was compared against the use of a circular TD. A circle was chosen as its rotational and reflectional symmetry means that orientation differences in terms of sense, size and position could never occur. Therefore, no environmental reference frame effect could result by superimposing a ground target with a circular TD. If this is so then subjects should perform with greater accuracy when using the circular TD as opposed to the square or no TD.

Method

Four male and eight female volunteer subjects participated in this experiment. All were employees at the Defence Research Agency. In order to be able to distinguish between the four tanks used in the experiment, subjects were given a presentation sheet displaying each tank in silhouette form from the flank. Subjects were also provided with formal instructions about the experiment and about using the DRA Workload Scales. Each subject had five minutes to study the presentation sheet. A practice run of seven stimuli presenting a combination of each tank, angle and target designator was then carried out. The three types of TD's used were square, circular or no TD at all. The tank status was either hostile or friendly (comprising of two friendlies, the Vickers Mk3 Main Battle Tank and the Challenger 1 Main Battle Tank and two hostiles, the Type 85-III Main Battle Tank and the Type 80-II Main Battle Tank) and were presented at seven different angles to the subjects eyes (0°, 30° left, 60° left, 120° left, 30° right, 60° right and 120° right). Each subject sat 75 cm from the screen with the index finger of each hand placed ready over the 'F' (for friendly) and 'H' (for hostile) keys. On presentation of the stimuli the subject made a response upon which the screen would go blank and the next stimuli would appear. This continued until all 28 runs had been completed. After each condition the subject made a workload rating before beginning the next condition. After the final condition the subject was asked to complete the general questionnaire. Reaction times and error score data was collected.

Results

The reaction time data was calculated from the moment the tank stimulus appeared on the screen to the moment a response was made by the subject. All response times were recorded in milliseconds. A two factor ANOVA was performed on the reaction times.

The two within-subjects measures were TD type and angles. No significant main effect for TD type was found (F=<1), however a significant main affect of angle was found ($F(3,33) = 89.1$, $p<0.01$). The reaction time differences between the four different angles (0°, 30°, 60°, 120°) increased as a function of the increase in the degree of bank.

The error score data was calculated from the accuracy of the subjects' responses. The number of errors were totalled for each condition. A two-factor ANOVA was performed on the error score data. Due to TD type almost meeting the significance level of 0.05, TD type was re-analysed using a one-way ANOVA and the error data log transformed in order to conform to the requirements of parametric analysis.

The two within-subjects measures were TD type and angles. A significant main effect of TD type was found ($F(2,22) = 3.62$, $p<0.05$) showing fewer error being made when using the circular TD.

Table 1. The means and Standard Deviations for the TD type error score data.

	Mean	Std. Dev.
No TD	0.675	0.126
Square TD	0.655	0.118
Circle TD	0.521	0.222

Post-hoc analysis was carried out on this main effect. The circular TD was found to be significantly different from the square and no TD. The square and no TD did not significantly differ from each other. A significant main effect of angle was found ($F(3,33) = 14.7$, $p<0.01$) showing more errors being when the tanks were presented at greater angles of rotation. No significant effect of TD type was found for any of the four workload components. No significant effect of TD type was found for any of the questions asked in the general questionnaire.

Discussion

The possible environmental referencing framing effects caused by using a square TD was tested with the proposal that using a circular TD would reduce these effects. Data collected from the experimental trials supports this hypothesis. Firstly, the significant differences between angles primarily shows that environmental referencing is a potential influencing factor within this particular area of application. If there was no significant difference between the angles then orientation of the tank within the TD would not be important and therefore, neither would the issue of reference framing. The greater the degree of bank (or rotation of the tank target inside the TD), the greater the errors made and time taken to respond to the stimuli. Such a finding can be attributed to the intrinsic model of recognition which states that shapes tested vertically should be faster to recognise than shapes tested at an oblique or horizontal position (Weiser, 1981, Julez, 1971).

The significant difference in error scores between the TD types, when considered in relation to the insignificance of reaction times for the TD's, provides support for a possible environmental reference framing effect (Rock, 1972). Table 1 illustrates the significant difference in accuracy arising from using the circular TD as opposed to using the square or no TD. Using the circular TD led to fewer recognition errors being made, and as the Newman-Keuls analysis shows, subjects were as likely to make as many errors using the square TD as they were using no TD at all. In addition there was no time trade-off as the reaction time data shows that it did not take significantly longer to make these responses.

Although the post-hoc tests questions the possibility of errors resulting from the reference framing effect of the square TD, (as clearly there was no frame of reference in the control condition) the fact that subjects performed better with the circle shows that there must be some form of referencing dynamic occurring. One could argue that a circle is as little use as no TD when it come to referencing as a circle is rotationally and reflectionally symmetric and, therefore, has no fixed points of reference. However, it may be possible that in the absence of a TD subjects were referencing the tank target against the next environmental cue which would be the boundaries of the screen. Therefore, it is possible that the perceptual advantage of the circle is that it removes the disruptive effect of any other type of environmental frame.

The lack of significant workload data suggests that the DRA Workload measure may be not be sensitive enough to the small differences between the TD types. It could be argued that subjects did not really notice any particular workload differences between conditions and that, equally, TD type does not influence workload. However, from an analysis of the general questionnaire there certainly seems to be a perceived problem with the square, as subjects stated a general preference for the circle as the square "tended to distract their attention" (subject 7).

Experiment Two

Introduction

In the light of subjective comments and to further examine the perceptual effects caused by using a square TD, the proximity of the TD box to the ground target has been considered in the second experiment. Discriminating tank features may be disrupted by the close proximity of the TD to the target, especially if the target and TD share similar featural characteristics such as sharp horizontal lines and angles. As Pomerantz's theory of configural superiority shows, features placed in close context can produce configural effects. Since the ground target and the TD are currently presented within one degree of visual arc, there is a strong possibility that the features from both the target and the TD box are configured to create a unitary shape rather than two distinctly different ones. In a situation where identification of critical detail is of upmost importance, the possibility of disruptive configural effects is an important factor to consider.

In addition to this potential problem there is an added concern that whenever the TD box and the target are displayed at different angles to one another as a result of banking, the pilot may potentially be perceiving a different object each time, thus causing recognition errors. In order to explore the possibility of a configural effect on perception, the second set of experiments compares response times and errors from large and small TD boxes. The large TD box lies outside the operators field of focal attention to ensure that the TD and target features are not combined. If proximity is an influencing factor then subjects should perform with greater accuracy when using the large TD.

Method

Six male and six females voluntarily participated in this experiment. All subjects were from the University of Surrey. Subjects were presented with formal written instructions and the tank presentation sheet. The procedure followed that of experiment one except that only the two types of TD were used (small and large) and that both TD types were randomised within one set of trials. There were 56 stimuli presentations in the trial. Only the general questionnaire was completed after the experiment.

Results

The reaction times were calculated and analysed as in experiment one. No significant main effect for TD type was found ($F=<1$). A significant main affect of angle was found ($F(3,33) = 61.2$, $p<0.01$) showing an increase in time taken to respond as a function of the increase in the degree of bank.

The error score data was calculated and analysed as in experiment one (two-factor ANOVA). No significant main effect for TD type was found ($F=<1$).
A significant main effect of angle was found ($F(3,33) = 7.9$, $p<0.05$), again showing an increase in the number of errors being made as a function of the increase in the degree of bank.

A significant effect of TD type was found for question 2 in the general questionnaire ($F(1,11) = 6.87$, $p<0.05$).

Discussion

The possibility that the proximity of the TD to the target which may be causing recognition problems was tested by removing the TD so it lay outside of the operators field of focal attention. However, analysis of data from experiment two strongly suggests that this should be rejected.

The main effect of angle for reaction time and errors was found to be significant and indicated the same pattern as proposed in the intrinsic model, i.e., the greater the rotation from the vertical, the greater the reaction time (Weiser, 1988). Reflecting the main effect found in experiment one the reaction times and errors, when using the small square TD, increased as a function of increased rotation of the tank. The fact that this is true for both TD's implies that there is still a strong effect of environmental referencing upon recognition, regardless of proximity - a possibility proposed in the discussion of results from experiment one.

The lack of significance for TD type suggests that no disrupting configural effects occurred as a result of having the TD in close proximity to the target. Therefore, the possibility of focal attention leading to errors in recognition should be discounted. If the subjects' responses had been affected due to combining features from the TD and the target as a result of focal attention (Treisman and Gelade, 1980, Pomerantz et al. 1977), then the reaction time and error scores should have increased as a result of using the small TD. Despite a lack of significant objective data for TD type, one of the subjective rating taken from the general questionnaire was found to show a significant difference between the small and the large TD. In response to question two, subjects rated using the large TD as more difficult in discriminating the tank's status when the aircraft was banked. This, perhaps, stresses the importance of having some type of symbology in close proximity in order to quickly reference the orientation of the tank. However, this does not mean that the smaller square is a more functional design of symbology with which fewer errors are made. Indeed, seven out of the twelve subjects later in the questionnaire stated a preference for the larger TD on the basis that it "did not clutter the tank and distract attention away from its features" (subject 10). This again raises the possibility that although a small square TD may create a distraction, something is needed within a relatively close degree of visual arc to the target in order to provide a reference frame. Error data from experiment one supports this assertion.

Conclusion

Results from both experiments highlight the importance of further investigation into the impact of environmental reference framing as created by the TD symbology. Findings from experiment one indicate that the current square TD will cause recognition problems for operators when the aircraft is moving through various angles of bank. Further to this, whilst the possible perceptual drawbacks of focal attention have been discounted, experiment two has highlighted the importance of the operator having some close form of reference frame. Therefore, in considering the findings from both experiments, a small circular TD is proposed as a more suitable symbology design. A circular TD, unlike the current square TD, is less likely to produce any disruptive environmental referencing effects, yet at the same time still provides the necessary frame for positive recognition of a ground target.

References

Briggs, R. W. & Goldberg, J. H. 1995, Battlefield recognition of armored vehicles, *Human Factors*, **37(3)**, 596-610.

Dror, I. 1992, Visual mental rotation: Different processes used by pilots, *Proceedings of the Human Factors Society 36th annual meeting*, 1368-1372.

Dudfield, H. J. & Hardiman, T. D. 1994, A review of HUD and HMD symbology: Part one, *DRA technical report*.

Foskett, R. J., Baldwin, R. D. & Kubala, A. L. 1978, The detection ranges of features of armored vehicles, *U.S. Army Research Institute Report 78-A37*

Humphreys, G. W. 1983, Reference frames and shape perception, *Cognitive Psychology*, 15, 151-196.

Johnson, R. M. 1981, An information processing model of target acquisition, *Proceedings of the Human Factors Society - 25th annual meeting*, 267-271.

Pomerantz, J. R., Sager, L. C. & Stoever, R. J. 1977, Perception of wholes and their component parts: Some configural superiority effects, *Journal of Experimental Psychology: Human Perception and Performance*, **3(3)**, 422-435.

Rock, I. 1973, *Orientation and Form*, (New York: academic Press), 1-137.

Rock, I., Halper, F., & Clayton, T. 1972, The perception and recognition of complex figures, *Cognitive Psychology*, **3**, 655-673.

Treisman, A. M. & Gelade, G. 1980, A feature Integration theory of attention, *Cognitive Psychology*, **12**, 97-136.

Wiser, M. 1981, The role of intrinsic axes in shape recognition, *Third annual conference on cognitive science*, 184-186.

Human Factors Assistance to the Specification of a Mission Planning System

Adrian Bryant, Iain MacLeod, Ian Innes

Aerosystems International,
West Hendford, Yeovil,
Somerset, UK. BA20 2AL

The specification of a system is normally performed under the tenets of some form of engineering. Human Factors has usually little impact on initial system requirements specification and is normally only evoked at later stages of the system design life-cycle. Indeed, the earliest specification of human requirements for a system are usually considered as a series of constraints on system design rather than as a functional contribution. The more complex the system the more important it is that its effectiveness is assessed considering all of the functional contributors to system performance. It is argued that the human performance contributions to a system can be usefully specified alongside engineering functions at an earlier stage of the system design process than hitherto.

Introduction

The specification of a system is normally performed under the tenets of some form of engineering, whether that be Systems Engineering or Integrated Logistics Support (ILS). Human Factors has little impact on system specification and is usually only evoked at later stages of the system design life-cycle. Indeed, the specification of human requirements for a system is usually considered as a series of constraints on system design.

Too often system requirements and function sets are specified by the use of methods depending on engineering logic. The completed function set is then usually examined by considering whether human or the machine capabilities and performance can best handle a function of groups of functions. However, as the considered functions have been defined by engineers, it is unlikely that they will encompass human cognitive functions related to such as human understanding and decision-making.

Moreover, functions are latent properties of a system until evoked. In contrast, system performance is concerned with the dynamic use of system functions as tasks and the efficiency with which both machine and human tasks are integrated and directed towards common goals. If human cognitive functionality within a system is not addressed in system requirements specification, then human system related tasks cannot be addressed through design until later, and often too late, in the design life-cycle.

Study Objectives & Utility

The more complex the system the more important it is that its effectiveness is assessed considering all of the contributors to system performance. This paper will argue that the human performance contribution to a system can be usefully specified at an earlier stage of the system design process than hitherto. For more detailed argument in this area see MacLeod, I.S. and McClumpha, A. (1995) and Final Report on ASTOR HE Study (1994). Furthermore, if a sub system is to be designed and developed in the support of an already designed system, it should be possible to specify the sub system HF issues from the onset. This is especially true of a Mission Planning System (MPS) that collects, filters, collates, and disseminates pertinent information required for the operation of an existing aircraft.

The MPS Problem Area

The systems that MPSs serve are becoming increasingly complex and technology driven, engineered to be capable of actions and computer based handling of data at rates beyond the capabilities of the aircraft operators to assimilate and comprehend. Regardless of this complexity, the aircraft crew are still required to direct and control overall system performance as this is still beyond the capabilities of engineered systems. For further discussion on the changing nature of human tasks with complex systems see the ASTOR HE Study. Changes to aircrew roles incurred by changes in technology require an associated improvement in the designer's appreciation of aircrew cognition and system complexity.

Efficient aircrew direction and control of an aircraft requires pre mission planning and rehearsal if it is to be effective in its maintenance of the mission aims. The quality of the information used in MPSs must be sufficiently understood by the system operators to allow judgements to be made on the application of that information to the planning process. For example, good operator knowledge on the source and recency of information is always important.

However, an understanding on the form of the information and the purpose of its presentation is also important. For a discussion on some perspectives for appraising information forms, purposes, and limitations see MacLeod, I.S., 1995. As an example, maps are of prime importance to the conduct of aircraft related mission planning. The application of modern technology to the benefit of the aircrew use of map information, both in the air and on the ground, is possible if progressed sensibly. Such sensible progression should approach the efficacy of aircrew aids and assistance, possible through the application of the technology, before being seduced by the myriad of imaginative possibilities available under the technology regardless of its application (Taylor, R.M. & MacLeod, I.S., 1995).

The one true tenet of any proactive pre-mission planning is that plans will need to be changed to equate or overcome real events during the actual conduct of the mission.   Thus reactive planning must also be performed during a mission.   Such planning is normally catered for by effective proactive planning for mission contingencies, the use of a suite of appropriate tactics (MacLeod, I.S. & Taylor, R.M., 1993), the use of Standard Operating Procedures (SOPs), and the application of aircrew skills and   expertise.

To aid the understanding of what constitutes an MPS, Figure One is a generic high level schematic of an MPS considering mission planning   equipment.

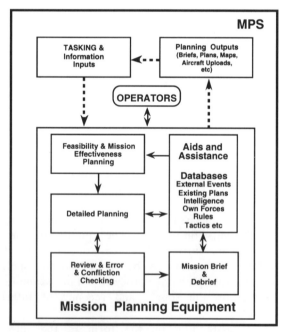

Figure One: A Generic MPS

Needs of the MPS Designer

The MPS designer needs a good grasp of the concepts of operation of the associated aircraft, that aircraft's equipments and performance, the aircrew roles, and the information needed by the aircrew for the performance of the mission.   The specification of the MPS should be based on customer requirements and system functional requirements derived through the designer's understanding of the above and extensive liaison / iteration with customer representatives, especially Subject Matter Experts (SMEs).

This paper suggests that the following issues should be addressed during the requirements definition for MPSs:

1) Definition of human and machine functional requirements for the MPS - these functions to encompass the type of input information required for the MPS operation and the outputs from the MPS required for aircraft operation.

2) Knowledge of expected MPS performance.

3) Knowledge of formal aircraft operating procedures and the working / organisational mores of the aircraft controlling authorities.

4) Knowledge of the expected roles and competencies of the intended system operators.

5) Knowledge of the optimum forms of system automation and control considering the above points and the optimum forms of control for the MPS operator at the Human-Machine Interface (HMI).

From a consideration on the above issues, the designer can then consider how best to design the MPS to aid the operator in the mission planning task.

A Route to MPS System Specification

A route used by Aerosystems International for the current specification of an aircraft MPS will be briefly discussed, this addressing all the points raised in the previous section. The particular MPS system will not be named.

The initial system specification represented the customer requirements for the MPS and was primarily based on the obvious mission information requirements of the associated aircraft system considering its envisaged missions. This specification was used to scope the level of design work to be performed by the winning contractor and a contract was awarded on that basis.

However, realising that the initial specification contained possible ambiguities, and bearing in mind the increasing awareness of the customer on the capabilities of the aircraft, an iteration on the specification is being conducted to ensure that, within the limits of liability on the contract, the customer gains the best design for their money.

The process has involved frequent liaison with customer SMEs, the compilation of a Concept of Operations biased towards the use of the MPS, a firming of the functional requirements specification considering the operators contributions, an iterative approach to the early design of the MPS, and a planned HF

programme applicable both to the MPS prime contractor and the
sub contractors.   Considering the customer requirements, HF work
will play an important role in this MPS project, this from the
completion of the initial system requirements specification to the
final customer acceptance of the system.

Importantly, the HF programme has been integrated into the
mainstream Systems Engineering (SE) work of the MPS design and
development.   To ensure this integration, the HF work from
concept of operation & detailed specification to Task Analyses is
being incorporated into the SE data base to ensure that HF
findings are not lost, that system design trade-offs considering HF
are recorded, that HF requirements are maintained, and that all
HF work is traceable to source.   The toolset being used is the
Requirements Driven Design (RDD) toolset made by the Ascent
Logic Corporation.

Assistance Afforded by ASCC Advisory Publication 61/116/11

A good and recent guide to the application of HF to MPSs is the
ASCC Advisory Publication 61/116/11.   This document gives an
overview of what an engineered MPS system should be capable of
doing but proffers very little specific guidance on what aircrew
cognitive functional contributions and needs are required within
the mission planning process.   Rather, the needs of the operators
in the manipulation of the engineered MPS functionality are
addressed, and the ways of HF addressing the best use of that
functionality.

The ASCC 61/116/11 states that it is intended as a guidance on
the HF requirements for MPSs in order

> " .. to achieve a common standard for functionality and
> usability at the mission-planning aid (MPA) operator
> equipment interface, and to improve interoperability and
> performance effectiveness."   ASCC 61/116/11

The publication gives the operational objectives of a MPS as:

- The reduction of the turnaround time between missions;
- The reduction of aircrew in-flight navigation workload;
- The production of better mission plans than previously;
- The provision of timely mission information to
  interested parties;
- To increase the probability of the detection of mission
  planning errors.

The ASCC publication also gives a high level coverage of the
generic purposes and functions of mission planning.   This
document provided a very useful 'checklist' to validate the overall
scope of the engineering functionality specified for the subject
MPS.

Conclusion

The above discussion has indicated that HF work can be planned to contribute to systems design from the early specification of system functional requirements to the final customer acceptance of the product. This planned contribution largely negates the need for the traditional 'Allocation of Functions', an allocation based on an engineering performed logical derivation of system functionality that, by its very basis, cannot address the importance of aircrew cognition to the direction, control and ultimate performance of a system. Instead, this article has discussed an attempt to understand the complementary human functionality and tasks required to assist the analysis of anticipated mission events through the mission planning process. This form of operator based assistance to system analysis is currently essential for anticipation of the effects of diverse influences from the mission environment on the planned conducted of the mission.

Further, only by recruiting in a rigorous way the existing knowledge and skills of SMEs, through appropriate and explicit methods designed to iteratively capture the cognitive aspects of system use throughout the design life-cycle, can it be expected that the performance of an MPS will meet the desired goals.

References

- ASCC Advisory Publication 61/116/11 dated 30 April 1996, "Human Factors Aspects of Mission Planning Systems".
- *Final Report on ASTOR HE Study,* undertaken under UK MOD(PE) contract SLS 1b/132. (AeI 1539K/1/TR.1-2 dated December 1994).
- MacLeod, I.S. & McClumpha, A (1995) *An Engineering Psychology Approach to Conceptual Design,* Contemporary Ergonomics 1995, Taylor and Francis.
- MacLeod, I.S. & Taylor R.M. (1994) 'Does Human Cognition Allow Human Factors (HF) Certification of Advanced Aircrew Systems?' in Wise, J.A., Hopkin, V.D. & Garland, D.J. (Eds) *Human Factors Certification of Advanced Aviation Technologies,* Embry-Riddle Aeronautical University Press, Florida.
- MacLeod, I.S. (1995), *Is effective aircraft display design driven by technology or to support aircrew task performance?,* Proceedings of the Electrical Research Association (ERA) Avionics Conference, Heathrow, November.
- Taylor, R.M. and MacLeod, I.S. (1995), *Maps for Planning, Situation Assessment and Mission Control,* Journal of the Royal Institute of Navigators, May, Cambridge University Press.

IDENTIFYING THE ENEMY: THE USE OF SHAPES TO DENOTE ALLEGIANCE IN THE FAST-JET COCKPIT

D.G. Croft [1], S.J. Selcon [2], C.S. Jordan [2], H. Markin [2] and I. Davies [1]

[1] *Department of Psychology, University of Surrey*

[2] *Human Factors Group, Systems Integration Dept,*
Air Systems Sector, DERA, Farnborough

Military cockpit displays specify the number of nearby aircraft, their allegiance and headings. We assessed the utility of shape coding (hostiles = diamonds or triangles) for such displays. Observers specified how many hostiles were present in displays with varying numbers of aircraft. For both symbols, search time increased with the number of aircraft (serial search), but triangles were detected faster than diamonds. Next, possible effects of direction indicators, and of the counting task, were explored. Now, triangles were only faster than diamonds when direction indicators were used. With no direction indicators search was parallel for both symbols. However, with direction indicators search for diamonds was serial, but parallel for triangles. Implications for display design are discussed in terms of Feature Integration Theory (Treisman, 1991).

Introduction

As the technology in modern fast-jet cockpits becomes more complex, the workload of the pilot also increases. For this reason it is imperative that cockpit displays provide information that can be correctly and rapidly interpreted. One requirement of fast-jet Head Down Displays is to provide pilots with allegiance and heading information for other aircraft. If allegiance information cannot rapidly and accurately be interpreted during combat the pilot may fire on friendly or neutral aircraft, or may be shot down himself. In obtaining such information from displays the pilot is performing a visual search operation. In this paper we apply general theories and findings from visual search tasks to guide the design of effective cockpit displays. In particular, we focus on the use of shape coding to improve visual search in such displays.

Perhaps the most recognised model of visual search is Treisman's Feature Integration Theory, or FIT (Treisman, 1991). This theory postulates that visual input from a display will be processed in two successive stages - a preattentive parallel stage, followed by a more rigorous serial stage. The parallel processing stage consists of a set of '*feature maps*', each of which will encode a particular feature present in the stimulus (e.g. a line of a particular orientation; a specific colour etc.). However, the parallel stage cannot provide information relating to the 'conjunction' of features. It is the serial stage of processing, which requires attention to be focused on each element of the display in turn, that provides information about how features are conjoined. If a target is defined by a separate feature (one not possessed by the distractors), then the resulting 'feature search' will be parallel (search times will be unaffected by the number of display elements), and the target will 'pop-out' (Treisman and Souther, 1985). If a target is defined by a conjunction of features, then the 'conjunction search' will be serial, with focal attention needed to interrogate the separate feature maps and to recombine these features into objects of a particular shape, size, colour etc. FIT was modified by Treisman and Gormican (1988) to allow for the possibility that feature searches can be serial when targets and distractors are very similar. In such a case, it is believed that the distractors would excite the target feature map, with net activity levels confusing the decision as to whether a target is present or not.

In this study we examined the utility of a newly proposed symbology set for fast-jet cockpits. The shape coding used to designate 'friendly', 'unknown', and 'hostile' aircraft was assessed in terms of its ability to produce pop-out, as compared with an alternative set of symbols.

Experiment 1

Experiment 1 investigated the possible difficulties that would arise with the new symbology, as a result of using diamonds (for hostile aircraft) and squares (for unknowns). Search performance for diamonds was compared with that of triangles using a visual search 'counting task'. The *total number of aircraft* symbols displays contained (6, 9, or 12), and the *number of aircraft depicted as 'hostiles'* (one, two, or three), were manipulated to investigate whether pop-out would occur.

Method

Subjects. 12 subjects, aged between 20 and 35 years of age were used. None of them had any flying experience and all had normal or corrected to normal eyesight.

Apparatus. Stimuli were presented on a 10" Macintosh LC monitor. Responses were recorded via the computer keyboard, which was masked to cover all but the three response keys; one key for each of one, two, and three hostile targets. The stimuli were static, monochrome displays measuring 9.5cm x 11.5cm, viewed from a distance of 50cms, and were representative of a generic modern fighter aircraft radar display. Hostile aircraft were either diamonds or triangles, depending on the experimental condition. Unknown aircraft were depicted as squares. The direction of travel for aircraft was indicated by a short line emerging from the shape (the direction indicator). Each display contained either one, two, or three hostiles, and either 6, 9, or 12 aircraft symbols.

Procedure. Ss were required to indicate how many hostiles were present in each display. Half the Ss began with the trials containing triangular hostiles, whilst the other half started with diamond shaped hostiles. For each shape, Ss began with a practice session, followed by three experimental trial blocks, corresponding to the 6, 9, and 12 aircraft display formats. Each experimental block contained 15 trials - five trials for each level of the hostile aircraft variable. The order in which the 15 trial blocks were presented for each shape was balanced, and the order of the 15 trials within each block was fully randomised. After completing the practice session and the first 45 trials, the procedure was repeated for the remaining shape (90 trials in total). Stimuli were presented until a response was made, with reaction times being measured from stimulus onset.

Results

The data were analysed using three-way ANOVAs (*symbol shape* by *total number of aircraft* by *number of hostile aircraft*), and Newman Keuls Range Tests. Here, we report only the major findings. All differences were significant at at least the p<0.05 level.

    Response Times. The two most pertinent effects were first, *symbol shape*: responses were faster for triangles (1223.2 ms), than they were for diamonds (2505.8 ms). Second, the *total number of aircraft*: for both diamonds and triangles, an increase in the number of distractors was accompanied by an increase in RTs, implying serial search (see Figure 1).

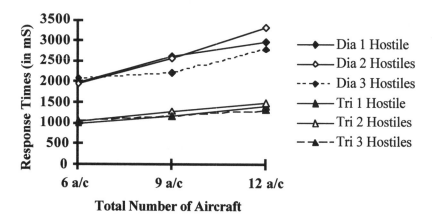

Figure 1: Mean RTs for one, two, and three triangular and diamond shaped hostiles as a function of the total number of aircraft presented.

    Error Scores. In general, the error score data mirrored the search time data: there were more errors for diamonds than for triangles, and errors increased with the number of aircraft.

Discussion

Visual search was faster and more accurate for triangles than for diamonds with the same distractors. In addition, both shapes demonstrated a monotonic increase in RTs as a function of the number of distractors. As previously mentioned, FIT distinguishes between those targets defined by a single feature and those defined by a conjunction of features. Both shapes operating as targets for this experiment were distinguished from distractors on the basis of a single feature. For triangle targets this feature was the number of sides possessed by triangles, as compared with squares. Since the diamonds proposed by the new symbology set were essentially rotated squares, the number of sides, internal angles etc. could not serve as distinguishing features. In this case, the distinguishing feature was a rotation of 45°.

Both targets for this experiment should, theoretically, afford parallel feature searches. Why then were they both serial? Treisman and Gormican's (1988) 'target - distractor similarity' hypothesis can possibly explain why searches for a diamond target among the very similar square distractors was conducted serially, but cannot offer an adequate explanation for triangles. The most likely explanation for the serial search performances is that of Quinlan & Humphreys (1987). They postulate that if Ss are asked to verify the presence of two targets, rather than one, search performance may in some instances be serial, even for feature searches. Although the 'counting' task was intended to emulate the reality of a pilot in combat being presented with multiple targets on radar, it differs considerably from the type of task used in other visual search studies. Also, that symbols had direction indicator lines, cannot be ignored; they were not simple stimuli like those used in other visual search experiments. It is possible that the lines disrupted the orientation information for diamonds, but did not interfere with the distinguishing feature for triangles. This could explain the performance differences between the two target shapes. To address these issues, we conducted experiment 2.

Experiment 2

Two main questions have been raised: first, would the same results be obtained for symbols without direction indicators? Second, were the results an artefact of the counting task employed for Experiment 1? To address these questions we replaced the counting task with a simple 'present/absent' identification task, and removed the direction indicators from half the stimuli. In addition, the *geometric shape* used to represent hostile aircraft (diamond or a triangle), the *total number of aircraft* symbols (6, 9, or 12), and the *target status* (present or absent) were manipulated. Unless specified, the method for this experiment was the same as for Experiment 1.

Method

Subjects. 16 subjects, aged between 19 and 32 years of age were used. None had participated in Experiment 1, and all had normal or corrected to normal eyesight and no previous flying experience.

Procedure. Ss had to identify the presence/absence of a hostile aircraft for each trial using a two-option, forced choice keyboard response. The experiment was run as four blocks; the blocks were derived from the factorial combination of *geometric shape* and *direction indicator status*. Each block consisted of 24 trials - four trials for each of the six combinations of total number of aircraft and target status. Practice was given prior to each trial block. The order in which the four experimental blocks were presented was balanced using a 4x4 Latin Square, and within each trial block the order of the 24 stimuli was randomised. After completing the practice session and the 24 trials for the first block, the procedure was repeated for the remaining three conditions (96 trials in total).

Results

A four-way ANOVA (*symbol shape* by *total number of aircraft* by *direction indicator status* by *target status*) was employed for both RTs and error scores. Post-hoc tests were either Newman Keuls Range Tests, or t-Tests adjusted with the Bonferroni correction. Unless reported otherwise, all tests were at the $p<0.01$ level of significance.

Response Times. All main effects were significant: *symbol shape* (F=29.263, d.f.=1,15; $p<0.001$), where Ss responded faster for triangular hostiles (768ms) than for diamonds (1233ms); *total number of aircraft* (F=44.709, d.f.=2,30; p<0.001), with RTs increasing with the number of display elements; *direction indicator status* (F=45.566, d.f.=1,15; p<0.001), with Ss responding faster when aircraft symbols had no direction indicators (685.3 ms), than when they did have them (1315.3 ms); and *target status* (F=99.781, d.f.=1,15; p<0.001), with Ss identifying the presence of hostiles faster than they could identify their absence. The most pertinent results come from the 'positive' trials (trials where a target was presented), and were revealed by a number of significant interactions (see Figure 2). From Figure 2 it can be seen that when there are no direction indicators there is little difference between triangles and diamonds, and no effect of the number of aircraft (parallel search). On the other hand, when there are direction indicators present, performance with diamonds is considerably slower than for triangles. Further, performance for diamonds increases with the number of aircraft (serial search), whereas there is no such increase for triangles (parallel search).

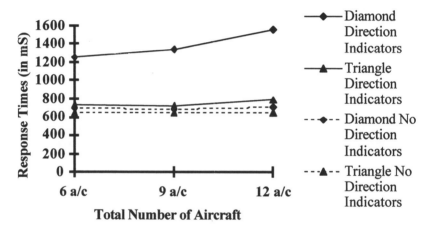

Figure 2: Mean RTs for diamond and triangle positive trials, with and without direction indicators, as a function of the total number of aircraft presented.

Error Scores. The error score data again mirrored the search time data: there were more errors when direction indicators were present than when they were not, particularly for diamonds.

Discussion

Regardless of whether trials were negative or positive, when direction indicators were removed there were no differences between shapes for RTs or error scores. In addition, visual search for both shapes was parallel. Although this finding seems unremarkable, it is of importance, in that a single line attached to each symbol can alter the way displays are searched.

The most consequential finding of this experiment was that when direction indicators were used and trials were positive, visual search was serial for diamonds, but parallel for triangles. Contrast this with the serial search performances obtained with the counting task used in experiment 1, and it becomes clear that the choice of experimental task is crucial when evaluating symbology sets. Visual search for both shapes should have been parallel, since both targets had a unique distinguishing feature (orientation and shape, respectively). Why then did triangles 'pop-out', when diamonds did not ? It is plausible that, for diamonds, direction indicator lines obscured the distinguishing feature i.e. orientation. The diamonds used were rotated squares, and the addition of a line to all symbols may have interfered with the only feature which allowed Ss to distinguish diamonds from squares. For triangles, the difference in the number of sides for distractors and the target would remain unaffected by the addition of an extra line. Alternatively, the addition of the line may have destroyed the configural properties of the diamond, but not the triangle.

General Discussion

This study has successfully demonstrated the negative impact on visual search performance that can result from: first, the choice of target and distractor shapes; second, minor arbitrary modifications to simple geometric shapes; and third, the experimental task used to explore search performances of different shapes. Although we have only scratched the surface of this complex research area, we can use the findings of this study to generate some display design guidelines for symbology sets:
(i) When utilising shape coding it is essential that all shapes are as dissimilar as possible (e.g. rotated square targets and square distractors should never be used simultaneously).
(ii) The addition of lines, numerals, letters etc. to simple geometric shapes may have a negative impact on visual search performance for some shapes and not others.
(iii) The primary task of the display operator may vary and, therefore, it is imperative that proposed symbology be tested for all tasks in which the operator will engage.

References

Quinlan, P.T. and Humphreys, G.W. 1987, Visual search for targets defined by combinations of color, shape and size: An examination of the task constraints on feature and conjunction searches, *Perception & Psychophysics*, **41**, 455-472.

Treisman, A. 1991, Search, similarity, and integration of features between and within dimensions, *Journal of Experimental Psychology: Human Perception and Performance*, **17**, 652-576.

Treisman, A. and Gormican, S. 1988, Feature analysis in early vision: Evidence from search asymmetries, *Psychological Review*, **95**, 15-48.

Treisman, A. and Souther, J. 1985, Search asymmetry: A diagnostic for preattentive processing of separable features, *Journal of Experimental Psychology: General*, **114**, 285-310.

METHODOLOGY

USING EPISTEMIC LOGICS TO SUPPORT THE COGNITIVE, ERGONOMIC ANALYSIS OF ACCIDENT REPORTS

Chris Johnson

Glasgow Accident Analysis Group,
Department of Computer Science,
University of Glasgow,
Glasgow, G12 8QQ, Scotland.

johnson@dcs.gla.ac.uk
http://www.dcs.gla.ac.uk/~johnson

Many of today's accident reports suffer from a number of limitations. In particular, it can be difficult to extract critical events from the mass of background detail in natural language accounts of human 'error' and system 'failure'. This, in turn, makes it difficult for readers to identify the evidence that supports the recommendations and findings in an accident report. Logic provides a means of avoiding these limitations. Formal proofs can be constructed to demonstrate that particular findings are actually consistent with the reported evidence. If such a proof cannot be derived then additional evidence must be found to support the analysis that is presented in an accident report. Unfortunately, conventional logics only provide limit support for this approach. In particular, they cannot be used to document the reasons **why** human 'error' jeopardises the safety of an application. This paper, therefore, demonstrates that epistemic logics can be used to formalise the implicit judgements that accident reports make about operator knowledge and motivation.

Introduction

Accident reports are intended to ensure that major failures do not recur in similar systems. It can, however, be difficult for readers to identify critical events from the mass of background detail that is presented in these lengthy, natural language documents. This, in turn, prevents readers from extracting the evidence that supports particular findings. This paper, therefore, argues that logic can be used to focus upon critical sequences of operator 'error' and system 'failure'. An important benefit of this approach is that formal proof techniques can be used to demonstrate that the necessary evidence is actually available to support the findings and recommendations in a report. The intention is not to replace the natural language accounts in accident reports but to ensure that the use of prose does not lead to fallacious arguments or implicit assumptions.

Formal notations are increasingly being used to support the development of safety-critical systems (Johnson 1996). Unfortunately, they, typically, cannot be used to support accident analysis. In particular, they cannot easily be exploited to document the causes of operator 'error'. This is an important limitation, given the central role of human failure in most major accidents (Reason, 1990). The following pages, therefore, demonstrate that epistemic logics can be used to support the cognitive, ergonomic analysis of accident reports.

Case Study

The Air Accident Investigation Branch's (1996) report into an incident involving a Boeing 737-400 aircraft near Daventry on the 25th February 1995 is used to illustrate our argument. This report provides an appropriate case study because it typifies the way in which cognitive analysis can be lost amongst a mass of contextual detail. Briefly, the incident took place when the aircraft was climbing after leaving East Midlands Airport en route for Lanzarote Airport. The crew noticed a loss in the volume of oil and a consequent drop in the oil pressure. They, therefore, diverted to Luton Airport where both engines were shut-down during the landing roll.

The subsequent enquiry found that the aircraft had undergone Borescope inspections to survey the state of the turbine sections in both of its engines during the night prior to the incident. This was normal practice after 750 operating hours. Borescope inspections involve the removal of high-pressure rotor drive covers. When the inspection was completed, these covers were not refitted and this, in turn, enabled oil to drain from both engines during the flight. Fortunately, no-one was injured and the aircraft was undamaged.

First Order Logic and Accident Analysis

Accident reports are, typically, large and complex natural language documents. They not only record the events that actually contributed to an accident but also the mass of contextual detail that helps to build a more complete picture of the failure. Unfortunately, it can be extremely difficult to identify critical events from contextual detail (Johnson, McCarthy and Wright 1995). For example, the following citation is taken from the AAIB (1996) account of the Daventry incident:

> "He (the Controller) assessed that his remaining Inspector and the Licensed Technicians should, if properly organised, be sufficient to supervise the required overnight Base Maintenance workload. So, he offered to take over the Borescope Inspections personally if the Line Maintenance engineer could take over moving the 737-500 from the Ramp to Base. This offer was accepted and the transfer of the task was noted, at 22:00hrs, in the Airline's Line Maintenance log" (Section 1.1).

This extract contains background information. The competency of the technicians to perform routine line maintenance plays no further part in the report, neither does the task of moving the 737-500. The previous extract also describes a critical event in the sequence leading to the accident. In particular, the transfer of the task to the Controller was significant because Borescope Inspections were not a frequent component of his job. First order logic provides a means of stripping out contextual details to focus in upon these critical events:

$$task(controller, borescope\_inspection, 22:00:00). \qquad (1)$$
$$\neg\ task(line\_engineer, borescope\_inspection, 22:00:00). \qquad (2)$$

The first clause states that the Controller has the task of performing the borescope inspections at 22:00:00 hours. The second clause states that the line engineer does not have the task of performing those inspections at 22:00:00 hours.

The previous clauses represented the precise time of the hand-over. Unfortunately, analysts frequently do not know the exact timings for all of the events in the lead up to an accident. For instance, the AAIB report does not state precisely when the Line Engineer informed the Controller of the state of the Borescope inspections:

> "The Line Engineer gave a verbal brief to the Base Controller of where he had got to in the preparation of engines for the inspection. This brief included the facts that the aircraft was 'dead' (i.e., without electrical or hydraulic power and had no-one else working on it); that one engine (the No. 1) had been fully prepared for inspection and the inspection paperwork pack was with the aircraft in T2" (Section 1.1).

Such imprecise timings can be represented in first order logic using the $\exists$ operator (read as 'there exists'). Informally the following clauses state that there exists a moment at which this communication took place. The terms *aircraft_dead* and *no_1_ready* refer to the exact utterances made by the engineer. These are not cited in full because the report does not provide the exact words that were spoken by the operator. This again illustrates how first order logic can be used to represent the events leading to major accidents at a suitable level of abstraction:

$$\exists t: communicate(line\_engineer, controller, aircraft\_dead, t). \qquad (3)$$
$$\exists t: communicate(line\_engineer, controller, no\_1\_ready, t). \qquad (4)$$

The first clause states that the line engineer informed the Controller that the aircraft was dead at some point during the accident. The second clause states that the line engineer informed the Controller that the Number One engine was ready at some point during the accident.

It is important to emphasise that accident reports, typically, contain two different types of information. The previous citations capture factual observations based upon the evidence that was presented to the board of enquiry. In contrast, accident reports also contain hypotheses or theorems that analysts propose about the causes of an accident. For example, the AAIB report concluded that:

> "It was in not following these (Aircraft Maintenance Manual and Boeing Task Card) procedures that the performance of the Controller and the fitter had its most serious

shortcomings. Having not done a 750 hour Borescope Inspection for about a year, it is surprising that the Controller did not feel it necessary to use the Task Card to remind himself of the complete procedure. A possible reason for him not having done this, before first going to the remote T2 hangar, may have been that he expected Task Cards and AMM extracts to be part of the normal workpack, in line with normal Base Maintenance practice" (Section 2.10.1).

This suggests that the accident occurred because the Controller and the fitter did not consult the Aircraft Maintenance Manuals and the Boeing task cards before going to hangar T2. First order logic can be used to explicitly represent this hypothesis. For example, the following clause states that there is the potential for an error if any person, p, does not consult the appropriate sections of an Aircraft Maintenance Manual and the Boeing Task Cards. The formula might be refined to indicate a real-time constraint between t and t'. This would specify the minimum period within which an individual must consult the relevant procedural information in order to reduce the likelihood of an operating error:

$\forall t, t', p: potential\_procedural\_error(p, borescope\_inspection, t) \Leftarrow$
$\qquad \neg consult(p, aircraft\_maintenance\_manual, borescope\_inspection, t') \land$
$\qquad \neg consult(p, task\_cards, borescope\_inspection, t') \land before(t, t').$ (5)

This states that any person, p, is likely to commit a procedural error at any time, t, if they do not consult the task cards and the Aircraft Maintenance Manuals at any time prior to t.

An important benefit of the logic notation is that formal proof techniques can be applied to determine whether the previous analysis is actually supported by the available evidence. For instance, the following proof rule states that if we an inference of the form Q is true given P and we know that P is indeed true, then it is safe to conclude that Q is true::

$\forall t: P(t) \Rightarrow Q(t), \ \forall t': P(t') \vdash \ \forall t': Q(t')$ (6)

If we know that Q is true at all times P is true and we know that P is, indeed, true at all times then we can conclude that at Q is also always true.

This can be directly applied to our example. From clause (5), we know that there is potential for a procedural error if any operator fails to consult the task cards and the Aircraft Maintenance Manuals. This gives us a rule of the form, $\forall t: P(t) \Rightarrow Q(t)$. In order to prove that our account of the error that took place is accurate, it follows that we must establish that the Commander did indeed fail to consult the task cards and the Aircraft Maintenance Manuals. In clause (6), this is formalised as $\forall t': P(t')$. In the evidence presented to the enquiry we find that:

"When he got to T2 he looked through the workpack for the job. Although it was not familiar to him, being Line Maintenance paperwork, and rather less than he was used to because there were no Boeing task cards attached, he considered it unnecessary to draw any additional reference material" (section 1.1).

The following proof starts with a formalisation of the observation that the Controller failed to consult the Aircraft Maintenance Manual and the Boeing task cards at all times during the accident. From this, it is possible to construct a formal argument in support of the informal hypothesis given in the citation from section 2.10.1 of the AAIB report:

$\forall t: \neg consult(controller, aircraft\_maintenance\_manual, borescope\_inspection, t) \land$
$\qquad \neg consult(controller, task\_cards, borescope\_inspection, t)$ (7)

$\forall t, t': potential\_procedural\_error(controller, borescope\_inspection, t) \Leftarrow$
$\qquad \neg consult(controller, aircraft\_maintenance\_manual, borescope\_inspection, t') \land$
$\qquad \neg consult(controller, task\_cards, borescope\_inspection, t') \land before(t, t')$
$\qquad$ [By instantiation of *controller* for p in (5)] (8)

$\forall t: potential\_procedural\_error(controller, borescope\_inspection, t)$
[Application of (6) to (8) given (7), *before(t, t')* is true given quantification of (7)] (9)

> This sequence of proof steps illustrates that the controller was ill-prepared to perform the borescope inspections at all times during the incident given that he failed to consult the Aircraft Maintenance Manual and the Boeing task cards.

The previous section has shown that formal notations offer a number of benefits for accident analysis. They can represent the critical observations that can be lost amongst a mass of background information They provide precise and concise means of representing the temporal properties that determine the course of an accident. Formal proof techniques can also be used to examine the validity of informal argument structures that support the findings of natural language reports. Unfortunately, these approaches cannot be applied to analyse the underlying causes of operator 'error' during major accidents. For example, clause (5) merely states that an error is likely if a individual fails to consult appropriate documentation during a Borescope inspection. It does not state the reasons **why** the controller failed to consult the manuals and task cards before leaving for hanger T2. In order to capture such motivations, first order logic must be extended with some means of representing the cognitive factors that influence operator behaviour.

Epistemic Logic and Accident Analysis

A number of different approaches might be used to introduce cognitive analysis into formal models of major accidents. For example, Duke, Barnard, Duce and May (1995) have recently used the Interacting Cognitive Subsystems (ICS) model to support the formal design of interactive systems. In general terms, ICS models cognition as a flow of information through different levels of mental representation. A number of practical reasons make it difficult to use this approach to support accident analysis. It is often impossible in the aftermath of a major accident to determine the precise demands on an operator's various cognitive subsystems.

In anticipation of current research into the ICS model, a number of alternative approaches might be recruited to introduce cognitive constraints into formal representations of accident reports. For instance, epistemic logics have been developed to reason about an individual's knowledge over time (Barwise and Perry, 1983). In contrast to the full ICS model, these notations do not capture the many subtle changes that occur within a user's various cognitive subsystems. Epistemic logics can, however, represent the ways in which an operator's knowledge affects their actions during major accidents. Although this is a more limited approach than that proposed by Barnard et al, epistemic logics do provide a convenient means of expressing some of the propositional and implicational information within the ICS model. For example, clause (5) stated that there is the potential for a procedural error if an operator fails to consult the Aircraft Maintenance Manuals and the Boeing task cards. The previous citation from Section 2.10.1 of the report stated that 'A possible reason for him (the Controller) not having done this, before going to the remote T2 hanger, may have been that he expected Task Cards and AMM extracts to be part of the normal workpack'. Epistemic logics provide a means of formalising this assertion about the Controller's expectations. The following clause states that the Controller did not consult the task cards or the Aircraft Maintenance Manual at any time before he went to T2 because he believed that this documentation was already available at hangar T2. The AAIB report gives the Controller's time of arrival at T2 as 21:40:

> $\forall t: \neg\ consult(controller, task\_cards, borescope\_inspection, t) \Leftarrow$
> $\quad K\_controller(at(task\_cards, hangar\_t2, t), t) \wedge before(t, 21:40:00)$ \hfill (10)
>
> $\forall t: \neg\ consult(controller, aircraft\_maintenance\_manuals, borescope\_inspection, t) \Leftarrow$
> $\quad K\_controller(at(aircraft\_maintenance\_manuals, hangar\_t2, t), t) \wedge before(t, 21:40:00)$
> \hfill (11)
>
> Clause (10) states that the controller didn't consult the task cards at any time before 21:40 if he knew that the task cards were available in the T2 hangar at that time. Clause (11) states that the controller didn't consult the Aircraft Maintenance Manuals at any time before 21:40 if he knew that the manuals were available in the T2 hangar.

It is important to emphasise that the Controller's knowledge about the location of the documentation, represented by the $K\_controller$ clauses, is entirely independent of the actual position of the task cards and the manuals. The importance of such cognitive, or rather epistemic, formalisations is that they explicitly represent the temporal duration of the controller's beliefs. In other words, (10) and (11) draw attention to the fact that the Controller's belief that the documentation was available in hangar T2 could not have lasted beyond 21:40 when he, himself, arrived in the hangar. After this time, we must seek alternative reasons why he failed to consult the manuals and task cards about the Borescope inspection. Such an alternative explanation is important because the Controller **did** consult the documentation about other matters:

"Having done this, he (the Controller) then returned to Base Maintenance to check the turbine damage which he had found during the inspections against the limits laid down in the Aircraft Maintenance Manual...The Controller then inspected the No.1 engine and after telling the fitter, who was working on the No.2 engine, that he would be back soon, went again to Base Maintenance..." (section 1.1).

This suggests that any human factors analysis must focus upon the reasons why the Controller did not consult the appropriate documentation after he had discovered that it was not available in the inspection paperwork pack. Unfortunately, only five paragraphs, filling half a page, are devoted to this topic in a report of well over fifty pages. Leaving this issue to one side, the section in question explains the Controller's failure to consult the documentation after 21:40:00 as follows:

"...as a result of changes in the social and working climates they have been augmented by non-task specific health and safety warnings... As a result of this continual amendment process, the thread of the task descriptive text in the manuals has been, in some measure, obscured. The effect is clear in the manuals produced by all manufacturers, not any one in particular. The result of that is that the recognition and abstraction of text directly relevant to and descriptive of the maintenance task becomes more time consuming. As a result, there is an increased likelihood that isolated pieces of important information, which are surrounded by highlighted warnings and other general information, will be missed" (section 2.10.2).

The implicit theory or hypothesis in this statement is that the Controller failed to consult the task cards and manuals about the Borescope inspection procedure after reaching hangar T2, at 21:40:00, because they knew that they could not easily find the information that they wanted in these documents. This analysis is made explicit in the following clauses:

$\forall t: \neg\ consult(controller,\ task\_cards,\ borescope\_inspection,\ t) \Leftarrow$
$\quad K\_controller(\neg\ supports(task\_cards,\ borescope\_inspection,\ t),\ t)$
$\quad \wedge\ before(21:40:00,\ t)$ $\qquad\qquad\qquad\qquad\qquad\qquad (12)$

$\forall t: \neg\ consult(controller,\ aircraft\_maintenance\_manuals,\ borescope\_inspection,\ t) \Leftarrow$
$\quad K\_controller(\neg\ supports(aircraft\_maintenance\_manuals,\ borescope\_inspection,\ t),\ t) \wedge$
$\quad before(21:40:00,\ t)$ $\qquad\qquad\qquad\qquad\qquad\qquad (13)$

Clause (12) states that the controller didn't consult the task cards at any time after 21:40 if he knew that the task cards did not support the Borescope inspection task at that time. Clause (13) states that the controller didn't consult the Aircraft Maintenance Manuals at any time before 21:40 if he knew that the manuals did not support the Borescope inspection task.

Further evidence must be sought in order to confirm the hypotheses that are explicitly represented in (12) and (13). There must be some means of determining whether or not the Controller actually did believe that the documentation would not support their task. Without such supporting evidence, the previous clauses ought to be re-written to reflect the true status of the analysis. That is to say, they represent the analysts' beliefs about the Controller's beliefs about the documentation:

$\exists t',\ \forall t: K\_analyst($
$\quad K\_controller(\neg\ supports(task\_cards,\ borescope\_inspection,\ t),\ t),\ t')$ $\qquad (14)$

$\exists t',\ \forall t: K\_analyst($
$\quad K\_controller(\neg\ supports(aircraft\_maintenance\_manuals,\ borescope\_inspection,\ t),\ t),t')$
$\qquad\qquad\qquad\qquad\qquad\qquad\qquad\qquad (15)$

The accident analyst knows that the Controller knows that the task cards and Aircraft Maintenance Manuals do not support the Borescope inspection task.

Unfortunately, the AAIB report does not present any detailed evidence in support of the analysts views about the Controller's opinion of the documentation. This is not surprising. It is difficult to accurately assess an operator's attitude to their system after an accident or near-accident has occurred. On the other hand, the general nature of the hypothesis cited from section 2.10.2 does suggest that further analysis is urgently required to determine whether other maintenance staff share this low opinion of their documentation. Evidence in other domains suggests that a range of measures must be adopted if such attitudes are to be effectively countered within an industry (Wickens, 1984). Our analysis raises further questions. In particular, the Controller's reluctance to consult Borescope inspection procedures does not explain why he did 'check the turbine damage which he had found during the inspections against the limits laid down in the Aircraft Maintenance Manual' (Section 1.4). Documentation about damage levels supported the users' tasks while documentation about Borescope inspection procedures did not:

$\exists t: K\_controller(supports(aircraft\_maintenance\_manuals, turbine\_damage\_limits, t), t)$ (16)

The Controller did at some time believe that the Aircraft Maintenance Manual did support the task of identifying the limits of admissible turbine damage.

Conclusions

This paper has argued that accident reporting techniques suffer from a number of limitations. In particular, it can be difficult for readers to extract critical information from the mass of background detail in these documents. One consequence of this is that the recommendations and findings of accident reports often seem to be unsupported by the detailed evidence that they present. We have shown that logic can be applied to identify these weaknesses. Formal proof techniques can be applied to determine whether informal arguments are justified in terms of the available evidence. Unfortunately, first order logic cannot easily be used to reason about the causes of human 'error'. We have, therefore, shown that epistemic extensions to this notation can be applied to identify the high-level cognitive factors that influence operator behaviour. A by-product of this analysis has been to reveal the slender nature of the evidence that is often provided to support claims about the causes of human failure during major accidents.

Acknowledgements

Thanks are due to the members of the Glasgow Accident Analysis Group and the Glasgow Interactive Systems Group. This research is supported by the UK Engineering and Physical Sciences Research Council, grants GR/L27800 and GR/K55042.

References

Air Accident Investigation Branch, *Report on the Incident to a Boeing 737-400, G-OBMM near Daventry on 25th February 1995*. Published on behalf of the United Kingdom Department of Transport. Aircraft Accident Report 3/96 EW/C95/2/2. 1996.

J. Barwise and J. Perry. *Situations And Attitudes*. Bradford Books, Cambridge, United States of America 1983.

D. Duke, D. Duce, P. Barnard and J. May. *Systematic development of the human interface*. In APSEC'95- Second Asia-Pacific Software Engineering Conference 1995.

R. Fagin, J. Halpern, Y. Moses, and M. Vardi. *Reasoning About Knowledge*. MIT Press, Boston, United States of America 1995.

C.W. Johnson, Impact of Working Environment: *A Formal Logic for the Integrated Specification of Physical and Cognitive Ergonomic Constraints*. Ergonomics (39)3:512-530, 1996.

C.W. Johnson, J.C. McCarthy and P.C. Wright, *Using A Formal Language To Support Natural Language In Accident Reports*. Ergonomics, (38)6:1265-1283, 1995.

J. Reason, *Human Error*, Cambridge University Press, Cambridge, United Kingdom, 1990.

C.D. Wickens, *Engineering Psychology And Human Performance*, C.E. Merrill Publishing Company, London, United Kingdom, 1984.

USE OF MICROSAINT FOR THE SIMULATION OF ATC ACTIVITIES IN MULTI-SECTOR EVALUATIONS

Hugh David

Eurocontrol Experimental Centre
91222 Bretigny-sur-Orge, France

MicroSAINT provides an 'empty' network model, which may be used to investigate the practical value of proposed innovations in ground-based ATC systems involving multiple operators. A generic model was developed, representing Centre, Sector and Individual levels. Initially, an orthodox allocation of staff - one Planning Controller (PC) and one Executive Controller (EC) was simulated for six sectors, and approximate workloads were evaluated. These models were then modified to investigate the workloads associated with six ECs working with one, two or three PCs. A final model was developed using six ECs, and two PCs with one 'Super-controller' controlling traffic flow. The advantages and disadvantages of MicroSAINT are discussed.

Introduction

In a survey performed by Carlow Associates (Fleger et al 1993) for the US Army Human Engineering Laboratory MicroSAINT was cited as the most frequently used computer tool for the analysis of human workload. Fontanelle and Laughery (1988) describe a Workload Assessment Aid (WAA) which predicts workload, and allows re-allocation of tasks between crew members, developed as part of the MANPRINT effort, and based on MicroSAINT. The Armstrong Aeromedical Research Laboratory, at which MicroSAINT was originally developed (as a mainframe package called SAINT), now uses MicroSAINT extensively, as do several other US military research centres. The Australian Department of Defence Aeronautical Research Laboratory uses MicroSAINT to analyze crew workload in military operations (Manton and Hughes 1991) This work is particularly interesting because they consider not only the simple time-line approach, but more detailed cases where operators may be engaged on visual and auditory tasks at the same time. There is a considerable literature on the choice between simple and multiple resource models, summarised in Damos (1991). Although for most practical cases the simpler models are already sufficiently complex, a study by Nijhuis (1993) suggested that data-link could actually reduce system capacity, if it competed for manual/visual resources instead of using the speech/audio channels of traditional control. The WAA mentioned above takes multiple resource theory into account, as does CREWCUT, (Hahler et al 1991) and the CAATS(Macleod, Farkin and

Hellyer 1994). Laughery and Hegge(1989) describe the use of MicroSAINT, in conjunction with multiple resource theory and psychophysical knowledge of performance deterioration to estimate the time to carry out military team activities and how they might be affected by fatigue. Hone and Barber (1993) used microSAINT to model error correction dialog in speech recognition

MacLeod, Biggin, Romans and Kirby (1993) used MicroSAINT for the predictive workload analysis of helicopter crew, with particular emphasis on error analysis..

GENERIC Model

EUROCONTROL has developed a generic ATC model, capable of modelling ATC systems can be seen at the Centre, Sector or Working Position levels.

Centre Level

At the CENTRE level, traffic is generated, and distributed to the sectors through which it will go. In this simulation model, we are not mainly concerned with the details of traffic flow, so we specify routes in terms of times in successive sectors, and a probability of taking that route. We generate traffic at a rate which can be set to increase in successive runs, until some part of the system reaches saturation. As each aircraft is generated, a record is created for its progress through the hands of the PCs and ECs of the sectors through which it will pass This is updated as successive activities are completed.

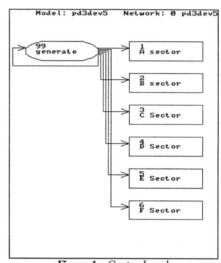

Figure 1. Centre Level

Sector level

At the SECTOR level, the system generates the activities required for the PC and the EC. It also records the passage of the aircraft through the sector. ATC differs from most systems simulated because we assume that the aircraft will fly through the sectors at the time appointed, and that if the coordination between sectors has not been completed, the

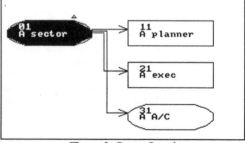

Figure 2. Sector Level

consequence is a late handover, not a delayed aircraft. At each stage of the aircrafts' progress, the system is programmed to verify that the necessary planning procedures have been completed. Queues of events form because some events, such as the hand-over by the executive controller to the next sector's executive controller, cannot occur before a specified time, because several tasks may arrive for a controller at the same time, or because he has not had time to complete his current tasks. Only when an activity has not been completed by the time it is due does it become important.

Planning Controller

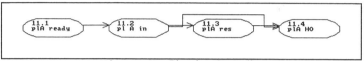

Figure 3. Planning Controller

Executive

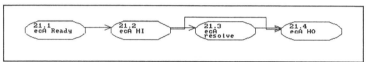

Figure 4. Executive

The diagrams representing the flow of tasks for Planning Controller and Executive are very similar, but they have important differences in practice. The Planning Controller normally operates well in advance of the arrival of the aircraft in the sector, and treats one aircraft completely before going on to the next. In contemporary systems, he compares times and heights at fixed positions to try to find a clear path for the aircraft. He usually treats an aircraft once, before going on to the next. The Executive is more closely bound to current events, and normally treats several aircraft in turn, before returning to the next activity for the first aircraft. He cannot, for example, hand over an aircraft until it is close to the sector boundary, even if he has nothing else to do (It is important to note that tasks such as conflict resolution are not in fact carried out in these models - all that is simulated is the time associated with them.)

Specific Application

In a typical application of this model, we examined the consequences of modifying the traditional way of manning sectors. In a region of European airspace, there is a set of six sectors arranged, topologically, as three high (North, Central and South) by two wide (East and West). The traditional manning would be to have one Planning Controller and one Executive in each sector. New systems were being proposed which would considerably ease the burden of planning, and it was suggested that it might be possible to have three Planning Controllers (one each for North, Central or South), two Planning Controllers (one for East and One for West), or even one Planning Controller for all six sectors.

A 'clone' was made of the basic GENERIC model to explore each possible case (three, two or one Planning Controllers). This involved modifying the identity of the Planning Controller so that three pairs of sectors, two sets of three or all six sectors had the same Planning Controller, and the workloads involved in transfers between sectors with a common Planning Controller were eliminated. These models were then run with systematically increasing traffic, and graphic images were produced to show how workloads, and the numbers of 'late' events of various types (Planning incomplete on entry, executive not aware of aircraft, late hand-in or hand-out) in each sector. (Each run consisted of 20 runs of MicroSAINT, and took up to thirty-six hours on am IBM PC clone using a 386 processor chip.)

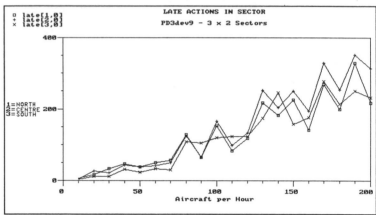

Figure 5. Late Actions

Figure 5 shows an example of the data presentation available from MicroSAINT. In this instance, the increase in 'late actions' as traffic increases when three PCs each handle two sectors. (Generally, contrary to expectation, there was no clearly-defined threshold beyond which late actions occurred. This may be because most action times were exponentially distributed rather than fixed.)

Similar diagrams were derived for one, two and six PCs operating as above, showing the numbers of 'late actions', the maximum numbers of aircraft present in sectors and the notional workloads associated with them. Examination of these showed that, although the overall workload was optimal when two PCs handled the East and West sectors, the traffic load was extremely unbalanced - most North-South traffic passing through the East Sectors.

A final model was therefore derived employing the two Planning Controllers for East and West and a notional 'Super-controller' whose sole task was to direct the traffic evenly through the East and West sectors. The work of the 'Super-controller' was realised in the traffic generation module in the top level Centre model, switching aircraft according to the number of aircraft present or warned to the East or West.

A relatively low workload for the 'Super-controller' produced a well-balanced workload for the two Planning Controllers, and this configuration was therefore recommended for further investigation by Real Time Simulation.

Discussion

The advantages of the Micro-SAINT modelling process are that it divorces the simulation process from the laborious compilation of detailed traffic samples, and allows a tighter control of 'human variability' and learning effects than is normally practicable in Real-Time (Human-in-the loop) simulation, or in the observation of real events. It can be used, for example, to investigate the effect of changing from a telephone or intercom based coordination between sectors, (which requires the attention of both controllers involved) to a computer moderated dialog (where each controller may attend to the task when free to do so).

The disadvantages are that realistic values for times to complete tasks must be available, and that the normal critical review of procedures carried out by participating controllers is not available. (Many formally defined procedures turn out to be incomplete, inconsistent or even impossible as soon as they are put into practice. A major value of Real-Time simulation is that these problems are identified and solved at the start.) Potentially disastrous programming or procedural errors can only be identified by detailed study of results and rigorous testing of models. Particularly careful consideration must be given to the ordering of priorities where tasks are queued. For example, use of a first-in first-out priority rule for tasks such as hand-overs will overestimate the number of late handovers if the time in the sector is very variable, since a controller would give priority to the handover closest to its due time, not to that which had been waiting longest.

Practical disadvantages are that MicroSAINT, except for the most simple applications, requires elementary programming skill, of the order of BASIC, and willingness to think rigorously and test models exhaustivly.

MicroSAINT is essentially intended for the on-line investigation of systems, and is provided with graphic tools for that purpose. MicroSAINT for Windows, the most recent version, provides a facility for producing animated images of the events simulated. MicroSAINT for DOS, an earlier version which is used at EEC, provides a symbolic version, from which the images in this paper have been derived (with some difficulty - a 'frame-grabber' was used to make PCX files of the images, which were then converted to monochrome BMP files using PAINTBRUSH, and inverted, moved and re-sized using WordPerfect.)

Conclusions

In this exercise, a predefined MicroSAINT model "GENERIC" of the time-limited operations of air traffic controllers was modifed to provide a series of models reflecting different allocations of functions. These models made it possible to explore the probable workloads, and the extent to which controllers would be able to carry out their tasks in the time available for them, in a relatively short time,and without the effort required for the construction of detailed traffic samples, or the running of Real-Time simulations.

MicroSAINT is a useful tool for the exploration of complex human-in-the-loop systems, provided that its limitations are recognised. It is particularly valuable where systems consist of large numbers of essentially similar events, whose individual characteristics are well-known, and whose individual interactions can be well defined.

Ergonomists who expect to be concerned with complex systems should be aware of MicroSAINT and the toolsets in which it can be embedded.

Acknowledgements

The Author wishes to thank the director of the Eurocontrol Experimental Centre for permission to publish this paper. This paper represents the opinions of the author, and should not be taken to represent Eurocontrol policy.

References

Damos Diane (ed),1991, *Multiple Task Performance*, Taylor and Francis, ISBN 0-85066-757-7

Fleger Stephen A, Permenter Kathryn E, Malone Thomas B ,1988, *Advanced Human Factors Engineering Tool Technologies* , NTIS AD-A195 252 ,

Fontanelle G. and Laughery K.R., (1988) A Workload assessment aid for Human Engineering Design, *Proceedings of the Human Factors Society 32nd Annual Meeting.*

Ford C.M., Manton J.G. and Hughes P.K., 1990, A Worked Example of Job Simulation using MicroSAINT, Aircraft Systems Technical Memorandum 125, Aeronautical Research Laboratory, Department of Defence, Melbourne, Australia.

Hahler Beth, Dahl Susan, Laughery Ron, Lockett John and Thein Brenda, 1991, CREWCUT - A Tool for Modelling the effects of High Workload on Human Performance, *Proceedings of the Human Factors Society, Thirty-Fifth Annual Meeting* , pp 1212-1214

Hone K. S. and Baber C., 1993 Using Task Networks to Model Error Correction Dialogues for Automatic Speech Recognition, *Contemporary Ergonomics 1993*, Taylor and Francis, London ISBN 0-7484-0070-2

Laughery Ron and Hegge Frederick,1989, A Method for Mapping Laboratory Performance Research to the Effects on Military Performance, *Proceedings of the Medical Defence Bioscience Review*, June 1989 , pp 765-772

Macleod I.S., Biggin K., Romans J. and Kirby K.,1993, Predictive Workload Analysis - RN EH101 Helicopter, *Contemporary Ergonomics 1993,* Taylor and Francis, London ISBN 0-7484-0070-2

Macleod I.S., Farkin B. and Helyer P., 1995, The Cognitive Activity Analysis Tool Set (CAATS), *Contemporary Ergonomics 1994*, Taylor and Francis, London ISBN 0-7484-0203-9

Manton J.G. and Hughes P.K., 1991, Aircrew Tasks and Cognitive Complexity, Aircraft Systems Technical Memorandum 150, Aeronautical Research Laboratory, Department of Defence, Melbourne, Australia.

Nijhuis, H.B.,1993, Workload in Air Traffic Control Communication, *Contemporary Ergonomics 1993,* Taylor and Francis, London ISBN 0-7484-0070-2

USER DATA IN PRODUCT DESIGN
OUTSIDE OF THE DEFENCE INDUSTRY[1]

Ronald W. McLeod   Ph.D.,C.Psychol.,M.ErgS

Nickleby & Co. (Scotland) Ltd.,
239 St. Vincent Street,
GLASGOW G2 5QY

A key issue in the MoDs' Human Factors Integration (HFI, formerly MANPRINT) initiative, is how to specify the characteristics of the target users of equipment in a way which is useful to designers. This paper describes a study investigating what use organisations outside the defence industry make of information on user characteristics in the product design process.

Twenty four individuals (from an initial survey of 187 non-defence organisations) took part in structured telephone interviews.

Little systematic use is made of user specifications in product design outside the defence industry. Those who do use them have similar experience to those involved in the design of defence equipment in terms of the types of user information which are available and used. Visual imagery may be a powerful means of conveying such information.

There is considerable confusion in the use of terms such as "designer" and "engineer" and in the assumptions about who does what in the design process.

Introduction

The Ministry of Defence (MoD) requires a programme of "Human Factors Integration" (formerly MANPRINT) to be conducted as part of the process of developing defence equipment. The MoD are expected to provide industry with information about relevant characteristics of the people who will use, operate and maintain the equipment once it is in operational use. These specifications are

---

[1] This work was undertaken for the Defence Evaluation Research Agency (DERA), Centre for Human Sciences (CHS), as part of the MoD's Corporate Research Programme in Human Factors. The contribution of all those who assisted with the telephone interviews reported herein is gratefully acknowledged.

referred to as the 'Target Audience Description' (TAD).   The TAD might, for example, include information on the intended users ranks, specialises, medical condition, general intelligence, body size and training history. The TAD is important in helping designers take account of characteristics of the user population.

The MoD are seeking to improve the quality and utility of Tads.   The study reported herein investigated how non-defence related industries incorporate information about user characteristics in the product design process.

Method

The study was based around structured telephone interviews covering a range of organisations and product types outside the defence industry.   Interviews were conducted using a standard pro-forma covering the following topics: qualifications and experience of the interviewee; the type of products designed and who uses them; typical ergonomic considerations; the use of user data in design; and the design process.

Results

Of the total sample of 187 organisations contacted, there was a response rate of 35.3% (66 replies).   Of these, 36 individuals indicated their willingness to participate in a telephone interview.   Twenty four of these were eventually interviewed (it was not possible to schedule the remaining twelve in the timescale of the study).

Respondents were predominantly involved in Ergonomics-related activities, in a broad range of market sectors and industries. The majority were involved in R&D, Engineering or Design with an average of over 17 years in their profession. Most had experience in areas other than their current professional field.   Most were involved in transportation, telecoms, IT, finance, power generation or aerospace.   Interviewees were generally very experienced with good awareness of trends in Ergonomics and product design

Most of the interviewees were involved in the design of computer systems and consumer products, although products such as vehicles, seats and medical equipment were covered.   Products were intended for use about equally by the general public and specific user groups.   Specific user groups were characterised by aspects such as particular groups of employees or professions (such as police, air traffic controllers) or their physical characteristics (such as age, gender, or physical abilities).

The greatest number of respondents indicated that their products were intended to be capable of being used on a 'walk-up-and-use' basis, although user manuals and training courses were also widely assumed to be required.

Ergonomic Issues in Design

Figure One summarises the range of ergonomic issues identified as typically being relevant to the types of products with which the interviewees were involved.

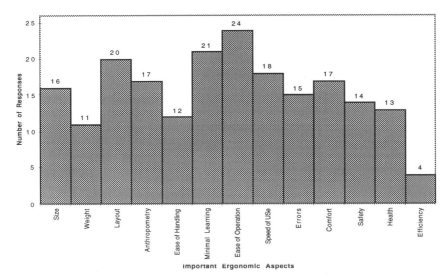

Figure One. Summary of Issues Identified as being Important in Product Design

The Use of User Data in Design

Twenty one of the respondents said they had been involved in projects where information about the user population was specified in advance. Of these, fourteen felt it was not common practice to specify these characteristics, while the other seven said it was common practice. Figure Two shows the distribution of characteristics which these respondents had seen specified in advance. Specifications were generally prepared in textual format supported with diagrams. In a few cases mannequins and computer models were also reported as being used.

Twenty of the respondents said they personally had a role interpreting information about the characteristics of the target users in the design.

Figure Three summarises who was typically involved in preparing these specifications. The data contained in them was derived from sources such as reference books, market research, standards, and personnel records as well as user representatives. Eight of the respondents felt that these sources were sufficiently up to date for their purposes, while eleven said they felt they were not up to date. Seven felt that the sources were sufficiently specific to the user populations they were interested in, while twelve said the sources tended not to be sufficiently specific.

The majority of interviewees felt user data was usually accessible and comprehensible. Those that did not tended to suggest that people using the data either do not understand the information, or the way in which the data was presented was not "user friendly."

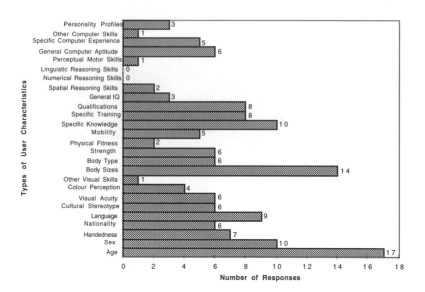

Figure Two. Summary of Types of User Characteristics Specified

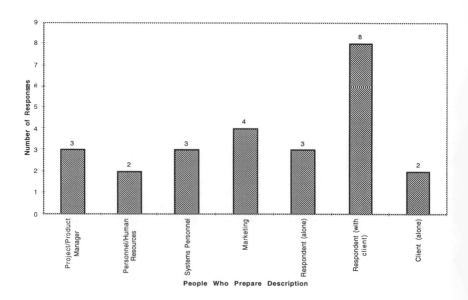

Figure Three. Summary of Who Prepares User Specifications

How Effective is the Information in Improving Design Decisions ?

Less than half of the respondents had any feeling for whether specifications of the user population actually led to improved product designs. Seven respondents mentioned data on anthropometry/biomechanics as being most effective. Three mentioned cognitive data as being most useful, although a further three identified cognitive data as being least useful. Some respondents felt that information about users computer experience is generally not useful because the nature and relevance of such experience often becomes outdated very quickly. Concern over the usefulness of personality profiles was expressed because of the uncertainty as to how the data is used.

What Other Types of Data Could be Effective ?

Interviewees were asked whether they could suggest other types of information which they felt would potentially be helpful to designers. The value of studying experience with existing similar designs was suggested. Three respondents felt that an overall characterisation of the target users could be particularly valuable. Two of these suggested the use of visual images, such as drawings or sketches to represent the target user. The other respondent suggested providing a "day in the life" snapshot of the target users. One interviewee discussed a product development programme for a new kitchen product. The target user description included drawings of prospective users, such as the housewife cooking while tending to the kids and family pet, the scholarly professor with an aversion to cooking, the professional couple with little time for meals, etc. There is a type of insight in this representation method that differs from the tangible and explicit data currently included a TAD.

The Design Process

Ten of the interviewees said they tended to be involved at all stages of the design process. Nine said they were generally involved at the beginning and/or middle stages. The remaining five said they tended to become involved late in the process. Most of these stated a preference to be included earlier on, although many of the projects they were involved in precluded an "early" role. They therefore tended to have a remedial, or reactive, presence in the design program.

Eighteen of the respondents said that designers in their organisation tended to have some form of obligation to demonstrate that their designs incorporate information about the target users. The other six respondents said that their designers had no such obligation.

Interviewees were asked whether they had any feeling for the proportion of a design budget which might be specifically allocated to addressing ergonomic issues. Seven of the respondents felt unable to offer a view. A further seven provided estimates ranging between 2 and 10% of the total budget: five gave general responses that it was "a small amount", or "very little". One participant suggested 15 - 20%. One respondent pointed out that even if an explicit budgetary figure was given, the reality is that the "hidden" budget is considerably higher because many tasks in a project address ergonomic issues, though they may not be recognised as such.

Every respondent said their organisation adopted some type of procedure to ensure that designs were suitable for their target users. Procedures cited included Market Research, Focus Groups, Prototyping & Usability Testing and Auditing against Legislation and Standards.

The term "designer" was used throughout by many of the interviewees. Clearly a "designer" working on an IT system is substantially different from a "designer" working in a product development programme for an industrial design firm. The same applies to the description "engineer". Recognising this confusion is important because it effects how data should be presented in the target user specification. Ideally, personnel preparing the data for a target user specification will allow for the characteristics of the personnel who are expected to make use of the data.

Summary and Conclusions

The majority of interviewees had personal experience specifying user characteristics in the design process. This was not considered to be common practice in product design however. The study found little evidence of any formal or structured process used outside the defence industry equivalent to the use of TADs in defence procurement. Approaches tended to be project specific, often depending on the initiative of the individuals involved. There was a tendency to feel that available sources of published information on user characteristics were either out of date or not sufficiently specific. The principal exceptions were where information was available directly from user groups or from specific market research.

There is an important difference in design strategy between defence and non-defence related organisations. Without the constraints of fixed-price competitive tendering, designers in non-defence organisations are able to develop more collaborative relationship with the users of their products, and involve them much more closely in the design process than is possible in the defence industry.

There is considerable confusion in the use of terms such as "designer" and "engineer" and in the assumptions about who does what in the design process.

There may be considerable potential in the concept of a "visual snapshot" type of user specification.

AGRICULTURAL ERGONOMICS

A CASE FOR DECISION SUPPORT TOOLS IN CROP PRODUCTION?

Caroline Parker

HUSAT Research Institute
The Elms, Elms Grove
Loughborough, Leics LE11 1RG

In recent years there has been an increasing interest in the potential of decision support tools for crop production. Farmers have less control over influences on their product than other industrial managers, and access to and use of up-to-date, accurate information is critical to the success of a farming enterprise. Attempts to provide such tools in the past 10 years have met with limited success. The paper discusses reasons for failure and explores the possibility that the adoption of a user centred design approach would overcome previously observed problems. The case for decision support systems is examined in the light of results from two user requirements exercises. The paper concludes that the industry has a genuine need for the technology and outlines the main features it should concentrate on to succeed.

Introduction

There has been an increasing interest in the potential of decision support systems (DSS) in the crop producing agricultural and horticultural industries[1] in recent years Crop producers have an enormous need for information when making the simplest of decisions, e.g. simply selecting a variety of a given crop requires knowledge about current market requirements, prices, the nature of the target field, the yield characteristics of the variety, the likely interaction between field and variety, and the disease pressure in the area. Access to and use of up-to-date and accurate information is therefore critical to the success of a crop producing enterprise.

Farmers have less control over external influences on their product than many other industrial managers, and for smaller producers a bad attack of a pest or disease can threaten the viability of the enterprise. While many effective

---

[1]To the outsider all farming activity seems to be the same but in fact there are a number of critical differences between horticulture (production of fruit, flowers and vegetables) and agriculture (the rest, e.g. cereals, oil seeds and animals) , most notably one of scale.

pesticides exist which remove the danger of loss, they are costly and their use is increasingly restricted by legislation.

For some time now, expert systems and decision support systems have been employed in other industries, particularly in manufacturing and finance. The use of rule based, model based or even neural network based support modules as part of everyday monitoring, decision support or diagnostics systems have proven their worth as the annual DTI awards have shown. However the success of decision support technology within the agricultural/horticultural industries is by no means certain and indeed attempts to provide decision support tools in the past 10 years have met with little or no success. A list of publicly available agricultural decision support systems, over 20 in total, does not contain a single name in common use on the farm (Parker, 1996a).

The reasons for failure may be many e.g. poor interface design, failure to answer the right questions, asking for inputs the user cannot easily provide, absence of affordable supporting technology(meteorological monitoring devices), targeting the wrong areas for support, etc, and are examined in more detail in (Parker, 1996b). One common feature in many of these earlier attempts however is the absence of a user-centred approach to design.

This paper describes part of the work conducted in the past 3 years in the application of user-centred design principles to the design of DSS, the focus of the paper is on the initial user requirement surveys and their findings. The research has been conducted as part of two distinct projects, GRIME (Graphical Integrated Modelling Environment) a MAFF Open Contract project based in horticulture and DESSAC (Decision Support Systems for Arable Crops) an Agricultural LINK project funded by MAFF and the Home Grown Cereals Authority.

The first aim of user centred design is to ensure that the users needs serve as the focus of the design process. In the context of the current problem therefore it is essential that research should initially be aimed at: firstly, determining if the industry needs decision support tools and who the likely users will be; secondly, if the industry has the necessary technological and informational components (i.e. computers and weather monitoring equipment) required to support it; and thirdly what the industry believes to be the most critical and cost-effective uses of DSS technology.(Preece, 1993). This paper describes the research methods employed to answer these and related questions and discusses the results in the context of ongoing work in the area.

Method

Two methods were adopted, face to face interviews and postal survey, providing both scope and depth to the data collection.

Interviews

Informal and largely unstructured interviews with two or three key figures in each industry were used to get a broad feel for the problem domain. The basic aims of both surveys were the same: to gather information on the decision process, the people involved in it, areas of current difficulty and on the computer ownership and technological awareness of the people concerned. In the horticultural survey 15 face to face interviews took place, 4 of which were with consultants; in the arable survey 32 farmers and 12 consultants were interviewed. Sampling was based on mailing lists made available by industry bodies.

Postal Survey

The questionnaires were designed to be as concise as possible whilst still making it possible to answer the 3 key questions outlined in the introduction. The horticultural survey was 2 pages long and the arable 4 pages. The survey was targeted at winter wheat growing members of a major arable research centre (Morley) as this group contained a large number of arable farms of a variety of sizes. Two hundred and fifty questionnaires were distributed in the horticultural survey and seven hundred and fifty were distributed in the arable survey. While the questionnaire differed in their content and size they contained sufficient similarities to permit comparison.

Results

Interviews

The potential users of decision support systems within both industries appear to be consultants and the farm managers/owners of medium to large enterprises and of co-operatives.

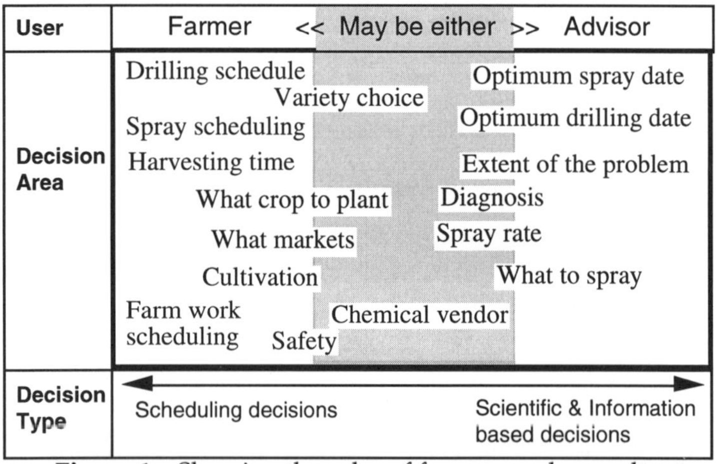

User	Farmer << May be either >> Advisor
Decision Area	Drilling schedule Optimum spray date Variety choice Spray scheduling Optimum drilling date Harvesting time Extent of the problem What crop to plant Diagnosis What markets Spray rate Cultivation What to spray Farm work Chemical vendor scheduling Safety
Decision Type	◄────────────────────────────────► Scheduling decisions Scientific & Information based decisions

Figure 1: Showing the roles of farmers and consultants.

It can be seen from figure 1 above that the farmer has the stronger role in any activity which involves the scheduling of time or resources e.g. planting, harvesting. The consultant plays the strongest role in those activities relating to the application of information based and scientific knowledge: i.e. the need for insecticides, fungicides and herbicides and the best rates of application.

Technological readiness

The results of these surveys appeared to support the view that the industry was ready for the type of support offered by computer based DSS's i.e. most of those interviewed either owned a computer or were planning to get one. There was however a distinct lack of access to the more complex types of meteorological data required by some models and an equal unwillingness to collect it.

Does the industry need DSS's and what will they use them for?

Most of those interviewed were of the opinion that any system that offered them the possibility of producing a better quality product for less cost or for reducing some of the uncertainty involved in the decision they took was worth having, providing it didn't require any special inputs.

The work context and environment in which the software will be used

The market and political context in which farm based decisions are made are too complex to be covered in detail here, (Parker, 1994) and (Parker, 1995) provide some specific coverage of the issues .

In brief the two main causes for the increased need for decision support on the farm are: increasing pressure from consumers to reduce pesticide inputs and increasing financial pressures requiring the reduction of variable costs. The main environmental influences on the software are: time - all potential users are extremely busy during the peak season and have very limited time to make use of computers; availability - unless computers become pocket based potential users will only be in the vicinity of a computer at certain restricted points in the average day; restricted access to field measures - most users will be unable to provide the detailed level of sampling required by some biological models because they lack time and resources.

Postal Surveys

The return rate for the surveys was 40% for the horticultural survey and 20% for the arable survey.

Computer usage

Two thirds of the horticultural growers and consultants who responded owned a computer, and 75% of the arable farmers and all of the arable consultants said they had at least one computer.

The questionnaires asked people what things they used their computer for, he most common uses were financial (e.g. accounting, 86% horticultural growers, 71% arable; gross margins, 76% arable farmers), farm records (73% arable farmers; 59% horticultural growers,) word processing (82% horticultural growers, 91% consultants),

Who are the users?

Almost 50% of the horticultural growers and 83% of the arable farmers said they used the software themselves as did all consultants owning a computer. Nearly 50% of the horticultural and 20% of the arable producers who don't own a computer were thinking of getting one in the near future.

Meteorological Data

Sixty-four percent of the grower sample and 85% of the arable sample make use of meteorological data either gathered themselves, from radio/TV or from the Met Office. The most common type of data obtained by growers, farmers and consultants is rainfall, min/max, straight temperatures and wind direction.

Decision support tool functionality

Horticultural growers and consultants were particularly interested in decision support tools concerned with pest and disease behaviour and control (78% and 67% for growers and 71% and 76% for consultants). Around a half

(51% and 47% growers and consultants respectively) of respondents were interested in tools concerned with nutrition.

Arable respondents were asked which of a list of facilities they would find useful and to what degree they felt those facilities represented a saving of time and/or money. The most popular facilities for farmers and consultants were: access to market information, local trials results; variety information; early pest/disease warnings; and anything concerned with the application of chemicals and nutrients and calculating the costs/savings of their use. ...

Discussion

The purpose of this research was to discover whether the horticultural and agricultural sectors wanted and could make use of decision support technology as a means of bringing together and presenting the vast amount of information on which it relies.

Prior to the current surveys research had suggested that computers were not being used in the industry. Gibbon, in a study examining the failure of uptake of computers in England discovered that many would not purchase computers because: they were too expensive; the size of the farm was too small to justify their use; or simply because the expertise to use them did not exist in the enterprise (Gibbon, 1992). However the results described here suggest that computer use has increased dramatically and is continuing to expand in both the horticultural and agricultural sectors.

Unlike most other industries decision making in crop production is heavily influenced by the weather and many of the simulation models under development are based on very specific measures of local conditions e.g. leaf wetness. Unfortunately these surveys make it very clear that the technological infrastructure for these models does not currently exist. Unless models are developed which can extrapolate from regional data to the field in question, they are unlikely to be of much use to the average grower, farmer or consultant. While meteorological equipment remains costly the need to buy in new technology or employ someone to gather information specifically for the model will prevent its widespread use. The problem of limited and expensive meteorological data is one that dogs many initiatives in the industry and a concerted effort is being made by a number of groups to address the issue and find solutions.

Key decision makers in crop production (growers, farmers, farm managers and consultants) are rarely in one place for very long. Unlike their industrial counterparts, it is possible that these people may not spend long enough in an office type environment to make use of decision support tools. The results of the surveys suggest that, although time is very limited, growers and farmers are using computers for other things. While location and time are most certainly going to impact on the use of DSS's in the short term, the key to encouraging use will be to ensure that the data the decision maker needs is quickly accessible and offers sufficient cost-benefits to make it worth the additional use of time.

Summary

It has been suggested that the failure of previous DSS's in the industry is largely due to the absence of a user centred design methodology. These surveys were conducted as the first stage of a user centred approach to the development of crop production support software. The conclusion of the research is largely positive. The industry does need the technology and the computer technology to support them is increasingly available. On the negative side however the meteorological technology required by many of the models underlying the decision support tools is not widely available and is unlikely to become so in the immediate future. This suggests that the successful decision support tools will be those which do not rely on field specific weather data.

The main users of the technology are likely to be the managers of larger farm enterprises, farm co-operatives and consultants. The main difference between the two sectors is in target of the decision making, in horticulture the main aim is to maintain quality, which often implies zero pest/disease presence in the crop; in the arable sector the main aim is to produce the highest yield for the least inputs and therefore maximise profits. The support which decision support tools provide these sectors must therefore reflect this difference.

The support systems for which this research was conducted will be released on the market over the next 2 to 3 years. The user centred design methodology will continue to be employed in both projects. A full evaluation process involving laboratory and on-site use is being planned for the cereal decision support tools for 1998-2000.

References

Gibbon, J & Warren, M. F. 1992. Barriers to adoption of on-farm computers in England. *Farm management* 8(1 , Spring): 37-45.

Parker, C.G. 1994. G.R.I.M.E. Results of the First Phase User Requirements Analysis. Report to Ministry of Agriculture, Fisheries and Food. 1.0.

Parker, C.G. 1995. Decision processes for Winter wheat. User Requirements for future decision support systems. Postal survey of User requirements. DESSAC. User Needs Working Group Report UNG/1.3.

Parker, C. G. 1996a. Compiled list of Agricultural and Horticultural Decision Support System's. agmodels-l@unl.edu,

Parker, C.G., Phelps, K., Reader, R. & Hinde, C. 1996b. Why computer-based models need more than just good science. *Applied Biology* 46:

Preece, J. (Ed). 1993. *A Guide to Usability: Human Factors in Computing*. The Open University. Addison Wesley.

Acknowledgements

The author would like to acknowledge the support of MAFF and the HGCA in funding this work and also the support of her project partners namely: ADAS; HRI International, IACS Rothamstead; Loughborough University Computer Studies Department.; Morley Research Centre; Optimix Computer Systems Ltd; and the Silsoe Research Institute.

PARTICIPATORY ERGONOMICS WITH SUBSISTENCE FARMERS

D H O'Neill

International Development Group
Silsoe Research Institute
Silsoe, Bedford MK45 4HS

The use of "bottom-up" rather than "top-down" approaches is currently in favour to increase users' acceptance of the outputs of research and development projects. This paper presents the findings of a farmer participatory research project to improve the handtools used by subsistence farmers in industrially developing countries. The participatory techniques applied to identify the problems faced by farmers and their families are outlined and the ergonomics methodologies selected as the most appropriate are presented. The "bottom-up" approach was successful in both identifying problems and conducting on-farm research to solve them. Various research activities, mainly on land preparation and crop care, were undertaken which have resulted in farmers adopting new equipment. The achievements demonstrate the success of the participatory approach and, furthermore, have given the farmers the confidence to apply ergonomics principles to other problems they may face.

Introduction

Several projects have addressed issues relating to the application of ergonomics in agriculture in industrially developing countries (reviewed by Rogan and O'Neill, 1993). However, the efforts have been disparate and as a consequence there has been very little impact. Most of the work involved has laboratory research and development, with researchers aiming to solve problems as they perceived them - the "top-down" approach. All the crop production tasks in subsistence agriculture are labour-intensive but the most demanding are land preparation and weeding (Rogan and O'Neill, 1993) and these are also associated with high levels of drudgery.

There is now accumulating evidence (Chambers, Pacey and Thrupp, 1989; Gamser, Appleton and Carter, 1990; Haverkort, van der Kamp and Waters-Bayer, 1991) that unless the target population, in this case subsistence farmers, is involved in the research and development, the impact will be minimal. This is because problems which have not been addressed from the farmers' perspective are unlikely to yield solutions appropriate to the farmers' real needs. The subject of this paper is a research project which aimed to improve farmers' tools and equipment by involving farmers themselves, first in the identification of their problems and then in the conduct of the research to alleviate them. A further attraction of this "bottom-up" approach is that it enables farmers to share ownership of the project outputs (better tools and equipment), with the increased likelihood of their adoption.

The Participative Approach - techniques and methods

There is an increasing number of examples demonstrating the effectiveness of the participative approach, both in ergonomics (Kogi and Sen, 1987; O'Neill and Haslegrave, 1990) and in research to improve subsistence agriculture (Appleton, 1994). However, there is virtually no recorded evidence of the participative approach being used in ergonomics projects to improve subsistence agriculture. One of the primary tasks, therefore, was to adapt participatory techniques found to be successful in subsistence agriculture to ergonomics research and development.

The use of participatory approaches is now widely advocated but, being a relatively recently established form of research, the recommended techniques are being continually developed and refined. Various approaches have become available, each serving a slightly different purpose and each with its own advantages and disadvantages. These include Rapid Rural Appraisal (RRA), Participatory Rural Appraisal (PRA) and Participatory Technology Development (PTD). Similar activities can be identified in all the approaches and the ones adopted in this project are shown in Table 1 (from McCracken, Pretty and Conway, 1988).

Table 1 Participatory techniques selected for the Handtools Project

Technique	Comments
Secondary data review	Examination of relevant information already existing
Direct observation	This may provide a good alternative to direct, and maybe awkward, questions (eg size and contents of a dwelling will reflect income/wealth)
Semi-structured interviews	Do not administer questionnaires and make as few notes as possible Do not have "official looking" documents and try not to work from prepared sheets Be prepared to digress on interesting points Make full use of "key informants" (eg village leaders, teachers etc)
Analytical games	Asking farmers to rank or rate aspects of their work or life
Stories and portraits	Discuss or describe difficult or delicate situations by creating stories or by reference to a "third" party
Diagrams	Produce, with members of the village, diagrams in space (ie maps) and time (ie calendars, work routines) A picture can save 1000 words and illustrate emphases in perception
Workshops	Aim to involve all stakeholders in discussion of problems and solutions; particularly try to avoid marginalisation of weaker groups

The overall structure of the project was divided into five phases and was unconventional in that any laboratory investigations were to be carried out after the field trials. The normal approach is to test various options in the laboratory and then evaluate the most promising in the field. In this case the plan was to identify the best options for farmers in (their own) field trials and then to find out, through laboratory tests, the scientific reasons behind the farmers' preferences. This approach is most closely associated with, but not identical to, PTD. Table 2 shows the five phases of the project and how they compare to the PTD approach, as described by van der Bliek and van Veldhuizen (1993).

Table 2 The Handtools Project in the context of PTD

	"Cluster of activities" (van der Bliek)	Stages of Handtools Project
1	Getting started	Formation of research groups and RRA training
2	Identifying options and making choices	Identifying tools and tasks requiring improvement
3	Improving and innovating	Devise and fabricate new designs
4	Spreading out	On-farm evaluation
5	Sustaining the process	Promotion of farmer-adopted improvements (including lab. tests for scientific support)

The conduct of ergonomics research and development on-farm in subsistence agriculture is severely constrained by the circumstances. The ergonomics techniques and methodologies selected as being the most appropriate and the most likely to yield meaningful results are given in Table 3. Some further validation of these approaches for developing country applications may still be required.

Table 3 Ergonomics techniques selected for the Handtools Project

Technique	Comments
Subjective assessment	Rating, ranking, interviews,questionnaires
Body maps (eg for discomfort, fatigue)	Simply indicate areas Rank areas (worst etc) Rate areas on a pre-determined scale (eg 0-5)
Postural evaluation	Diagrams, photographs, video Anthropometric measurements

Workload	Ratings of perceived exertion (Borg scale)
	Heart rate, for
	    work stress / heat stress
	    fatigue / recovery
	(Oxygen uptake)

Work study	How tasks are performed The organisation of tasks

Research Activities

After a year of contact with four villages, we held Workshops to discuss whether there were productivity constraints associated with farmers' tools and equipment (and thus within the broad remit of the project) and, if so, how to proceed. There was no shortage of ideas and suggestions on what might be done from not only the farmers but also the artisans responsible for making and maintaining the tools and equipment. Broadly speaking, there were many shared problems but each village formulated its own set of priorities - in other words, the things that mattered most to them. So, in the spirit of the participatory nature of the project, each village produced its own agenda. These were accepted without trying to identify a common goal, or goals, in the expectation that the findings would be transferable between the villages and, ultimately, would be made more widely available as options for farmers in similar farming systems. The main research activities were:

i)   the development of improved sickles
ii) modifications to country ploughs to improve stability and handling characteristics
iii) the development of push-pull weeders
iv) evaluation of the benefits offered by a threshing machine
v)   identification of mechanisation options and the ergonomics and economic implications
vi) the safer use and management of agro-chemicals.

Achievements

The most significant achievement, which has had a positive influence on all the research activities, was an attitude change in the participating farmers. Few, if any, of the farmers had appreciated when the project started that they might be able to increase their productivity and reduce their drudgery by analysing the design of their tools and how they used them. However, they quickly grasped the concepts of ergonomics and, having done so, reacted very constructively with ideas for improvements, especially for their own situations. The prevailing attitude, that being a farmer means that life has to be hard and they therefore have to suffer, was undermined. They appreciated that the application of ergonomics could not remove all the hardships, but that the opportunity to alleviate at least some of them was now available to them.

Progress with the research topics was varied and some of the activities are still ongoing. The progress includes the design and adoption of improved tools and equipment, a greater understanding of the ergonomics issues and clarification of the farming systems constraints. Brief details are given below.

i)   Research into sickle design has established that the farmers' preferred design is the outcome of a specific production process and that use of this design results in a 12% increase in productivity whilst the farmers have a lower working heart rate. Measurements of wrist angles, particularly adduction, may bring about further improvements.
ii) The instability of the country plough was caused by excessive wear and the introduction of a liner has overcome this. Other benefits include reduced (animal) draught requirement, reduced maintenance costs and the conservation of scarce hardwood resources.
iii) Development of a push-pull weeder was largely obviated when a local blacksmith was found who made these implements close to the farmers'

specification. Effort has since been directed to developing this (wet-land) weeder to dry-land conditions.

iv)   The benefits associated with a pedal-powered threshing machine were not fully evaluated as the machine broke after only a few days use. Comments on the design of the user interface and how the women have taken advantage of the time savings attributable to use of the machine (tending garden crops and animals) have provided useful insights into its potential benefits.

v)   Regarding mechanisation, the project helped sponsor the purchase of a power tiller, after much deliberation on the economic viability. The users reported numbness of the feet and tremor/tingling of the hands, which deserve further investigation. The occasionally high workloads, notably when moving between terraces, suggest that appropriate work-rest routines should be proposed.

vi)   The research into the use of agro-chemicals, particularly pesticides, revealed serious ignorance of dermal absorption, the commonest means of ingress (Ambridge, 1988) and hence poisoning. Medical tests undertaken within the project indicated that the incidence of poisoning/intoxication is increasing. Current training methods on the use of pesticides were found to be inappropriate and the findings are contributing to better targeted training procedures.

The local collaborators have shared in the success of these simple and inexpensive, but effective, approaches and so are encouraged to apply them subsequently.

Concluding Remarks

The use of participative and ergonomics methodologies, which can not necessarily be prescribed but which are appropriate to the context, has elicited the full collaboration of farmers and their families. Farmers, once exposed to the philosophy of ergonomics, especially through participatory experiences, may be predisposed to pursue ergonomics improvements themselves, without outside intervention.

The sharing of information and decision-making have benefitted the farmers and project scientists. The farmers have derived real benefits and the scientists have achieved impact through farmer adoption.

The research activities have produced positive and sustainable results.

Acknowledgement

The support of the Overseas Development Administration (ODA), London for funding this work is acknowledged. The views expressed are attributable to the author and not necessarily to the ODA.

References

Ambridge, E.M., 1988, Pesticide management in relation to user safety. In Prinsley, R.T. and Terry, P.J. (Eds) *Crop Protection for Small-scale Farms in E and C Africa - A Review.* (Commonwealth Science Council) 98 105.

Appleton, H., 1994, Ownership through participation. *Appropriate Technology* **21**(1), 1-4.

van der Bliek, J. and van Veldhuizen, L., 1993, Developing tools together. *Proceedings of an International Symposium "Participatory Technology Development: Innovation through Dialogue",* Amsterdam 10 December 1993 (TOOL, Netherlands).

Chambers, R., Pacey, A. and Thrupp, L.A., (Eds), 1989, *Farmer First: Farmer Innovation and Agricultural Research*, (Intermediate Technology Publications, London).

Gamser, N.S., Appleton, H. and Carter, N., (Eds), 1990, *Tinker, Tiller, Technical Change: Technologies from the People*, (Intermediate Technology Publications, London).

Haverkort, B., van der Kamp, J. and Waters-Bayer, A. (Eds), 1991, *Joining Farmers' Experiences: Experiments in Participatory Technology Development*. ILEIA Readings in Sustainable Agriculture, (Intermediate Technology Publications, London).

Kogi, K. and Sen, R.N., 1987, Third World Ergonomics. In D.J. Oborne (ed), *International Reviews of Ergonomics*, 1 (Taylor and Francis, London) 77-118.

McCracken, J.A., Pretty, J.N. and Conway, G.R., 1988, *An Introduction to Rapid Rural Appraisal for Agricultural Development*. (International Institute for Environment and Development, London).

O'Neill, D.H. and Haslegrave, C.M., 1990, The application of ergonomics in industrially developing countries. In E.J. Lovesey (ed), *Contemporary Ergonomics 1990, Proceedings of the Ergonomics Society's 1990 Annual Conference*, (Taylor and Francis, London) 418-423.

Rogan, A. and O'Neill, D.H., 1993, Ergonomics aspects of crop production in tropical developing countries: a literature review. *Applied Ergonomics* 24(6), 371-386.

ERGONOMIC ISSUES IN TRACTOR ACCIDENTS: THE HITCHING SYSTEM

Lisa Cooper and Caroline Parker

LUTCHI Research Centre, Loughborough University,
Loughborough, Leics. LE11 3TU

HUSAT Research Institute,The Elms, Elms Grove,
Loughborough, Leics. LE11 1RG

Farming has a notoriously high accident rate and recent statistics from the Health and Safety Executive (HSE) suggest that tractor accidents are on the increase. While HSE concentrates on operator error, it seems likely that other factors play a significant role. This paper tests the hypothesis that training, design and management factors are equally responsible. Hitching, a significant cause of non-fatalities was selected as the focus of this investigation. Semi-structured interviews and a task analysis were used to identify potential errors and Task Analysis For Error Identification and Contributory Factors in Accident Causation, were used to classify the information. The paper concludes that other human factors issues will prevent the uptake of recommendations. Potential and legislative solutions are discussed.

Introduction

Next only to construction, in 1993/94 the rate of fatalities in the agricultural sector was found to be the second highest of the five industrial sectors HSE (1995). Seventy five per cent of UK agricultural machinery market consists of tractors (Cracknell; 1994) and as such is a major cause of most fatal and non-fatal accidents. The HSE studied over 1000 tractor related accidents from 1987 - 1993 based on the RIDDOR (Reporting of Injuries, Diseases and Dangerous Occurrences Regulations, 1985) investigations by HSE inspectors. They found that, 'Moving accidents caused the most accidents by far, with operator error and unsafe systems of work being the predominant factors.' (McCalister, 1995) The HSE concluded that training was inadequate, and launched the 'Tractor Action Pack'. This consists of an educational video, booklet and poster on safe tractor usage and conveys the importance of tractor safety by shocking its audience into following safe procedures through graphic reconstruction of real life accidents.

In order to investigate the issues associated with tractor operation a simple procedure such as hitching an implement to a tractor was targeted for more in depth investigation. Hitching has been found to be a significant cause of non-fatal tractor accidents, (Farmer's Weekly, 1995) (Fatal Injuries Report, 1994/95) (McCalister, 1995) According to the 1994 Labour Force Survey (as cited in the Fatal Injuries Report, 1994/95) hitching totalled 19% of all accidents researched by the HSE from 1987-1993. The report states that the task of hitching 'has many contributory factors and obviously required a major input aimed at most levels of staff' (McCalister, 1995). As farmers are also notoriously poor at reporting accidents, particularly non-fatalities and it is estimated that a mere 28% of non-fatal injuries are actually reported, with less than 10% of non-fatal injuries to self employed farmers being reported (HSE, 1995).

While the HSE concentrates on operator error, it seems likely that other factors play a significant role in tractor based accidents, as Rassmussen (1987) states, "There are many causes of error when investigating an accident... component fault, operator error, manufacturing error, or design error." Why are 'operators' making 'errors', and why are 'unsafe systems of work' occurring in this industry as suggested by McCalister, (1995)?

Methodology

In order to cover all possible avenues of error leading to accident causation in this context an accident causation framework was selected from the literature. The CFAC (Contributing Factors in Accident Causation model - as proposed by Sanders & McCormick (1993) and the TAFEI (Task Analysis For Error Identification) Baber & Stanton(1994) provided a sound base on which to select relevant stakeholders from each area, to build semi-structured interview formats, to effectively categorise and analyses the results and to put the example 'unsafe procedure' of hitching in context by highlighting all relevant issues for later discussion.

Subjects
Fourteen interviewees took part in this investigation as identified by a stakeholder analysis. This constituted 10 farm managers/farmhands from 7 farms, 2 designers, a sales representative and a lecturer at an Agricultural College.

Data Collection
Data collection took the form of semi-structured interviews on site. Each interview took approximately one hour to complete and the information was recorded and later transcribed using a dictaphone. Observation and ethnographic methods of data collection were also employed.

Data Analysis
A hierarchical task analysis and time line analysis of the hitching procedure were used and the CFAC framework allowed categorisation of the interview results for later analysis and discussion under the headings of Management, Physical Environment, Equipment Design, The Work Itself, Social and Psychological Environment, Co-worker/worker, Unsafe Behaviour.

Results & Discussion

Management

'Supply of any resource is best allocated by demand. Demand or safety should come from the worker and it doesn't happen'

Half of the farm managers interviewed made a concerted effort to ensure safe working procedures. However, failure to comply with safe working procedures was a common complaint from the management. Production pressure is common to all arable farms during harvest. This requires optimum machine and man power at the end of September to the end of October. Mixed farms also have to deal with young cattle born at around the same time. Keeping costs down often results in keeping recruitment of extra workers to a minimum. This often results in dairy and cattle workers, untrained in tractor operation, helping out on mixed farms, and long shifts which could lead to higher risk of error as described in accident causation theories. Management structure is similar to the Theory J organisation (ouchi, 1981 - as cited in Huczinsci and Buchanan, 1991). A structure characterised by low staff turnover, non-specialist workers, and flat hierarchy (a 'them' and 'us' attitude)

Physical Environment

Although areas for animals and machinery tended to be designated, manure, oil and other 'slippery' substances were easily transferred onto the boots of farmers, a potential hazard when controlling the clutch during hitching. Noise also hindered effective communication during the procedure as highlighted through time line analysis, HTA and observation.

Equipment Design

Controls and displays within cab design have become far more complex and varied not only among different tractor makes but also among like tractors of differing ages. An array of multi-functional, multi coloured knobs and switches serves to confuse irregular tractor operators such as the farmer's wife or dairy man. Ford, John Deere and Massey Ferguson were by far the most popular makes and upgrading of farm machinery was restricted primarily to arable farms, upgrading one of the tractors every couple of years to cope with harvest. Only two of the seven farms in the sample used a single make of tractor, the others owned up to three different makes ranging from the youngest tractor of 2 years old to the oldest at 36. Compatibility, in terms of cab design and link arm design, compounded the problems associated with the design of implements; these too varied in terms of pin diameter and width of the implement arms. Often an assistant is required to adjust the width of the link arms to match those of the implement arms while the tractor operator reverses the tractor. This exposes him/her to risk, from poor visibility, possibly due to the positioning of the pick up hitch, smashed mirrors, poor PTO guarding and absence of rear window wipers. This factor combined with a noisy environment inevitably causes problems. Finally all tractors have controls to adjust the width of the link arms. These are situated at the rear of the tractor; a major area of risk. Some controls to make minor adjustments to link heights also require the engine to be running during operation.

Worker/co-worker

'..how often do we see brand new machines being used only to find out the instruction manual is still in its sealed cellophane pack' Roger Kendrick speaking at the

Tractor Action Pack launch in 1995

Farming is a male dominated industry, while there is an abundance of literature on poor manual usage in general, workers attitude towards potentially unsafe machinery is *'They get used to it..'*. While lack of time may be largely to blame, other reasons could be responsible e.g. Zeitlin (1994) links unsafe behaviour with the risk/benefit ratio; the perceived benefits being time, convenience, self image and social acceptability. It has also been found that 45% of men enjoy the sense of danger. The interview data collected as part of this investigation supported these claims for taking shortcuts and avoidance of safe procedures.

Unsafe Behaviour & The Work Itself

How do all of the above issues influence unsafe behaviour when carrying out the work (or in this case when hitching an implement to a tractor?) The answer to this question was investogated by observation, task analysis, and a time line analysis . This data together with that collected during the interviews made it clear that each the issues mentioned previously, while insignificant on their own, provide a cocktail for disaster when mixed. The best way of highlighting this is to state each of the HSE guidelines for hitching and judge whether this is realistically possible in 'real life' situations. This has been covered in some detail in Cooper (1996). The HSE guidelines are as follows:
- Avoid incompatible hitch operation
- Only use controls from the operating position
- Never stand between the tractor and other machines
- Never stand with feet under or near the drawbar
- Ensure that skids and other supports are well maintained
- Communicated clearly if assistance required
- SAFE STOP

Design issues such as lack of standardisation, poor positioning of controls, poor visibility, hitching design arrangement, mismanagement, noisy environment, time pressures and many others as highlighted in this study, not only create the need for these guidelines but makes it extremely difficult to adhere to some, if any, of the above. The issues all serve to inflict an unsafe procedure which can only be dealt with by employing shortcuts to save time, frustration and inconvenience, hence exposure to risk and unsafe behaviour, Zeitlin (1994).

General Recommendations

After the collation of the results, recommendations were put forward. These are as follows:

Management
- A high degree of inspector presence and spot checks on unsafe systems of work needs to be implemented.
- Awareness of tractor accidents needs to be raised by stressing financial and criminal costs of mismanagement.
- More competitions which reward good management practice should be run by the HSE in conjunction with current competitions for safe tractor use among employees.

Equipment Design
- Safety of tractors should be upheld by providing tractors with tests similar to car M.O.T.'s.
- The hydraulics on the back of tractors should be removed or repositioned. Thus reducing exposure of second operator to risk from the back wheel of the tractor.
- The adjustment of the width should be hydraulically, not manually, controlled from within the cab or from another safe position.
- All tractor link arms should be standardised preferably with the open hook (waltersheid end) arrangement.
- The clevis hitch should be made obsolete, only hooks should be used for drawn implements.
- Guiding frames should be marketed more thoroughly to farmers. This would help with the correct adjustment of the link arms.
- All implement arms should be standardised where possible, or at least be upgraded to class 2. This would also eliminate the need for different size coupling balls.
- More effort should be made to colour code automatic pick up hooks, perhaps illuminace paint could be used to heighten visibility.
- More space should be allocated to the cab region of tractors to give room for swivelling chairs. This would optimise visibility of the hitching system.
- All tractors should have rear window wipers fitted as standard not optional.
- New legislation needs to be formulated concerning the design of tractor controls. This should be an international, if not a European, standard ensuring that all manufacturers have compatible control systems.
- Mirrors which are shatterproof are a possibility in enhancing vision.
- Mirrors should be attached to buildings wherever possible to enhance vision of reversing vehicles.

Unsafe Behaviour
- Using two people to reverse machinery should be encouraged but guidance on how to do this safely should be incorporated into training courses.

Conclusion

Accident research has put forward many theories concerning risk taking behaviour and human error. The approach taken by the HSE, to combat the problems of tractor accidents, has been solely directed at this unsafe behaviour among farm workers. But as Rasmussen (1987) states, *'There are many causes of error when investigating an accident. They are identification in terms of component fault, operator error, manufacturing error or design error.'* These factors seemed to have been overlooked by the HSE research. This may lead to their recommendations and the other contributory factors to compound each other and exert further pressure. These pressures serve to aggravate worker's abilities to perform this relatively unsafe task. The full report highlights the problems concerning areas of management, equipment design, etc. which are imposed on already overloaded workers. Until other players in the industry such as manufacturers and farm managers have been targeted by the HSE, problems will continue to be attributed to 'operator error'. Employees are already faced with a difficult job under considerable time pressures and in poor working conditions, and although training is important, it seems to be more important that this 'high risk' industry provides a safer working environment.

Recommendations for Further Research
- More detailed investigation of hitching accidents needs to be conducted. For example through further interviews with hitching accident victims to identify the exact sources of error.
- Boot design or clutch design also requires evaluation. There are various possibilities, for example a hand operated clutch situated conveniently inside the tractor cab would reduce the incidents of feet slipping off the clutch while hitching or a simple lip or curvature added to the pedal design.
- The possibilities of positioning the pick up hitch above the PTO as seen among French and other European makes of tractor, is another alternative and if engineering problems could be overcome this would be far safer.
- Lowering the levels of external noise in the operation of tractors would aid communication between individuals.
- Further research is needed to investigate the possibility of designing a method of automatically cutting out the engine when the operator is absent, perhaps in the future the possibilities of robotic operators may become a viable alternative.
- Finally re-conceptualisation of the hitching system and implement design, could be carried out by looking at other hitching systems in use by other vehicles such as lorries and caravan design.

References

Anon (1995) 'Farm deaths up' Farmer's Weekly. (Mar -Apr 1995)

Baber, C.. & Stanton, N.A. (1994) Task Analysis for error identification: a methodology for designing error-tolerant consumer products, Ergonomics, 1994, 37, 11, pp1923-1941

Cooper, L. (1996) Ergonomic Issues in Tractor Accidents. Submitted in part fullfillment of B.Sc Hons. in Ergonomics at Loughborough Unviersity.

Cracknell, John, (1994) Factors influencing the mechanisation of UK agriculture since 1972. The Agricultural Engineer, 49, 3,   pp81-84.

HSE (1994-1995) Fatal injuries investigated by HM Agricultural Inspectorate. HMSO Press.

HSE (1995) Tractor Action leaflet: a step - by - step safety guide. The Health and Safety Executive

McCallister, J. (1995) An investigation into farming accidents. Internal Document. Health and Safety Executive.

HEALTH AND SAFETY

PROCESS-BASED TOOL TO EVALUATE OCCUPATIONAL SAFETY AND HEALTH PROGRAMS

Ali Alhemoud*, Ash Genaidy* and Michael Gunn**

University of Cincinnati, Cincinnati, Ohio, U.S.A.
***City of Cincinnati Employee Safety/*
Environmental Compliance Division, Cincinnati, Ohio, U.S.A.

A valid and reliable tool to assess the effectiveness of occupational safety and health systems in the workplace is needed. The review of the literature indicated that there is a limited amount of research on the evaluation of comprehensive safety and health programs. There is also a need for the development of a valid and reliable instrument that would begin to incorporate workers feedback and encourage them to participate in the development and application of their workplace overall program. An instrument was designed to analyze the corporate safety and health climate and contained three processes. Content validity of the instrument was established by a panel of three expert judges. Field testing of the survey instrument was carried out in cooperation with the management and workers of the City of Cincinnati Department of Public Works.

Introduction

A self-evaluation strategy by which employers as well as employees can work together to identify and systematically attempt to solve safety and health problems at their workplace is needed. A potential method of assessing safety and health problems is through the use of self-administered surveys (Peterson, 1989). The survey is intended to analyze the corporate culture-climate of safety and health program and organizational culture. A survey of reports and research studies indicated that there have been few studies concerned with workers evaluations of their workplace safety and health programs (Cohen et al., 1975; Acthyl, 1983; Chung, 1984; Zohar, 1980; Brown and Holmes, 1986; Bailey and Peterson, 1989; Hayes et al., 1994). Although most of these studies pointed out the importance of factors which should be included in the design and evaluation of safety and health programs, most of them did not assess the validity or the reliability of their instruments.

Model Development

The model developed in this study consists of three major components: the organizational culture, the management process, and the safety and health surveillance process (figure 1). Figure 2 presents the model for evaluating the third component (safety and health surveillance process) and shows the six necessary steps required to evaluate any safety and health process.

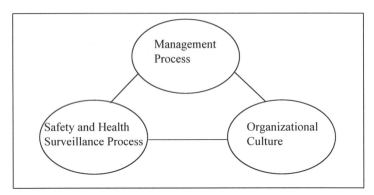

Figure 1. Evaluating corporate safety and health system

Instrument Validity

Instrument validity is the determination of the content representativeness or content relevance of the elements/items of an instrument by the application of a two-stage (development and judgment) process (Lynn, 1986). The development stage of content validity was established first by a thorough review of the literature and assistance from experts in safety and health to identify all measures of the instrument. A panel of three expert judges were asked to assist in determining whether the items and the entire instrument are content valid. Feedback from the members proved useful in clearing up item terminologies and formulating the final form of the survey instrument. Occupational safety and health items on the instrument were rated using a modified version of Borg's category scale of physical exertion (Borg, 1982). The scale used in this experiment deviates slightly from the Borg model, in that it is bipolar. Perception scores on all items using the scale vary from very strongly agree (+7) to very strongly disagree (-7).

Pilot Study

Pilot of the survey instrument was conducted on 15 workers from the City of Cincinnati Department of Public Works. Those workers were randomly selected from four divisions (Highway Maintenance, Sanitation, Facilities Management, and Traffic Engineering). At the end of the session, participants were asked about the length of the questionnaire "was it too long?", its content "was it too wordy?", and interpretation of the items "were they clear?". No revisions were suggested by the participants.

Instrument Reliability

Reliability deals with whether or not the instrument can measure the same traits consistently upon repeated measurements. Cronbach's alpha for measurement of internal consistency was used to analyze the respondents' scores. Table 1 presents the reliability for the seven subscales (excluding "safety and health goals" since it was based on frequency counts) of the instrument. Mishel (1989) suggested that the criterion level for coefficient alpha in a new scale to be about .70 or above.

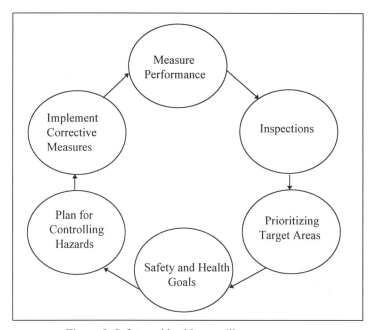

Figure 2. Safety and health surveillance process

Method

A total of 229 workers were selected from the divisions of Public Works. An analysis was made to examine the joint effect of these two factors (worker differences and division differences) on each safety and health process. A three-by-four factorial (three worker categories and four divisions) MANOVA was used to determine the linear combinations of the variables that best separate the factors (worker and division). Then, stepwise discriminate analysis was performed as a follow-up procedure to determine whether those variables are significantly different for the factors.

RESULTS

Tables 2 shows MANOVA and discriminant analysis for all processes of the model. Management process results indicate that divisions' responses differ significantly on the set of variables (division main effect, multivariate $F=.87$, $p<.032$,) and the following variables were noted to be most significant- violations disciplined and setting goals. Regarding the organizational culture, table 2 shows that only one variable was indicative of differences in worker perception (worker main effect, multivariate $F=.91$, $p<.001$)- organization cares for workers' safety, health, and welfare. No significant

Table 1: Reliability of the Survey Instrument

Subscale	Number of Items	Coefficient Alpha	Mean	(SD)
Management Process	5	.66	3.2	1.7
Organiz. Culture	2	.91	1.4	3.7
Safety and Heath Surveillance Process				
Measure Performance	4	.76	3.0	2.7
Inspection	6	.92	3.4	3.2
Prioritizing Targets	2	0	4.2	2.1
Plan to Implement	4	.96	3.7	3.3
Corrective Measures	6	.89	4.2	2.3
	----	----		
Total Scale	29	.96		

differences in worker and/or division responses were noted for measure performance process. The process of inspections, however revealed that significant differences exist in division responses on two variables (division main effect, multivariate F=.81, p<.003)- houskeeping and safe practices. Prioritizing target areas process used a log-linear model since its measured in terms of frequency data. Data analysis showed that no model fits the data indicating that worker and/or division responses did not contribute to any differences in the priority levels. Data analysis on the other hand, revealed that response frequencies on safety and health goals differed among workers and divisions (X^2=37.84, p<.01). The process of planning for controlling hazards revealed no statistical differences in worker and/or division responses, and corrective measures implemented process revealed significant differences between workers on two variables (worker main effect, multivariate F=.87, p<.018)- complaints and needed PPE.

DISCUSSION

Validity and Reliability

Literature review revealed that most workplace safety and health audit surveys either did not undergo any validity or reliability tests to determine if they adequately measure perceptions regarding safety and health or they did not comprehensively assessed the entire safety and health system (lacked a theoretical basis). Measurements of validity and reliability are paramount to any newly designed questionnaire. A comprehensive review of the literature and assistance from professional safety and health administrators provided for the formulation of a valid instrument aimed to comprehensively evaluate safety and health systems at the workplace. Field testing of the instrument provided good evidence of its internal consistency (reliability- Cronbach's alpha of .7 or higher).

The Model

The survey instrument contained three processes: the management process, the organizational culture, and the safety and health surveillance process. Furthermore, six measures were incorporated to assess the safety and health surveillance process. All items within a given process or measure were correlated as evidenced by a rejection of Bartlett's test of sphericity which justifies grouping items within their respective processes and measures. According to Peterson (1989), "To analyze the corporate [safety]

Table 2: MANOVA and Discriminant Analysis of the Model

Source of variation	Variable	Ms Between	Uni- Variate F	P<	SDFC
Management Process					
Division-	Setting Goals	41.18	2.63	.052*	-.55
Differences	Workers' Suggest.	31.33	2.41	.068	-.38
	Reward/Incentives	2.83	0 .15	.929	.54
	Violations Discipl.	34.02	2.89	.036*	.73
	Rules Enforced	5.07	0.41	.741	-.04
	Supervisory Train.	5.34	0.27	.848	.11
Organizational Culture					
Worker-	S & H Goals	93.49	4.88	.009*	-.04
Differences	Organization Cares	189.62	9.76	.000*	1.03
Measure Performance					
Worker-	Policy Posted	36.45	3.84	.023	.376
Differences	Accidents Investigated	28.59	2.82	.062	.522
	Rec. Accidents Posted	59.47	3.72	.026	.458
	Non-Rec. Posted	25.44	1.41	.248	-.076
Inspections					
Division-	**Daily**				
Differences	Safe Practices	31.21	3.28	.022*	.552
	Equipments	24.69	1.39	.246	.442
	Housekeeping	69.47	4.23	.006	-.611
	Routine/Scheduled				
	Equipments	41.94	2.37	.072	-.272
	Emergency Systems	43.32	2.37	.072	-.511
	Environmental	18.52	.95	.420	.121
Plan for Controlling Hazards					
Worker-	Awareness Training	35.91	2.73	.067	.095
Differences	PPE Training	45.37	4.35	.014	.524
	Design assessment	58.33	4.63	.011	.333
	Reporting	64.66	4.23	.016	.241
Corrective Measures Implemented					
Worker-	Orientation	30.93	2.71	.069	.065
Differences	Meetings	29.67	2.26	.107	-.627
	Guards Placed	38.31	3.02	.051	.086
	PPE Provided	63.24	5.81	.004*	.474
	Medical Facility	54.31	4.35	.014	.271
	Complaints	174.71	9.16	.000*	.734

Note: SDFC= Standardized Discriminant Function Coefficient.

climate, it must be divided into two distinct areas- climate of the safety program and corporate climate". He further adds "A perception survey if properly constructed is a better measure of safety performance and a much better predictor of safety results". The developed instrument was designed to analyze the corporate safety and health climate from all angles- climate of the safety and health program, quality of the management systems, and the organizational culture.

Survey Administration
    The developed instrument was administered to assess the effectiveness of Cincinnati's Department of Public Works safety and health program. Four divisions of Public Works were given the survey and responses were obtained from all workforce (field employees, supervisors, and managers). Most of the subjects selected attended the survey administration sessions suggesting that they were interested in participating. An initial examination of data demonstrated varying response levels to questions. Average reported scores concentrated in the range -2 to +7 (weakly diagree to very strongly agree) which imply an increased effort by the department to initiate safety and health actions.

References

Achtyl, J. (1983). *A survey of workers' perception of safety and health promotion.* (Masters Theses, University of Cincinnati).

Bailey, C.W., & Peterson, D. (1989). Using perception surveys to assess safety system effectiveness. *Professional Safety*, 34 (February), 22-26.

Borg, G.A. (1982). Psychophysical bases of perceived exertion. *Medicine and Science in Sports and Exercise*, 14, No. 5, 377-381.

Brown, R.L., & Holmes, H. (1986). The use of factor-analytic procedure for assessing the validity of an employee safety climate model. *Accident Analysis and Prevention*, 18, 445-470.

Chung, J.Y. (1984). *Comparison of safety programs and employee attitudes within University of California facilities management departments.* (Doctoral Dissertation, University of California, Santa Barbara, 1984). Dissertation Abstracts International, 46, 8509426.

Cohen, A., Smith, M., & Cohen, H. (1975). Safety program practices in high versus low accident rate companies- An interim report (questionnaire phase). *(DHEW publication No. 75185). NIOSH.*

Hayes, B.E., Johnson, P.R., Strom, S.E., Langlie, J.M., & Trask, J.M. (1994). Development and evaluation of the work safety scale (WSS). *Paper Presented at the Annual Conference of the society for Industrial and Organizational Psychology*, Nashville, TN.

Lynn, M.R. (1986). Determination and quantification of content validity. *Nursing Research*, 35, 382-385.

Mishel, M.H. (1989). Methodological studies: instrument development. *In P.J. Brink, & M.J. Wood (Eds.), Advanced Design in Nursing Research.* Newbury Park, CA: Sage.

Peterson, D. (1989). Techniques of Safety Management: A Systems Approach. New York: Professional and Academic Publisher.

Zohar, D. (1980). Safety climate in industrial organizations: theoretical and applied implications. *Journal of Applied Psychology*, 65(1), 96-102.

A COST EFFICIENT METHOD FOR PREVENTING ACCIDENTS IN ELECTRICAL DISTRIBUTION SYSTEMS AND PERIPHERAL EQUIPMENT

Ercüment N. DİZDAR[1], Hüseyin CEYLAN[2], Mustafa KURT[3]

[1] I. E. Dept., K.K.Ü., Kırıkkale, Turkey
tel: +90 318 357 24 58 / 120, e-mail: dizdar@kku.edu.tr fax: +90 318 357 24 59

[2] E. E. E. Dept., K.K.Ü., Kırıkkale, Turkey
tel: +90 318 357 24 58 / 127, e-mail: ceylan@kku.edu.tr fax: +90 318 357 24 59

[3] I. E. Dept., G. Ü., Ankara, Turkey
tel: +90 312 210 23 10 / 28 54, e-mail: dizdar@kku.edu.tr fax: +90 312 230 84 34

In this study, a low cost ergonomic method is developed to decrease the accident risk in substations which are important parts of electrical distribution systems. In this approach, the accident risk which may occur as a result of formation of spark was minimized by deviating the direction of electric field through formation of a magnetic field around the substation. This approach is not so costly since it can easily be realized by integration of a magnetic circuit on the present system. It was proven that the productivity of this low cost method developed is higher than the present methods.

Introduction

The facilities for production and distribution of electricity have a great importance nowadays. Because the present technology depending heavily on the usage of electrical energy is very sensitive even to very little problems in these facilities. In fact any fault may occur in such places may terminate production activities of industrial organizations fed by from the units where such fault occurred within the time pass set up to start up of such facilities, may interrupt the daily life and may cause great financial losses in such organizations. Moreover such faults may frequently result in accidents finished with death and injury of human-being. Domestic and industrial regions consuming electricity need substation centers of various capacity depending on quantity of electricity that they consume. Substation centers (transformer centers) are one of the most important units of power transmission systems. This unit has vital functions for operation of electrical transmission network. Because this center is unique alternative that present technology has in adjustment of electrical energy produced for transmission and consumption of electrical energy by the end users. For this reason the faults and accidents may occur in substation centers may terminate power transmission systems totally, the measures to be taken for prevention of the faults and accidents in these units are very important. The most important factor causing problems in substations is high temperature. Because the substations are heated by the current passing through the coils inside. This situation necessitate cooling of substations. However since the liquidated gas used in solution of such problem is of flammable and explosive nature the accident risk is very high. For this reason as a result of any accident, there may occur fires and explosions in substation centers due to sparks released on the cooling gas. Throughout researches have been carried out it is tried to solve the problem by enlarging substations cabins. This solution, on the other hand, brings a very expensive reconstruction cost. As an alternative, in this study, the accident risk which may occur as a result of formation of spark was minimized by deviating the direction of electric field through formation of a magnetic field around the substation. This approach is not so costly since it can be simply realized by integration of a magnetic circuit on the present system. It was proven that the productivity of this low cost method developed is higher than the present methods.

Electrical Distribution Systems And Substations

The facilities for production and distribution of electricity have a great importance nowadays. Because the present technology depending heavily on the usage of electrical energy is very sensitive even to very little problems in these facilities.

Electrical distribution systems consist of transmission lines and transformer centers. Electricity produced in energy producing centers are transmitted to the end users by transmission lines. During transmission of electricity, a part of energy is lost in transmission lines. The lost in this energy is proportional to the current and resistance of transmission lines. That is, the energy is given as a heat to the environment by warming coils, by increasing current in transmission lines. As a result, in order to minimize the loss of energy in transmission lines, the transmission must be done in low current-high voltage form of electricity. For this purpose, transformers are used at the beginning and end of transmission lines. Transformers are the electrical equipment used for adjustment of voltage and current of electricity.

Accident Risks In Electrical Distribution Systems

Any fault may occur in electrical distribution system may terminate production activities of industrial organizations fed by from the units where such fault occurred within the time pass set up to start up of such facilities, may interrupt the daily life and may cause great financial losses in such organizations. Moreover such faults may frequently result in accidents finished with death and injury of human-being.

As stated before, electrical distribution system consist of transmission lines and transformer centers. The important part of the system is transformer centers. Because, the transformers are expensive and the accident risk in this units is higher then rest of the system. Any fault in transmission lines can easily be detected and problems can be solved cheaply and in a short time. But, accidents in transformers are more important and costly. Because, lots of the time this accidents cause transformer completely out of services and time elapsing to repair this unit is very long.

Problem And Method

The most important factor causing problems in substations is high temperature. Because the substations are heated by the current passing through the coils inside. This situation necessitate cooling of substations. However since the liquidated gas used in solution of such problem is of flammable and explosive nature the accident risk is very high. For this reason as a result of any accident, there may occur fires and explosions in substation centers due to sparks released on the cooling gas. Under the light of many researches made in this century, we know that field emission (emission of electrons from metal surface under the action of high electric fields at room temperature) plays an important role at breakdown initiation and also to be the essential of at the stage of maintaining and development of the succeeding discharge process. This effect is also appears in high voltage transformers. A strong electric field at the surface of transformer caused by any reason, causes emission of electrons from the surface of the transformer. This emitted electrons strikes the surrounding metal cabin and cause the emission of other electrons. This situation continues consecutively. After a while, this free electrons cause formation of a spark in the medium and results with a breakdown. Since the cooling gas used between transformer and surrounding metal cabin is flammable and explosive, this situation constitutes a great problem for transformer centers.

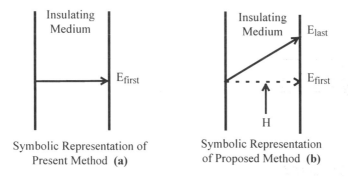

Symbolic Representation of Present Method **(a)**

Symbolic Representation of Proposed Method **(b)**

Figure 1. Symbolic Representations of Present and Proposed Methods

In this study, for the solution of this problem, the direction of electric field (E) is deviated through formation of transverse magnetic field (Figure 1).By this way, the probability of striking the emitted electrons from transformer surface to the across metal surface is decreased. As a result, the probability of formation of spark is decreased. Applied magnetic field (H) is externally produced by a simple magnetic circuit.

Experimental Work

Apparatus

In order to realize the proposed method, transformer and surrounding metal cabin is modeled as a point to plane electrode configuration. This point to plane electrode configuration is put to the Plexiglas chamber as shown is Figure 2. The two sides of chamber is closed with ordinary glass. The diameter of the chamber is 8.5 cm and the distance between electrodes is 1 cm. In order to compare efficiency of proposed method, the experiments are done firstly without the application of extra magnetic field (i.e. present method) and then the same experiments are repeated with the same parameters for the situation of existence of extra magnetic field. To produce the extra magnetic field, the circuit given in Figure 3 is designed. Calculation related with the magnetic circuit is given in Appendix. All the experiments are done in a darkroom in order to follow the formation of spark sensitively.

Experiments
Experiments with present method (Without extra magnetic field)
        Firstly experiments were performed for the present method. High voltage (DC) is applied to the terminals of electrodes shown in Figure 2. And applied DC high voltage

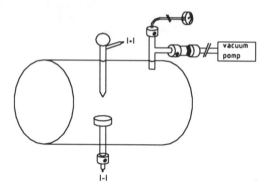

Figure 2. Point to Plane Electrode Configuration.

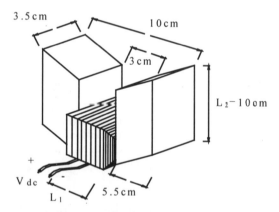

Figure 3. Magnetic Circuit used in the experiments.

increased until the formation of a spark between electrodes. The voltage value at which spark is formed is recorded as breakdown voltage.

Experiments with proposed methods (Application of extra magnetic field)
        A magnetic field perpendicular to the electric field is produced by applying a DC voltage to the magnetic circuit shown in Figure 3. And the same procedure for the present method is followed. And the applied DC voltage is increased, the procedure

given above repeated. By this way, for different magnetic flux density, the breakdown voltage is determined.

Experiments are performed in free air and in vacuum for the same parameters. The experiments performed for point to plane electrode configuration is repeated for point to point electrode configuration.

Results And Conclusions

The data obtained from experiments for different parameters are given in tables below:

Table 1. Breakdown voltages for <u>free air</u> and <u>point to plane</u> electrode configuration. Distance between electrodes 1 cm. The first row is for the present system and the other rows are for the situation of externally applied magnetic field.

DC voltages applied to the magnetic circuit	Currents on the conductors of magnetic circuit	Externally applied magnetic field	Breakdown Voltage
0 volt	0 ampere	0 tesla	3 kilovolts
2.72 volt	1 ampere	$3.86*10^{-3}$ tesla	3.6 kilovolts
4.93 volt	2 ampere	$7.72*10^{-3}$ tesla	3.6 kilovolts

Table 2. Breakdown voltages for <u>vacuum</u> and <u>point to plane</u> electrode configuration. Pressure of vacuum in the chamber is 26 inches. Distance between electrodes 1 cm. The first row is for the present system and the other rows are for the situation of externally applied magnetic field.

DC voltages applied to the magnetic circuit	Currents on the conductors of magnetic circuit	Externally applied magnetic field	Breakdown Voltage
0 volt	0 ampere	0 tesla	600 volts
2.72 volt	1 ampere	$3.86*10^{-3}$ tesla	680 volts
2.93 volt	2 ampere	$7.73*10^{-3}$ tesla	700 volts
7.80 volt	3 ampere	$11.59*10^{-3}$ tesla	700 volts
10.57 volt	4 ampere	$15.45*10^{-3}$ tesla	710 volts

Table 3. Breakdown voltages for <u>vacuum</u> and <u>point to point</u> electrode configuration. Pressure of vacuum in the chamber is 26 inches. Distance between electrodes 1 cm. The first row is for the present system and the other rows are for the situation of externally applied magnetic field.

DC voltages applied to the magnetic circuit	Currents on the conductors of magnetic circuit	Externally applied magnetic field	Breakdown Voltage
0.0 volt	0 ampere	0 tesla	4000 volts
7.0 volt	5 ampere	0.208 tesla	4100 volts
13.0 volt	10 ampere	0.417 tesla	4100 volts
18.0 volt	15 ampere	0.626 tesla	4250 volts

The following results can be extracted from this tables. Field emission plays an important role at breakdown initiation. By using an external magnetic field, the direction of the motion of free electrons between electrodes can be chanced. By using a suitable magnetic circuit, the breakdown voltage of the system can be increased. That is why, the accident risk which may occur as a result of formation of spark was minimized. This approach is not so costly since it can be simply realized by integration of magnetic circuit on the present system.

References

Ceylan, H. 1994, *Investigation of Breakdown Phenomena In Vacuum And In Free Air*, Graduation Project, (G. Ü., Gaziantep, Turkey).

Dizdar, E. N. Ceylan, H. and Kurt, M. 1997, A low cost approach for reducing accident risk in substations of electrical systems, *IAE'97*, (Tampere, Finland).

Feng, H. and Weihan W. 1990, *Electrical Breakdown of Vacuum Insulation at Cryogenic Temperature'*, IEEE Trans. On Electr. Insulation, **25**, 3.

Isa H. 1991, *Breakdown Process of a Rod-to-Plane Gap in Atmospheric Air Under DC Voltage Stress*, IEEE Trans. On Electr. Insulation, **26**, 2.

Kuffel and Zengel, *High Voltage Engineering.*

Miller, C. H. 1991, *Electrical Discharges In vacuum*, IEEE Trans. On Electr. Insulation, **26**, 5.

.... 1989, *Application of a Magnetic Field to Analyze The Prebreakdown Current in Vacuum Insulation*, IEEE Trans. On Electr. Insulation, **24**, 6.

Appendix

Related Formulas

Calculation Of Magnetic Field Intensity In Magnetic Circuit

$$H = B/\mu \tag{1}$$

where H magnetic flux intensity, B flux density and μ permeability of the material

$$\varphi = B*A \tag{2}$$

where φ magnetic flux and A area

$$\mu = \mu_0 * \mu_r \tag{3}$$

where μ_0 absolute permeability ($4\pi*10^{-7}$ H/m) and μ_r relative permeability

$$\oint H * dl = N * I \text{ (Ampere's law)} \tag{4}$$

where dl infinitesimal length, N number of conductor and I current

$$\Re = \frac{l}{\mu * A} \tag{5}$$

where $\Re$ reluctance and l length. If we apply Ampere's law to our magnetic circuit given in Figure 3: $N*I = H * l_1 + H * l_2 + H * l_{air} + H * l_2$

By using (1), the above equation can be written as,

$N*I = B/\mu * l_1 + B/\mu * l_2 + B/\mu * l_{air} + B/\mu * l_2$

By using (3), $N*I = B/\mu_0*\mu_r * l_1 + B/\mu_0*\mu_r * l_2 + B/\mu_0*\mu_r * l_{air} + B/\mu_0*\mu_r * l_2$

By rearranging,

$$N*I = B / \mu_0 * [l_1/\mu_r + 2*l_2 / \mu_r + l_{air}] \tag{6}$$

where in our case $l_{air} = 6.5$ cm, $l_2 = 10$ cm, $l_1 = 3$cm, $\mu_r = 6000$ (for commercial iron with 0.2 impurity) and N= 200 turns. If you put this numbers in equation(6) then

$$I = 258.8 * B \tag{7}$$

For a specific value of current, B can be calculated from equation(7) and by using equation (1) H value can be determined for a specific current value.

HEALTH AND SAFETY PROBLEMS IN COMPUTERISED OFFICES: THE USERS PERSPECTIVE

Randhir M Sharma

Division of Operational Research and Information Systems
The University of Leeds
Leeds
LS2 9JT

Computers are everywhere, it seems that whatever we do, we cannot escape from them.From the workplace to the supermarket to schools the computer has found a place for itself in all of our lives. Whenever a product becomes an integral part of our lives concerns are raised over the effects on us, our bodies, our environment and those around us. The car is a prime example, nobody thought twice about emissions, lead content or recyclability in the early days of the car, it is only relatively recently that we have seen a sudden increase in concern. However perhaps it is too early to be sure that the blame we attribute to workstations is really justified. The purpose of this work was to obtain an understanding of the level of familiarity amongst typical users of these problems and if problems exist, which factors users feel are to blame. In short this paper summarises problems experienced by users and the components of their working environment that they blame for their problems.

Introduction

From simple measurable complaints like headaches and stiff necks, to more serious medical disorders such as eye strain and repetitive strain injury through to complex social problems like stress, unemployment and work alienation, people are paying the price of rapid computerisation at home and work(Bawa, 1994). The purpose of this paper and the surveys that it describes is quite simply to determine whether or not we are paying the price. In a similar survey carried out whilst an undergraduate the author was alarmed at the widespread disregard for the subject of health and safety amongst computer users. It was not considered to be serious at all, it was seen as trivial. Users seemed unaware of the numerous cases highlighted in the press. The large library of information on this topic including hundreds of case studies did not concern them. Perhaps it was that these books, journals and reports had no relevance to real computer users. Should we be concerned or do these problems really not affect us? In a famous quote Judge John Prosser once referred to those people who complain of problems after using computers as "Eggshell personalities who need to get a grip on themselves....The RSI epidemic is a form of mass hysteria rewarded and reinforced by the compensation system." (case of Mughal vs Reuters, 1993) The wide range of opinions on this subject and widespread confusion regarding terminology do not aid its progress towards being accepted as a problem which users will either accept or take precautions against. Within the School of Computer Studies at Leeds two members of staff are using voice recognition systems because they can no longer use keyboards comfortably, they are *"normal"* users, if they can be affected, surely it can affect the rest of us.

The purpose of the survey was firstly to establish the scale of the problem, secondly to identify those components held responsible by users for their problems and finally to assess the importance of particular working habits. Two hundred and fifty six replies were obtained, the replies came from a wide range of computer users. One hundred and sixty one replies came from the School of Computer Studies at Leeds. Respondents included undergraduates, postgraduates, academic and non academic staff. Ninety five of the replies were obtained from workers working for a large national newspaper in India. The respondents varied significantly in the tasks they were doing and the number of hours per week they were using workstations. This large variation in the roles and tasks performed by users allows a better simulation of computer users as a whole. The results obtained from this questionnaire are relevant to those using computers in all walks of life, not just programmers and data entry clerks who tend to be the focus of most surveys of this type.

Questions Asked

One of the most important questions was the number of years that users had been using workstations. This question was asked in order to see whether or not there was a correlation between the number of years a workstation was used and the incidence of health problems. Another question was the number of hours per week that users used their workstations. It has been shown that the greater the amount of VDU use per day, the greater the amount of pain experienced by users. Fahrbach and Chapman (1990) compiled statistics for head, neck, shoulders and back pains and Evans (1985) showed that increased usage affected eyesight.

The distance between a computer user and the screen is a factor which is given a lot of attention. Sitting too close to the screen encourages poor posture and is often held responsible for visual problems arising from the use of VDU's. There is widespread debate concerning the ideal distance between the user and the screen. This in the main stems from a change in viewing angle. Earlier recommendations advocated the monitor being some 20 degrees lower than the line of sight, now the common consensus is that the monitor should be in the line of sight thus removing the risk of neck and shoulder pains as a result of the flexion required to view a monitor which is below the line of sight. The average of all viewing distance recommendations is 20 inches (50 cm) (Godnig and Hacunda,1991)

The time spent working between breaks was also important, regular breaks are recommended by most specialists.Without adequate pauses to rest, the mind loses its clarity and sharpness, and the body suffers wear and tear. You become less productive and more prone to injury(Choon-Nam Ong, 1995). Although there is no specific guideline for the time interval between breaks, most texts do recommend a 10 - 15 minute break every hour. One particular study showed that increasing the number of breaks actually sped up work and improved quality (Stigliani, 1995). The Health and Safety Regulations (Display Screen Equipment) 1992 contained specific guidelines for breaks. Regulation 45c states "Short, frequent breaks are more satisfactory than occasional, longer breaks:eg, a short 5 -10 minute break after 50-60 minutes continuous screen and/or keyboard work is likely to be better than a 15 minute break every two hours".

Stretching is another recommendation that frequently crops up. Stretch before you start work just as athletes do before vigorous exercise, concentrate on your arms, neck and shoulders but stretch your whole body too for general flexibility (Stigliani, 1995). By comparing the incidence of pains in arms and wrists, back pains and neck and shoulder pains amongst users who stretch and those who do not, it should be possible to see whether stretching does reduce the risk.

Another issue that was addressed was that of user familiarity with the topic of health and safety helping to prevent problems. Are those who are aware of the subject less likely to suffer than those who are not? The author was convinced that those who are well acquainted with the possible problems are more likely to take preventative action. The questionnaire asked respondents whether they were familiar with the terms CTS, RSI and WRULD. Although RSI is a very controversial term which we can argue is not really as relevant as WRULD, it is still regarded by computer users as an acronym for muscular problems associated with the use of a keyboard. The definition is vague and questionable but having knowledge of the term is an indication of awareness of the topic of health and safety. The results were then split into groups depending upon the level of knowledge and compared in terms of the incidence of problems.

Finally users were also asked for their opinions on how best to distribute Health and

Safety information so that it would be read. If users are to learn about these problems it is essential that the information is presented to them in a manner that they want and are comfortable with.

Results

The largest single group of users were undergraduates within the School of Computer Studies. In addition to undergraduates there were also responses from postgraduates and staff. The replies obtained in India came from users in a variety of roles within the newspaper, they included journalists, accountants, graphic designers and secretarial staff. In response to the number of years spent using a workstation, the largest group of respondents (126) had been using a computer for 5 years or less, this figure was influenced by the large number of undergraduates taking part in the survey. 83% of the respondents worked for 10 hours or more per week with a workstation.

Using the criteria that a safe break was one which was taken somewhere between 45 and 60 minutes only 37 (15%) of respondents took breaks within this safe time frame 219 (85%) users exceeded this. The figures for the distance between the user and the screen were also quite surprising, using the distance of 50 cm as a safe distance only 35 (13.7%) of the respondents were sitting a safe distance from the screen, 221 (86.3%) were sitting at a distance which is deemed unsafe.

Only 9% of respondents stretched before working. The small sample size made it impossible to draw any accurate conclusions from the data which provided very similar results for both users who did and did not stretch.

The results in the chart below illustrate the responses to the question asking which problems were suffered by users.

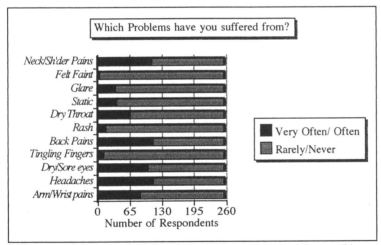

Figure 1. Problems suffered by users and the frequency of these problems.

Quite clearly the five key problems identified were neck and shoulder pains, back pains, dry sore eyes, headaches and wrist pains. The results indicate a frequency of between 30 and 40% for these problems. If this chart is modified to combine very often, often and rarely to form a new group of users - those who have suffered at some time this figure rises to between 60 and 75%. This clearly suggests that a large percentage of computer users have at sometime experienced these problems as a result of using a workstation.

The next step was to look at these five most frequent complaints and see if the users who made these complaints had anything in common in terms of their working habits. The habits looked at were the number of years they had worked, the distance they sat from the screen and the frequency of the breaks that they took.

Headaches

Those respondents who suffered from headaches most frequently sat closer to the screen, and worked the longest amount of time between breaks when compared with their colleagues who did not suffer as frequently. They also worked a longer number of hours. There was no apparent connection with the number of years worked. Next those factors were looked at which were highlighted as being responsible for these problems. The key factor highlighted was heating and ventilation. First impressions were that this result had been influenced by the replies obtained in India. Further interrogation of the spreadsheet revealed that this was not the case. Workers in India take the hot weather for granted, as a result of this all offices are equipped with both air conditioning and fans. The majority of complaints came from the School of Computer Studies. Since this survey was carried out a new air conditioning unit has been installed in the main laboratory used by undergraduates in response to complaints about excessive heat. The other point to raise is that in order to solve the problem we need to be clear as to which factors are responsible for the problems. It is not unreasonable to have expected poor quality monitors to be the factor pinpointed by users as the cause of their problems. The result obtained illustrates that it is not always the most obvious factor which is responsible for these problems, it also illustrates the importance of the working environment. The causes of certain problems may not be workstations but the environment in which we use them.

Pains in Arms and Wrists

Those respondents who suffered very often from pains in arms and wrists also shared many of the characteristics of those who suffered very frequently from headaches. They sat closer to the screen and took breaks after longer periods of time. They also worked the longest number of hours per week. However in this case there was a correlation between the number of years worked and the frequency of these problems. The greater the number of years worked the more they suffered. One of the noted characteristics of these sort of problems is that they have a cumulative effect. The results obtained from the survey support this claim and also stress the importance of prevention being better than cure. The factors highlighted for these problems were poor chairs, poor furniture and poor keyboards respectively. Poor chairs and poor furniture which do not allow a user to use a workstation comfortably were identified as the most important factors ahead of keyboards which the author had expected to be the most troublesome factor.

Back Pains

Once again the same three factors surfaced, users who suffered most sat closest to the screen, worked a longer number of hours between breaks and also worked more hours per week when compared with those users who did not suffer as frequently. There was no identifiable relationship between the number of years worked and the frequency of back pain. As expected the key factor identified here was poor chairs, followed very closely by poor furniture in general.

Neck and Shoulder Pains

The results obtained here were also very similar to those obtained for the previous problems. Users who suffered most sat closest to the screen, worked longest between breaks, and the worked the largest number of hours per week. No correlation was found between the number of years worked and the frequency of these problems. The key factor identified as being responsible for these problems was poor quality chairs. Surprisingly heating and ventilation was in second place followed by poor quality furniture. The surprising prominence of heating and ventilation led the author to the conclusion that fatigue associated with excessive temperatures within offices and laboratories was responsible for encouraging poor posture whilst working.

Dry Sore Eyes

The results obtained for this problem did not follow the pattern of previous results.

The survey did not replicate results found in other surveys. There was no apparent connection to either of the three factors which were so prominent in the other cases. The conclusion drawn from this was that individual differences in visual quality were more likely to be responsible for these problems than the working habits of users. However one surprising result was found by looking at the factors blamed by users. Heating and ventilation was the factor identified as being most responsible. The statistics showed that there was a huge gap between the blame attached to this factor and any other indicating that this particular factor was considered to be primarily responsible for the visual problems experienced by users.

The issues of whether users who are well informed suffer less or whether responsibility is solely in the hands of service providers who must provide a working environment which is free from potential risk was explored. Respondents were asked about three specific terms, CTS, RSI and WRULD. Only a handful of respondents were aware of all three terms. The results obtained from those respondents who were not aware of any of the terminology were compared with those who were familiar with one of the three terms or a combination. The results indicated that those who had knowledge of the subject worked fewer hours per week and took breaks more frequently. There was however little to choose between the two with regard to distance between user and the screen. The key differences however were found by looking at the frequencies of problems. The table below illustrates the results obtained for the five key problems highlighted earlier in the paper.

Table 1. Mean Problem Frequencies

	No Knowledge	Some Knowledge	
Pains in Arms and Wrists	2.39	3.22	
Headaches	2.05	2.88	
Back Pains	2.16	2.90	$p<0.05$
Dry/ Sore Eyes	2.77	2.61	
Pains in Neck and Shoulders	2.21	3.04	

Respondents were asked to indicate the frequency of their problems on a scale of 1 to 4 with 1 indicating very often and 4 never. The respondents were grouped into those who knew nothing about the subject and those who knew something. The figures in the table are the means of the two groups. With the exception of dry sore eyes which as mentioned earlier is probably more related to individual differences in vision, those who were familiar with the subject appeared to suffer less. This quite clearly suggests, as with many other problems, that education may the best prevention technique. 80% of respondents felt that they did not know enough about health and safety. Following on from this respondents were asked which techniques they thought should used in order to get the information to them. The current policy employed both within the School of Computer Studies and in India placed most of the emphasis on the user. They had to use their own initiative. Users were asked to suggest mechanisms which they felt would be most effective. The chart on the next page contains the five most preferred techniques. Altogether fifteen different techniques were suggested by respondents. The lack of relevant literature is highlighted by the responses obtained. Relevance to the user is a priority, most users have neither the time or inclination to read books and journals before using a computer. Information should be both concise and clearly presented in a such a manner that it does not become a burden to the user. The fact that users felt that more information and literature is needed suggests that the information currently available is either not easily accessible or is not in a format which will encourage users to read it. The various techniques put forward by users also indicate that a more varied approach needs to be taken. Different users have different needs, and it is important that these are given due consideration.

The results given so far have illustrated that problems do exist and when you consider the variance of the types of use, they also illustrate that it is not only particular types of users who suffer, anyone and everyone can suffer. It is essential therefore that users are educated and are aware of the apparent risks of using a workstation, only by doing this will they be in a position to take preventative measures. In order to educate users, the methods by which information is circulated have to be those which users will read and more importantly understand.

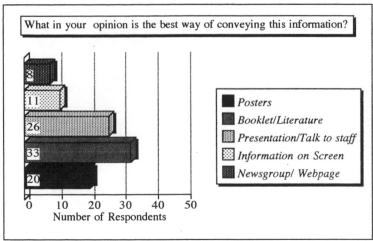

Figure 2. Preferred techniques

Conclusion

The purpose of the work described in this paper was to obtain an understanding of
the users perspective. The results obtained indicate that users are suffering and that
these problems are not just confined to a handful of individuals. The importance of
taking frequent breaks and sitting a safe distance from the screen has been shown.
Users have highlighted those factors which they believe are causing them problems and
in several cases the key factors are not those which would have been expected. Most
importantly the results have shown that those users who have some degree of
knowledge regarding the problems are less likely to suffer. Worryingly 80% of
respondents felt that they did know enough. Users have indicated which techniques
they would like to see employed to educate them. In order to prevent such problems it
is important to understand and identify risks associated with the technology, but it is
more important to have a full understanding of the user. Understanding and educating
users is essential if we are to eliminate these problems. Users need to be be aware not
only of the problems but also prevention techniques. The delivery of this information
has to consider the needs of individuals to ensure that it stimulates interest. Without
careful consideration, we risk endangering the very people we are people we are trying
to help...... the USERS.
    The next stage of this work is to look at several of the techniques highlighted above
in order to find the most effective.

References:

Bawa J. 1994, *The Computer Users Health Handbook: Problems Prevention and Cure.*
(Souvenir Press)
Bentham P. 1991, *VDU Terminal Sickness: Computer Risks and How to Protect
Yourself.* (Green Print)
Choon-Nam Ong, 1990, *Ergonomic Intervention for Better Health and Productivity.*
in Promoting Health and Productivity in the Computerised Office, edited by S. Sauter.
(Taylor & Francis,)
Evans J. 1985,*VDU Operators Display Health Problems.* Health and Safety at Work
Fahrbach , P.A & Chapman L. J., 1990, *VDT Work Duration and Musko-skeletal
Discomfort.* AAOHN Journal , 38(1)
Godnig & Hacunda. 1991, *Computers and Visual Stress: Staying Healthy* (Abacus)
HSE 1992. *Display Screen Equipment Work: Guidance on Regulations.*
Huws U. 1987, *The VDU Hazards Handbook: A workers guide to the effects of new
technology.* (Calverts Press (TU) Workers' Co-operative)
London Hazards Centre 1993, *VDU Work and Hazards to Health.*
Stigliani J. 1995, *The Computer Uses Survival Guide* (O'Reilly & Associates, Inc)

VIRTUAL ERGONOMICS

ARGUING IN CYBERSPACE: THE MANAGEMENT OF DISAGREEMENT IN COMPUTER CONFERENCING

Fraser Reid[1] and Rachael Hards[2]

[1]*Department of Psychology*
[2]*School of Computing*
University of Plymouth
Plymouth,
Devon, PL4 8AA

We report an experiment in which pairs of undergraduate students used keyboard-based conferencing software to resolve disputes on two controversial issues under conditions either of time scarcity or time abundance. Analysis of discussion content revealed differences between issues in the availability of a compromise position. Where compromise was possible, argument in time-scarce conditions focused on the positional concessions necessary to reach agreement within the constraints of the conference session. Where compromise was less readily available, time scarcity led to an increase in value-based argument. We conclude that temporal constraints profoundly affect the way in which disputes are interpreted and collaboratively managed in computer conferencing.

Styles of discussion in computer conferencing

Research on the use of computer technologies for human communication has revealed important differences between face-to-face and computer-mediated communication, and a reasonable degree of consensus now exists on the appropriate organizational uses of this technology (McGrath & Hollingshead, 1994). How actual workplace conditions affect computer-mediated communication is less well understood, however, and some conditions may yield contrary indications. This paper examines the effects of a pervasive yet often neglected task constraint--time scarcity--on the management of disagreement in synchronous, keyboard-based computer conferencing.

Recent studies of face-to-face meetings have shown that time scarcity focuses conversational argument on the goal of achieving consensus, reducing the amount of unique and unshared information discussed, and discouraging the systematic evaluation of alternative solutions and proposals (Parks & Cowlin, 1995). We have demonstrated elsewhere (Reid, Ball, Morley & Evans, in press) that these effects are further amplified in keyboard-based computer conferencing. Firstly, time pressure discourages complex argument in computer-mediated discussion, and increases direct pressure for consensus. A discussion style characterised by frequent decision proposals, emphatic statements of position, a cursory evaluation of evidence, straw polling, compromise, and acquiescence is the typical result. Secondly, consensus will be sought through appeals to norms and values, particularly when the problem involves an element of judgement. Norms and values function in this context as shared rules for evaluating the validity of claims, and allow relatively simple argument structures to be deployed, especially when facts are absent or skimpy (Antaki, 1985). In our study, we found that both of these argument structures were more prevalent in computer-mediated than face-to-face discussions, but *only* towards the end of the experiment when participants were under pressure of time to complete the discussion task. We reasoned that time scarcity is an important moderating variable, combining with the inherent restrictions of keyboard-based communication to present an increasing obstacle to conversational argument, to which participants adapt by adjusting their conversational goals and communication style (Reid, Malinek, Stott, Evans, 1996),

To test the generality of this analysis, we needed to examine the effects on conversational argument of directly varying time constraints in computer conferencing. We hoped to show that discussion style was not a property of computer-mediated argument *per se*, but a result of the interplay between time limits and communication modality. Specifically, we expected time-scarce computer conferences to be more consensus-driven than time-abundant conferences, with fewer arguments based on inference or evidence, more frequent appeals to norms and values, and a greater focus on contrasting positions and the concessions necessary to reconcile them.

The conferencing experiment

To investigate this hypothesis, we carried out an experiment at the University of Plymouth in which pairs of newly-arrived undergraduate students used desktop conferencing software to resolve disputes on two controversial issues under conditions either of time scarcity or time abundance. The conferencing software (MacConference; Kirkpatrick, 1994) connected networked Apple Macintosh computers together to allow synchronous, interactive keyboard-based "chat" sessions. Each computer screen was divided into two conference windows, which scrolled to display time-stamped messages as they were entered. Each message was labelled with its sender's code name. All communication between conference partners took place through the conferencing system.

A pretest survey showed that opinions for and against two statements--*All experiments on live animals should be banned*, and *In view of the risks of long-term injury, boxing should be banned*--were evenly distributed within the undergraduate group. Piloting also suggested that a 30 minute MacConference session was more than adequate for two people to engage in a full exchange of arguments and opinions relevant to these statements. Halving this time allowance would impose significant time pressure on these debates. However, an additional and arguably more realistic arrangement is to allocate the full 30 minutes for participants to engage in *two concurrent debates*, so that attention must be switched back and forth between the two statements, two sets of arguments, and two discussion partners.

Accordingly, 124 undergraduate students were grouped into sets of opposing pairs on the basis of their pretest opinions concerning the animal experiments and boxing statements, each set being assigned at random to one of three conditions: *condition C60*, in which participants took part in two consecutive 30-minute conferences, each on a different statement and with a different partner; *condition C30*, in which participants discussed the two statements in two consecutive 15-minute conferences, each with a different partner; and *condition P30*, in which 30 minutes was allowed for participants to engage in two parallel conferences on different statements with different partners. Statement order was counterbalanced over pairs. Before commencing their first conferences, participants completed a check questionnaire on their opinions, and were trained in the use of the conferencing system. They were also instructed to reach agreement on each of the statements by changing their partners' viewpoints, rather than by compromise or capitulation. An experimenter was present throughout each session to give timed announcements of

minutes remaining for each conference, to supervise the changeover between
conferences in the C60 and C30 conditions, and finally to administer a post-
experimental questionnaire comprising a number of rating scales and other check
questions.

Analysing argument profiles

After eliminating conferences based on unstable, weakly held, or unopposed
opinions, 40 animal experiments conferences and 41 boxing conferences remained.
Transcripts of these conferences were analysed by adapting Vinokur, Trope and
Burnstein's (1975) coding scheme to the present task. This involved first dividing
each message into units corresponding to single propositional statements, and then
identifying instances of the three kinds of argument relevant to our hypothesis:
arguments referring to *outcome utilities* (O arguments), or the consequences, costs,
benefits, probabilities of success or failure associated with the action proposed;
arguments referring to *values inherent to the action itself* (V arguments), or the
moral or social utility of the proposed action regardless of its outcomes; and
statements of position (P arguments), or explicit suggestions for resolving the
dispute. Four additional categories were used to classify other non-argument
statements. All 81 transcripts were initially coded by the first author. A random
subsample of 18 transcripts was then coded blind by the second author. A
comparison of coding decisions on this subsample revealed 85% agreement on unit
identification, and a significant kappa coefficient of agreement (κ=.70) on the
assignment of units to statement categories.

O, P, and V arguments accounted for over 80% of the 2,277 statement units
identified in the two sets of transcripts (787, 680, and 369, respectively). To
compensate for differences between conditions in the time available for discussion,
raw argument frequencies were first expressed as proportions of the total number of
statements recorded for each conference, and then normalised by applying an arcsin
transformation. The resulting *argument profiles* show how participants distributed
their effort over argument categories, regardless of the total volume of
communication. Mean argument profiles (with standard error bars) are shown for
each statement separately in Figure 1.

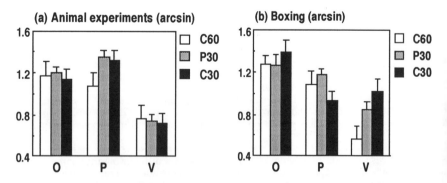

Figure 1. Argument profiles for the two discussion statements.

We expected the argument profiles of the time-scarce C30 and P30
conferences to contain proportionally fewer arguments based on inference or
evidence (O arguments) than the time-abundant C60 conferences, but instead to
show more frequent appeals to norms and values (V arguments), and a greater
emphasis on reconciling opposing positions (P arguments). In fact, *a priori* contrasts

between the time-scarce and time-abundant conditions found no difference in the prevalence of O arguments for either discussion statement. However, P arguments were more prevalent in time-scarce animal experiments conferences ($F(1,38)=4.70$, $p=.04$), and V arguments were more prevalent in time-scarce boxing conferences ($F(1,37)=7.83$, $p=.008$), *but only these conferences.* We did not find evidence for a uniform argument profile across the two discussion statements.

The opportunity for compromise

Curiously, our hypothesis appeared to fit some aspects of the data, but different aspects depending on the discussion statement at issue. To investigate this, we set about comparing the two sets of conferences on each of the measures we had to hand. The results of these comparisons are shown in Figures 2 and 3. What we found was a clear difference between discussion statements in the opportunity for a compromise response, and this appeared to explain the different patterns of argument deployed by conference participants.

Firstly, we examined the argument profiles generated by the two conferences, irrespective of time limits (Figure 2a). The profile for the boxing statement resembled that for the sample as a whole, with O arguments outnumbering P arguments, and P arguments in turn outnumbering V arguments (Tukey HSD, $p<.001$ in both cases). The profile for the animal experiments statement differed, however: here O and P arguments were more numerous than V arguments (HSD, $p<.001$ in both cases), but were equally prevalent. In relative terms, the boxing issue appeared to generate *less positional argumentation* than the animal experiments issue.

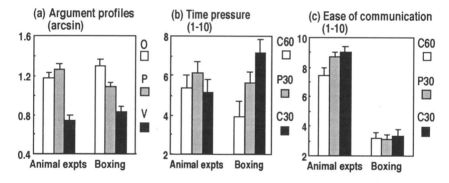

Figure 2. Comparisons between the two discussion statements

Secondly, we examined responses to the two statements in the post-experimental questionnaire. Factor analysis of the rating scales contained in this questionnaire revealed three factors common to both statements, which together accounted for 62.9% of the variance. *Post hoc* tests on the aggregate scores of scales loading significantly on these factors (scores range from lowest=1 to highest=10) revealed differences between conditions on factors representing perceived time pressure (Figure 2b), and ease of communication (Figure 2c). Interestingly, differences in time pressure were not reported by participants in the animal experiments conferences: *only in the boxing conferences* were the time-scarce conditions judged to be more time-pressured than the time-abundant C60 condition (HSD, $p<.05$ in both cases). Conversely, only in conferences about the animal experiments issue was ease of communication rated highly, and *more so in the time-scarce than time-abundant conditions* (HSD, C30 $p=.04$, P30 $p=.10$).

Thirdly, we inspected each transcript to determine whether conference partners reached agreement on the issue in question, and whether this was achieved by one partner acquiescing to the views of the other (with or without evidence of a genuine change of opinion), or whether conference partners discovered and agreed upon a compromise position not included in the original statement. The striking finding here was that conference partners were more than twice as likely to reach agreement in the time-scarce C30 condition than in either of the remaining conditions (Figure 3a), *but only in the animal experiments conferences* ($\chi^2(2$, N=41)=6.76, *p*=.03). Although corresponding rates of agreement for the boxing conferences in this condition were extremely low (18.2%), no significant difference was detectable among conditions for this statement.

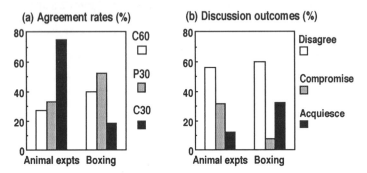

Figure 3. Further comparisons between the two discussion statements

Finally, although conference partners found agreement difficult to reach overall (Figure 3b), where agreement was reached on the animal experiments issue, *a compromise was the most likely outcome* ($\chi^2(2$, N=41)=11.90, *p*<.01). In contrast, acquiescence was the most common form of agreement in the boxing conferences ($\chi^2(2$, N=40)=16.55, *p*<.01). Inspection of these transcripts soon revealed the reason for this. For the animal experiments issue, a compromise position--to ban animal experiments for certain specifiable purposes (such as testing cosmetics) but to retain them for other purposes (such as testing life-saving drugs)--was readily available to conference partners. This was, without exception, the accepted middle-ground for this statement. Few such opportunities for compromise arose in disputes over the banning of boxing: where they were proposed, compromises on the boxing issue more often took the form of conditions to be met if a ban were *not* to be agreed (e.g. to improve safety measures in boxing tournaments), rather than as a redefinition of the options available to conference partners, as was the case with the animal experiments issue.

Managing disagreement in computer conferences

We can summarise these results as follows: time scarcity impacts on style of discussion in computer conferencing in different ways, depending on how conversational disputes are managed within the discussion. Our data suggest that where the opportunity for compromise is recognised--as in the debates on animal experiments--participants focus argument on the positional alternatives and concessions (P arguments) necessary to reach agreement within the combined constraints of the electronic medium and the limited time available. The greater the pressure of time, the more intense this focus is likely to be, and the more readily will a compromise be viewed as an acceptable means to resolve the dispute. When viewed retrospectively--as in the post-experimental questionnaires in the present study--

conferences of this kind are likely to be perceived as relatively uncomplicated, allowing agreement to be reached relatively easily with time to spare.

Where a compromise response is less readily available--as in the boxing debates--time scarcity amplifies the constraints of the electronic medium to create a significant conversational problem. These debates will be perceived as difficult to conduct, with insufficient time to express one's thoughts, persuade one's partner, or resolve disagreement. Under these difficult conditions, argument takes the form of assertion and counter-assertion, and appeals to norms and values (V arguments) will be deployed as shorthand surrogates for reasoned argument. If agreement is reached, it will be by one party's acquiescence rather than by their persuasion.

Some of our results are less readily explained, however. Our failure to find differences in the prevalence of O arguments between time-scarce and time-abundant conditions is perplexing, as are inconsistencies in the patterning of results for the two time-scarce conditions, C30 and P30. Clearly, conference partners in this latter condition are able to distribute their attention unevenly between the two parallel conferences, and this appears to have attenuated the impression of time scarcity in this condition.

Whatever the means by which disagreement is managed in computer conferencing, our results do confirm the general hypothesis that discussion style is a result of the interplay between communication modality and other constraints on participants, rather than a consequence of computer mediation *per se*. We believe it more profitable to view argument in this context as a way of collaboratively managing the practical problems presented by the existence of disagreement (Jacobs, 1987). The shape and substance of the arguments presented in computer-mediated disputes are worked out collaboratively, within the constraints of argumentative discourse, the interpretation jointly placed on the issue under dispute, and the practical opportunities afforded by the conferencing session. What we have shown is that it may be possible to predict preferences for particular argument structures from a knowledge of the temporal constraints on the conference, and the issue under dispute.

References

Antaki, C. 1985, Ordinary explanation in conversation: Causal structures and their defence, European Journal of Social Psychology, **15**, 213-230.

Jacobs, S. 1987, The management of disagreement in conversation. In F.H. van Eemeren, R. Grootendorst, J.A. Blair & C.A. Willard (eds.), *Argumentation: Across the Lines of the Discipline*, (Foris Publications Holland, Dordrecht)

Kirkpatrick, H.W. 1994. *MacConference 2.0*, (MacAnswers Consortium, New York)

McGrath, J.E. and Hollingshead, A.B. 1994, *Groups Interacting with Technology: Ideas, Evidence, Issues, and an Agenda*, (Sage Publications, Thousand Oaks, California)

Parks, C.D. & Cowlin, R. 1995, Group discussion as affected by number of alternatives and by a time limit. Organizational Behavior and Human Decision Processes, **62**, 267-275.

Reid, F.J.M., Ball, L., Morley, A.M. & Evans, J.StB.T. in press, Styles of group discussion in computer-mediated decision making, British Journal of Social Psychology.

Reid, F.J.M., Malinek, V., Stott, C.J.T., & Evans, J.StB.T. 1996, The messaging threshold in computer-mediated communication, Ergonomics, **39**, 1017-1037.

Vinokur, A., Trope, Y. & Burnstein, E. 1975, A decision-making analysis of persuasive argumentation and the choice-shift effect, Journal of Experimental Social Psychology, **11**, 127-148.

BUILDING A CONCEPTUAL MODEL OF THE WORLD WIDE WEB FOR VISUALLY IMPAIRED USERS

Mary Zajicek and Chris Powell

School of Computing and Mathematical Sciences
Oxford Brookes University, Gipsy Lane Campus, Headington,
Oxford OX3 OBP, UK
Tel: +44 1865 483683 Fax: +44 1865 483666
Email: mzajicek@brookes.ac.uk

There has been a discernible shift on the World Wide Web from basic text pages to the use of more complex information structuring techniques such as lists, tables and frames, and multimedia elements such as imagemaps, video and virtual reality. These advances make the interface more information rich for the sighted user, but make life more difficult for the visually impaired user who can only 'read' the textual part of the interface using speech synthesis. The aim the project is to present as much **real information** contained in the Web page to the visually impaired user as possible. The experimental work described in this paper was aimed at determining how best to sort, group, and annotate the information contained in a Web page to promote the most effective conceptual model in the user.

Introduction

This paper describes experiments conducted using a helper application WebChat, developed within the project, to find an optimum method for representing the contents of a World Wide Web page for visually impaired users using speech. The modality of WebChat changed in order to incorporate the different features described below and evaluate them for effectiveness and usability.

Currently available screen reading products allow the visually impaired user to 'read' the screen using speech synthesis or Braille display hardware. Most of these applications are only capable of reading one line of the screen at a time and the document is accessed in a sequential manner. They are also general purpose and not specifically designed for Web use. However as graphical interface technology develops uninterpreted screen reading becomes less cost effective or effectual as has been shown by the following authors, Laux, McNally, Paciello and Vanderheim (1996), Petrie, Fabrizi and Homatas (1993), Stephanidis, Savidis and Acoumianakis (1995).

Our system provided access to conceptually different parts of the Web page using a function key system and different synthesised voices to indicate conceptually different features of text such as headings, links and multimedia objects.

Existing facilities for visually impaired access to the Web

Web pages are created by the Internet Browser (Netscape Navigator, Mosaic or Internet Explorer are examples) from HyperText Markup Language (HTML) code. HTML incorporates an increasing number of multimedia elements and multimedia presentation can be further enhance by incorporating Java Applets. Software such as WebChat must run alongside the Internet Browser and intercept the HTML code to 'interpret' it for speech output.

Problems with inaccessible HTML code

Screen readers and WebChat itself rely on HTML code being written in a standard and accessible manner. Sadly many Web sites provide code which is far from accessible as described in O'Brien 1996. It is argued that Web sites should be legally bound to provide only standard accessible HTML in the same way that buildings must be accessible for disabled people.

Attempts have been made to improve the access of vision impaired users to HTML code as described in Polovina (1996). The International Committee for Document Design (ICADD) was set up to establish a standard Document Type Declaration by which textual information can be successfully transformed for print disabled users into the more accessible ICADD format. This format includes the contextual information that would enable conversion into the existing alternative formats such as Braille, but in a way that can now be interpreted by alternative output software. Alternatively the W3 Access for Blind People (W3ABP) approach exploits the Web's Proxy Server Technology. Web pages can be read via the W3ABP's server. The server automatically changes the format of the Web page, adding certain contextual information, so that it can become meaningful to a user who needs a screen reader.

How we presented the pages

The aim of these experiments was firstly to investigate ways of orientating visually impaired users to the contents of a Web site and then to present as much real information contained in the Web page as possible. The focus of the work `is on determining how best to sort, group, and annotate elements of the Web page to this end.

For sighted users orientation to the contents of a Web site is cognitively demanding. We make decisions like 'what it is about?', 'does it really contain the information I am seeking?' as quickly as we can by 'scanning' the page to build up a working picture of its contents. Sighted users were observed in active scanning i.e. consciously searching the text (passive scanning will be discussed later) and it was found that they tended first to look for headings, images and links to orientate themselves to the page. If these did not provide a sensible picture they then investigated paragraphs in the main text. The aim was to provide functionally equivalent facilities for active scanning for visually impaired users.

M. Zajicek and C. Powell

The Web page was presented conceptually as a collection of textual objects (consisting of heading and accompanying paragraph) and links. These were ordered hierarchically within heading and sequentially down the page. As HTML headings allow for six levels, in a well designed document they describe a tree structure. This feature is used by WebChat to subset the document and give the user the ability to examine only the subtrees of interest. Navigation of the Web page was achieved by easily locatable keys. By use of the Select headings menu (F7), a heading is selected. The sub-document described by the subtree associated with that heading can then be examined by the user with the shifted F2 to shifted F5 keys described below.

The tables below show the functionality of the keys used.

Table 1. Control Keys

Function Key	Normal Function	Shift Function
F1	Read document title	Load a URL
F2	Read all links	Read links under selected heading
F3	Read Headings	Read headings under selected heading
F4	Read entire document	Read document under selected heading
F5	Read entire dictionary	Read dictionary of words under selected heading
F6	Select link menu	
F7	Select heading menu	Search for a user-specified word
F8	Select bookmark menu	Search for interesting words

Table 2. Menu Keys

Key	Function
Esc	Cancel any menu and return to ready state, Stop any speech
Spacebar	Pause/resume speech
Page up	Go to 10th link/heading/bookmark before current position
Page down	Go to 10th link/heading/bookmark after current position
Home	Go to first link/heading/bookmark
End	Go to last link/heading/bookmark
Cursor up/right	Go to previous link/heading/bookmark
Cursor down/left	Go to next link/heading/bookmark
Return	Select the current menu item

The links (i.e. anchors) and headings (H1 to H6) in a document are extracted from the document and placed in the Link and Heading lists respectively. These lists are available to the user as an aid to navigation and summarisation.

The Menu keys above provide a keyboard-operated menu system where feedback to the user is vocal (Fig. 1). For example, pressing F6 selects the links menu. The first link is read out. Pressing the down cursor key causes the second link to be read out. Once the required link is located using the cursor and page keys, pressing Return will load the associated document (equivalent to clicking on an anchor in a standard web browser). Pressing escape will cancel the menu and return WebChat to its ready state. The Heading and Bookmark menus are used in the same way.

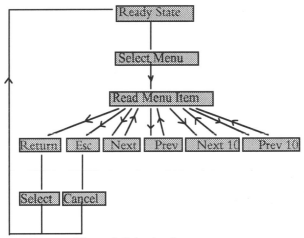

Figure 1. Selecting from a menu.

Dictionary/Audio scanning

The dictionary consists of words from the document after *noise* words (and, the, to, where etc.) have been removed. Additionally, stem-stripping is used to normalise words before insertion into the dictionary. This avoids the duplication of words where the root is the same, but the suffix is different (e.g. browsing, browsed). The dictionary has two functions :

1) To provide a word index into the document for performing word searches.
2) To provide a summary of the document (passive scanning).

The word index can be used by the user to look for a specific word in the document (shift F7), or to scan for any words which match those in the user-maintained *interesting word* database (a list of words which the user commonly searches for, i.e. those which relate the users interests).

The summary can be used to read-out to all the non-noise words in the document. This reduces the time taken to review a document (experiments show around 50% less), enabling the user to scan the document, actively or passively, for anything of interest. This audio scanning attempts to provide the non-sighted user with the audio equivalent of visual scanning.

Site orientation using function keys

Users could build a model of the information content of the page by 'trying out' the lists of headings, subheadings and links simply by pressing the function keys.

Evaluation of Web page structure presentation

Inevitably the time taken by the user to orientate themselves to the information content of a Web site was significantly more than the time taken using visual scanning. Therefore evaluation of effectiveness was based purely on information gathered rather than time taken. Three aspects of Web page presentation were evaluated using six user subjects for each aspect.

(1) Function key evaluation - Users were observed using the function keys and asked about the functionality they offered. It was found that the system worked well. The facility for locating and selecting a heading and reviewing its associated text was particularly time saving. However users could be disorientated if they became distracted due to the serial nature of speech and the lack of refresh. This is a common problem with speech driven systems which will be addressed in the next version by incorporating a message tagging system.

(2) Web site orientation - Users were asked to familiarise themselves with a whole (small) Web site ready to answer questions on 'what it was about' rather than questions about specific information. They were asked general questions designed to determine the extent of their conceptual model of the information available at the site. It was found that the six subjects had built a conceptual model of the information content of the page comparable to that of a sighted user after scanning.

(3) Information retrieval - Users were required to use the site as an information source and were given the task of retrieving specific information. It was found that the design of the Web site and length of sentences were major factors in the level of success, mainly due to the uninteresting nature of synthesised speech.

Conclusion

It is hoped that the results of the work described above will be relevant for both those designing Web pages and those providing extensions to the HyperText Markup Language (HTML) which forms the Web page. By exploring the presentation of data in a different context we also hope to provide insights for those presenting information to sighted users.

It cannot be stressed enough that without goodwill on the part of Web page providers who use accessible code, visually impaired users will not be able to use helper applications.

The work was funded by the Higher Education Funding Council Executive under the Widening Access to Higher Education initiative.

References

Laux, L., McNally, P., Paciello, M., Vanderheim, G., 1996, 'Designing the World Wide Web for People with Disabilities: A User centred Design Approach', *Proceedings of ASSETS'96, The Second Annual ACM Conference on Assistive Technologies*, ACM, New York

O'Brien, S., 1996, *'Multimedia and hypertext'*, Ability: The Journal of The British Computer Society Disabled Group, Issue 18

Petrie, H., Fabrizi, P., Homatas, G., 1993, 'Report on user requirements for the accessibility of graphical user interfaces by blind people', *Report to the CEC for TIDE Pilot Action Project 103: Graphical User Interfaces for Blind Persons, London, Royal National Institute for the Blind.*

Polovina, S., 1996, *'Weaving the Web: Designing for Universal Access'*, Ability: The Journal of The British Computer Society Disabled Group, Issue 18

Stephanidis, C., Savidis, A., Acoumianakis, D., 1995, 'Tools for User Interfaces For All', *Proceedings Second TIDE Congress, Amsterdam*, IOS Press

THE DEVELOPMENT OF A VISUAL TEST BATTERY FOR VIRTUAL REALITY USERS

Peter A. Howarth and Patrick J. Costello

Visual Ergonomics Research Group (VISERG)
Department of Human Sciences
Loughborough University
Leicestershire LE11 3TU   England, U.K.

Immersive virtual reality systems, using binocular headsets, represent a new step in human-computer interaction. We have recently evaluated changes to the visual system which occur as a consequence of headset use. The rationale underlying the choice of objective measures for the test battery is discussed.

Introduction

Virtual Reality (VR) systems represent a revolutionary development in human-computer interaction techniques allowing the user to step into a computer-generated Virtual Environment. There are various implementations of VR, the most sophisticated being 'immersive' systems in which the virtual world is viewed through an enclosing binocular headset (HMD). A number of recent studies have reported that wearing these headsets can adversely affect the user, bringing about symptoms similar to those reported by users of driving and flight simulators (e.g. Regan, 1995; Regan and Price, 1993; Howarth and Costello, 1996a). Although some symptoms, such as nausea and disorientation, may be caused by sensory conflict rather than because the visual system is affected directly, users also report eyestrain, tired eyes, and blurred vision - symptoms which are more likely to have an ocular genesis. We have recently completed a study for the U.K. Health and Safety Executive (HSE) of the visual effects of immersion in Virtual Environments, and as part of this work we developed a battery of objective vision tests. This battery was designed specifically to reveal ocular changes caused by wearing a headset, and the rationale underlying its development is discussed here.

A major issue is that tests selected should measure visual parameters that are expected to change over the period of immersion. These can be determined by analysis of the visual display, and by consulting previous research to identify particular areas where changes in visual performance have been documented. The display analysis has been broken down into three sections, dealing with the spatial, temporal, and binocular aspects of the stimulus.

The expected time course of any changes is an additional consideration which is common to all three of these aspects. Previous research that has identified visual changes under immersive conditions suggests that such changes may only be experienced for a short time after immersion is completed before there is a return to initial baselines. Similarly, we found that, when short immersion periods were involved, changes to the visual system were only short term: the majority of subjects who experienced visual changes after a 20 minute immersion returned to initial levels within a matter of minutes of exiting the immersive environment.[1] This is in contrast to symptom changes, which can last for longer and can re-appear after the subject has apparently returned to normal.

Spatial aspects

An HMD will, in general, contain either one or two display screens and two optical systems. The screen can either be a CRT or an LCD display, depending upon the HMD type. In either case, the image seen is made up of a number of small pixels, the borders of which may be obfuscated by "depixelation". The display resolution is a function of the pixel density, and generally is poor when compared with a normal VDU. Word-processed text, read easily on a 15" CRT monitor, is hard to resolve on low-cost HMDs manufactured for the home games market (e.g. Virtual i-glasses). However, this poor resolution is not, necessarily, a bad thing because the accommodation of the eye need not precisely match the stimulus distance (the depth of focus of the eye increases with decreasing target clarity, obviating the need for accommodative precision).

The angular size of the display is likely to be of critical importance in determining the extent of any virtual simulation sickness brought about by sensory conflict between the visual and vestibular systems (e.g. Howarth and Costello, 1996b). However, in terms of physiological effects on the visual system itself, the angular subtense of the screen is unlikely to be of serious consequence.

Refractive Error

Refractive error refers to a condition in which the optical components of the eye do not directly form a clear image on the retina. When uncorrected, refractive errors can lead to symptoms such as headache, blurred vision and eyestrain.

Our own measurements of display screens have shown that, generally, the image of an object produced within an HMD will be at an optical distance further than a metre from the eye. This finding holds whether the image seen is *geometrically* at a far or a near distance. At first sight, this would not appear to be an optical stimulus which might be expected to cause refractive error changes. However, it has long been known that viewing through optical instruments causes the eye to accommodate, leading to instrument myopia (see Ong and Ciuffreda [1995] for a review of this area) and so it is reasonable to expect HMD use to bring about myopic changes in a young population.

A study of immersive VR use for an assembly task by Miyake, Akatsu, Kumashiro, Murakami, Hirao, and Takayama (1994) revealed that HMD wearers experienced a myopic shift in the order of -0.2 to -0.35 Dioptres after 36 minutes of VR work. In our own experiments for the HSE (in which three different VR systems were evaluated) the changes in refractive error found were small, and could not be distinguished from normal day-to-day variation. However, subjects were immersed for a

---

[1] Although this would tend to suggest that any tests used should be quick to administer, allowing visual functions to be measured before any decay takes place, one must then consider whether these changes, if short-lived, have real *practical* significance.

shorter (20 minute) period than used by Miyake et al. (1994) which could account for the difference in the results found.

Visual Acuity (VA)

Visual acuity (VA) is a measure of the eye's ability to resolve fine detail and is dependent upon the person themselves, the illumination level and the contrast between target and background. It is widely regarded as the most basic metric of visual performance: reading black letters on white charts is still the method used by optometrists to measure VA. However, most of the world is not high-contrast, and so this clinical method does not relate well to everyday situations. Accordingly, we chose to assess vision using both low (18%) and high (90%) contrast letter charts of the type designed by Bailey and Lovie (1976). This approach allowed VA to be measured under sub-optimal visual conditions (low contrast) and also at a point near the optimum performance level (high contrast) to provide a more extensive assessment of vision.

The 'noise' in these measures is large compared with the size of expected effects, and their usefulness is dependent upon having adequate subject numbers. In our study we measured 41 subjects in total, and did not find any statistically-significant acuity changes for either high or low contrast targets over the twenty minute immersion period.

Contrast Sensitivity (CS)

The need for an initial analysis of the visual stimulus is well-illustrated by consideration of early studies which investigated CS changes amongst VDU users. In these studies, a broad range of spatial frequencies was tested and no adaptation was found. However, given a knowledge of the spatial frequency spectrum of text displayed on a screen, one would not expect a broad deficit to occur. When the stimulus was Fourier analysed, and subjects were tested at the spatial frequencies present in the stimulus, then adaptation *was* seen (Greenhouse, Bailey, Howarth and Berman, 1992).

The display within an HMD will generally be made up of a number of individual pixels, as it is on a VDU. However, it is likely that the image produced within the HMD will change its size and position frequently, unlike, say, the text produced on a VDU by a word-processor, and as a consequence the visual system will be exposed to a *constantly-changing* spatial spectrum. It is thus unlikely that *frequency-specific* adaptation will occur, and so CS was not included in our test battery.

Colour Vision

Although one might expect to see some colour vision changes, in the same way that VDU use can be shown to affect colour vision mildly and temporarily, there is no reason to suppose that any changes would differ from those seen with other displays. Although the range of colours presented to the user in a virtual environment is greatly restricted when compared with the range in the real world, this restriction in itself would not be expected to be detrimental to the users, in terms of inducing adverse *physiological* changes, and so colour vision tests were not included in our test battery.

Temporal aspects

In this section we need to consider the regeneration of the image, and the two main concerns are how often the screen is refreshed, and the lag between a head-movement and the subsequent alteration of the screen image position.

Flicker Perception

When one considers the temporal stimulus provided by the screen one might expect to find either a reduction in sensitivity at the refresh frequency, if the modulation is high, or no change in temporal sensitivity, if it is low or absent. The expectation will depend upon the technology involved in the display presentation. However, as it does not differ from that associated with the use of any electronic display, evaluation of temporal sensitivity was not included in the battery.

Time lag

The lag between a head movement and the update of the screen image produces a complex stimulus when considering the sensory conflict between the visual and vestibular systems. One might reasonably expect nausea to result, along with possible compensatory behavioural changes. The physiological effects one might expect to see involve modification of the vestibulo-ocular reflex (see Peli, 1996), but evaluation tests for these changes are complex and as such were inappropriate for inclusion in a test battery. This is an area which remains to be explored.

Binocular aspects

Most HMDs include two optical systems, one for each eye, and in considering how the brain copes with the two images we need to look at not only the sensory aspects of the joining together of the images, but also the sensory-motor aspects of eye position.

Heterophoria

Heterophoria is a condition of motor imbalance of the eyes where the active position of the eyes coincides with the fixation position, but the passive *fusion-free* position (seen when one eye is covered and visual feedback is prevented) deviates from it. There are two possible causal factors of heterophoria change in users of VR headsets: refractive error changes (see above) and prism adaptation caused by optical misalignment (see Howarth, 1997). If the equipment were *correctly* aligned, one might expect esophoric (inward) changes to be seen, accompanying instrument myopia. If the equipment were *incorrectly* aligned, the form of the heterophoria change would depend upon the direction of misalignment of the screen, the optical system, and the eye.

Heterophoria was measured for both distance and near vision during the HSE study, and changes were seen after the use of each of the three HMDs. These changes were not consistent, however, (see Howarth and Costello, 1996a) which can be explained by the differences between the optical systems of the HMDs.

Fixation Disparity

This test, at first sight, is similar to heterophoria in that it shows where the eyes are pointing, but it is a measure of very small changes in eye position when fusion is *allowed*. Although the test provides information about the fidelity of the ocular fusion system, many subjects with a fixation disparity are asymptomatic (Sheedy and Saladin, 1978) and the measure cannot be considered, in isolation, as a good objective measure of visual strain. We found little change in fixation disparity after HMD wear.

AC/A Ratio

When the eyes accommodate they automatically converge at the same time: this is termed accommodative-convergence (AC), and the amount of change per unit of accommodation (A) is termed the AC/A ratio. When the eyes view virtual images at different geometric distances from the eyes the normal relationship between accommodation and convergence could be disrupted (see Howarth, 1996), and, also, if accommodative ability were altered as a result of using the equipment, one might expect to see a change in the accommodation-convergence (AC/A) ratio.

The AC/A ratio was assessed by measuring heterophoria under different accommodative conditions. These were achieved by using ophthalmic lenses to alter the accommodation stimulus to the eye. The assessment revealed oculomotor changes after the use of each of the three VR systems, but, as with heterophoria, these changes were dependent upon the HMD used, and were not consistent across the systems.

Stereopsis

Absolute distance perception, and distance judgements, are essentially psychological issues, whereas relative distance judgements (which are needed for tasks such as threading a needle) are physiological issues. Although one might expect distance perception itself to be affected by immersion, one would not expect any *physiological* changes to occur within the stereoscopic visual system. Hence no measures of stereoscopic ability were included in the test battery.

The Value of the Approach

Throughout this paper, the approach has been adopted that tests should be performed on the basis of a rational *a priori* analysis of the stimulus presented to the eye. The value of this approach can be illustrated by considering results of the heterophoria changes seen during HMD wear (Costello and Howarth, 1996).

Optometrists and opticians will usually take great care whilst fitting spectacles (particularly those with higher power prescriptions) to ensure that the centre of each lens is positioned directly in front of the eye. At first sight, one would expect that the distance between the eyes (the inter-pupillary distance, IPD) would be a crucial issue because of the prismatic distortion introduced by not viewing through the centre of the high-power HMD lenses. However, analysis of the optic paths (Howarth, 1997) shows that if the images are at infinity, or at a far distance (Peli, 1996), then prismatic effects are dependent not upon the subjects IPD, but primarily on the distance between the two screens. Heterophoria changes seen after the use of an HMD are shown in figure 1,

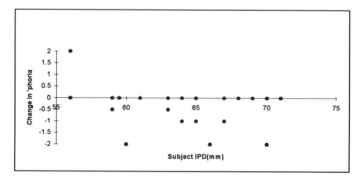

Figure 1. Change in distance heterophoria for subjects with different IPDs. (Virtual i-glasses HMD). A negative change indicates an exophoric shift.

taken from Costello and Howarth (1996). The headset used had no IPD adjustment, and from the measures of the image position described earlier one would expect that heterophoria changes (and any consequential visual discomfort) would be independent of IPD, which is the opposite of suggestions previously made (Regan and Price, 1993; Kolasinski, 1996). The distribution of data points suggests there is little relationship between the measures, and although there is a negative correlation between them it is weak: $y = - 0.0805x + 4.7375$ and $R^2 = 0.1485$.

These changes are to be expected, given the configuration of the optical system used in the HMD, and are a consequence of the hardware design rather than because a virtual environment was used. The result is as predicted from our analysis of the optical stimulus provided to the visual system, illustrating the value of our approach.

References

Bailey, I.L. and Lovie, J.E. 1976, New Design Principles for Visual Acuity Letter Charts *Am J Optom Physiol Opt* **53**, 740-745

Costello, P.J. and Howarth, P.A. 1996, *The Visual Effects of Immersion in Four Virtual Environments*, VISERG Report 9604 (HSE Contract 3181/R53.133)

Greenhouse, D.S., Bailey, I.L., Howarth, P.A. and Berman, S.M. 1992, Spatial Adaptation to VDT text *Ophthal Physiol Opt.* **12,3** 302-306

Howarth, P.A. 1996, Empirical Studies of Accommodation, Convergence, and HMD use *Proceedings of the Hoso-Bunka Foundation Symposium*, Tokyo, Dec.3 1996

Howarth, P.A. 1997, Oculomotor Changes within Virtual Environments *Submitted for publication*

Howarth, P.A., and Costello, P.J. 1996a, Visual Effects of Immersion in Virtual Environments: Interim Results from the UK Health and Safety Executive Study *Society for Information Display International Symposium Digest of Technical Papers, Volume XXVII*, San Diego, 12-17 May 1996, 885-888

Howarth, P.A., and Costello, P.J. 1996b, The Nauseogenicity of Using a Head-Mounted Display, Configured as a Personal Viewing System, for an Hour *Proceedings of the F.I.V.E. Conference* Pisa, Italy, Dec. 19-20 1996

Kolasinski, E.M. 1996, Prediction of Simulator Sickness in a Virtual Environment *PhD Thesis*, Dept. Psychology, University of Central Florida, Orlando, USA.

Miyake, S., Akatsu, J., Kumashiro, M., Murakami, T., Hirao, H., and Takayama, M. 1994, Effects of Virtual Reality on Human *Proceedings of The 3rd Pan-Pacific Conference on Occupational Ergonomics* Nov. 13-17, 1994. Seoul, Korea

Ong, E. and Ciuffreda, K.J. 1995, Nearwork-Induced Transient Myopia - a Critical Review *Documenta Ophthalmologica* **91**, 57-85

Peli, E. 1996, Safety and Comfort Issues with Binocular HMDs *Proceedings of the Hoso-Bunka Foundation Symposium* Tokyo, Dec. 3 1996

Regan, E.C. 1995, An Investigation Into Nausea and Other Side-Effects of Head-Coupled Immersive Virtual Reality *Virtual Reality* **1,1** 17-32

Regan, E.C. and Price, K.R 1993, Some Side-effects of Immersion Virtual Reality: An Investigation Into the Relationship Between Inter-pupillary Distance and Ocular Related Problems *APRE Report 93R023*

Sheedy, J.E. and Saladin, J.J. 1978, Association of Symptoms with Measures of Oculomotor Deficiencies *Am J Optom & Physiol Opt* **55,10** 670-676.

ORGANISATIONAL ERGONOMICS

The Management of Stress

Taylor Bourne

Glasgow Caledonian University
Department of Risk and Financial Services
Britannia Building
Cowcaddens Road
Glasgow, G4 0BA

Absenteeism and reduced work performance resulting from stress cost British industry in excess of £6 billion annually. In addition, it has been estimated that stress-related illness directly causes the loss of 40 million working days each year. Under recent legislation employers have a duty to assess and prevent workplace stress as a health hazard.

Radical changes to employment practices combined with smaller workplaces and more pressure to perform have resulted in stress being an inevitable part of organisational life. When the costs associated with lost production and potential liability are taken into account it makes economic sense for employers to consider monitoring and prevention.

This paper explains what causes stress and outlines how it can be managed and reduced.

Definition of Stress

The word "stress" is used to describe a wide range of feelings of ill-health, frequently arising from a very active pace of life. Stress may also be thought of as the 'fit' between people and their environment.

Stress arises when the physical and emotional demands on the person do not match their resources to cope with them. When people have demands placed upon them that are beyond their ability to cope they experience stress and the body reacts with a primitive 'flight or fight' response. This involves increased heart rate, the release of adrenaline and other hormones into the blood stream, increased blood sugar levels, and tensing of the muscles.

The individual's response to stress is subjective and may vary considerably depending on health, social, domestic and employment factors. Stress has both positive and negative sides. For example, on the positive side a certain level of stress may be required and enjoyable in order to get going. In the workplace some pressure can be a good thing. The task and challenges we are faced with at work can be beneficial and stimulating. Stress can raise levels of energy, drive and productivity, qualities that are required for successful performance. However, there is a limit to the amount of stress with which people can cope. If workplace pressure is too high, the resulting stress can be harmful to business performance and also the health of the workforce.

Harmful stress has been defined by Jee and Reason (1992) as 'an excess of demands on an individual beyond their ability to cope.' For example, it may be that a task is too difficult or challenging for an individual's capabilities. Conversely, just as too much demand can result in stress, so can an absence of stimulation and challenge. In other words, the task may not be challenging enough. Table 1, HSE (1995), illustrates how an employee copes with different amounts of pressure.

Table 1

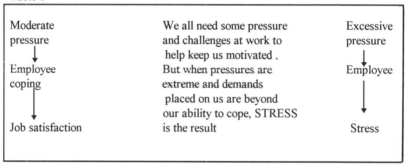

People's ability to cope with stress varies, but the idea is not to think of one worker as 'weaker' than another. Instead it should be as GMB (1994) propose 'how can the fit between the worker and the work environment be improved?'

The Causes of Stress

It is difficult to forecast what will cause harmful levels of stress. Stress means different things to different people. A situation which is intolerable to one person may be stimulating to another. A person's response will depend on their personality, experience and motivation and how much help they receive from others including managers, colleagues, family and friends.

People may experience stress from a variety of sources including their home and work environments. Table 2, GMB (1994), shows the main causes of stress at home.

Table 2 The main causes of stress at home

- The death of a partner or close relative
- Divorce or separation
- Organising a marriage
- Change in the health of a family member
- Pregnancy
- The death of a close friend
- Taking on a large mortgage or other large debts
- Moving house
- Arranging holidayChristmas/family get-together.
- The unfair division of household tasks or single parenthood can be another source of stress which is mainly experienced by women.

From the individual's perspective it is important that they know what causes stress in their home environment and how this stress can affect their health. This will put them in the position to be able to modify their lifestyle so that they can decrease the amount of stress they experience and to escape ill health.

From an employer's perspective, stressors at home are outside the scope of their responsibility. However, good employers may want to know about them, because employees may become more susceptible to stress at work as a result of them.

According to the HSE (1995) harmful levels of stress are most likely to occur where:

- pressures <u>pile on top of each other</u> or are <u>prolonged</u>
- people feel <u>trapped</u> or unable to exert any control over the demands placed on them
- people are <u>confused</u> by conflicting demands made on them

In terms of the workplace the above situations will probably have more of an affect on clerical and manual staff than higher management. GMB(1994) have highlighted the following as the main causes of work related stress:-

a. <u>The work environment</u>
- noise/vibration
- poor lighting
- poor ventilation
- incorrect temperature
- dust
- toxic fumes and chemicals
- overcrowding
- open-plan offices
- badly designed furniture
- poor maintenance
- poor canteen or rest facilities
- poor childcare facilities

b. Job design
- exposure to violent or traumatic incidents
- boring or repetitive work
- the under utilisation of skills
- too much or too little supervision
- job isolation
- constant sitting
- change or the pace of change
- working with VDUs and other machinery

c. Contractual problems
- job insecurity
- low pay
- shift work
- excessive hours
- flexitime and annualised hours
- insufficient meal breaks
- lack of adequate rest breaks

d. Relationships
- harassment and discrimination
- bad relations with supervisors
- working with the public
- customer or client relationships
- impersonal treatment at work
- lack of communication.

The effects of stress

A little stress is good for the body and alerts the mind. However acute, constant or prolonged stress can be harmful to the mind and the body. As stress starts to take its toll physically, emotionally and behaviourally, a variety of symptoms can result. Tables 3 and 4, GMB (1994), list the main symptoms and diseases associated with stress.

The symptoms experienced by each individual will vary. In any group of workers undergoing high levels of stress some will suffer less harm than others. There will be others who will display hardly any effects. Nevertheless all of the symptoms listed in the tables are known to be the result of or exacerbated by stress.

Table 3 Short term stress symptoms and diseases

Physical	headaches, muscle tension, backache, neck ache, chest pain, palpitations, indigestion, poor sleep, nausea, dizziness, skin rashes, eye problems, period problems, impotence, excessive sweating, rapid weight loss or gain.
Behavioural	relationship strains, poor work performance, absenteeism, heavy smoking, heavy drinking, use of tranquillisers, abuse of other drugs, indecision, unusually impulsive, unusually emotional, accidents, restlessness.
Emotional	anxiety, irrationality, boredom, tiredness, fear and panic, nightmares, loneliness, cannot concentrate, low self-esteem, depression.

Table 4 Long term stress symptoms and diseases

Physical	headaches, hypertension, high blood pressure, poor general health, asthma, peptic ulcers, dermatitis, diabetes, gynaecological problems, gastro-intestinal problems
Behavioural	marital breakdown, family break-up, social isolation, alcoholism, other addictions, suicide.
Emotional	insomnia, neurosis, chronic depression, chronic anxiety, nervous breakdown, mental breakdown

The legal position relating to stress

In the United Kingdom the legislation most pertinent to the control of stress at work is as follows:-

- Under Section 2 of the Health and Safety at Work Act, 1974 employers are required to ensure the health, safety and welfare at work of all employees
- The Management of Health and Safety at Work Regulations 1992 require employers to assess the risks to health and safety in their workplace and to put in place suitable avoidance and control measures.

Research has shown that stress is a major cause of work-related ill health and it increases the probability of accidents at work. Ill health arising from stress at work should be handled in the same way as ill health arising from other, physical causes present in the workplace. The significance of this is that employers have a general duty to ensure that the health of their employees is not placed at risk through excessive and sustained levels of stress arising from sources in the workplace. Stress is a workplace health hazard and should be borne in mind by employers when carrying out risk assessments. Therefore employers should identify the risks of stress arising and put in place measures to prevent or decrease it.

The prevention and control of stress

Occupational stress has been widely publicised. Despite this, the majority of employers only possess a vague awareness of the problems associated with stress in the workplace. A number of myths and problems with stress have to be overcome. First of all, the myth that stress is a problem only for high flying executives has to be dispelled. Secondly, and perhaps even more difficult to overcome is the concept that stress is an individual's own problem. Going hand in hand with this idea are arbitrary management decisions on whether one worker is 'weaker' than another and a bias towards personal management.

Those few employers who do accept stress as a problem generally seek solutions in terms of educating their workers to 'manage' stress instead of endeavouring to deal with the problem at source. GMB (1994) believe the only truly effective way of preventing harmful stress at work is to tackle its root causes. They propose that stress should be dealt with in the same manner as any other potential health and safety problem. Firstly, the hazard must be identified (establish the causes of stress in the workplace) then, in order of preference:-

- eliminate the hazard (remove the causes of stress in the workplace)
- control the risk (reduce the amount of stress experienced by individual workers or groups of workers)
- provide the means for protecting individuals (training in stress management techniques)

The HSE (1995) provide advice and guidance to employers on how to deal with stress at work. In relation to attitudes to stress, employers need to set in motion procedures to ensure that the problem is understood and taken seriously by managers, staff and the organisation as a whole. Regarding the job itself, this needs to be 'do-able' and should be matched with the abilities and motivations of the person in it. Management style should be such that it avoids inconsistency, indifference or bullying.

The HSE also list the following things which can help reduce or remove unnecessary stress:
confidence, competence, consistency of treatment, good two-way communication combined with
some flexibility, scope for varying working conditions, open attitude by managers, treating people fairly, providing staff with the necessary skills, training and resources.

Action on stress can be cost-effective and the HSE list the following formula for pay off of reducing stress:

reduced stress	=	better health	+	reduced sickness absence	+	increased performance and output	+	better relationship with clients	+	lower staff turnover

References

GMB, 1994, *Stress at Work (A GMB Guide)*, (GMB, London).
HSE, 1995, *Stress at Work (A Guide for Employers)*, HS(G) 116, (HSE Books, Sudbury, Suffolk).
Jee, M and Reason, L 1992, *Action on Stress at Work*, (Health Education Authority, London).

A REVIEW OF METHODS FOR SAFETY CULTURE ASSESSMENT: IMPLICATIONS FOR ERGONOMICS

Richard Kennedy and Barry Kirwan[1]

Industrial Ergonomics Group
School of Manufacturing & Mechanical Engineering
University of Birmingham
Edgbaston Birmingham B15 2TT

A review of the currently available safety culture assessment methods showed that there is often only a limited treatment of ergonomics in these approaches. As ergonomics is known to have a significant effect on health and safety, suggestions are made on how safety culture assessment approaches may better address ergonomic factors.

1. Safety Culture and Risk Potential

The concept of 'safety culture' evolved after a number of major accidents in the 1980's, in particular the Chernobyl nuclear accident. These accidents were not attributable to engineering failures alone and so safety culture was used as an explanative term for the wider social aspects of the organisation [such as individual and group attitudes, values, beliefs and work practices] which effect system safety. Safety culture is significant as it will be the dominant influence on the risk potential of the system (Kennedy and Kirwan, 1995). Therefore approaches and techniques for assessing safety cultural factors have needed to be developed. Although safety culture will determine the system risk more than basic ergonomics, it will however influence how much ergonomics is built into a system and whether error / incident information is analysed causally and acted upon. Therefore this paper reviews the approaches of the currently available safety culture assessment methods and asks whether the assessment of ergonomic factors is built into their questioning frameworks.

2. Safety Culture Assessment Methods

A review of the safety management literature identified 15 methodologies for safety culture assessment [for a definitive review of the area and references for the techniques see Hale and Hovden, 1996]. The methodologies are shown in Table 1.

---

[1] ATMDC, National Air Traffic Services Ltd, Bournemouth Airport, United Kingdom.
The contents of this paper are the opinions of the authors, and do not necessarily reflect those of their respective companies.

Table 1. Review of methods for safety culture assessment

Technique & Developmental Origins	Approach & Rationale	Inputs / Outputs
British Council 5 Star System: Developed to be applied in a range of industries including mining, manufacturing and offshore oil and gas.	A qualitative approach which assesses the adequacy of the procedures and workplaces to manage safety.	The audit consists of a checklist of weighted questions which determine the star rating given to the organisation. A 5 star rating (91-100%) is excellent and a 1 star (40-50%) represents basic safety standards.
CHASE [Complete Health and Safety Evaluation]: Developed for use in a number of industries and to different sizes of companies.	A qualitative approach which concentrates upon the identification and control of risk within the organisation.	Weighted questions are divided into 12 different sections and require 'yes' or 'no' answers. The overall score allows the organisation to identify problem areas and those needing improvement.
INSAG [International Nuclear Safety Advisory Group]; ASCOT [Assessment of Safety Culture in Organisations Team]; ACSNI [Advisory Committee on the Safety of Nuclear Installations]: Developed as prompt lists for use in nuclear installations.	Three similar qualitative approaches attempting to identify the safety culture elements that influence organisational activity	ASCOT requires discussions with government and regulators, senior managers of the organisation and plant visits. In all three of the approaches, the organisational levels assessed are individual, management, regulatory or government, corporate and supporting organisations. The output is a collection of 'indicators' or 'factors' that effect the safety culture and in turn how work is carried out.
ISRS [International Safety Rating System]: Developed for use in the mining industry and subsequently applied to the chemical processing industry.	A qualitative approach in which accidents are considered to be caused by a lack of management control.	ISRS comprises 600 questions arranged into 20 main subject groups. The system provides five stars as classification indices of achievement. A plant with a 1 star rating requires major improvement, whereas a 5 star rating is given for excellent performance.
MANAGER [MANagement Guidelines in the Evaluation of Risk]: Developed for use in the process industries, in an attempt to incorporate organisational factors into PSAs.	A quantitative approach which reviews management controls and their effects on the frequency of occurrence of major events.	A qualitative as well as quantitative technique, consisting of 114 questions divided into 4 categories. The scoring system is based on the norm for the industry. Responses are categorised as 'better than average' 'average' or 'worse than average' to calculate the management [modification] factor.
MORT [Management Oversight Risk Tree]: Developed as an accident investigation tool.	A qualitative approach in which accidents are considered in terms of the failed 'safety function' of the organisation using a normative model of safety management.	Details of the accident and the system are collected via detailed interviews and analysed via a generic safety management fault tree. MORT delivers specific failures which caused the accident, or safety management failures which allowed it to happen. Means of rectifying the situation are usually apparent for the analyst.

NOMAC *[Nuclear Organisation and Management Analysis Concept]:* Developed for use in Nuclear Power Plants (NPPs) in the USA.	A qualitative approach in which the components of organisational structure are linked to key management and supervisory roles.	Claimed to be able to explain processes in NPPs via the use of functional analyses, questionnaire of safety culture attitudes and observation. The consistency of taxonomy use is claimed to allow the ability of the method to discriminate organisational activities both within and between different NPPs.
OSTI *[Operant Supervisory Taxonomy and Index]:* Developed for use in organisations where managerial activities are important (if not crucial).	A qualitative approach in which safety management adequacy is judged through assessing managers activity.	Seven categories of supervisory behaviour give an idea of the proportion of time spent carrying out certain types of task which can effect safety performance. Interventions concerning revisions of the time spent on each category can be suggested.
PRIMA *[Process Risk Management Audit]:* Developed for use in the chemical industry to assess loss of containment and the factors which effect the frequency of these events.	A quantitative approach which uses a 'model' of accident causation (socio technical pyramid) to underpin a quantitative audit scheme.	The audit contains eight audit areas which can produce qualitative and quantitative outputs on the adequacy of the safety management. In the quantitative assessment, a modification factor is produced which is used to modify the generic failure rate data used in the risk assessment.
PRISM *[Professional Rating of Implemented Safety Management]:* Developed from UK safety legislation and best practice in safety management in a number of industries.	A qualitative approach which reviews effectiveness of safety management against best practices.	The audit consists of interview and inspection and is split into 10 basic topics, each addressing a major element of safety management, and 4 line management levels. The scoring system is 'yes' or 'no' with a scale of 0 to 3 where judgement is required. Quantitative results and qualitative recommendations are provided.
SCHAZOP *[Safety Culture HAzard and OPerability study]:* Developed for use in nuclear installations although can be applied to other industries.	A qualitative group based approach which is used to identify the influence of safety culture factors on safety management processes.	The SCHAZOP guide and property words are applied to a graphical representation of the safety management process. Safety management vulnerabilities are identified and influencing factors suggested.
TRIPOD: Developed by research at Leiden and Manchester Universities for use in the oil industry (has been modified for the aviation and rail industries).	A qualitative approach in which the adequacy of the 'organisational health' of the company is assessed.	TRIPOD gives details of Generalised Failure Types (GFTs) and their associated indicators. It uses interviews and subjective ratings of operations personnel to judge the significance of failure types and where safety effort should be directed.
WPAM *[Work Process Analysis Model]:* Developed for use in USA NPPs for inclusion in Probabilistic Safety Assessments (PSAs). It uses much of the method and modelling of PSAs.	A quantitative approach in which organisational factors are explicitly incorporated into PSAs.	An organisational factors (OF) event tree is constructed for each key work process and an OF fault tree for each unsafe act and breach of the barrier. An organisational factors matrix is defined for each key work process. The information is used to identify tasks which allow the events to occur and to identify common organisational failures that may contribute to the recurrence of the identified events.

From Table 1, the four dominant trends in safety culture assessment method development are identified as being the following:

• *Quantitative integration into risk assessments* [MANAGER, PRIMA and WPAM]. These approaches attempt to derive a safety culture numerical factor which is used to affect the total risk determined by the risk assessment.

• *Consolidation of stand-alone scored audit approaches* [5* System, CHASE, INSAG-ASCOT-ACSNI, ISRS, PRISM]. These approaches address the organisational contribution to risk qualitatively, outside of the risk assessment framework. An overall measure of the acceptability of the system and areas requiring improvement is derived.

• *Identification of fundamental general problem areas* [NOMAC and TRIPOD]. These approaches identify general problem areas and associated indicators but utilise a notional framework of the organisation to guide the data collection.

• *Modelling and analysis of safety management system activities* [MORT, OSTI, SCHAZOP]. These approaches attempt to model or show the dynamics of safety management activities and the qualitative influence that they have on the overall system risk.

3. Linking Safety Culture and Ergonomics

Safety culture assessment approaches are tending towards evaluating safety culture in the form of the human error effect on system risk. In many cases, the safety culture assessment will raise issues about the ergonomics adequacy of the system. However, although many companies may have an ergonomist, they are often not used for addressing these concerns. This is a safety culture problem in itself as safety culture will determine whether the ergonomist is brought into a project or whether they are blocked out. Therefore an alternative way of making sure that the ergonomics issues are addressed is to integrate ergonomics more effectively into safety culture assessment approaches. How this can be done will now be suggested.

4. Safety Culture Assessment and Ergonomics

Some of the assessment approaches, such as MORT, incorporate assessments of the adequacy of the ergonomics of a system, and / or the degree to which human error concerns have been evaluated and addressed (e.g. via Human Reliability Assessment). However, the approaches generally tend to deal with ergonomics in a piecemeal fashion. They ask some good questions but miss out many others due to: i) the breadth of ergonomics itself; ii) the necessity to restrict the audit set size to a manageable size; and, iii) not to give ergonomics disproportionate significance compared to other areas. Whilst such an approach may detect problems, it is entirely conceivable that they may fail to detect inadequacies even at a general level. This is because ergonomics, by its nature, tends to be testable only in a detailed contextual way. It is often the detailed interactions between system and user, and the way the interface deals with these interactions, that makes the difference between a 'good' and 'poor' interface. A superficial ergonomics review can be worse than no review at all, as it might say that the interface is fine when it is not. The opposite situation could also arise where radical change of an interface which may look poor [but is actually very supportive of operator needs] may be suggested.

Given that audits most often need to be carried out in a matter of days, and that there may be 10-20 areas of audit of which ergonomics may be only one, it is at first sight difficult to see how the issue may be resolved. However, given the contextual nature of the audits themselves, an approach can be borrowed from safety management and Total Quality Management (TQM). Theory and practice in these domains suggests that *'what is at the top of company's priorities is what counts'*. This accounts for the fact that when asked, all companies will currently state that safety is their number one priority. Whether this is actually true is another matter, but the truth of the statement can be verified by determining in practice whether safety is prioritised via such factors as budget and resource allocation and projects on safety issues. This 'higher-level review' approach can be applied to ergonomics. Therefore, rather than asking key detailed questions such as whether the desk height is 720mm, looking for footrests and the absence of visible flicker on VDU screens etc.[2], questions such as those in Table 2 could be asked.

Table 2. Possible questions for the ergonomics component of an audit

<table>
<tr><td colspan="2" align="center">**Audit Question**</td></tr>
<tr><td>1</td><td>Is there a Human-Machine-Interface (HMI) design guide that the company works to, with mandatory and advisory recommendations on ergonomics design aspects?</td></tr>
<tr><td>2</td><td>Is the HMI design guide applied to all projects?</td></tr>
<tr><td>3</td><td>Does the company have a model of the system user?</td></tr>
<tr><td>4</td><td>At what stage does ergonomics enter the project design life cycle?</td></tr>
<tr><td>5</td><td>Are there people with approved ergonomics qualifications within the company doing ergonomics work? What level in the organisational company structure are they at? If not, what contractor assistance is used, and who and at what level is the company officer dealing with such issues?</td></tr>
<tr><td>6</td><td>What proportion of the company R&D budget is allocated to ergonomics and human error issues? Who decides this proportion and on what basis?</td></tr>
<tr><td>7</td><td>Are all incidents analysed for their human error contribution? Are these errors classified causally and are trends computed? What is then done with such information? Is such information used to inform the R&D programme? Cite examples of a human error problem that has been identified and successfully resolved.</td></tr>
<tr><td>8</td><td>How is the ergonomics work linked to the safety and reliability work?</td></tr>
<tr><td>9</td><td>What is the relationship of the company with the human factors inspector or regulator?</td></tr>
<tr><td>10</td><td>How does the company keep up to date on ergonomics issues and advancements?</td></tr>
</table>

The questions in Table 2 are therefore operating at a high level, questioning fundamentally whether the company is dealing with ergonomics matters strategically and proactively, whether it is acting in a tactical way, or whether it is ignoring the key issues partly or even entirely. The questions shown in Table 2 are appropriate to a company with significant ergonomics needs. However, they could be 'scaled down' for other companies. Human error is generic across all industries: *all companies should protect themselves from human error, if not for safety reasons, then for financial reasons.*

---

[2] *These type of questions obviously need to be asked when carrying out a specific audit to comply with Display Screen Equipment Regulations.*

The safety management assessment of whether ergonomics is receiving adequate (and appropriate) attention in a company can also be addressed via a more open-ended set of three questions based on the Hale et al's (1997) three-level model of safety management (planning, procedures and execution levels).

• **Planning**: the analysis of ergonomics needs for the company which requires analysis of and action on the results of the following:

- *the prevalence and consequences (risk) of human error in operations;*
- *the importance (criticality) of the operator (and maintainer) roles;*
- *legal requirements and applicable standards;*
- *the state of the art in comparable companies / industries.*

• **Procedures**: the formalisation of recurrent tasks and available data:

- *detailed human factors design guidance data and principles;*
- *formal methods descriptions of task analysis methods used;*
- *interface design evaluation approaches;*
- *Human Reliability Assessment (HRA) approaches;*
- *incident analysis approaches and taxonomies etc.*

• **Execution**: evidence in the form of:

- *reports (including R&D strategies and programmes);*
- *changed designs;*
- *budget accounts;*
- *staff in post and publications;*
- *participation of ergonomists in design and safety team meetings.*

Both these question sets are analysing the extent to which ergonomics has been integrated into the company safety management infrastructure. These approaches may therefore fit more neatly into conventional safety audits. They may also be more effective, within the structure of safety audits, at determining the priority of ergonomics within the company's operating philosophy.

5. Conclusions

This paper has reviewed a number of methods for organisational safety culture assessment. Currently the assessment of ergonomics is often neglected or is treated in a piecemeal way in many of these approaches. Therefore ideas for assessing ergonomics as part of an overall audit approach have been suggested.

References

Hale, A.R., Heming, B.H.J., Carthey, J. and Kirwan, B. 1997 (in press), Modelling of safety management systems. *Safety Science.*

Hale, A. and Hovden, J. 1996, Management and culture: the third age of safety. A review of approaches to organisational aspects of safety, health and environment. Paper presented at a the *Worksafe International Symposium on Industrial Injuries.* Sydney, Australia, February 26-27.

Kennedy, R. and Kirwan, B. 1995, The failure mechanisms of safety culture. Paper presented at the *International Topical Meeting on Safety Culture in Nuclear Installations.* Vienna, Austria, April 24-28.

ORGANISATIONAL VARIABLES IN CALL CENTRES: MEDIATOR RELATIONSHIPS

Kevin Hook
The Decisions Group, Mitre House
Canbury Park Road, Kingston, KT2 6LZ

Lana Matta
Goldsmiths College,
University of London, SE14 6NW

This study explored the role of perceived satisfaction of the environment as a mediator variable in the relationship between seven objective organisational variables and individual affective outcomes (stress and job satisfaction). Eleven organisations were involved in the study (n = 566) focusing specifically on telephone call centres. Using analysis of covariance, mediated, partially mediated and direct effects were observed. Practical implications are discussed. Theoretically, the results highlight the validity of developing a more complex conceptual link between the environment and its impact upon individuals.

Introduction

Researchers have viewed the work environment as conceptually divided into objective characteristics and the individual's experience of the environment. For example, Archea (1977) makes the distinction between properties and attributes. Properties are the physical, ergonomic or architectural characteristics of a work setting (e.g. size of office, air conditioning, lighting); attributes are individuals' attitudes and perceptions of those properties.

Objective physical characteristics or properties of the work environment can be seen to have direct and indirect effects on the individual (Wineman, 1986). One attempt that has been made to understand the relationships between direct and indirect effects has been to explore moderator and mediator relationships (Carlopio & Gardner, 1996; 1992). By examining the relationship between job level and affective outcomes (e.g. job satisfaction, intention to turnover), they illustrated the significant effects of employee perceptions of the workplace as mediators of affective outcomes. They highlight the need for a richer conceptualisation of both work and physical setting characteristics that influence employee perceptions.

In addition, Carlopio & Gardner (1992) claim that existing research has examined a limited set of objective characteristics of the physical work environment and their relationships to possible outcome variables. This view has also been postulated by Marans & Spreckelmeyer (1986), who claim that there has been a lack of carefully developed conceptual links between physical environmental attributes and individuals' response to those attributes.

Therefore, the current study is designed to explore how individuals' perceived satisfaction with the workplace mediates the relationship between seven distinct objective physical characteristics of the workplace and affective outcomes in terms of job satisfaction and stress.

Diagram one illustrates the model that will be evaluated in the current study, following Ferguson & Weisman (1986), with the underlying assumption that one's perceived satisfaction with the work environment will mediate outcome variables.

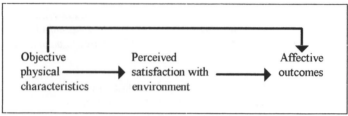

Diagram One: Model to be tested.

The specific work environments selected for the current study are telephone call centres. Growth in direct forms of customer communication have led to the development of groups of individuals whose sole purpose is taking telephone calls from customers or enquirers. The call centre is not a typical, traditional office environment. Employees work predominantly with a VDU or computer, talking to customers by phone up to 80% of their time. The office may be open twenty four hours a day, and employees are constantly available to talk to customers. Participants in the current study were drawn from eleven call centres across the UK. All were of the same job level (call centre agents) undertaking the same job function. Seven objective physical characteristics relevant to the call centre environment were chosen.

1) *Do agents have their own desk.* As many call centres run multiple shifts, it is a common practice for individuals not to have their own desk. In the current sample, three of the eleven organisations followed this practice.

2) *Use of partitions.* Research on partitions has been used as a measure of office openness and enclosure (Ferguson & Weisman, 1986; Oldham & Rotchford, 1983). Call centres generally are located in open-plan offices, with a large number of people in the same office. Partitions around each workstation are common, often made from sound absorbent material. From the existing sample, three distinct groups emerged; (a) those without any partitions between desks (three organisations); (b) those with "cut-away" partitions that are no more than 18 inches in height, thus allowing some direct eye contact with colleagues (four organisations); and (c) those with "full" partitions that are above 18 inches in height both in front and to the side of the individual, virtually encasing the individual in a booth (four organisations).

3) *Number of workstations per group.* Workstations in call centres are grouped together to optimise available space; a consequence being that colleagues become an immediate source of social support. In the current study, three different workstation configurations were observed; (a) groups of four, sitting in a square (four organisations); (b) groups of eight, typically sitting in two rows of four workstations with their backs to each other (four organisations); and (c) groups of ten, in a variety of different configurations (three organisations).

4) *Type of lighting.* Previous research has highlighted the impact of inappropriate lighting on performance, comfort, satisfaction and emotional strain (Sutton & Rafaeli, 1987; Ellis, 1986; Knave, 1984). In the current sample, two types of lighting were used; (a) traditional louvered overhead fluorescent strip lighting (nine

organisations), and (b) up-lighting, designed to minimise glare from PC screens (two organisations).

5) *Type of desk.* Ergonomic furniture has been suggested to affect a range of physical and mental outcomes (Carlopio & Gardner, 1992). From the current sample, two types of desks were observed to be used (a) traditional rectangular desks (seven organisations), or (b) ergonomically designed desks which are curved, allowing the user to sit comfortably at the desk with the PC directly in front, (four organisations).

6) *Purpose built.* A relatively unexplored variable in previous research relates to whether or not the office environment was specifically designed or refurbished for its current purpose. In the context of the current study, seven organisations were in purpose -built environments, whilst four organisations were using offices which were not specifically designed to be call centres.

7) *Number of offices.* The number of discrete open-plan offices that were used as call centres by the organisation. A total of six of the organisations sampled used a single office, the remaining five organisations used more than one office.

Method

Sample

Data was collected from 566 call centre agents from eleven different organisations across the UK. The age range of the sample was 17 to 60 years with a mean age of 29.96 years. The sample was 39.5% male and 60.5% female. Employees had worked in their current department for between 1 month and 308 months, with a mean of 26.51 months.

Procedure

Department managers were approached, the study purposes and procedures were explained, and their call centre's participation was requested. A three page questionnaire was sent to all call centre agents (the individuals taking calls), including an explanatory note assuring them of confidentiality and anonymity. Completed questionnaires were returned in a sealed envelope. The research team then visited each call centre, and the objective organisational variables were collected at this stage.

Measures

The following measures were used in this study. Scale reliabilities (Cronbach's alpha), items per scale, the number of valid responses and means for each scale are shown in table one.

Perceived satisfaction with the environment. A variation on the Human Factors Satisfaction Questionnaire (Carlopio, 1986) was employed as a measure of physical work environment satisfaction. Using factor analysis (VARIMAX rotation), two orthogonal factors emerged: perception of workstation and work organisation (15 items), and perception of ambient environment & facilities (12 items).

Anxiety-Stress Questionnaire: Developed by House and Rizzo (1972) to measure the existence of tensions and pressures growing out of a job. It contains three sub-scales: job induced tension, somatic tension and general fatigue. Two items were dropped from the original seventeen item version due to adverse comments from respondents regarding items' face validity.

Job Satisfaction: A specially constructed eleven item questionnaire, covering intrinsic and extrinsic factors relating to job satisfaction.

Table one: Scale reliabilities.

Scale name	Items	Cronbach's alpha	N	Scale mean
1. Perceptions of workstation & work organisation	15	.793	483	44.93
2. Perceptions of ambient environment & facilities	12	.424	397	38.87
3. Job Induced Tension	7	.539	531	1.32
4. Somatic Tension	4	.915	560	1.34
5. General Fatigue	4	.925	537	1.26
6. Job Satisfaction	11	.691	542	32.74

Results

Initial correlations between background variables, mediator variables (the two perceived satisfaction with environment variables), and outcome variables (job induced tension, somatic tension, general fatigue and job satisfaction) are shown in table two below.

Table two: Correlation matrix.

	1.	2.	3.	4.	5.	6.	7.	8.
1. Age								
2. Gender	.1987**							
3. Tenure	.4415**	.1998**						
4. Perceived satisfaction with workstation & work organisation	-.1690**	-.0886	-.4202**					
5. Perceived satisfaction with ambient environment	-.0015	.0302	-.2766**	.7403**				
6. Job Induced Tension	.1476**	-.0206	.2062**	-.3876**	-.3886**			
7. Somatic Tension	.0634	.0308	.2190**	-.4178**	-.4098**	.6016**		
8. General Fatigue	.0808	.0258	.2800**	-.4254**	-.3363**	.4111**	.4697**	
9. Job Satisfaction	-.0602	-.0394	-.3300**	.7176**	.6227**	-.4231**	-.4172**	-.

** $p < .01$

As shown above, tenure correlates significantly with all four outcome measures. Therefore, to control for it effect statistically, tenure was included as a covariate in all subsequent analyses. Summary results are reported for each of the seven organisational variables. The approach used follows Carlopio & Gardner's (1995) method of isolating the mediating effect of variables using analysis of covariance (ANCOVA). A two-stage analysis was adopted; analysis of covariance using organisational variables as main effects and tenure as a covariate; followed by a further ANCOVA with the two perceived satisfaction with the environment variables as additional covariates (details of analyses can be provided by the first author).

If the hypothesis of perceived satisfaction with environment acting as a <u>mediator</u> was to be supported, a significant main effect (p < .05) would be found at step one; at step two, the covariates would be significant predictors of outcome, while the main effects should disappear. A main effect found both at stage one and stage two of the analysis would indicate a <u>direct effect</u> of the organisational variable on affective outcome variables. A main effect found at stage one and a trend found at stage two (p < .1) indicates a <u>partially mediated relationship</u>, where some of the variance in outcome can be accounted for by the mediator variables, but the influence of the objective physical characteristics still has some direct effect.

In an attempt to give an holistic view of the analysis, table three summarises the results of the various ANCOVAs across the seven organisational variables, for each of the four outcome variables.

Table three: Summary of analyses.

Organisational Variable	Job Induced Tension	Somatic Tension	General Fatigue	Job Satisfaction
1. Own desk	*Mediated*	✗	✗	✗
2. Use of partitions	*Direct*	*Mediated*	*Mediated*	*Mediated*
3. Number of workstations in group	*Direct*	✗	*Direct*	*Mediated*
4. Type of lighting	*Partially mediated*	*Mediated*	✗	*Mediated*
5. Type of desk	*Partially mediated*	*Mediated*	*Mediated*	*Mediated*
6. Purpose built	✗	✗	✗	*Mediated*
7. Number of offices	*Direct*	✗	✗	*Direct*

✗ = No direct effect observed at stage one

Discussion

The current study had two principal objectives; to explore the role of perceived satisfaction with the environment as a mediator variable in the relationship between quantifiable organisational variables and individual affective outcome variables; and to review the specific impact of organisational variables once the influence of mediator variables have been accounted for.

With regard to the first objective, as can be seen from table three, eleven significant mediator relationships were observed, across all four outcome variables. In these instances, significant differences in outcome variables due to organisational variables disappeared once environment satisfaction variables were included as covariates. In addition, two partial mediator relationships were recorded, where significant main effects were reduced to a trend ($p < .1$) upon the inclusion of mediator variables. This provides clear support for the mediating role of perceived satisfaction with the environment (Carlopio & Gardner, 1992; Oldham & Rotchford, 1983).

Five direct effects were observed, where mediator variables did not alter the relationship between organisational variables and outcome variables. It can be seen, therefore, that direct, partially mediated and fully mediated relationships can exist.

Focusing on the second objective of the study, a number of tentative conclusions can be drawn. Direct effects for three variables were observed after statistical control of mediators (partitions, number of workstations in group, number of offices).

Firstly, the use of either full partitions or no partitions seems to results in similar high levels of job induced tension. Given the potentially monotonous nature of call centre work, producing an environment which facilitates social contact whilst still providing a sense of personal space would appear an optimal arrangement.

Secondly, looking at the number of workstations in a group, analysis of the means shows that groups of eight were significantly higher in general fatigue and lower in job satisfaction. No conceptual reason for this result can be given; it may be an artefact of the typical workstation arrangements observed in those call centres with eight workstations per group - that agents were in straight lines with their backs to each other.

Having the call centre split over more than one office seemed to result in higher levels of job induced tension, but also higher levels of job satisfaction. Perhaps having more than one office produces smaller, more cohesive groups of agents (therefore, higher job satisfaction) but also negatively influences perceptions of privacy, crowding and distractions which may influence job induced tension.

Partially mediated effects were observed for two additional organisational variables (lighting and desk type), showing that the use of up-lighting and ergonomically designed desks was associated with significantly lower job induced tension.

Due to the design and analyses used in the current study, it is impossible to infer causality in any of the relationships discussed. There are also potential dangers of common method variance which may have contaminated results. However, findings are broadly consistent with the results of earlier research (Carlopio & Gardner, 1995,1992; Oldham & Rotchford, 1983).

To conclude, the present study highlights the utility of using mediator variables in developing more complex conceptual links between the environment and its impact on employees. In addition, this study was relatively uniquely in keeping the job type and function constant across organisations - increasing its internal validity but reducing generalisability. Further research of this type is recommended.

References

Archea, J. 1977, The place of architectural factors in behavioural theories of privacy, *Journal of Social Issues*, **33** (3), 116-137.

Carlopio, J. 1986, The Development of a Human Factors Satisfaction Questionnaire, *Human Factors in Organisational Design and Management II*, 559 566.

Carlopio, J. and Gardner, D. 1992, Direct and interactive effects of the physical work environment on attitudes, *Environment and Behaviour*, **24** (5), 579-601.

Carlopio, J. and Gardner, D. 1995, Perceptions of work and workplace: Mediators of the relationship between job level and employee reactions, *Journal of Occupational and Organisational Psychology*, **68**, 321-326.

Ellis, P. 1986, Functional, aesthetic, and symbolic aspects of office lighting. In J.D. Wineman (ed.), *Behavioural issues in office design*, (Van Nostrand Reinhold, New York).

Ferguson, G. and Weisman, G. 1986, Alternative approaches to the assessment of employee satisfaction with office equipment. In J.D. Wineman (ed.), *Behavioural issues in office design*, (Van Nostrand Reinhold, New York).

House, R.J. and Rizzo, J.R. 1972, Towards the measurement of organisational practices: Scale development and validation, *Journal of Applied Psychology*, **56** (5), 388-396.

Knave, B. 1984, Ergonomics and lighting, *Applied Ergonomics*, **15**, 15-20.

Marans, R. and Spreckelmeyer, K. 1986, A conceptual model for evaluating work environments. In J.D. Wineman (ed.), *Behavioural issues in office design*, (Van Nostrand Reinhold, New York).

Oldham, G.R. and Rotchford, N.L. 1983, Relationships between office characteristics and employee reactions: A study of the physical environment, *Administrative Science Quarterly*, **28**, 542-556.

Sutton, R. and Rafaeli, A. 1987, Characteristics of work stations as potential occupational stressors, *Academy of Management Journal*, **30** (2), 260-276.

Wineman, J.D. 1986, *Behavioural Issues in Office Design*, (Van Nostrand Reinhold, New York).

HUMAN SUPERVISORY CONTROL:
WHAT OF THE FUTURE?

Melanie Ashleigh and Neville Stanton

Department of Psychology
University of Southampton
Highfield
Southampton
SO17 1BJ

The question of how developing technology will be able to facilitate the human machine interface into the next millennium and beyond is very much one which is becoming the focus of human factors research. This paper speculates on possible future scenarios for system control, based on a conceptual framework of Human Supervisory Control and how technology has forced changes in the way we control systems. A brief discussion of these issues will be tackled before giving a brief synopsis of current research in the field of virtual reality and multi-media environments. Finally, conclusions will be drawn as to possible future speculative ideas for Human Supervisory Control.

Introduction

As many organisations are streamlining their businesses and adopting centralisation policies, this has had a considerable affect on both the workers themselves and their working practices. In domains such as telecommunications, manufacturing plants and energy transmission companies, organisations have taken steps to reduce the number of control centres, and those remaining have meant fewer people with more responsibility to service the same systems.Re-organisational strategies has forced re-location, and often disruption to the workers and their families.Reductionism also implicates concern for safety issues, as errors in human supervisory control can have potentially disastrous consequences, which can impact upon the lives of many people, beyond those making the initial errors. Therefore, findings from research into human supervisory control is becoming even more widely applicable and an important area of ergonomic research.

There is a high likelihood, when speculating about future systems, that one may fall victim of the 'Tomorrow's World' phenomenon i.e. hindsight shows that the predictions were wide of the mark. Yet we feel that the speculation may serve as a useful framework to guide research effort, even if we revise the prediction many times as the decades pass. Without a research framework to guide us, we will lack focus. In contrast to many other technology-based predictions, we have opted for a human-centred approach concentrating on two orthogonal factors.

- Demographics of control (either centralised or decentralised).
- Representation of information (either physical form or functional abstraction).

Whilst there are many other factors that could be taken into account, the two representations we have selected to discuss are:- a decentralised control room utilising virtual reality, which describes the physical form and a centralised control arena which takes a functional representation, utilising multi-media communications. Not only do these very polarised views offer us marked contrasts in the form that future human supervisory control could take, they also challenge conventional control room technology, which as Buxton (1990) argues, currently does not seem to utilise all aspects of human functioning, Therefore only from careful analysis and prospective research into the human element of HSC can representative future technology be developed which will capitalise upon human attributes as well as meet the higher demands placed upon the operator in Human Supervisory Control.

Brief history of Human Supervisory Control (HSC)

As HSC has evolved over the last century, so technological developments have led to dramatic changes in the nature of work practices and behaviours, (Kragt, 1994). The first revolution was to automate parts of the process so that people were able to supervise larger areas of plant. The second revolution was to centralise the controls and displays into a single control room, again enabling workers to supervise larger areas of plant. The third revolution was to put all the information at people's fingertips via information technology, further reducing the personal requirements. Arguably, virtual environments constitute a fourth revolution in process control. This could return human supervisory control to monitoring (virtual physical components or could lead to the development of new forms of representation. HSC has been described as involving tasks where *"a person intermittently gets information from and gives instructions to a computer which in turn continuously controls a physical process by commanding machine actuators and reading machine sensors"*. (Sheridan p.149, 1988). The role of the operator has drastically changed, from overt physical effort to covert mental manipulations, many of the activities surrounding the control process, do not in fact involve many actual control actions per se. Umbers, (1979), cited in Baber (1991) from his research estimated that, control actions only occurred 0.7 times per hour; arguing that HSC was largely a cognitive task requiring little physical action. The changing work patterns and multi task role the control room operator has, has forced researchers to try and reach a better understanding of the human component in the human-machine interface.

Levels of Control

To understand the demand placed upon human operator's, researchers have developed cognitive models of HSC. (Lind, 1983; Rasmussen & Lind, 1981; Rasmussen, 1984; 1986). Rasmussen's (1983; 1986) 'levels of abstraction' framework, is a particularly helpful in explaining the dichotomy between physical local control and remote functional control. Future scenarios could follow the two extremes of this continuum. One by means of virtual reality, representing the physical level, the other representing the functional level of abstraction, utilising multi-media communication systems.

The 'levels of abstraction' model is a hierarchical representation which characterises the different stages of human decision making in supervisory control; describing how the operator moves cognitively from concrete physical appearance of system components to goal seeking functional purposeful objectives whilst interacting with the system. As systems have become more complex and layered, design technology in developing current interfaces have necessarily had to compromise between the physical form and functional purpose. However, the operator in his/her supervisory capacity may shift cognitive control from level to level, as required by the demands of the situation. Rasmussen argues that the requirement to convert process objectives into physical plant manipulations, puts a complex cognitive overhead on the operators task. He suggests that functional displays could reduce this demand substantially.

Physical Form Level of Abstraction

From this perspective, complex system technology today has enabled the operator to control the whole system via SCADA (system control and data acquisition) systems. This allows the operator access to large amounts of information including system support which are displayed in a variety of formats, via the VDU. Driven by window/menu based systems, mimic displays offer a graphical representation of the system which reflect the topography of the physical process plant, whilst sequence and alphanumeric displays can assist in procedural and monitoring tasks. Although this has provided greater flexibility to the human operator in terms of viewing different parts of the process at different levels and at different times, the amount of information that can be presented at one time is limited according to the amount of screens. Accessing information is quicker, although processing that information can increase cognitive workload, (Wilson & Rajan, 1995). Most control room tasks involve both discrete and continuous control, where operators are constantly having to fluctuate between different levels of processing. They must not only deal with the complexity of the plant/process and monitor surveillance systems, but also have the task of navigating their way through the interface.

Functional Purpose Level of Abstraction

Relating HSC to this higher level of the model at the other end of the spectrum, may be useful in trying to think about a generic design for future systems; a design based on the functional relationship of components and subsystems within the bigger system. For example, research work on plant process by Praetorius and Duncan (1991), has shown that operators are able to control a process with limited knowledge of the plant. This is based on the assumption that most complex plants are made up of interrelated parts that can be controlled with a knowledge of system dynamics and mass/energy flow functions. In their model, a water reactor plant is depicted as mass and energy flow functions using identical, yet limited symbols, throughout the plant at every subsystem, or component level. Results of experimental work, simulating fault scenarios and using this functional flow representation of the plant, showed that naive engineering undergraduates, who had no knowledge of the actual plant could accurately diagnose faults, their consequences and why the faults occurred after only a few hours training. The symbols used in the flow function model do not have to bear any resemblance to the physical plant components as their appearance size shape, location etc. is irrelevant to the mass and energy flow function. When participants were interviewed after the experiments and shown the conventional plant diagram, they expressed the view that the conventional model would not have helped them to understand the dynamics of the system in carrying out the fault diagnosis, in fact may even have served to be more confusing.

Two Futures - Virtual Reality and/or Multimedia

Virtual Reality

Virtual Reality (VR) environments are now being created in design technology so that people can now become totally immersed in a simulated world. Although VR technology is still in its infancy and therefore presents innumerable limitations (e.g. similar to the first generation of telephones, televisions and computers), progressive technology is gradually enabling a more interactive mutuality between human and systems. Computer simulations are already successful in creating learning environments to examine how complex systems work under various conditions, such as vehicle testing (Pacejka, 1991, cited in Smets et al, 1995). Other kinds of simulation have initiated 'what if' scenarios where in the aviation industry a concept termed 'telepresence' (Sheridan 1989) has been created where virtual worlds are simulated, which has proved invaluable in pilot training.

Further work is currently being carried out with creating improved CAD systems using VR, (Smets et al 1995). This can be seen as promoting better design in manufacturing, where an object in the virtual world is tangible. For example in vehicle design one would be able to *"virtually walk around it, open its doors and experience and evaluate is spatiality without having to build the physical model."* (Smets et al 1995, p192) This means that the design can be more rigorously evaluated aesthetically and ergonomically before the real product is made. Another important advantage is that designers have access to direct manipulation through VR technology, something that is currently lacking, not only in current CAD packages, but in all human-machine interaction. The work of Smets et al is based on Gibson's ecological perception theory, which briefly purports that perception is by means of directly interacting with our environment, both with and on its surfaces, with other animals and objects. Through the basic properties of the environment which are specified through ambient light arriving at any given observation point at any time. It is the relationship of the objects in our environment to the ground surface that directly provides us with perceptual information. We also learn to perceive things correctly by their affordance to us - what they offer us, whether for good or bad. For example we learn to use tools by their structure and shape and how they can help us interact directly with our world. (See Gibson, 1986). Smets related this principle to the VR CAD system, which allowed designers to physically adjust the shape size and location of prototypes by means of direct manipulation with the 'virtual' tools.

It is worth contemplating that in the not too distant future, these type of virtual-world environments may be applicable in other industries such as process control, where through telepresence operators will be able to exist in a virtual plant where they can physically interact with other members of their team, no matter how far apart they are. Geographical barriers would be removed, where operators could even be located at home and with a simple headset or glasses, could be transported to the virtual control room. This would enable increased social interaction with more members of a distanced team simultaneously, which would help to reduce cultural differences. Becoming embodied in the physical environment and controlling processes via direct manipulation, would allow the operator more contextual knowledge, representing the actual situation and therefore be more meaningful. Hence where technology can match the operators mental image, there is less likelihood of human error, thus facilitating enhanced control.

Multi-Media Interaction

Technological advances may make it possible to develop more natural interactions in HSC. These advances include automatic speech recognition, speech synthesis, gesture and pen recognition and liquid crystal displays. If we are to design future HSC systems around the people who operate them, the technology might be rather different than the form in which it exists today. Humans communicate freely with each other using speech. Technological development has led to commercially viable automatic speech recognition (ASR) and synthesis systems (Baber & Noyes, 1993). Current applications exist in industry, offices, telecommunications and the military. There has also been serious consideration of use in HSC applications (Baber, 1991). Contemporary ASR systems require trained operators working with a restricted dialogue set. This hardly represents natural speech, as in human-human communication. Speech synthesis systems are similarly restricted. To overcome the inhuman sound, digitised speech has been used in some applications (e.g. the speaking clock). Despite these limitations, one might speculate that future ASR and SS systems might be more naturalistic.

Research into the use of speech in control room systems suggests that it may be potentially useful (EPRI, 1986, cited by Baber, 1991). Certainly the benefits of speech are apparent, including breaking through attention, eyes free/hands free operation, omnidirectionality, little learning required and reduction in visual clutter (Stanton, 1993). Speech would free the operator of HSC from a desk. allowing them to move around the control room at will. However, Stanton (1993) cautions that speech needs to be paired with some form of permanent visual display because of its transitory nature. Since the advent of VDU technology, and the subsequent removal of large panel displays, operators in HSC have been left with the legacy of a window of the process they are supervising. This has, on occasion, led to some problems in operation of systems (Stanton, Booth & Stammers, 1992); Stanton & Baber, 1995).

Liquid Crystal technology is reaching a stage where it is possible to imagine large wall based displays which will replace the need for desktop VDU's. Some attempt has been made in this direction with the use of video-walls,(a bank of VDU's clustered together to give the impression of a large screen) in the French electricity supply and distribution industry. Indeed the desk is only present to support the VDU and keyboard, and if they are removed, so may the need for a desk. The control room of the future may more closely resemble the bridge of the Star Ship Enterprise, than a traditional control room layout! The use of the QWERTY keyboard and mouse as interaction devices has already been questioned (Buxton, 1990) and speech has been offered as an alternative. However, speech may not be appropriate for all forms of interaction as Carey (1985, cited in Baber, 1991) pointed out. He specified five generic operations, which could be considered for the purpose of identifying appropriate input/output media.. Using a combination of ASR, directional pointing and pen input one could perform input operations such as specifying and locating objects, specifying and carrying out actions. Output operations such as presenting objects and their location, presenting numerical values, confirming actions and presenting results would require a combination of semi-permanent visual display both of text and graphics , together with speech synthesis. The use of pointing and pen input (Frankish et al 1994) combines method for object selection that are seen outside the control room and enhance the naturalness of the interaction. For example, if one wished to view an object in more detail, one would simply point at the object on the LCD wall display and request a "status report." The detailed information on the object would be displayed together with a verbal response. If one wished to change the setting on an object, one could simply point to the object in question and make a verbal request. The system would respond by repeating the command, to which the control operator would respond to actuate the command.

Rather than the physical levels of interaction, a functional image could represent plant dynamics at the level of abstraction closer to peoples mental models of the process control system. As previously mentioned, Praetorius & Duncan (1991), describe a method of representing a system in terms of 'mass' and energy flow' functions. Generic components of any system can be reduced to a simple symbol set comprising six flow functions.

- Source - where the energy or mass comes from.
- Sink - where the energy or mass goes to
- Balance - to keep energy or mass in balance
- Storage - to store energy or mass
- Transport - to move flow.
- Barrier - to prevent flow.

The role of the operator would be to manage this functional system, i.e. to keep the system in balance within specified limits. It might be possible to represent any process control system in this manner. Thus it would be functions that the operator would be concerned with, rather than individual physical objects. Praetorius & Duncan (1991), claim that this functional representation enables opeartors to diagnose novel faults quicker and more easily than traditional physical representations.

Conclusions

As researchers are continually looking to understand HSC, the introduction of smart knowledge based systems into process control has been discussed (Hollnagel et al, 1988). One may anticipate process plants becoming completely computerised and the humans will be unnecessary. However, as Baber argues *"it would be impossible to design a plant that would be able to cope with all the potential faults that can occur."* (p.19 Baber, 1991, He points out that to emulate human creativity in the function of problem solving for example would be very difficult - even harder to capture the whole of human knowledge and experience. Therefore the human operator, even though his/her role may have drastically changed, and continue to do so according to the type of technology implemented, one must assume that humans are still an essential part of the control in HSC. Therefore, we should focus on developing better ways of adapting technology to fit the tasks of control.

In considering these two visions of the future in this paper, we have attempted to capitalise on human abilities, in a way which is not currently realised in the human machine interface. It is possible that many more ways of creating future interfaces will evolve as technology develops even further. The future is something we can only speculate on; technology may well be able to provide an amalgam of these two visions, or neither. Whilst we cannot know the exact future of HSC, one thing we can be certain of is that the future is unlikely to be exactly as we predict. Whilst control and display technology may change dramatically over the next 30-40 years, human beings are highly unlikely to evolve further. Therefore, we argue, a human-focused view of the future is likely to yield maximum benefit. If we can chart human capabilities, limitations, preferred methods of working etc., then we may select the technology that is appropriate for them. In any future design, the emphasis therefore should be on the users in supporting their tasks, rather than introducing technology for its own sake.

However, there are concerns that in reducing one set of problems, we may be gaining others with regard to diffusion of responsibility associated with physical isolation of individuals in the virtual world. Virtual control rooms, although physically different, but functionally equivalent to conventional systems they are unlikely to be socially equivalent. Therefore prospective research should be aiming to critically evaluate the performance of teams in virtual control systems before any future technological vision into HSC can become a reality.

References

Baber, C, 1991 *Speech technology in Control Room Systems*. (Ellis Horwood, London)

Baber, C. & Noyes, J.M. (1993) *Interactive Speech Technology*. (Taylor & Francis, London)

Buxton, W. (1990) "There's more to interaction than meets the eye: some issues on manual input" In: J. Preece & L. Keller (eds) *Human-Computer Interaction*. (Prentice Hall,Hemel Hempstead)

Frankish, C.; Morgan, P. & Noyes, J. 1994. Pen Computing in context. In C. Baber & N.A. Stanton (eds). *Designing Future Interaction*. The Ergonomics Society: Loughborough.

Gibson James, J, 1986. The Ecological Approach to Visual perception. (Lawrence Erlbaum Associates, Inc., New Jersey)

Hollnagel, E. Mancini, G & Woods, D.D. (1988*) Cognitive Engineering in Complex Dynamic Worlds* (Academic Press, London)

Kragt, H 1992 Introduction to enhancing industrial performance. In H. Kragt (ed.), *Enhancing Industrial Performance* (Taylor & Francis, London)

Lind, M 1983 A systems modelling framework for the design of integrated process control systems. Proceedings of *IASTED Symposium ACI 83 on Applied Control and Identification*, (Copenhagen Denmark)

Praetorius N., & Duncan, K.D. 1991 Flow representation of Plant processes for fault diagnosis. Behaviour & Information Technology, **10**, 41-52.

Rasmussen, J & Lind, M 1981, Coping with complexity. *Proceedings of 1981 European Annual Conference on Human Decision Making and Manual Control, Delft.*

Rasmussen, J 1986 *Information Processing and Human-Machine-Interaction - An Approach to Cognitive Engineering* (Amsterdam: North Holland).

Rhinegold, H. 1991 *Virtual Reality*, London: Secker & Warburg.

Sheridan T, B, 1988 Human and Computer roles in supervisory control and telerobotics: musings about function, language and hierarchy. In L.P. Goodstein, H.B. Andersen, S.E. Olsen, (eds.), *Tasks, Errors and Mental Models* (Taylor & Francis, London)

Smets, G.J,F., Stappers, P, J, Overbeeke, K.J. & Van Der Mast, C. 1995 .Designing in virtual reality: perception-action coupling and affordances. In Karen Carr & Rupert England, (eds.), *Simulated and Virtual Realities, elements of perception.* (Taylor & Francis, London)

Stanton, N. A., Booth, R. T. & Stammers, R. B. 1992 Alarms In Human Supervisory Control: A Human Factors Perspective. *International Journal of Computer Integrated Manufacturing* **5**, 81-93

Stanton, N.A. 1993 Speech-based alarm displays. In: C. Baber & J. Noyes (eds). *Interactive Speech Technology*. (Taylor & Francis, London)

Stanton, N.A. & Baber, C. 1995 Alarm initiated activities: an analysis of alarm handling by operators using text-based alarm systems in supervisory control systems. *Ergonomics, 38,* 2414-1431

Wilson, J.R. & Rajan, Jane. A. 1995, 'Human-machine interfaces for systems control'. In J.R. Wilson & E.N. Corlett, (eds.), *Evaluation of Human Work: A practical ergonomics methodology*

CO-OPERATIVE WORKING IN THE AUTOMOTIVE SUPPLY CHAIN

S.M. Joyner, C.E. Siemieniuch

HUSAT Research Institute
Loughborough University, Leics. LE 11-3TU

We describe some generic aspects of supply chains in the automotive domain, and discuss the requirements to support teamworking in this domain. Key issues are the federated control structure, the consequent impoverishment of information, and the need for trust to overcome this. CSCW issues include delays, conferencing, hardware, quality of data, quality of service, and security

Introduction

Since 1989 HUSAT has been working in collaborative projects with funding from the European Union involving most of its automotive assemblers and some of their suppliers on the problems of tele-co-operation in the product introduction process. We report some generic findings from three of these projects: Project R1079 CAR - 'CAD/CAM for the automotive industry in RACE', Project R2112 SMAC - 'Suppliers and Manufacturers in Automotive Collaboration', both in the RACE Programme and Project AC070 TEAM - Team-based European Automotive Manufacturing'.

Classes of supply chains

Five generic classes of supply chains have been defined (Wortmann 1991; Hoekstra and Romme 1992; Wortmann 1992):

- Make-and-ship-to-stock: products are manufactured and distributed to stock points which are spread out and located close to the customer.

- Make-to-stock: end products are made and held in stock at the end of the production process and from there are sent directly to many customers who are scattered geographically.

- Assemble-to-order: only system elements or subsystems are held in stock in the manufacturing centre, and the final assembly takes place on the basis of a specific customer order.

- Make-to-order: only raw materials and components are kept in stock, and each order for a customer is a specific project.

- Purchase-and-make-to-order: no stocks are kept at all, and purchasing takes place on the basis of the specific customer order; furthermore, the whole project is carried out for one specific customer.

This paper is concerned with the automotive domain, which is currently characterised by the third of these classes, and we therefore restrict our discussion to this class. Fig. 1 illustrates a supply chain structure likely to be found in this industry in the near future.

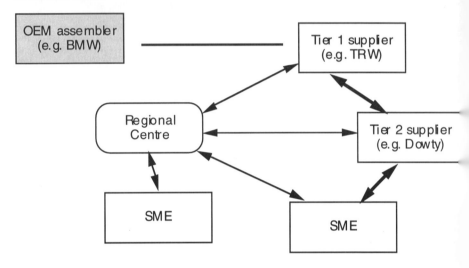

Fig. 1   Example of a supply chain in the automotive industry, from the TEAM project

Issues of federated control.

Contrary to unitary organisations, supply chains do not have an hierarchy of control, but must operate under federated control, by agreement between entities (or nodes) in the supply chain. In principle, a node can, if it wishes, restructure itself or disconnect itself from the cluster as it wishes. Consequently, co-operation occurs by negotiation, the acceptance of common policies (see Dobson (1988, 1991); Martin and Dobson (1991), for an interesting series of discussions), and, most importantly, the establishment of trust. It is clear that in this environment an efficient supply chain cannot exist without provision for, and extensive use of, communications. Nodes must be able to communicate at several levels of discourse:

- the transactional level - there must be provision for goods and/or payments to be transferred between them;
- the operational level - there must be provision to co-ordinate and control the transactions (shopfloor and distribution and accounting issues);
- the policy execution level - negotiate targets, develop products together with their necessary resources, create delivery arrangements, etc.;

- the strategy level - define role and level of participation in supply chain, discuss market research information, and set other policy issues (e.g. define the type and scope of the contracts between nodes).

Effects of federated control on information availability and co-operation.

The consequences of this federated system of control are that legal requirements, security and confidentiality requirements restrict the classes, timing, destinations, and quantity of information at the four levels discussed above that can be passed within the supply chain - hence, the supply chain as a whole necessarily has a sparse information environment (Sinclair and Siemieniuch, 1995, 1996). In particular, there is imperfect information (e.g. lack of timely information about designs due to slow dissemination procedures); excluded classes of information (e.g. about an alternative supplier's products); and inappropriate communications facilities and hardware and software to utilise information (e.g. 'legacy' systems - IT systems created in earlier years with unique characteristics which have been 'patched' for purposes not originally envisaged and which are no longer fully documented or understood).

It is obvious that in such an environment (which is typical of many supply chains), improvements to IT systems are essential. However, this will not necessarily overcome the problems of sparse information due to legal, organisational and cultural reasons. Closer relationships between suppliers will in principle enable more preview and planning information to be provided (i.e. widening the information window), but there must also be reliance on personal relationships, the use of alternative communication channels (e.g. telephone, dinners, etc.), and, as the basis of all of this, on trust in the integrity and performance of others.

The importance of trust.

Where one must operate with incomplete information, and use what is passed along the supply chain, one must have trust in the both the people and the companies with whom one deals, if co-operation is to flourish. This relies on a number of things: a common understanding of terms and their usage (Kanoi 1991; Mackay, Siemieniuch et al. 1992); the establishment of common goals and shared benefits; and integrity in relationships. The latter in particular requires (a) a clear commitment to ethical behaviour; (b) operational roles, procedures and the organisational structures in which they are embedded must enable integrity to be demonstrated; and (c) there must be some freedom and empowerment for individuals to take individual initiatives to ensure efficient operation in an environment of imperfect information. The negative side of these issues can be seen in one common characteristic of many classic supply chains, particularly in the construction industry; the final product constructor creates a supply chain to tender for a contract on a given timetable. The suppliers conform to this, but for various reasons there are delays, price renegotiations, and contract rewordings. The result is that smaller suppliers find themselves severely squeezed over design time and costs, and do their best to pass these on. The environment created by this is distrust, with information hoarding, and continuous litigation to move costs around the supply chain, utilising any nooks and crannies in the contracts. The are inevitable delays in completion of contracts, cost over-runs, and denial of responsibilities for any difficulties. In such an environment it is difficult to achieve any new efficiencies and competitive edge, particularly if the sharing of intellectual property rights are involved.

Required characteristics of jobs for co-operative working in design engineering.

Following the discussion above, there are a number of characteristics that a firm should exhibit in its behaviour with respect to others in the supply chain. These are listed below; however, it should be noted that these characteristics will only be evident if the jobs in the firm are designed to exhibit them, and the firm has implemented the appropriate policies and procedures:

- Transparency about goals, problems, and ways of working.
- Willingness to share any benefits accruing from improvements or windfalls.
- Respect for commercial confidentiality.
- Room for personal relationships, to be built up over time.
- Speedy, efficient, and correct execution of promises.
- Recognition of the 'favour bank', and its tacit rules, as a means of overcoming unexpected problems and difficulties in the supply chain, usually arising from the sparse information environment.

It follows, then, that the following human and organisational issues must be addressed:

- The organisational design should reflect the needs of the firm's role in the various supply chains in which it participates. Typically, this is interpreted as a process-based team structure, involving 'business groups', etc.
- The people who commit themselves (and hence the firm) to promises must be empowered to execute them (i.e. they must be given appropriate responsibility, authority, and access to resources). Note the importance of the four levels of discourse in this; an individual is unlikely to be empowered to undertake all four of these, implying that there must be good internal communications between different levels in the hierarchy to ensure that the firm's responses are seen to be appropriate and integrated.
- The people executing the promises must have access to appropriate information and knowledge - i.e. appropriate applications, supporting tools and training in their use
- The processes and procedures involved in the execution of promises must be clear, and be capable of being controlled by the people executing the promises
- The people involved must have appropriate skills and expertise to execute their tasks
- While the people involved must be accountable for their errors, there must be appropriate, sensitive treatment of these errors by those in more senior positions.

Supporting co-operative working for the product introduction process in the supply chain.

In the projects mentioned earlier, and in some additional projects, over 100 semi-structured interviews have been held with stakeholders concerned with the operation of supply chains. Hidden within these interviews were the generic issues discussed above, concerning the systems aspects of supply chains. The more evident and most frequently-mentioned, general issues were as follows (Joyner, Siemieniuch et al., 1996).

- Difficulties in identifying parts - suppliers and OEMs tend to use individual numbering systems, which points to the need for a secure, open-access product data model.
- Drawings not of satisfactory quality, or missing - CAD models are not always correct, and may follow local rules in their structure, leading to translator

difficulties. Compliance with standards such as STEP will ameliorate the problems.

- The need for solid geometry and supporting information, and the need for fast, reliable transmission of this - to allow others to explore options as soon as possible, and to discover constraints. If information is sparse, then that which is transmitted should be capable of maximum elaboration.
- Ensuring that design information is up-to-date - procedures for validation and version control may delay release of important design changes, which have already been discussed with others.
- language difficulties (particularly with foreign design data) and incomplete information on both the design and on the supplying company - as discussed above, several levels of discourse are required for adequate transmission and the establishment of trust.
- Design problems are often filtered through Sales or Marketing personnel - ostensibly to maintain integrity and responsibilities, but causing delays and reinterpretations of problems. Internal reorganisation into project-based teams and removal of functional demarcations will be necessary to alleviate these problems.
- Reliance on face-to-face meetings - it has been estimated that on average a one-hour meeting consumes 4 hours of otherwise useful time; more importantly, problems are held over until a meeting can be arranged. Properly-supported CSCW applications are the best answer to these problems.
- The effects of delays in the transmission of design information on lead times - delivery dates don't change, whatever the delay; hence the importance of project scheduling and planning tools, as discussed below.
- Incompatibility of systems and applications - causing problems of data interpretation, and double-entry of data. Legacy systems and old hardware are particular problems, for which investment represent the only viable solutions.
- Suppliers willing to invest, but only small amounts of money available - usually on internal networking, moves from mainframes to PCs, and external communications.

Analysis of both the systems issues and the direct issues led to the following requirements for supporting teamworking along the supply chain.

- Provision of tools for project management. There should be support for all groups in the supply chain to input scheduling information, with secure, open access to the master project schedule. Whatever tools are used, there should be interoperability among them.
- Rapid exchange of information. EDIFACT, STEP, and the standards being proposed in the CALS initiative will do much to enable this, but there are fundamental translator problems still to be solved, as well as requirements for physical networks to be extended both nationally and internationally.
- Provision for static and dynamic visualisation of data. This is predicated on product data models involving solid geometry and properties, and is essential for matching tools, component designs, and production facilities.
- Shared database facilities. This should cover not only product data models, but also national and international standards, legislative requirements, test procedures, contract histories, etc. Within this category are Product Libraries for standard parts, produced by suppliers of these parts.
- Multipoint-multi-user conferencing facilities. These should be at the desk (but not necessarily all desks), and should include voice, face-to-face video, and remote live video feed. The conferencing facilities should act as a distributed platform, allowing applications to be opened for multiple users during the conference (e.g. whiteboard, CAD, project planning, directories). It should be possible for people to join and leave the conference as necessary. Finally, there must be easy means for recording the essential points of the conference.

- Security. Companies must protect their core engineering competence and components; this conflicts with the need for external access to design data, and easy-to-use security systems. Furthermore, these systems should be predicated on the ownership of data by individuals, to enhance security even though this implies human resource management complexities.
- Hardware considerations. The key issues here are interoperability, standards and quality of service. The growth of the internet and of intranets will fulfil many of these requirements, though there are still issues of internal system management and of quality of service. A key issue to be addressed is mutual assistance within the supply chain to achieve the necessary investments to allow fast, flexible and agile responses to problems and demands.

Finally, we repeat the point made earlier; fulfilment of these technical requirements will not by themselves ensure a flexible, agile and appropriate performance by the company in the supply chain; this requires due attention to organisational structures, policies, procedures job designs, reward structures, and motivation. Without these, and the development of an appropriate culture as implied by these, little will be achieved at the cost of a great deal of effort.

References

Dobson, J. 1991, Information and denial of service. *Database Security and Prospects V.*, (North Holland, Amsterdam)

Dobson, J. E. 1988, Modelling real-world issues for dependable software. *High integrity software.* ed.: C.T. Sennett. (London, Pitman) pp 274-316.

Hoekstra, S. and Romme, J. 1992, *Integral logistic structure - developing customer-oriented goods flow* (McGraw-Hill, New York)

Joyner S.M., Siemieniuch, C.E. et al. 1996, User and organisational requirements. Report no. HUSAT/TEAM/WP1/DRR003, for ACTS Project AC070.

Kanoi, N. 1991, Manufacturing modernisation - Sony's approach, First International Manufacturing Lecture, Inst. of Manufacturing Engineers.

Mackay, R., Siemieniuch, C. E. et al. 1992, A view of human factors in integrated manufacturing from the perspective of the ESPRIT-CIME programme. *Proc. 2nd Information Technology and People Conf. (ITaP '93)*, Moscow, Northwind Publications, and Publ. International Centre for Scientific and Technical Information, 125252, 21-B Kuusinen Str. Moscow.

Martin, M. and Dobson J. E. 1991, Enterprise modelling and security policies. *Database security, IV.* status and prospects., (Elsevier Science Publishers D.V. Amsterdam).

Sinclair, M. A., Siemieniuch, C. E. et al. 1996, Concurrent engineering and the supply chain, from an ergonomics perspective. In S. Robertson, ed.: *Proceedings of the Annual Conference of the Ergonomics Society*, University of Leicester, (Taylor & Francis, London)

Sinclair, M. A., Siemieniuch, C. E. et al. 1995, A Discussion of Simultaneous Engineering and the manufacturing supply chain, from an Ergonomics Perspective. International Journal of Industrial Engineering 16(4-6): 263-282.

Wortmann, J. C. 1991, *Factory of the future: towards an integrated theory for one-of-a-kind production.* ESPRIT-91, Brussels.

Wortmann, J. C. 1992, "Production management systems for one-of-a-kind products." Computers in industry 19: 79-88.

HANDS AND HOLDING

What is the optimum surface feature? A comparison of five surface features when measuring the digit coefficient of friction of ten subjects.

George E. Torrens

Department of Design and Technology,
Loughborough University,
Loughborough,
Leicestershire. LE 11 3TU

This study begins to identify appropriate featured surfaces to be used on product interfaces. The paper discusses the results of comparing five different featured surfaces when measuring digital pulp friction of ten young adult subjects, five male and five female covering the anthropometric and body type scale. The surface features used were dry and smooth in surface texture. The measurement of digit friction performance are shown as the friction force generated when a (known) normal force is applied through the digit. Suggestions will be made about which surface features appear to be most suitable for different subjects, i.e. male and female.

Introduction

The aim of this pilot evaluation is to provide ergonomists and designers with more information to place within the complex structure of interactions that occur at the hand and handle interface. The main use of a hand is to provide a stable connection between a person's body and the object which they are using to perform a task (Torrens 1996) In this comparative assessment the focus of interest is the intermediate level of interaction of the hand with an object. To assist in the study of the complex interaction between hand and object the author has grouped the aspects of scale of mechanical interaction under three headings: micro interaction between skin and the fine features of a textured surface, intermediate interaction where the skin and soft tissues are combined to interlock with the fine and coarse features of a surface texture, and gross interaction which involves the overall shape of the object and the grip pattern and muscle strength used to hold it. During the grip of an object all three levels of interaction take place. This study concentrates on the scale of the surface features that may affect the mechanics of grip at the intermediate level of interaction involving the glaborous (hairless) surface of the hand. The cognitive aspects of the hand and object interaction will be considered once the basic mechanics of the interaction have been defined and validated.

The features referred to in this study are the marks on the surface of the object material that make up the overall texture. The scale of the features on the textured surfaces used was based on complementary work studying skin friction by Bobjer, Johansson, Piguet (1993) and Buchholtz, Frederick, Armstrong (1988). The roughest surface texture, (texture 4), described by Bobjer *et al* had rectangular sectioned features, 0.5 mm in depth on a pitch of 2 mm. The peak or asperity was 0.5 mm wide, (25% of the overall surface area available). The features were larger than features described by Buchholtz *et al*. The roughest texture described by Buchholtz *et al* was 320 American grade sandpaper. The roughest Bobjer *et al* texture may be taken as the limit of micro interaction, the interaction between the texture and skin alone. From an earlier study (Torrens, 1996) it was suggested the optimum featured texture from mechanical grip (shown as coefficient of friction) might be 4 mm from peak to trough on a pitch of 8 mm The shape of the asperity in the authors earlier study was a 1 mm radius. The smallest scale of the features used on the textures in this study were larger than those described by Bobjer *et al* during the measurement of dynamic coefficient of friction. The largest scale of features within the textures used were larger than the optimum estimated by the author in an earlier study. In this study five surface textures were compared. All the features were based on a square section, repeated over the surface. A square section was chosen to offer equal amounts of peak and trough area to the finger. The dimensions of the features were: Texture 1, 2.5 mm square, on a pitch of 5 mm, Texture 2, 3 mm square, on a pitch of 6 mm, Texture 3, 4 mm square, on a 8 mm pitch, Texture 4, 5 mm square on a 10 mm pitch and Texture 5, 6 mm square on pitch of 12 mm.

Method of assessment

A test rig was used to measure static finger frictional force during this assessment as described earlier (Torrens 1996). The frictional force was displayed as the coefficient of static friction. During a gripping action the concern of the subject is to avoid slippage. If slippage occurs more grip force is generated by muscles in the forearm and intrinsic muscles of the hand . Based on the previous statements static friction is more important than dynamic friction in many everyday task performances.

A number of other measurements were taken to link the subjects within currently available ergonomic and anthropometric information. A Stadiometer was used to measure stature, maximum finger diameter was measured over the distal index finger joint, right hand only, using a circle template, elbow to finger tip length and hand breath across the finger knuckles were measured using a large set of callipers, finger width, breadth, and hand depth over the finger knuckles were measured using a set of digital callipers. Room temperature was monitored and each subject's finger temperature taken at the start, midway and at the end of the testing. The measurements were based on those described in PEOPLESIZE ™ and British Standards Institute (1990).

The friction tests were undertaken using the second digit of the right hand. The right hand is mainly the dominant hand and the index finger (second digit) is involved in most prehensile grip patterns. The five friction measurements were separated by the taking of anthropometric measurements and filling in a questionnaire about lifestyle. Each subject spent at least three minutes resting before the first finger friction performance test and a similar time between each test. The subject was asked to put the second digit of their right hand into the finger friction test rig. The orientation of the subject's finger and hand position during testing was constrained by a number of foam barriers under the hand and forearm and a clear plastic cover plate above the hand. The barriers helped the subject to stay within the plane of orientation chosen for this test. The subject was asked to point their index finger, (second digit), towards the back of the test rig keeping the finger straight but relaxed. The subject's finger was positioned over the featured surface, with the tip of the subjects finger just at the point where the featured surface began. From the instructions given to the subject their hand was naturally pronated.

The subject was asked to lower their hand until their finger was in light contact with the test surface. They were then asked to look at a mark on an oscilloscope screen in front of them. The mark had been pre-set using a 1 Kg weight placed on the test rig carriage. They were asked to press down with their finger (only) until they could bring a moving line on the oscilloscope up to the same level as the fixed mark on the screen. The subject was then asked to pull their finger back towards their body, while maintaining the position of the moving line level with the mark on the screen. They were also asked to maintain the finger, hand and upper limb position taken up during this act, as shown in Figure 1. The subjects had an opportunity to practice the test using their middle finger (third digit), right hand. The index finger was then used for the documented test. There was a number of random variations in the order of the tests due to assessing a number of subjects at the same time who were at different stages of the assessment. The oscilloscope was connected directly to a printer and so there was a delay in processing the subjects.

Figure 1. Shows finger friction test rig in use.

Between friction measurements the subjects were asked to provide information about age, gender, hand dominance and occupation. They were also asked whether they were involved in any sport or past time that involved physical exertion related to their hands. The described procedure was repeated using five different surface featured for ten subjects, five male and five female. The following results were obtained.

Results

The stature of the subjects was found to cover the most of the anthropometric scale of the United Kingdom population. The smallest female matched a 27 th percentile UK female and the tallest male in the study matched a 96th percentile UK male from the PEOPLESIZE ™ database. The distribution of stature is shown in Table 1, where subjects one to five are female. Most of the male and female subjects fell within the normal limits of the Body Mass Index (Garrow 1979), where the combined height and weight provide an index. Finger dimensions were found to generally follow stature when compared with the PEOPLESIZE ™ anthropometric database.

The peak static friction reading was taken from each graph produced during the finger performance test. There were to lines on each test graph, one indicating friction force, the other the maintenance of normal force (1 Kg) by the subject. A reading was taken from the highest point on the printed graph before the line began to fall, indicating slippage. Signal noise was taken into consideration using operator judgement. The distance between the highest point and the datum line was measured using a digital calliper to reduce operator error. Each interval of 1 mm corresponded to 66.7 grams weight, or 0.654 Newtons force. The normal force was taken nominally to be 1Kg weight, 9.81 Newtons force. The variation of graph line produced on each normal force documented was measured and the error, due to the subject being unable to accurately maintain the normal force whilst pulling their finger back across the texture, was estimated to be ± 10% of the overall value. The coefficient of static friction was calculated, normal force divided by friction force.

The lowest overall value, 0.28μ, was produced by a female, subject 4, when using surface texture 3. The maximum overall value, 1.44μ, was produced by a female, subject 2, using surface texture 5. The minimum value produced by a male, subject 6, was 0.48μ, when using surface texture 1. the maximum male value was 1.7μ, produced by subject 10 using surface texture 3. The result of subject nine using surface feature four was spoiled. Figure 2 shows a comparison between the average coefficient friction values from the ten subjects and the pitch of the features within the five different surface textures. Figure 3 shows the average coefficient of friction value for female results and male results when using the five different surfaces.

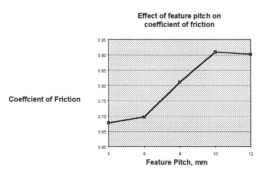

Figure 2. A comparison of average μ of ten subjects with feature pitch on surface texture

Discussion

A clear trend can be seen in Table 2 where the coefficient of friction average values are compared with the pitch of the features used on the textured surfaces. When compared, the difference between the results of the ten subjects using Texture 1 and Texture 4 were found to be consistently different, (P= 0.05), but may not be so relevant due to the large variation of subject performance results. The values have been averaged in Figure 3 and again clear trends can be seen. The female subjects produced higher coefficient of friction results than the males when using the finer featured surface textures 1 and 2. Males produced higher results with larger featured textures than females. Overall the larger featured textures produced higher coefficient of friction values. The high friction performance of the female subject 2 may be accounted for through methodological error. However, from the questionnaire it was noted the subject was involved in many racket sports activities and used her hands in practical workshop environment. Many of the subjects were office based and did not undertake a notable amount of manual work. There was a notable increase in performance between surface Texture 2, made up of 3 mm square section features, and surface Texture 4, made up of 5 mm square section. The trends shown in Tables 1 and 2 correspond to the author's findings from an earlier study presented last year at the Conference.

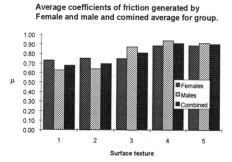

Figure 3. Average μ generated by male, females and combined subject group using five different textures

The lower coefficient of friction results from this study are not as high as the results from Buchholtz *et al*, (0.66μ, using 320 Grade Sand paper and a dry surface). However, the highest results from both male and female using the larger featured textures are over twice the value of the Buchholtz results. This suggests interlocking at the texture and hand interface that is more than skin related friction alone. No direct comparison can be made between the Bobjer *et al* findings and this study as Bobjer *et al* were concerned with dynamic friction and not static friction. It is interesting to note that Bobjer *et al* found an increased perception in discomfort when using the textures with ridges on a longer pitch. The variation of individual results may have been due aspects of the methodology that allowed variations during the assessments. The room in which the testing took place was cooler than expected, at 13° Centigrade, due to unusually cold weather. It was noted that one subject's finger temperature dropped by 9° Centigrade, from 27° C to 16° C, during the twenty to thirty minutes of the test. For the majority of subjects the temperature loss was 3° C. Moisture was not objectively measured which also may have affected the performance of the subject's digit.

It would seem large surface features are effective in optimising friction force that may be generated by the skin and soft tissue of the glaborous surface of the hand. It would seem the effectiveness of the larger textured surface was reaching a peak with the 5 mm square section on a 10 mm pitch. The finger tip length of the subjects ranged from 21.5 mm in the smallest female to 28.8 mm in the next tallest male, who had the largest hand dimensions. At this scale only two featured sections of the textured surface were being used to grip through the subject's digit. This placed a high localised pressure on two sections of the digit. It would seem high localised pressures increase the interlocking effect of skin and soft tissue of the hand with the object. The optimum scale would seem to be a 5 mm square section on a 10 mm pitch. . This scale corresponds to the estimated optimum surface feature scale described in earlier work of the author relating the scale of the fat lobules found in the digit based on anatomical sources such as Foucher (1991). Furthermore, at least two and ideally three features, (or asperities), should be used to interlock with the finger tissues. This may be one reason why females, with smaller fingers, performed better at lower scales of feature pitches

What is less clear was the effect of the surface features in relation to discomfort. The subjects used the textured surfaces for only a few seconds. There were no comments relating to discomfort noted. It would seem logical to assume there is a limit to the friction that may be generated by the hand before discomfort and tissue damage occur. Further studies are required to define the forces involved in the performance of specific tasks and link the results to forms that optimise the scale of a featured surface for the required grip whilst minimising discomfort over the time in use.

Acknowledgements

This study was in part funded by the Defence Clothing and Textiles Agency.

References

British Standards Institute, 1990, BS 7231: Part 1: 1990, Body measurements of boys and girls from birth up to 16.9 years, Part 1. Information in the form of tables.

Bobjer O., Johansson S.E., Piguet S., 1993, Friction between hand and handle. Effects of oil and lard on textured and non-textured surfaces; perception of discomfort, *Applied ergonomics*, **24**, 190-202

Buchholz B., Frederick L.J. and Armstrong T.J., 1988, An investigation of human palmar skin friction and the effects of materials, pinch force and moisture, *Ergonomics*, **31**, 317-325.

Foucher G.(ed.), 1991, *Fingertip and nailbed injuries*, (Churchill Livingstone, Edinburgh) 2-3.

Garrow J.S., 1978, Energy, balance and obesity in man, 2nd edition, (North Holland, Amsterdam)

Torrens G.E., 1996, A contribution to the understanding of the role of digital pulp in hand grip performance,. In Robertson S.A. (ed.), *Contemporary Ergonomics 1996, Proceedings of the Annual Conference of the Ergonomics Society*, (Taylor & Francis, London) 75-80

INDIVIDUAL VARIABILITY IN THE TRANSMISSION OF VIBRATION THROUGH GLOVES

Gurmail S. Paddan and Michael J. Griffin

Human Factors Research Unit
Institute of Sound and Vibration Research
University of Southampton
Southampton
SO17 1BJ England

An experiment has been conducted to determine intra-subject variability and inter-subject variability in the transmission of vibration from a vibrating handle through gloves to the palm of the hand. Two gloves were used in the investigation. One male subject took part in the intra-subject variability experiment and eight male subjects took part in the inter-subject variability study. The subjects held a horizontally vibrating handle with a push of 50 Newtons; no grip force was applied to the handle. The subjects were exposed to random vibration at frequencies up to 1200 Hz with a frequency-weighted vibration magnitude of 5.0 ms^{-2} r.m.s. (frequency weighting W_h). Vibration was measured at two locations: on the vibrating handle and at the palm using a palm-glove adaptor. A very large variability in transmissibility was seen between the subjects and between gloves. The application of the findings to the method of testing anti-vibration gloves is discussed.

Introduction

International Standard, ISO 10819 (1996) uses single figure transmissibility values averaged over three subjects to determine whether a glove can be considered as an 'antivibration glove'. The variability in transmissibility between gloves and between individuals has been partly addressed elsewhere (Paddan, 1996). However, the variability during repeat measures (i.e. intra-subject variability) for an individual has not been previously investigated. Such data are necessary when considering the causes of the differences in transmissibility between individuals (i.e. inter-subject variability) and the effects of other factors.

This paper presents intra-subject and inter-subject variability in the transmission of vibration through gloves. The method used in determining transmissibility is similar to that specified in ISO 10819 (1996). The data presented in this paper are taken from a larger study involving 10 gloves.

Equipment and Procedure

The experiment was conducted using an electrodynamic vibrator, Derritron type VP30 powered by a 1500 watt amplifier. A handle was attached to the vibrator such that the grip of the hand was vertical and therefore at right angles to the (horizontal) axis of vibration. The first resonance of the handle occurred at 1440 Hz. Strain gauges (copper nickel alloy foil type) were mounted on the handle so that both the push force (i.e. feed force) and the grip force on the handle could be measured.

Acceleration was measured at two locations: on the vibrating handle, and between the palm of the hand and the glove using a palm adaptor of mass 9.21 grams (ISO 10819 (1996) states a maximum mass of 15 grams). The accelerometers were of piezoelectric type (Brüel and Kjær type 4374) each with a mass of 0.65 gram. The acceleration signals from the two locations were passed through charge amplifiers (Brüel and Kjær type 2635) and then acquired into a computer-based data acquisition and analysis system (*HVLab*).

The subjects stood on a horizontal surface and held the vibrating handle with the right hand. The subjects held their forearms horizontal and at an angle of 45° to the axis of vibration. The handle was held such that the knuckle bones (the metacarpal bones) were at approximately right angles to the axis of vibration. The wrist was bent at an angle of 45°; there was no abduction at the wrist. The elbow formed an angle of approximately 90° between the forearm and the upper arm. There was no contact between the elbow and the body during the measurements. A push force of 50±8 N was applied during the measurements; no grip force was applied. A copy of the written instructions given to subjects is shown in the appendix.

Results from two commercially available gloves are presented, see Table 1. In accord with International Standard ISO 10819 (1996), the gloves were worn by the subjects for at least 3 minutes prior to the vibration measurements. The room temperature during the tests fluctuated between 22 °C and 25 °C (the standard specifies a temperature range of 20±5 °C) and the relative humidity varied between 29% and 46% (the standard specifies that the relative humidity shall be below 70%).

The experiment was approved by the Human Experimentation Safety and Ethics Committee of the Institute of Sound and Vibration Research. One male subject took part in the repeatability study (age 28 years; weight 78 kg; height 1.78 m). Eight male subjects participated in the inter-subject study (mean age 31.25 years; mean weight 75.0 kg; mean height 1.81 m). Each subject was exposed to the vibration two times: once with each of the two gloves. The subject who took part in the repeatability study was exposed to the vibration eight times for each condition.

Table 1. Gloves used in the experiment.

Glove number	Description
1	blue nylon lycra, yellow leather palm, hand pad, fingerless mitten, no wrist protection
2	yellow leather palm covering rubber-like material, full finger, full wrist protection

A data acquisition and analysis system, *HVLab*, developed at the Institute of Sound and Vibration Research of the University of Southampton, was used to conduct the experiment and analyse the acquired data. A computer-generated Gaussian random waveform having a nominally flat acceleration spectrum was used with a frequency-weighted acceleration magnitude of 5.0 ± 1.0 ms^{-2} r.m.s. at the handle. The frequency weighting used was W$_h$ as defined in British Standard BS 6842 (1987). The frequency range covered by the input vibration was 8 Hz to 1260 Hz. The waveform was sampled at 5480 samples per second and low-pass filtered at 1260 Hz before being fed to the vibrator. Acceleration signals from the handle and the palm adaptor were passed through signal conditioning amplifiers and then low-pass filtered at 1260 Hz via anti-aliasing filters with an attenuation rate of 30 dB/octave (5 poles) with a maximally flat response. The signals were digitised into a computer at a sample rate of 5480 samples per second. The duration of each vibration exposure was 5 seconds.

Analysis

Transfer functions were calculated between acceleration on the handle (i.e. the input) and acceleration measured at the palm-glove interface adaptor (i.e. the output). The 'cross-spectral density function method' was used. The transfer function, $H_{io}(f)$, was determined as the ratio of cross-spectral density of input and output accelerations, $G_{io}(f)$, to the power spectral density of input acceleration, $G_{ii}(f)$:

$$H_{io}(f) = G_{io}(f)/G_{ii}(f)$$

Frequency analysis was carried out with a resolution of 5.35 Hz and 108 degrees of freedom.

Results and Discussion

Transmissibilities between acceleration on the handle and on the palm adaptor for the subjects holding the handle while wearing the two gloves are shown in Figure 1. The data shown are for the 1 subject who participated in the intra-subject variability study and the 8 subjects who took part in the inter-subject variability study. It may be seen in Figure 1 that some of the transmissibilities for low frequencies are greater than unity; a value of unity would have been expected. The high transmissibilities at low frequencies are thought to be related to low coherency values for these data.

Very large variability is seen in transmissibilities between subjects for the two gloves tested. For glove 1, one subject (subject 4) showed a transmissibility below about 0.75 for frequencies above 300 Hz whereas most of the other subjects show values greater than 1.0. Phase for the transmissibilities presented in Figure 1 are shown Figure 2.

The transmissibilities for the two gloves are significantly different at frequencies above 192 Hz for the inter-subject variability study ($p < 0.02$, 2 tail, Wilcoxon matched-pairs signed-ranks test).

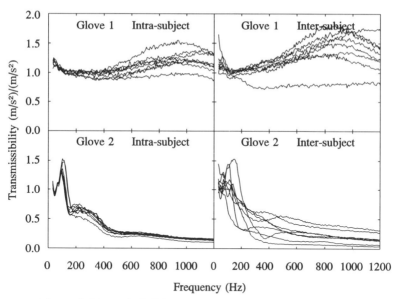

Figure 1. Transmissibilities between the handle and the adaptor
(5.35 Hz frequency resolution, 108 degrees of freedom).

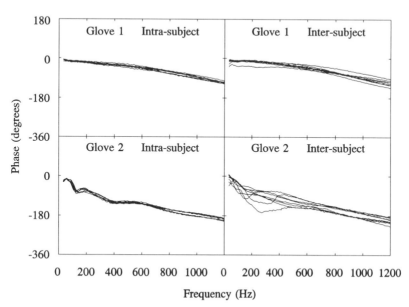

Figure 2. Phase between the handle and the adaptor
(5.35 Hz frequency resolution, 108 degrees of freedom).

Median and interquartile transmissibilities showing intra-subject and inter-subject variability for the two gloves are shown in Figure 3. A quantitative measure of the variability in transmissibility is required to compare the intra- and inter-subject variation. Relative variability, in this context, is defined as the inter-subject variation relative to intra-subject variation (Paddan and Griffin, 1994). The relative variability showed a maximum of 2.0 for glove 1 over most of the frequency range and about 10 for glove 2 over the high frequency range.

For the inter-subject variability, the interquartile ranges for the two gloves are approximately similar (about 0.3) at 1000 Hz for glove 1 and at 200 Hz for glove 2. However, the median values are different for the two gloves: 1.3 for glove 1 and 0.7 for glove 2. At these frequencies the interquartile ranges relative to the median values for the two gloves are 0.23 and 0.43: there is approximately twice as much variation for glove 2 compared to glove 1.

The variability presented here is of great importance in relation to the tests specified in ISO 10819 (1996). The results from three subjects (as specified in the standard) will vary according to the choice of subjects. The success or failure of a glove and whether it can be advertised as an "antivibration glove" could depend on the results from three subjects. The choice of subjects could be manipulated and used to an advantage without providing any improved protection to glove users.

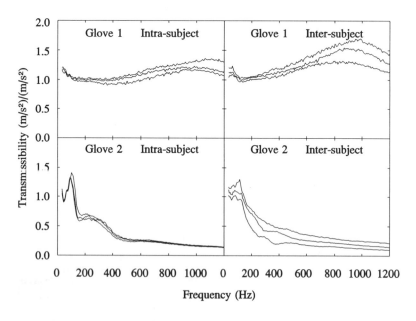

Figure 3. Median and interquartile ranges of transmissibilities between the handle and the adaptor (5.35 Hz frequency resolution, 108 degrees of freedom).

Conclusions

The variability in the transmission of vibration through gloves has been determined for eight subjects wearing two commercially available gloves. The variation between subjects was up to 10 times greater than the variation in repeat measures for one subject. The two gloves tested showed large differences in transmissibilities thus indicating the range of vibration characteristics that may be found in gloves.

References

British Standards Institution 1987, British Standard Guide to Measurement and evaluation of human exposure to vibration transmitted to the hand, *BS 6842*. London: British Standards Institution.

International Standards Organization 1996, Mechanical vibration and shock - Hand-arm vibration - Method for the measurement and evaluation of the vibration transmissibility of gloves at the palm of the hand. *ISO 10819 (1996)*.

Paddan, G.S. 1996, Effect of grip force and arm posture on the transmission of vibration through gloves. *United Kingdom Informal Group Meeting on Human Response to Vibration*, MIRA, Nuncaton, 18-20 September 1996.

Paddan, G.S. and Griffin, M.J. 1994, Individual variability in the transmission of vertical vibration from seat to head. *ISVR Technical Report No. 236*, Institute of Sound and Vibration Research, University of Southampton. November 1994.

Acknowledgements

This work has been carried out with the support of the United Kingdom Health and Safety Executive.

Appendix

Following are instructions that were given to the subjects taking part in the experiments on the transmission of vibration through gloves.

INSTRUCTIONS TO SUBJECTS
GLOVE VIBRATION: EFFECT OF ADAPTOR

The aim of this experiment is to measure the transmission of vibration through gloves to the palm and to the metacarpal bones of the hand.

Please stand and hold the handle with your right hand such that the upper arm is vertical and the forearm is horizontal. Ensure that your right arm (upper and lower) is not in contact with your body. You are to hold the handle such that the angle between your forearm and the direction of vibration is 45°.

If you are instructed to use the palm adaptor, then the adaptor should be inserted between the glove and the hand and positioned such that the transducer in the adaptor is in line with the direction of vibration. You will hold the handle and apply a push force of the 50 Newtons. Just prior to the start of each run, which the experimenter will indicate, you are to hold the handle with your right hand and apply the required push force as indicated on the visual display in front of you. This position is to be maintained throughout the vibration exposure.

You are free to terminate the experiment at any time.

Thank you for taking part in this experiment.

A discussion about the measurement of skin friction

W.P. Mossel

*Faculty of Industrial Design Engineering,
Dept. of Product and Systems Ergonomics
Delft University of Technology
Jaffalaan 9, 2628 BX Delft
The Netherlands*

The static skin friction of the right forefinger with stainless steel has
been measured for six subjects. Twenty-four measurements were taken.
The values of the measurements are put in both the linear and the
logarithmic friction models. In sixteen cases the logarithmic model fits
better than the linear model with a certainty of 99%. On one subject
eleven measurements were taken. The differences in these measurement
results are discussed. From the results of the measurements, the
limitations of the friction models in describing skin friction are
discussed.

Introduction

The fundamental law of friction is the law of Amontons. It describes the frictional
force F_f for non-lubricated surfaces as a linear function of the normal load F_n and the
coefficient of friction: $F_f \leq \mu . F_n$.

The independence of the frictional force from the size of the area of contact is
explicit in this equation. The coefficient of friction is among other things dependent on
the two materials. For soft surfaces and especially human skin the law of Amontons in
general does not apply. It is found (Bowden & Tabor, 1950) that for soft surfaces the
frictional force can be described as a non-linear function of the normal force. This is
supported by the measurements of Bobjer et al. (1993). Although the formula
$F_f \leq c . F_n^q$, developed by Comaish & Bottoms (1971) on the basis of measurements,
can not be used in general (Mossel & Roosen, 1994) it gives in principle a better
description of the skin friction. Based on the formula of Comaish & Bottoms and
formulas of Hertz a new formula to describe the skin friction has been developed
(Mossel & Roosen, 1994):

$$F_f \leq M . c_p . A_t^{1-q} . F_n^q \qquad \dots \dots (1)$$

in which A_t stands for the contact area and q for a dimensionless exponent smaller
than 1. M is a proportional factor. It has a dimension of $[(MPa)^{1-q}]$ when the forces are
expressed in [N] and the area in [mm^2]. c_p is the pressure distribution factor,

dimensionless and dependent on q. Formula (1) is further called the logarithmic model.

Measurements

To evaluate the logarithmic model it was decided that its reliability should be demonstrably better than the linear model. The rejection of the linear model is checked under the assumption of the logarithmic model. An error probability $\alpha = 0.01$ is chosen so that the result of the statement has a certainty of 99%. The rejection (or not) of the linear model is done on the basis of the Students-t table. All the measurements are done with the instrument described in (Mossel,1996) except the measurements A1, A2 and A3. The latter are done as described in Mossel & Roosen, 1994. (For the indication of the measurements, see the section Results and Discussion). In short, this means that the normal force F_n and the friction angle α are measured, excluding in the measurements A1, A2 and A3. The friction material was flat stainless steel. The measurements were done with nine subjects. The factor c_p is taken equally to 1 for the same reason as described in (Mossel,1996). The original intention was to find out if a value of q and M existed for human skin and stainless steel. The result of the group of five measurements on one subject gave the incentive to look for inter-individual differences and so the group of eleven measurements on one subject were conducted.

Results and Discussion

In table 1 (next page) the results of the measurements are collected. The measurements with the same letter are from the same subject. The measurements in which the desired certainty is smaller than 99% are printed italic. The measurements in which the q is found to be bigger than 1 are in conflict with the set-up of both the logarithmic model and the literature (Bowden & Tabor, 1950). They are printed in bold. The measurements A1-A4, B and C1 have been presented previously (Mossel,1996). There was a reasonable doubt about the quality of measurement A(4). The measurement instrument failed after this measurement and had to be repaired. The measurement A4 looked doubtful, not only for the value of q being bigger than 1 but especially for the very small value of μ compared to the other values of μ for this subject. These two facts combined with the

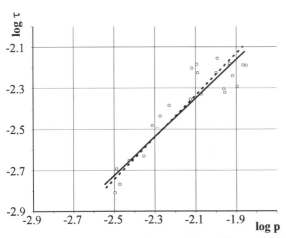

∘ measuring points — logarithmic model ·· linear model

Figure 1

Log τ as a function of log(p), subject C4

Table 1

Values for *q*, *M* and *μ* found in *n* measurements

subject	q [-]	M $[(MPa)^{1-q}]$	μ [-]	n [-]
A1	0.686	0.195	1.19	25
A2	0.564	0.086	1.03	35
A3	0.564	0.755	1.16	43
A4	1.314	1.627	0.38	25
A5	0.602	0.104	0.63	25
B	0.521	0.031	0.32	25
C1	0.632	0.119	0.63	25
C2	1.034	0.831	0.72	25
C3	0.844	0.321	0.64	25
C4	0.818	0.257	0.57	25
C5	1.065	0.719	0.54	25
C6	0.856	0.431	0.87	25
C7	0.815	0.361	0.88	25
C8	0.665	0.134	0.70	25
C9	0.694	0.140	0.62	25
C10	0.473	0.049	0.64	25
C11	0.864	0.262	0.51	25
D1	0.933	0.759	1.05	25
D2	0.813	0.605	1.50	25
E	0.815	0.221	0.47	25
F	0.862	0.180	0.32	24
G	1.204	0.704	0.27	25
H	1.302	2.040	0.48	25
J	1.364	1.428	0.25	25

possibility of a (partly(?)) failing measurement instrument gave rise to the doubts. On the other hand nothing was wrong in the measurement A5. The comparative low value of μ was the reason for looking for intra-individual differences. Therefore 11 measurements were done with one subject, namely C. In three measurements (C3, C4 and F) the certainty required to reject the linear model is not achieved. In figure 1 the measurement C4 is exposed. The figure shows the logarithmic model line and the linear model line being very close to each other. This small difference means that the linear model cannot be rejected with the certainty of 99%. The chance that the linear model is better than the logarithmic model is not lower than 1% but around 25%. For C3 and F these values are 21% and 5%. No clear reason has been found for the small difference between the logarithmic and linear models in these cases.

In six cases (A4, C2, C5, G, H and J) the value of *q* is bigger than 1. In figure 2 the measurement C5 is given. The lines for the logarithmic and linear model nearly cover each other because the value of *q* is very near to 1 (in case *q* = 1 they cover completely). Two measuring points are marked. These two measurements engender the anticlockwise turning of the line of the logarithmic model. In figure 2a the line of the logarithmic model should not be 'hollow' but 'domed'. The question is why the friction is suddenly so high whilst in the rest of the measurement it is rather stable. In table 2 the measured values of the normal force F_n and the friction angle α in the case of subject C5 are presented. The measurements took place in the order as presented. The two marked measurements of figure 2 are printed in bold. Each group of five measurements stands for a certain force: 0.5, 1, 2, 3, and 4 N. It is striking that the highest friction coefficient is the first measurement in the group (except in the case of the 4 N group). It could be the result of the action that before each start with a new force, the finger is cleaned with alcohol. When the finger is still a little damp in the

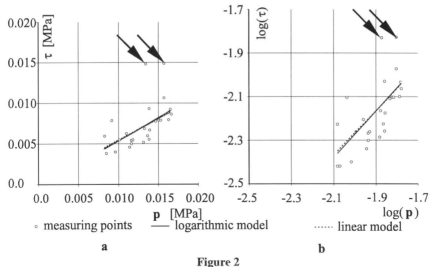

Figure 2

τ as a function of p and $\log(\tau)$ as a function of $\log(p)$, subject C5

bottom of the skin profile, the friction could be higher than in case of a completely dry skin. Because of the assumed linear *logarithmic* relation of the normal force (see next section) the size of the area of contact makes the area at a high force relatively small, so the pressure rises relatively quickly. This could be an explanation for the two extreme points in figure 2. For the measurements G, H and J this can not be a reason

Table 2

Measured values of F_n and α,

subject C5

F_n	α	μ		F_n	α	μ
[N]	[°]	[-]		[N]	[°]	[-]
0.377	40.6	0.86		**1.765**	**47.8**	1.10
0.247	35.7	0.72		**3.365**	**43.5**	0.95
0.275	24.2	0.45		3.365	34.2	0.68
0.455	22.5	0.41		2.824	27.7	0.53
0.561	28.4	0.54		3.012	27.3	0.52
0.777	29.7	0.57		4.110	29.3	0.56
0.965	25.0	0.47		4.235	27.5	0.52
1.035	25.0	0.47		2.541	27.9	0.53
0.973	23.3	0.43		2.008	34.0	0.67
0.894	22.0	0.40		3.890	26.1	0.49
1.678	27.4	0.52				
1.898	23.6	0.44				
2.055	25.7	0.48				
1.616	21.5	0.39				
2.008	21.7	0.40				

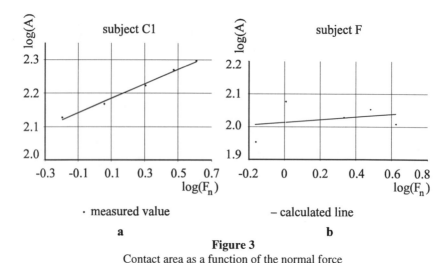

· measured value   − calculated line

a   b

Figure 3

Contact area as a function of the normal force

for q being bigger than 1. A list as table 2 for these measurements does not show striking differences at the start with a new force, but the influence of the area of contact in the calculation could cause the effect.

The area of contact

The scatter in all measurements is wide. A possible reason can be the effect of the area of contact in the calculation of the pressure (p) and the frictional stress (τ). The size of the area of contact will be larger when the normal force is higher. From this point of view the finger tip can be considered as a spring. It is not plausible that it will be a spring with a linear spring constant. The structure of the finger is too complex. In measurements it was found that the function between contact area and normal force could be described reasonably as:

$$\log (A_t) = a \ \log (F_n) + b \dots\dots\dots\dots\dots\dots\dots\dots\dots\dots\dots\dots\dots\dots\dots\dots\dots.(2)$$

In figure 3a a clear example of equation (2) is presented. Figure 3b makes clear that a straight line in cases likes this does not give a serious presentation of the connection between F_n and A_t. . For the measurements F, G and H this large scatter occurs, which could be a reason for q being more than 1 in the measurements G and H. For the measurement F this could be a reason for the certainty smaller than 99%. In the cases A4 and C2 the connection between force and area is like C1. No reason has yet been found to explain why q can be greater than 1, and nor can measurement J be explained by the foregoing approach. Maybe the only deviation in this case is the subjects skin still being black in the bottom of the skin grooves, even after cleaning with alcohol. (The subject had been busy with car repair and working in the workshop).

Table 4

Values of a and b, subject C

no	a	b
1	0.216	2.163
2	0.137	2.373
3	0.269	2.261
4	0.250	2.248

On subject C four measurements of the contact area were done. The results of the values of a and b from formula (2) are given in table 4. The differences are clearly visible; the reason not. It could be an effect of 'memory'. The finger maintains its impressed form, some time after the pressure is taken off. (Edin et. al., 1991) During the measurements this effect is avoided by spacing the measurements and doing the measurements with the lowest force first and the highest last.

Conclusions

In 16 cases from the 24 measurements the logarithmic model fits better than the linear model with a certainty of 99%. Nevertheless from the results presented in table 1 the problem arises that for a designer the values of q and M are seemingly dependent on the subject. A simple number is not enough to describe the friction between skin and stainless steel. The differences in the measurements C1 up to C11, the intra-individual differences, are not great, but noticeable. Amongst other reasons it could be that they are caused by differences in air humidity at the moment of measurement. The influence of humidity is known from literature (Highly et. al., 1977; Wolfram, 1983). It is plausible that humidity causes a change in the modulus of elasticity of the skin, if there is such a thing as modulus of elasticity of skin. Simplifying the skin *in vivo* by a mechanical model shows a modulus of elasticity but the question is if it can be expressed in a constant value useful for calculation (estimation) of skin friction. It is shown in the derivation of the logarithmic model (Mossel & Roosen, 1994) from the equation of Herz that the modulus of elasticity of the skin is hidden in the factor M. It is possible to make the modulus of elasticity explicit in the logarithmic model. The question is what value for the modulus of elasticity can be used? More research is needed.

References

Bobjer, O., Johansson, S. & Piguet, S. 1993, Friction between hand and handle. Effects of oil and lard on textured and non-textured surfaces; perception of discomfort, *Applied Ergonomics*, **24** (3), 190-202.

Bowden, F.P. & Tabor, D. 1950, *The friction and lubrication of solids*, (Oxford University Press, London), 5-32.

Comaish, S. and Bottoms, E. 1971, The skin and friction: Deviations from Amontons' laws and the effects of hydration and lubrication, *British Journal of Dermatology*, **84**, 37-43

Edin, B.B., Westling, G. & Johansson, R.S. 1991, Independent control of human finger-tip forces at individual digits during precision lifting, Journal of Physiology, **450**, 547-564.

Highley, D.R., Coomey, M., DenBeste, M. & Wolfram, L.J. 1977, Frictional Properties of skin, *The Journal of Investigative Dermatology*, **69**,303-305

Mossel, W.P. & Roosen, C.P.G. 1994, Friction and the skin, *Proceedings of the Ergonomics Society's 1994 Annual Conference*, (Taylor & Francis, London) 353-358

Mossel, W.P., 1996, The measurement of skin friction, *Proceedings of the Ergonomics Society's 1996 Annual Conference*, (Taylor & Francis, London), 69-74

Wolfram, L.J. 1983, Friction of skin, *Journal of the Society of Cosmetic Chemists*, **34**, 465-476.

VISUAL SEARCH

USER REQUIREMENTS OF VISUAL INTERFACES TO PICTORIAL DATABASES

Andrée Woodcock
Design Research Centre,
Derby University,
Mackworth Road,
Derby, DE22 3BL

Stephen A. R. Scrivener
Design Research Centre,
Derby University,
Mackworth Road,
Derby, DE22 3BL

Mark. W. Lansdale
Department of Human Sciences
Loughborough University,
Loughborough,
Leics., LE11 3TU

Querying of visual databases has relied predominantly on text based systems even though words do not always provide an appropriate or adequate means of describing visual artifacts. The DIVIUS (Depictive Interaction with Visual Information Using Sketches) interface allows users to describe, encode and query objects in a pictorial database, using a visual language derived from the database objects. Users can also indicate their levels of uncertainty regarding certain visual attributes of the query. This paper considers the nature of querying pictorial databases and the implications of this for the development of pictorial databases. It then outlines the system and presents preliminary usability results.

Introduction

A pictorial or visual database is a series of images which have been amassed for a particular reason. They can include photographs, drawings, paintings, slides, prints, or any approximately 2D, static item containing information in the form of a representational image, Shatford, 1986. Storage and preservation of such collections, especially when people need regular access to them can lead to degradation and loss of the original.

Visits to such diverse institutes as the Illustrated London News, Courtaulds Fabrics, Imperial War Museum and two hospitals (histology and ophthalmology departments) have shown that users interrogate databases for many reasons. Searches may be general or specific, precise or vague, and may be solved by the provision of one, or a group of images. The images themselves might be needed for instruction, communication, exemplification, sourcing, problem solving (e.g. medical diagnostics), evocation of moods, periods in history and general information. In some cases image retrieval can be mediated through text based methods, for example if the required items can be accessed via patient number.

DIVIUS has been developed to aid in the retrieval of objects which cannot always be accessed from their textual description. For instance when the retrieval relies on a users memory which cannot be translated into the terms of reference used in the system. For example, an author might wish to find photographs of soldiers with 'battle weary expressions' to illustrate an article s/he is writing, a doctor might 'remember' s/he has seen a similar cell pattern before, but cannot remember the patient number. If the items have not been specifically stored in terms of their visual attributes, the user must either search through all the photographs or rely on the archivists memory and ability to match the request with the items in the database. The role of the archivist in all searches is crucial - in interpretation of the request, knowledge of the database, and ability to physically locate information (not a trivial task). In such cases the categorisation and classification methods used to describe (index and retrieve) the visual material restrict the likelihood of mediating a successful search and limit the usefulness of the entire database.

Integral to the design of DIVIUS are the premises that text is undeniably a poor means of describing pictures and art, it is not always the most appropriate means of mediating interaction with a pictorial database: and secondly, that there is frequently a mismatch between the words used by the cataloguer and the query of the user. DIVIUS is a visual interface to a pictorial database which allows the information retriever and archivist to use a pictorial language, derived from the database images, to initiate searches and classify items in the database.

The domain and the visual language

After visiting several sites, the archives of Crown Derby were chosen as exemplifying many important attributes of visual databases: original artifacts are fragile, expensive and irreplaceable and not readily accessible to members of the public: various people are interested in looking/searching for particular cups: the artifacts need to be preserved; identification requires obtaining a satisfactory description of the object from the information retriever, translating this query into a visual memory, and then searching in the archives for likely matches.

Clearly computerisation of this type of archive would be beneficial. In the project we created a computer database from a small sample of this material (120 images), designed a visual language to describe the entries and developed an interface which could be used to describe the database contents (i.e. classification of the objects) and retrieve objects from it.

Textual descriptions of 700 cups were content analysed producing 9 features; body shape, fluting, handles, borders, rims, lid, reserves (miniature, framed painting - usually attributable), surface decoration and inside border; and their related attributes, which could be used to describe and distinguish between artifacts in the database.

Description of system

DIVIUS was developed in Delphi on a PC Pentium. The direct manipulation 'click and point' interface consists of a sketch area on which the schematic (similar to an identikit picture) is built up; a high level menu along the top of the screen from which the 9 high level features can be selected; a series of pop up menus for the selection of attributes (e.g. type of pattern); a control panel to move the schematic/template on the sketch area and system control features. Using these features a description of the cup can be constructed, for either encoding or retrieval. The final description contains approximately 40 items relating not only to features and their attributes but also to position, size and aspect ratio and positional uncertainty (for borders, handles and fluting). This information can be recorded in under 4 minutes (by experts and novices).

The query or encoding is built up step by step as a schematic representation (consisting of movable tiles) on the sketch area along with a high level signature (across the top of the screen). Classification of images takes place in the same manner as retrieval (i.e. using the same visual language). In querying the database, once a query has been completed to the users satisfaction it can be matched against the description of items in the database. At this stage users are given the option of including all or just some of their description for retrieval. For example, if someone knows they are poor at remembering aspect ratios, all of these attributes can be switched off. The final query is matched against descriptions of the objects in the database.

The match algorithm is described in Bartlett (1996). In outline, each attribute selected in the query has a match score calculated for it with respect to each stored object. The total overall score for a stored object is calculated as the average score of all the selected attributes of the query. The match algorithm exploits experimental results, Lansdale (in press) which show that users recall positions more accurately when they are close to natural anchor positions in an image. The DIVIUS matching algorithm assumes two such anchor positions, the top and bottom of the cup body, and weights the match score of a location feature as a function of its position with respect to a cue line.

Experiments have established that subjects are more confident about exact, or categorical, than inexact location recall. This suggest that a measure of user confidence in the location information provided might be used to select between exact and inexact matching mechanisms. This concept was implemented in DIVIUS by allowing users to define a confidence level associated with the location of features, such as borders. If the location of a corresponding database feature falls outside of this confidence window then a zero match is returned rather than a poor matching value.

Items which achieved a high level match are displayed in an intermediate representation, the match matrix, Lansdale and Edmonds (1992). A 16 level gray scale is used to show the nearness of the match on each attribute. An overall score (in gray scale) is also given. Items are rated according to the overall score, with the best match at the left hand side of the index. Any of the items can be selected to produce a thumbnail image of the item from the database. Inspection of these images has been shown to help in further refining the search

The design of DIVIUS incorporated features identified by previous research as being important to the design of pictorial interfaces e.g. filtering and browsing, Besser (1992), general hci, Shneiderman (1992, 1993), graphical representation and immediate display of results, visible limits on query ranges and interactive feedback.

Evaluation

DIVIUS was evaluated using 8 postgraduate engineers and ergonomists who, after 15 minutes training, completed a series of tasks using the interface which took 2- 3 hours to complete after which they rated the usability of the system using a questionnaire based on Shneiderman (1992). The evaluation was conducted to determine firstly the usability of the interface for encoding and retrieval; secondly, the appropriateness of the representation of visual information, and thirdly the potential for using an uncertainty measure. System performance was not measured directly at this time

Usability of the interface for encoding and querying

There is a strong association between the time taken to describe a cup and its complexity (as measured by the number of features it possesses). An encoding consisting of 40 items can be made in less than 4 minutes. Preliminary results (repeated encoding of 10 items) showed that the two 'experts' were consistent over time, and with each other. More research of a longitudinal nature needs to be undertaken to look at the stability of perceptual encoding over time.

One problem which did emerge for subjects was in combining information from two separate sources on the interface. Subjects focused their attention primarily on the schematic (and associated tiles) and overlooked information relating to the high level features (e.g. presence of inside border, lid, two handles), which was only present as a series of checked boxes. A solution would be to position all information appertaining to the query directly within the field of view, appending it to the schematic, as a label tile, after entry.

Overall subjects found it easy to encode cups from the photographs and were consistent with each other, especially in relation to the high level features, positional information and unambiguous attributes (such as type of reserve). Borders proved more problematic because they can be perceived and encoded in different ways. However there was consistency in the encoding of positional information and aspect ratios between subjects. Memory for positional and aspect ratio information was also good.

In developing a query subjects worked along the high level features first, especially those which did not require any further entries (e.g. a lid was either present or absent - once this information had been recorded, it did not have to be attended to further). They found the selection of features and match matrix easy and intuitive to use. When building a query from memory the interface actually functioned as an aid to memory. A more rigorous investigation is needed to determine which type of features actually degrade with memory, and whether these differences can be predicted and used for matching in the database. For example our results indicate that positional information does not degrade greatly.

The appropriateness of the representation of visual information

Subjects rated the system as usable overall and did not appear to have any problems in understanding and using the schematic representation effectively. Verbalizations which occurred during the course of experiments indicate that subjects converted the screen description into a mental 'image' of the target. Verbalizations such as 'it had a handle, but not that sort - it was bigger than that' suggest that visual memory was being accessed

The visual language itself was sufficient and comprehensible. However, the domain chosen was limited. When applied to describe cups outside of the period 1790-1810, and to cups produced by other manufacturers the palettes (visual descriptions of features) were found to be restrictive. This could be overcome by enriching the language with actual examples of items as well as their representation as a series of sketches. Sketches were also used to represent both concrete objects and abstract concepts. Subjects understood that a sketch of a flower represented a whole class of flowers which might be painted on the cups (e.g. roses, harebells), but found difficulty in applying the same rule to sketches representing complex concepts such as geometric border.

The potential for use of the uncertainty

Initially we were interested in looking at how psychological experimentation could be fed into system development, Lansdale, Scrivener and Woodcock (1996), especially with regard to positional uncertainty when querying databases from memory. Two ways of representing positional uncertainty were used, either the specification of the minimum and maximum positions within which the feature could occur or specification of the range within which the centre of the feature could lie. Subjects had the opportunity to use both methods to record uncertainty when querying the database from memory. Subjects found the facility to represent uncertainty desirable but neither of the methods implemented were particularly usable.

It is hoped to extend this work in three ways; firstly, to show the influence this type of information might have on the success of the query (i.e. it will widen the search criteria). Further investigations are required to show when uncertainty is used and to monitor the effects of its use in database searches; secondly to look at the representation of other uncertainties (e.g. users model of the encoder); and thirdly to look at other means of representing uncertainty at the interface (e.g. blurring of the image).

Conclusions

DIVIUS was developed as a predominantly visual interface to a pictorial database allowing both encoding (or cataloguing) and retrieval of database items through the same interface. The DIVIUS schematic could be interpreted by users to distinguish the object, and the interface allowed the input of visual and spatial information which formed the query. Evidence suggests that the structure of the interface aided recall. The intermediate representation of results in the form of an attribute list and match matrix enabled queries to be further refined. The ability to show uncertainty was seen as desirable.

In conclusion, we believe that DIVIUS represents a novel development in pictorial database systems for the following reasons:

1. It exploits visio-spatial (visual appearance) information in pictorial database encoding and retrieval. In particular, the user is able to construct a query that shares a visual resemblance with the pictures it is intended to represent;

2. The language for describing a picture combines both depictive and descriptive (or categorical) information in a single coherent representation - a sketch, Charles and Scrivener (1990);

3. The interface is designed such that the same method of description is used both in encoding and retrieval. This offers the potential for exploiting, in retrieval, the recall of visual memory acquired during encoding (in cases where the retriever is also the encoder).

We take the evaluation results as demonstrating the potential of pictorial databases that utilise visio-spatial information. A task analysis of pictorial database domains led us to conclude that what users do is strongly constrained by current technology, the content of pictures, and past and present conceptions of task. It would be of great interest to us to investigate how existing tasks evolve and new tasks develop when using a system such as DIVIUS. This could only be accomplished through a longitudinal study of extended field trials of the system.

Acknowledgements

Sharon Thomas and Sean Clark conducted site visits. The research was conducted in the Departments of Human Science and Computer Studies of Loughborough University, supported under the MRC/ESRC/SERC Joint Council Initiative, grant number G9200538.

References

Bartlett, R.A.,1996, *The Development of the DIVIUS Database*, Report, Loughborough University.

Besser, H., 1990, Visual access to visual images: the UC Berkeley image database project, Library Trends, **38**, 4, 787-798.

Charles, S. and Scrivener, S.A.R., 1990, Using depictive queries to search pictorial databases, *Proceedings of Human-Computer Interaction - INTERACT'90,* eds. D. Diaper et al., Elsevier Science Publishers B.V. (North Holland), 493-498.

Lansdale, M.W., 1996, Modelling errors in the recall of spatial position, MS submitted for publication.

Lansdale, M.W. and Edmonds, E.A., 1992, Using memory for events in the design of personal filing systems, IJMMS, **36**, 97-126.

Lansdale, M. W., Scrivener, S.A.R. and Woodcock, A., 1996, Developing practice with theory in HCI: applying models of spatial cognition for the design of pictorial databases, International Journal of Human Computer Studies, **44**, 777-799.

Shatford, S., 1986, Analysing the subject of a picture: A theoretical approach, Cataloguing and Classification Quarterly, **6**, 3, 39- 62.

Shneiderman, B., 1992, *Designing the User Interface; Strategies for effective Human Computer Interaction, 2nd edition,* Addison - Wesley, Boston.

Shneiderman, B., 1993, Dynamic queries; a step beyond database languages, Human Computer Interaction Laboratory, University of Maryland, September.

VISUAL SEARCH FOR LETTERS AND WORDS WITH DIFFERENT DISPLAY SHAPES

ALAN J. COURTNEY and PO YAN LAI

Department of Industrial and Manufacturing Systems Engineering,
University of Hong Kong, Hong Kong.

Display shape for visual displays has long interested researchers. This paper investigates visual search performance for various display shapes using either letters or words as search items, and considers the role of the visual lobe in determining search performance. A variety of array shapes were tested for each type of search item, and it was found that horizontally elongated arrays were searched faster than vertically elongated arrays. In general, search performance decreased as rectangular array shape gradually changed from horizontally elongated to vertically elongated.

Introduction

In a very early study, Paterson and Tinker(1940a) investigated the effect of width of line on speed of reading and found that there exists an optimal range of line width. Bloomfield(1970) studied search performance for three categories of display, namely, vertical, square and horizontal. More recently, Scott and Findlay(1990, 1993) investigated VDU user performance with various display array shapes. Subjects fared significantly better with an extreme landscape array. Visual search is a common and basic operation, so even a small improvement in search efficiency may have a large practical effect.

In a study with subtitled television programs, Praet, Verfaillie, De Graef, Rensbergen & d'Ydewalle(1990) found that the average number of fixations for one-line sentences (0.15 fixations per character) was larger than for two-line sentences (0.12 fixations per character). Pollatsek, Raney, Lagasse, and Rayner(1993) investigated search of multi-line displays but did not find any benefit from previewing the lines below fixation.

Paterson and Tinker(1940b) described a series of studies with variation in type size, line width, leading and color of print and background, which were all factors that affected speed of reading. By using undesirable variations of these factors they obtained a non-optimal arrangement which retarded speed of reading 22.2 percent in comparison with an optimal arrangement. In studying effect of width of line to speed of reading, they found that materials with width set in 19, 23, 27 and 32 picas (one pica is about 1/6 inch) were all read significantly faster than text in 14, 36, 40 and 44 picas. In another study, they varied line width by one pica steps from 17 through 27 picas. No differences in speed were found within this range.

Scott and Findlay(1993) analyzed eye movement data for letter search tasks. Their results showed that subjects tended to show larger saccades with more horizontally elongated arrays. There was also a trend towards shorter regressions with the horizontally elongated materials. Greater efficiency on more horizontally biased arrays was due to longer saccade amplitude. Their studies of search with arrays of different shapes demonstrated a superiority for horizontally elongated arrays over vertically elongated ones. They explained that horizontal arrays may be better matched to the visual lobe for text-like material. Also, the array shape may affect the size of scanning saccades, leading to more efficient scanning behavior.

The visual lobe is generally elongated horizontally(Courtney, 1985). Tinker and Paterson(1940) found that for 9 pica line width, words/fixation was only 1.4 compared to 1.6 for 19 pica line width. This may have been due to the inefficient use of the visual lobe at the both ends of lines; an effect which is relatively more significant in short lines.

Parafoveal preprocessing of the next word(Balota, Pollatsek & Rayner, 1985; Balota & Rayner, 1983; Inhoff & Rayner, 1980) will also favor horizontal displays. Since parafoveal preprocessing allows the first few letters of a word to be previewed, it might also cause the optimal landing position to be displaced to the right and allow faster gaze through the line. But another factor concerns return sweeps at the end of a line. Short lines allow more accurate returns. Long lines will result in some time being taken up adjusting for the landing point of return sweeps to the beginning of a line. Sometimes, readers may also need to get a clearer view of material or to reread it. Tinker and Paterson(1940), found that regression frequency for line widths 9 and 19 picas; and 19 and 43 picas are 22.5 and 26.2; 27.4 and 42.9 respectively. In the case of 43 pica line, regressions were chiefly at the beginning of lines.

For reading passages of about 20 words in a 3-line format, Everatt and Underwood(1994) showed that fixation duration data correlated significantly with passage reading time. The amount of time spent refixating a target word was found to be the main predictor of passage reading time.

The experiment here will be about the effect of layout of presentation on visual search performance specifically whether the array shape will have different effects for letter and word searches.

Method

Subjects

Four subjects, all undergraduates at the University of Hong Kong, volunteered to participate in this study. They were all male Chinese speakers literate in English and had normal or corrected-to-normal vision. The age range of subjects was 21 to 23. They were informed of the purpose of the experiment.

Stimuli and design

The stimuli were either arrays of lines of English alphabetic characters or arrays of lines of three letter words presented on a Cathode Ray Tube (CRT). Text was single-line spaced and 24 points(1 point = 1/72 of an inch) Times Roman in black on a white background was used to simulate experiments using projector or printed materials. Row by column dimensions of the arrays were 3×20, 5×12, 6×10, 10×6, 12×5 and 20×3. Thus, each array had 60 items. Letters or words were evenly spaced vertically and horizontally. Each subject saw all array conditions in a randomized design which was different for each subject. Subjects were seated 0.4 m from the display.

In any array there was only one target which was either 'a' or 'eat' which replaced one of the items in an array. Thus there were 59 distractors and one target in each array. The character sets used in the experiment were:

1. a   s   e   o   c
2. t   i   l   f

The character sets were chosen on the basis of the confusable character sets(Bouma, 1971). In the case of letter search, the target was 'a' and the distractors could be 's', 'e', 'o' or 'c'. In word search, the target was 'eat' and the distractors could be words like 'oet', 'eal' or 'ast', made up with the first two letters from group 1 and the third letter from group 2.

Apparatus and procedure

The subject sat in front of the CRT of a micro-computer with head fixed by means of a chin rest. The center of display was below the eye level and subtended 7° in the horizontal direction.

At the start of each presentation, there was a fixation cross in the top left corner of the search area. Subjects were ask to fixate the cross before each search in order to ensure that they started at the same point. The only input device provided to subjects was a computer mouse. For each search, when the subject was ready, the mouse button was pressed to start the search. The fixation cross was replaced by the stimulus array. When the target was found, the subject pressed the mouse button again to terminate the search. A pointer then appeared on the display and the subject had to move the pointer to the location of the target and select it by pressing the mouse button.

There were 3 practice searches followed by 8 trials for each array shape and there were six blocks of trials according to the six array shapes. Between any two blocks, there was a 1 minute break. Each block being all letter searches or all word searches. The sequence of the blocks were randomize and different for each subject in a counterbalanced design with two subjects starting with word searches and two starting with letter searches.

Results

In analyzing the results, the primary dependent variable was mean search time for an array. The search times for letter and word searches are shown in Fig. 1 plotted against array shape.

Subjects performed differently for letter searches and word searches. Letter searches were more affected by array shape than word searches. Search time for the 5×12 (horizontal) array being shortest with mean search time of 1.40sec. For the same number of items in each search, subjects searched 5×12 arrays 53% faster than 20×3 arrays. Results were analyzed using a two-factor ANOVA. For letter searches, there was an array shape effect for search time ($F(5,15)=3.178$, $p<0.05$). There was also a within-subjects effect for search time ($F(3,15)=3.175$, $p<0.1$). For word searches, there

was no array shape effect (F<1) and a within-subjects effect at only 0.2
(F(3,15)=1.97).

For word searches, each subject performed differently for different array shapes
(Fig. 2). Only subject 1 showed a search performance pattern similar to letter search
pattern. Subject 4 was the best search performer in both letter and word searches and
gave the least variable performance for word searches (search time between 5.5 and
10.2 sec). Subject 1 has the slowest search time for the 20×3 array (26.4 sec) and
fastest search time for 6×10 array (5.3 sec).

Table 1 shows mean search speed for each subject for both letter and word
searches. Subjects generally maintained their relative performance for the two types of
search. Subject 4 did the best for both letter and word searches while subject 1 did the
worst in both. Though subject 2 was better than subject 3 in letter searches, he was
slightly slower than subject 3 in word searches.

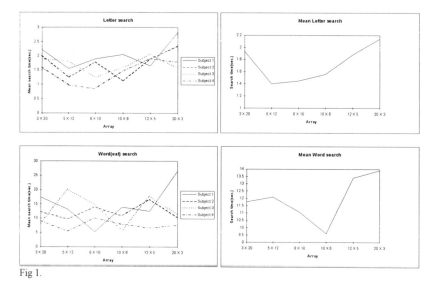

Fig 1.

Table 1. Search speed

Search speed(items/sec.)	Mean letter search	Mean word search
Subject 1	29.5	4.1
Subject 2	34.6	4.9
Subject 3	35.3	4.6
Subject 4	41.9	7.7
S.D.	5.1	1.6

Discussion

For the letter searches conducted here, there was a U-shaped function for search
time against array shape. Though with the exception of the 3×20 array, in letter search,
performance decreased as rectangular array shape gradually changed from horizontally
elongated to vertically elongated. The most horizontally elongated array (3×20), and

the most vertically elongated array (20×3), both showed poor search performance. One possible explanation is the under-utilization of the visual lobe. Although the visual lobe varies in size and shape from one task to another, it is likely that for a visual search task of this nature, it is roughly elliptical in form; its horizontal dimension being greater than its vertical. Therefore search areas cannot be tessellated without some degree of lobe area overlap. In addition, unless the array dimensions are in some multiple of visual lobe dimensions, there will inevitably be some area not utilized where lobes extend beyond the array dimension (Fig. 2).

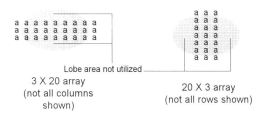

3 X 20 array
(not all columns shown)

20 X 3 array
(not all rows shown)

Fig. 2

Because the visual lobe has roughly the shape of an ellipse, it cannot be utilized as efficiently in vertical array as in horizontal array unless the width of the vertical array matches lobe width. This appears to be the case in the mean search time for the 3×20 array (1.9s) was shorter than that for the 20×3 array (2.1s) here. The result agrees with the finding of Scott and Findlay(1990) where search performance for an extreme horizontal array was superior to that for an extreme vertical array. The visual lobe should be a useful predictor of search performance for various array of items, like the letter searches used here.

Though letter searches showed a systematic effect for display shape, word searches did not. This may imply that searches involving higher levels of processing may reduce inter-subject differences. Another explanation is the effect of foveal task complexity. Ikeda and Takeuchi(1975) reported noticeable shrinkage of useful field of view as the primary foveal task increased in complexity. Chan(1993) also found that increase in foveal cognitive load affected peripheral target detection performance. For stimuli of the type used here, foveal attention probably played a larger part in processing items for word searches than it did for letter searches. Subjects were required to scrutinize items one by one for the word search and peripheral vision may not help much without some conspicuous feature present.

In a survey of printing practice among printers, Paterson and Tinker(1940) found no single preferred line width in double-column printing. American non-scientific journals showed variations in line widths from 12 to 22 picas, whereas American scientific journals varied the line widths for the most part between 13 and 18 picas. Single-column printing journals tended towards relatively long line widths i.e. 23 to 28 picas. Textbooks, on the other hand, varied between 19 and 24 picas. There was a strong tendency to limit line widths to a narrow range. For magazines, the typical line width was in the neighborhood of 24 picas or about 4 inches, whereas for books the typical line width was approximately 21 picas. There was no single preferred line width in printing. They also found that unless unusually long or short lines were used, reading speed was not affected to a significant degree.

Conclusion

Display shape clearly has an effect on visual search where foveal loading is not too high, such as in the letter searches used here. For high foveal loading tasks like word searches of the type used here, though performance was not significantly affected by display shape, the subjects still maintained their relative search performance. Both foveal and peripheral tasks can be useful to test search ability. Peripheral or low foveal load tasks especially can be enhanced with proper arrangement of search items.

References

Balota, D.A., Pollatsek, A., & Rayner, K. (1985). The interaction of contextual constraints and parafoveal visual information in reading. *Cognitive Psychology*, **17**, 364-390.

Balota, D.A., & Rayner, K. (1983). Parafoveal visual information and semantic contextual constraints. *Journal of Experimental Psychology: Human Perception & Performance*, **9**, 726-738.

Bloomfield, J. (1970). Visual Search. PhD Thesis, Nottingham University.

Bouma, H. (1971). Visual recognition of isolated lower-case letters. *Vision research*, **11**, 459-474.

Chan, H.S.(1993). Effects of cognitive foveal load on a peripheral single-target detection task. *Perceptual and motor skills*, **77**, 1993, 515-533.

Courtney, A.J.(1985). Development of a search-task measure of visual lobe area for use in industry. *Int. J. prod. Res.* **Vol. 23(6)**, 1075-1087.

Everatt, J., Underwood, G.(1994). Individual differences in reading subprocesses: Relationships between reading ability, lexical access, and eye movement control. *Langauge and speech*, **37(3)**, 283-297.

Ikeda, M. & Takeuchi, T. (1975). Influence of foveal load on the functional visual field. *Perception & Psychophysics*, **18**, 255-260.

Inhoff, A.W., & Rayner, K. (1980). Parafoveal word perception: A case against semantic preprocessing. *Perception & Psychophysics*, **27**, 457-464.

Paterson, D.G. and Tinker, M.A. (1940a). Influence of line width on eye movement. *Journal of experimental psychology*, **27**, 572-577.

Paterson, D.G. and Tinker, M.A. (1940b). *How to make type readable*. New York: Harper & Brothers publishers.

Paterson, D.G. and Tinker, M.A. (1944). Eye movements in reading optimal and non-optimal typography. *Journal of experimental Psychology*, **34**, 80-83.

Praet, C, Verfaillie, K, De Graef, P. & d'Ydewalle(1990) A one line text is not half a two line text. In *From eye to mind: Information acquisition in perception, search and reading*, R. Groner, G. d'Ydewalle, R. Parham (eds.). North-Holland.

Pollatsek, A., Raney, G.E., Lagasse, L. and Rayner, K.(1993). The use of information below fixation in reading and in visual search. *Canadian journal of experimental psychology*, **47(2)**, 179-200.

Scott D. and Findlay J.M. (1990). The shape of VDUs to come: A visual search study. *Contemporary Ergonomics* 1990, 353-358.

Scott D. and Findlay J.M. (1993). Visual search and VDUs. *Visual search 2*. Taylor & Francis, London.

EFFECT OF DEFECT PERCENTAGE AND SCANNING STRATEGY ON VISUAL SEARCH PERFORMANCE

Stefan W. H. Chow and Alan H. S. Chan

Department of Manufacturing Engineering
and Engineering Management
City University of Hong Kong
Tat Chee Avenue
Kowloon Tong
HONG KONG

In this experiment, the effects of defect percentage and scanning strategy on five different performance measures viz. correct acceptance percentage (CAP), correct inspection percentage (CIP), hit rate (HTR), inspection rate (IPR), and efficiency in improving the product (EIP) were investigated with simulated industrial inspection tasks. Analysis of variance (ANOVA) of the results showed that within the range of values tested, both defect percentage and scanning strategy were not significantly related to the search performance measures of the subjects. Moreover, the collected eye movement data showed that fixation duration was subject dependent and not significantly affected by the two factors tested. Scanning paths were found to exhibit some patterns even though subjects were instructed to adopt a random search strategy.

Introduction

Visual search is important in inspection task for improving product quality. Drury (1990) showed that while the decision making factor was general across different inspection tasks, the search factor did not correlate between tasks. So it is necessary to study the search process for understanding inspection process and improving performance. Search performance may be influenced by a large number of factors. Megaw (1979) suggested that, for example, visual acuity, eye movement, scanning strategies, age, lighting, inspection time, fault probability, product complexity, feedback and feedforward, training, etc. need to be considered when designing or modifying industrial inspection tasks. Amongst the techniques used for studying inspection performance, eye movement recording has been suggested as an important source of data for selecting remedies to improve performance and to establish standard time and inspection rate. It also provides information of the coverage of each item and scanning strategies of inspectors. Basically there are two major types of scanning strategies in performing visual search, viz. random scanning and systematic scanning. For random scanning, successive eye fixations within a search trial are independent of each other and possibly overlap each other. For systematic scanning, successive fixations do not overlap within a trial and coverage of subjects' visual lobe in the search field is exhaustive. Though mathematical models of visual search were established for simulating and optimizing inspection task, little is known on the effect of variation of scanning patterns on inspection performance measures. One of the objectives of this study was to compare various search performance measures under random and systematic scanning patterns.

In practice, it is not easy to carry out eye movement experiments with industrial inspection tasks. The large amount of data required and the setup of testing equipment usually prohibit eye movement studies in industries. Instead, simulation tasks with generation of testing stimuli by computer programming was usually used as an experiment tool (Stampe and Reingold 1995, Chan and Courtney 1994). In the laboratory, visual search has often been studied using a task with a single target appearing at a random location on a competing background (Chan and Courtney 1993). Under no time limit, subjects usually continue searching until the target is detected. But in practice, it is uncommon to always have one fault on an industrial product with the inspector continuing to search until detection. The defect percentage in industrial environment is comparatively much smaller than the 100% usually used in the laboratory. Fox and Haslegrave (1969) found that the level of detection of defectives increased with the probability of defect occurrence. Drury and Addison (1973) also found that hit rate fell with defect percentage. A recent study also showed that defect percentage (DP) is one possible critical factor (Chow and Chan 1996) that affected hit rate and correct inspection percentage. Moreover, there are also explicit or implicit time limits imposed in industrial inspection tasks by which subjects may stop searching by themselves based on experience and knowledge of task requirements, and not infinite searching until fault detection in the laboratory. Hence, the effects of defect occurrence percentage and scanning strategy on search performance was studied in this experiment with time limited inspection tasks, with the supplement of eye movement information provided by a eye movement tracker.

Method

Subjects

Four male subjects participated in this experiment with ages ranging from 22 to 26 years and a median of 24. They were all volunteer undergraduates of the City University of Hong Kong and had near acuity scores of at least eight (20/25 Snellen notation) on the Bausch and Lomb orthorator.

Equipment and Software

An NAC Eye Mark Recorder Model VI was used to record eye mark data through a controller onto a video tape. The NAC Data Processing Unit then transferred recorded eye mark data to the NAC Data Analysis Software for analysis. The responses of subjects on fault occurrence and location were recorded and analyzed by another computer program.

Stimulus

A computer program was written to produce simulated products (images) on a display of a personal computer. In each trial, it generated 18 grids, representing 18 components on a product. Each grid has 25 characters, representing individual features of a component. A grid consisting of all character "X"s was a normal component, and one with a "V" signified a defective component, and hence a defective product. An example of a stimulus is shown in Figure 1.

```
XXXXX XXXXX XXXXX XXXXX XXXXX XXXXX
XXXXX XXXXX XXXXX XXXXX XXXXX XXXXX
XXXXX XXXXX XXXXX XXXXX XXXXX XXXXX
XXXXX XXXXX XXXXX XXXXX XXXXX XXXXX
XXXXX XXXXX XXXXX XXXXX XXXXX XXXXX

XXXXX XXXXX XXXXX XXXXX XXXXX XXXXX
XXXXX XXXXX XXXXX XXXXX XXXXX XXXXX
XXXXX XXXXX XXXXX XXXVX XXXXX XXXXX
XXXXX XXXXX XXXXX XXXXX XXXXX XXXXX
XXXXX XXXXX XXXXX XXXXX XXXXX XXXXX

XXXXX XXXXX XXXXX XXXXX XXXXX XXXXX
XXXXX XXXXX XXXXX XXXXX XXXXX XXXXX
XXXXX XXXXX XXXXX XXXXX XXXXX XXXXX
XXXXX XXXXX XXXXX XXXXX XXXXX XXXXX
XXXXX XXXXX XXXXX XXXXX XXXXX XXXXX
```

Figure 1. Example of a defective product.
The grid (component) in the second row
and the fourth column contains a defective feature.

Procedure

There were two levels of defect percentage (DP) and two different scanning strategies (SS) tested in the experiment. A total of four batches of 50 simulated products for the four different conditions (Table 1) were tested in randomized sequence for each subject. Subjects were required to perform a zigzag horizontal search starting from the top left corner of the stimulus under the horizontal scanning condition. They were allowed to use any strategies they liked under the random scanning condition. A maximum of four seconds was allowed for each product inspection. Within the specified time limit for a trial, the subjects were required to visually inspect each product with the specified scanning pattern and report whether the product contained a defective component and where the defective component was. The actual search time required for each product inspection was recorded. The product was accepted if no response was provided within four seconds. No practice was given prior to the test.

Table 1. Combinations of DP and SS.

	Defect Percentage (%)	
	4(A)	16(B)
Random Scanning (R)	AR	BR
Horizontal Scanning (H)	AH	BH

Results

Performance Measures

There were five performance measures for the search task, viz., correct acceptance percentage (CAP), correct inspection percentage (CIP), hit rate (HTR), inspection rate (IPR), and efficiency in improving the product (EIP). The subjects' overall means, standard deviations and coefficients of variation for the performance measures are shown in Table 2. It was noted that the CAP and CIP were closed to 100% with small variations while

HTR, IPR and EIP varied greatly with coefficients of variation of 38.87%, 11.03% a 40.25% respectively.

Separate ANOVAs were performed on individual performance measure and t results are summarized in Table 3. Basically, neither defect percentage nor scanni strategy was found to affect any one of the five performance measures significantly. T DP x SS interaction was found non significant in all the five performance measures.

Table 2. Summary of statistics for performance measures.

Performance Measure	Mean	Std Dev	Coefficient of Variation (%)
CAP	99.85	0.60	0.60
CIP	97.88	3.30	3.38
HTR	80.47	31.28	38.87
IPR	0.27	0.03	11.03
EIP	79.53	32.01	40.25

Table 3. Summary of ANOVAs on the performance measures.

Performance Measure	Main Factor					
	DP			SS		
	F	DF v_1,v_2	Sig. of F	F	DF v_1,v_2	Sig. of F
CAP	1.00	1,12	NS	1.00	1,12	NS
CIP	2.73	1,12	NS	0.02	1,12	NS
HTR	0.01	1,12	NS	0.07	1,12	NS
IPR	0.76	1,12	NS	0.76	1,12	NS
EIP	0.03	1,12	NS	0.06	1,12	NS

Number of Fixation, Fixation Duration, and Saccade Velocity

Over 90% of number of fixations in the inspection tasks ranged from 5 to 12. There w no significant subject difference in number of fixations, and number of fixation was found r related to defect percentage and scanning strategy.

Over 85% subjects' fixation durations ranged from 100 to 1000 msec. Average fixati duration of the four subjects pooled from all task conditions were 290, 325, 380 and 405 mse and the subject factor was found significant ($p < 0.025$). Significant subject factor in fixati duration was also found in similar experiments by Chow and Chan (1996). Fixation duration w also found not related to the two main factors being investigated.

89% of saccade velocities ranged from 50 to 500 deg/sec, and 99% ranged from 0 to 5 deg/sec. No significant subject difference was found in saccade velocity, which was r significantly affected by variation in DP and SS.

Discussion

Performance Measures

As no incentives were provided for early termination of search, subjects alwa continued searching for a fault until stimulus removal, which was shown by an avera inspection rate of 0.272. Nevertheless, the overall hit rate was achieved at only a level 80.47%. A 99.85% correct acceptance showed that nearly all good products were accept Although a relatively low hit rate was reported, the CIP was still high at 97.88% as the number defects was small and the correct accept number was large. Significant individuals' differer was found in subjects' HTR ($p < .05$) and EIP ($p < .05$).

Although defect percentage (DP) was found significant to HTR in a similar experiment by Chow and Chan (1996) with variation of search time limits from 2 sec to 7 sec, it had no significant effect on all the five performance measures in this study. The significant DP effect in the previous experiment was believed due to testing of subjects under wide ranging search time limits, which led to significant DP x ST interaction, as well as significant DP effect.

Scanning strategy (SS) was shown non significant to search performance in this study, though subjects with random scanning had a marginally better performance in all measures than those with systematic scanning. During the test, it was interesting to note that all subjects used relatively "systematic" ways of searching, for example, looping circularly inwards or outwards when they were told to scan randomly with their own means, and no obvious random scanning paths were observed.

Number of Fixation, Fixation Duration, and Saccade Velocity

Other than fixation duration, there was no subject difference in number of fixation and saccade velocity. All the three subjects' eye movement parameters were reasonably constant irrespective of the variation in defect percentage and scanning pattern.

Conclusion

In conclusion, all the five performance measures were not significantly affected by defect percentage and scanning strategy within the range of values tested in this experiment. As revealed by the nonsignificant difference in performance between the stipulated horizontal and random scanning strategies, there were no advantages in asking subjects to adopt the horizontal scanning strategy for inspection. It was suggested by the authors that subjects should be allowed to use their own scanning strategies for inspection. Defect percentage and scanning strategy did not significantly affect the three eye movement parameters investigated. Subjects' fixation duration was found strongly subject dependent at various levels of defect percentages and scanning strategies.

Reference

Chan, H.S. and Courtney, A.J. , 1993, Inter-relationship between visual lobe dimensions, search times and eye movement parameters for a competition search task, *The Annals of Physiological Anthropology*, **12(4)**, 219-227.

Chan, H. S. and Courtney, A. J., 1994, effect of priority assignment of attentional resources, order of testing, and responding sequence on tunnel vision, *Perceptual and Motor Skills*, **78**, 899-914.

Chow, S. W. H. and Chan, A. H. S. 1996, Effects of defect percentage and search time limit on search performance. In *Proceedings of the 4th Pan Pacific Conference on Occupational Ergonomics*, 268-271.

Drury, C. G. 1990, Visual search in industrial inspection, in Visual Search, In *Proceedings of the First International Conference on Visual Search*, 263-276.

Drury, C.G. and Addison, J.L., 1973, An industrial study of the effects feedback and fault density on inspection performance, *Ergonomics*, **16(2)**, 159-169.

Fox, J. G. and Haslegrave, C. M. 1969, Industrial efficiency and the probability of a defect occurring. *Ergonomics*, **12**, 713-721.

Megaw, E. D. 1979, Factors affecting visual inspection accuracy. *Applied Ergonomics*, **10.1**, 27-32.

Stampe, D. M. And Reingold, E. M. (1995). Selection by looking: A novel computer interface and its application to psychological research, in J.M. Findlay (Editor) *Eye Movement Research*, pp. 467-478, Elsevier Science B.V.

Williams, L. G. 1966, Target conspicuity and visual search, *Human Factors*, **8**, 80-92.

HUMAN PERFORMANCE

VARIABLE WORK LOAD EFFECTS ON MONITORING ABILITIES WITHIN A FREIGHTLINE SIMULATION

Lee M. Crofts & Neil Morris.

University of Wolverhampton
School of health Sciences
Psychology Division, MC Block
62-68 Lichfield Street
Wolverhampton WV1 1DJ

The 'Freightliner' simulation examines the ability to keep track of vehicles within a model freight system. The current study builds on earlier Freightliner runs but, examines the effects of variable workload bias on human 'real-time' processing. Subjects were tested on their ability to keep track of messages about target vehicles, embedded in distractor traffic, when half the messages were presented in the first half of a session (50/50), 70% in the first half (70/30), and 30% in the first half. The results demonstrated higher cargo identification with 30/70 bias. However, depot visitation order recall, the 'real time' processing measure, was not affected by workload bias. The first five vehicle transactions were better recalled than the later transactions. The data are contrasted with other Freightliner runs and discussed in terms of work practices.

Introduction

Updating, monitoring, and reacting to constant streams of information requires 'real time' memory updating. This is especially the case within complex organisations such as freight distribution centres, right through to, military command and control centres. It is however, often difficult and expensive to highlight working practices that are not compatible with human cognitive abilities, once a system is up and running. For example, Harper (1974) commented that the command and control centre for the Los Angeles Fire Department could have been environmentally enhanced, with little cost, by applying ergonomic principles before construction commenced. To this end the use of 'real world' simulations, such as Freightliner, in combination with, sound ergonomic principles can highlight potential problems at an early stage.

'Freightliner' was developed to examine some aspects of command and control that were likely to be found throughout a range of environments. This would include, for example, military command and control centres, air traffic control, Police and Fire departments and freight services. The Freightline simulation requires the course of vehicles to be tracked by retaining in memory the cargoes and transactions of target vehicles. In many command and control situations ones own vehicles are not being tracked. An operative who follows the transactions of an unknown vehicle is in a position to create hypotheses about the intentions of the vehicles being tracked. This is likely to be very important in military and quasi-military organisations. It would usually be less important in the transport industry. However, the use of a freight simulation as a metaphor for such organisations, allows the examination of important aspects of command and control without specifying precise operational practices.

Memory updating is "the act of modifying the current status of a representation of schema in memory to accommodate new input" (Morris & Jones, 1990). For example, a friend moves and changes their phone number thus requiring you to update your memory to incorporate the new number. Human working memory (For a detailed overview of memory and working memory see, Baddeley & Hitch (1974), Baddeley (1981), Baddeley *et al* (1984), Baddeley (1986), Morris (1986)) is considered to have a very limited capacity. Considering the amount of information that is processed within the system, it is reasonable to suggest that it has a formidable processing rate. For example when the system is faced with multiple inputs how does one suppose it handles these inputs as we do? Morris (1991) has suggested that this problem can be understood relative to two points, a) the system itself has a multiple processor ability and that each processor has capacity limitations independent of the other. b) Working memory has real time, dynamic processing abilities and not simple static memory stores. Thus the dynamism of the system facilitates limited capacity at any moment in time, but allows a phenomenal amount of information processing to occur over time. New material enters the system and old material is lost, this is the essence of memory updating. These findings were reiterated in Freightliner studies conducted by Morris & Jones (1988). What is important is that according to Bjork & Landauer (1978) and demonstrated by Morris & Jones (Submitted) is that updating ones memory is not destructive. In other words it does not result in overwriting of earlier variable states. In examining these contentions various studies have been devised that could tap into memory updating, one of which is the Freightline simulation.

The current study is a replication, in part, of Freightline studies conducted by Morris & Jones (1988), and Morris & Jones (Submitted). In these studies they examined the effects of transcription on memory updating ability, and the effects of transcription with variable work load on memory updating. In the first study three tasks were devised. Each of these tasks consisted of sixty, six word sentences. Each sentence referred to a particular vehicle, a specific cargo and the depot at which it was to be loaded. The totality was sixty six word sentences referring to vehicle colour, cargo and depot, six vehicles, ten depots and sixty cargoes. Throughout a trial each vehicle would visit all ten depots but not carry the same cargo twice. Within the three texts the cargoes and depots were randomly assigned. The messages were then read onto a tape with a time spacing of twenty seconds with a tone preceding each message. Subjects were required to merely listen to the tape in the no transcription condition but pay attention to messages referring to the target vehicles (Black & White). In the partial transcription condition subjects were required to write down any messages that referred to the target vehicles, twenty in total, and instructed to ignore messages that referred to any other vehicles. In the full transcription condition, subjects were required to write down all messages, sixty in total , but told that they would only be tested on the target vehicles. An example of a message would be "Black vehicle, load tea, depot 9". After a message had been written down subjects were required to turn over the paper so they would have to rehearse the target message from memory. At the end of the trials the subjects were tested on depot visitation order and the cargoes carried by the target vehicles. Morris & Jones (1988) concluded that transcription does impair subsequent memory performance. This is important because the length of the task requires the subjects to store information in long term memory, thus errors are generated at the encoding stage. The study did find that memory updating is not destructive which was demonstrated by the primacy effects and smaller recency effects observed.

The second Freightline experiment Morris & Jones (Submitted) encompassed a variable work load element combined with transcription variables, e.g. 30/70 & 70/30, 30% refers to thirty percent of target messages appear in the first half of a trial and 70% in the second of and vies versa. The combination of transcription variables with work load variables resulted in a confounding variable. For example, 30% work load requires a grater number of distractor items presented which would not be a problem if either work load or transcription were examined independently. A more detailed description would make this point clearer. In the full transcription condition of the previous study subjects would have to write down all messages including the distractor items but the total amount of messages written down would be the same for all work load conditions, e.g. majority of target items presented in the first/second half of a condition would result in a reduced amount of distractor items. The subjects would therefore not have a true variable work load representation. This would not be the case however for the partial transcription condition because no attention would given to distractor items. A further problem of the previous study was that of time spacing between messages. The messages were presented at a rate of one every twenty seconds. This constant spacing between messages is not typical of a working environment and would facilitate an anticipation of message arrival. A third and final point to mention is that a base line was not required for the work load conditions but merely a reliance on previous data, this point requires some clarification. Morris & Jones (Submitted) examined a variable workload condition, 70/30 or 30/70 but did not run a base line 50/50 condition. This would have been a prudent experimental practice because reliance on previous base line data for comparison is unsound. It is the intention of the following study to negate the problematic nature of the previous study by using partial transcription under all three work load conditions, (50/50, 30/70 and 70/30) using a randomly assigned time spacing between messages and setting up an independent base line. It is hoped that by doing so a clearer examination of work load effects on memory updating ability will be achieved.

Method

Twenty four undergraduate students from the university of Wolverhampton volunteered to take part in this experiment. Twelve female and twelve male, none of which reported any auditory, visual or writing problems.

Three scripts were prepared each of which referred to a different work load condition, 50/50, 70/30, and 30/70, bias. Each script contained sixty six word sentences pertaining to vehicle colour, cargo and depot. A typical sentence would read, "Black vehicle, load sugar, depot ten".

The vehicles were assigned a number from 1-6 (six colours), the sixty cargoes (Cargoes were generated that fit the following descriptions, a; they were one word cargoes e.g. CEMENT, and, b; they were everyday items of a non-ambiguous nature e.g. DOORS, BEEF etc.) were assigned a number 1-60, and there were ten depots.

The numbers were then entered into a random number generator to enable a random assignment of vehicle colour, cargo and depot. Each script was then recorded onto a separate C90 audio cassette with randomised time spacing between each message The random time spacing between each message was achieved by removing the time taken to read sixty messages (300 seconds, 5 seconds per message) from the total time assigned for one condition (600 seconds) and then randomly assigning a time space between each message.

One Sony cassette player with electric adapter and two sets of headphones was used to play back the messages. A pen and note pad were provided for the subject to transcribe target messages. Four response sheets were provided, response sheet one pertaining to cargoes carried by target vehicles. Response sheets 2-4 pertaining to depot visitation order.

The subjects then listened to the tape containing the script for either condition 1, 2 or 3 (50/50, 70/30, 3070, respectively). The conditions were assigned to each subject using a simple Latin square design thus facilitation a counter balancing exercise. So, for example, subjects 1-6 would experience all six possible combinations of condition order (Possible condition orders = 1,2,3, / 3,1,2, / 2,3,1, / 2,1,3, / 3,2,1, / 1,3,2,) and so on for multiples of six subjects.

Subjects were instructed to write down only messages that referred to target vehicles, in this case, black and white. After writing down a message they were required to place the note paper face down in a box.

After completion of a tape, subjects were requested to fill in a response sheet pertaining to depot visitation order (forwards recall was not required) and a response sheet which listed all sixty cargoes carried by all vehicles. Subjects were required to signify only those cargoes carried by the black and white vehicles. If they could not remember all cargoes they were asked to make a guess, but only to select a maximum of twenty cargoes (The total number of messages referring to black and white vehicles). Subjects were then allowed a period of five minutes before a subsequent trial. Each subject was debriefed after completion of all three conditions.

Results

A one-way repeated measures analysis of variance was performed on the cargo item recognition scores. This analysis revealed a significant effect of work load bias upon recognition performance $F(2,46) = 13.07$, $p<0.0001$. Subsequent Newman-Keuls tests revealed significant differences between condition one (50/50) and condition three (30/70) $p<0.01$: Between condition two (70/30) and three (30/70), $p<0.01$: But not between condition one (50/50) and two (70/30), $p>0.05$. This would suggest that a work load biased towards the end of a session enabled a greater number of cargo items to be recognised, with the greater recognition deficit being produced in the 30/70 work load condition. This effect can be seen more clearly in table 1, where the scores are shown in comparison to those obtained by Morris & Jones (Submitted).

The depot visitation order scores were compressed to produced two scores per condition, first half and second half. This gave a total of six scores per subject. The scores were then subjected to a two-way analysis of variance with work load bias and serial position as factors. The analysis revealed no significant effect of work load bias on serial recall, $F(2,46) = 2.11$, $p>0.05$. There was no significant interaction between work load bias and serial position, $F(2,46) = 1.05$, $p>0.05$, but a significant difference between serial positions 1-5 and 6-10, $F(1,23) = 10.85$, $p<0.01$ was found (See Table 2), with the greater recall scores being achieved for serial positions 1-5.

Table 1. Mean recognition scores comparison

Work Load Bias	Morris & Jones (Submitted)	Current Study
50/50	12.94	12.708
70/30	11.72	12.208
30/70	10.42	15.375

In summary, facilitating a heavy work load towards the end of a task enabled subjects to achieve greater recognition for cargo items in the 30/70 biased condition. The 70/30 work load bias seemed to have a rather negative effect on subsequent recognition, with the middle ground being produced during the 50/50 work load bias. However, the difference between 70/30 and 50/50 was not significant. The effects of work load bias on depot recall order was however, unaffected by work load bias in any condition. The best recall results were produced for positions 1-5.

Table 2. Recall scores, means (Standard Deviations).

Work Load Bias	Serial Position 1-5.	Serial Position 6-10
50/50	3.167 (2.531)	2.208 (2.686)
70/30	2.792 (3.148)	2.250 (2.507)
30/70	3.875 (3.366)	2.375 (2.449)

Discussion

The cargo recognition scores demonstrate a clear effect of work load bias upon recognition abilities, especially when the work load is heaviest towards the latter part of a session (30%/70%) which implies a possible recency effect (See Watkins & Peynircioglu (1983) for expansive discussion on recency effects). This is contrary to the findings of the previous study by Morris & Jones (Submitted) who found work load biased towards the end of a session facilitated the worst recognition scores. This particular phenomena could be attributed to a procedural difference. Recall of depot visitation order however, appeared to show no such anomalies in the data. Recall performance for positions 1-5 was better than for positions 6-10 in all three conditions. This would suggest that recall (Real time processing element) is unaffected by work load bias which reinforces the results obtained by Morris & Jones, (1988) Morris & Jones (Submitted). They suggest that the possible reason for this effect is the reduced time that latter serial positions have in working memory.

What is clear however is that both the current study and the previous studies have highlighted the problematic effects that biased work loading could have on human memory updating abilities. As suggested by the data, creating heavy performance demands in the early part of a session has a negative effect on recognition abilities. This is not, however, the case with the recall task. The scores in this case were not effected by the biased work load. This could suggest that the processes involved with recall and recognition memory operate independently of each other within the memory system, each with its own capacity limitations. This point requires further research, however the implications of the current study's findings need to be addressed

What should be made clear is that simulations of this sort are very specific in the problems that they address. For example, in the current simulation subjects were required to transcribe target vehicle messages, recall their respective depots and the cargoes they carried. This is a very specific task and thus simulations of this sort should not be over generalised. The purpose of simulations such as this one is to highlight the effects that variable workloads could have on human real time processing abilities, and, to demonstrate that those effects can be predicted empirically.

Biasing work loads does have an effect of human performance, with an optimal scenario being produced with a heavy workload towards the end of a session. Obviously it may not be possible to facilitate such a scenario but it is essential that personnel in command are aware of the optimal performance scenarios

References

Baddeley, A. D. (1981) The Concept of Working Memory: A View of its Current State and Problematic Future Development. *Cognition.* **10**. 17-23.

Baddeley, A. D. (1986) *Working Memory.* Oxford: Oxford University Press.

Baddeley, A. D., & Hitch, G. J. (1974) Working Memory. In Bower, G. H. (Ed). *The Psychology of Learning and Motivation.* vol **8**. New York. Academic Press.

Baddeley, A. D., Lewis, V. J., & Vallar, G. (1984) Exploring the Articulatory Loop. *Quarterly Journal of Experimental Psychology*, **36A**, 233-252.

Bjork, R. A., & Landauer, T. K. (1978) On Keeping Track of the Present State of People and Things. In M. M. Gruneberg, P. E. Morris, & R. N. Skyes (Eds), *Practical Aspects of Memory.* London: Academic Press.

Harper, W. R. (1974) Human Factors in Command and Control For the Los Angeles Fire Department. *Applied Ergonomics,* **5**, 26-35.

Morris, N. (1986) Working Memory, 1974-1984: A Review of a Decade of Research. *Current Psychological Research & Reviews,* Fall 1986, vol **5**(3), 281-295.

Morris, N. (1991) The Cognitive Ergonomics of Memory Updating: Lovesey, E. J. (ed). *Contemporary Ergonomics*, 276-281.

Morris, N., & Jones, D. M. (1988) The Effect of Irrelevant Transcription on Performance on a Freight Service Simulation. *Contemporary Ergonomics.* 503-508.

Morris, N,. & Jones, D. M. (1990) Memory Updating in Working Memory: The Role of the Central Executive. *British Journal of Psychology*, **81**, 111-121.

Morris, N., & Jones, D. M.(Submitted).

Watkins, M. J. & Peynircioglu, Z. F (1983) Three Recency effects at the Same Time. *Journal of Verbal Learning and Verbal Behaviour*, **22**, 375-384.

Yntema, D. B. (1963) Keeping Track of Several Things at Once. *Human Factors*, **5**, 7-17.

LEVELS OF CONTROL IN THE EXTENDED PERFORMANCE OF A MONOTONOUS TASK

P. Tucker, I. Macdonald, N.I. Sytnik, D.S. Owens and S. Folkard

MRC Body Rhythms and Shiftwork Centre,
Department of Psychology,
University of Wales Swansea,
Swansea SA2 8PP

The work routine of the production-operator is partially simulated in a laboratory task that is performed over a full working day. Evidence is accumulated which indicates that two levels of cognitive control are operating within the single paradigm. Specifically, it is suggested that responses to the minority target stimuli are due to a more controlled form of processing than are the responses to the majority target stimuli. The findings suggest that in a task with these characteristics, the vigilance decrement and the development of a more automatic mode of response to target stimuli are resistant to the effects of practise and to the provision of additional motivation or feedback.

The industrial production operator's role is often that of a supervisor, confirming or correcting the operation of the system at irregular intervals. The majority of their actions will be routine responses to regular stimuli, but there will be a small proportion of unusual events for which a non-standard response is required, usually in the form of some kind of corrective action. These environments combine prolonged rehearsal of invariant stimulus-response relationships with the detection of rare and irregular events while monitoring an output over a prolonged period; i.e. the conditions associated with the development of 'automaticity' and with the maintenance of sustained controlled processing (i.e. vigilance conditions), respectively.

Automaticity occurs when individuals become highly practised in a repetitive operation, such that the amount of conscious awareness that is required to maintain performance gradually decreases to a point where their responses are largely outside voluntary control and require minimal attentional resources (Schneider & Shiffrin, 1977; Shiffrin & Schneider, 1977). Vigilance task performance is characterised by a decline in target detection rates over time (usually within the first 30 minutes) which stabilises over prolonged task exposure. According to Fisk and Schneider (1981), this is because vigilance tasks require sustained controlled processing. In contrast, no such decrement is observed in the performance of the repetitive tasks which involve automated processes (Schneider & Shiffrin, 1977). Performance in the latter is maintained for extended periods without showing a vigilance decrement because automatic responses involve no short term memory load and thus no participant effort (Schneider & Fisk, 1980a).

We contend that the job of the production-operator, as described above, combines elements of a 'controlled processing paradigm' and an 'automated processing paradigm', within the same task. The individual is required to react to the occurrence of relatively rare and usual events requiring unique responses, that are presented within a prolonged series of regular stimuli requiring routine responses. Strayer and Kramer (1994) examined the effects of combining controlled and automatically processed stimuli within a single paradigm. Responses that were assumed to be due to controlled processing were slower and less accurate than automatic responses.

The current research examines performance in this heterogeneous type of task environment, within the context of a prolonged vigilance-type task, in a laboratory setting. It is predicted that, in comparison with responses to non-target events, the slower and less accurate responses to target events will indicate that those responses are due to relatively controlled processing (after Strayer & Kramer, 1994). Furthermore it is hypothesised that the accuracy and latency of responses to target events will show vigilance decrements, while performance in relation to non-targets will not (after Fisk & Schneider, 1981).

Feedback and motivation within the monotonous task environment

In general providing feedback and additional motivation, either separately or in combination, can be expected to improve overall performance. Research into the effects of feedback and extrinsic motivation upon the vigilance decrement have produced inconsistent findings (see Davies & Parasuraman, 1982, for a review). Nevertheless, it seems likely that feedback or other forms of extrinsic motivation will have more effect upon performance that is subject to controlled processing than it will upon responses that are under 'automatic' control. This hypothesis is based on the assumption that motivation acts exclusively upon processes that are subject to intention. Motivation is therefore unlikely to impinge upon automatic processes that are less influenced by voluntary control (Hasher & Zacks, 1979) and are thus difficult to modify (Schneider & Shiffrin, 1977). Thus we predict that the accuracy of responses to targets will be enhanced by the provision of either feedback or additional extrinsic motivation, while there will be no such effects upon the accuracy of responses to non-targets.

Method

Eight participants performed the task on personal computers. A set of 4 'target' consonants were presented to the participants at the outset. An experimental trial comprised a single consonant being presented in one of four positions on the display screen, around a central point (upper left, upper right, lower left, or lower right). The correct response to the presentation of a non-target was to press the corresponding one of four keys labelled on the key-board. The spatial arrangement of the four keys corresponded approximately with the four presentation positions on the screen. If the consonant presented was a target the correct response was to press the space bar. A correct response resulted in the presentation of another stimulus, after a 500 ms interval. The stimuli were presented in blocks of 1700 trials. Two in every 20 contiguous stimuli were targets, but the precise positions within each 20 stimuli were random.

The task comprised four conditions: with feedback and extrinsic motivation, with feedback but without extrinsic motivation, without feedback but with extrinsic motivation, and without either feedback or extrinsic motivation. In the conditions which involved extrinsic motivation, the programme would count the number of erroneous responses made by the participant during a block of 1700 trials and add on an extra three trials for every erroneous response. Subjects were informed of the number of extra trials that they had incurred, prior to their presentation. In the conditions which involved feedback, an erroneous response resulted in the immediate presentation of an error message. If one of the four non-target response keys was pressed in response to the presentation of a target, or if the space bar was pressed in response to the presentation of a non-target, the message "WRONG KEY" was displayed. If the wrong one of the four non-target keys was pressed in response to the presentation of a non-target, the message "WRONG POSITION" was displayed.

The task was performed across five full working days. The first day was designated a practice day and comprised 3 blocks of 1700 trials in each of the four conditions i.e. 12 blocks in all. The remaining four days were experimental; one day spent in each condition. An experimental day's work comprised 14 blocks. The design allowed for the performance of a block approximately every half an hour, with a break of a few minutes between each block. The task was self-paced and participants were told that they had to complete a set number of blocks (6 on practice days, 7 on experimental days) before they could break for lunch.

The order of conditions was partially counterbalanced between participants, such that half of the subjects performed the two feedback conditions followed by the two no feedback conditions while the other half did the reverse; and also, half of the subjects performed the two extrinsic motivation conditions followed by the two no extrinsic-motivation conditions while the other half did the reverse.

Results

Errors

In practice only two categories of error featured in significant quantities. These two were 'target errors' (responding to the presentation of a target stimulus that occurs twice in every 20 trials, as though it were a non-target stimulus); and 'spatial errors' (or 'non-target errors'; i.e. choosing the wrong one of the four non-target response keys, to the presentation of an a non-target stimulus). Each are expressed as a proportion of the number of stimuli presented of that type. Means and standard deviations for the two types of error in the four conditions are shown in Table 1.

Table 1: Mean (and SD) of errors as proportion of the number possible errors

	Target	Spatial
Feedback & extrinsic motivation	0.154 (0.082)	0.016 (0.010)
Feedback, no extrinsic motivation	0.247 (0.106)	0.027 (0.106)
No feedback & extrinsic motivation	0.232 (0.106)	0.024 (0.018)
No feedback & no extrinsic motivation	0.356 (0.165)	0.048 (0.042)

A four-way (2 x 2 x 2 x 14) analysis of variance (ANOVA) examined the combined effects of Stimulus Type (target versus non-target; i.e. target errors versus spatial errors), Feedback (with feedback versus without feedback), Extrinsic Motivation (with extrinsic motivation versus without extrinsic motivation) and Block (1 to 14). There were main effects of Stimulus Type, $F(1,7)=47.03$, $p < .001$, Feedback, $F(1,7)=14.25$, $p < .01$, Extrinsic Motivation, $F(1,7)=12.20$, $p < .01$ and Block, $F(13,91)=10.99$, $p < .001$. There was a higher rate of target errors than spatial errors and the effects of including either feedback or motivation were to decrease error. Error rates increased across the duration of the day. There were significant interactions between Stimulus Type and Feedback, $F(1,7)=9.19$, $p < .05$, and between Stimulus Type and Extrinsic Motivation, $F(1,7)=9.19$, $p < .05$. The effect of including either feedback or motivation was to decrease target error rates, while there were no such effects of either feedback or motivation on spatial errors rates. There was a significant interaction between Stimulus Type and Block, $F(13,91)=5.38$, $p < .001$. This interaction is illustrated in Figure 1.

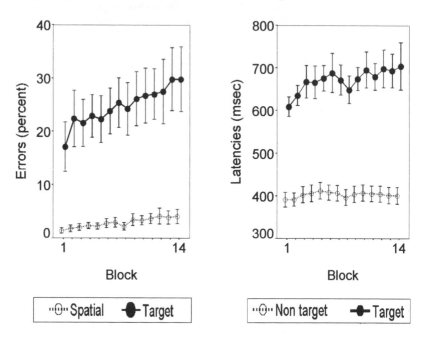

Figure 1. Mean error rates (+/- 2 standard errors).

Figure 2. Mean response latencies (+/- 2 standard errors)

Given that there were no interaction effects between experimental condition and Block, the opportunity was taken to conduct a second ANOVA upon the raw error data, in order to examine the effects of practice over the week. A three-way (2 x 4 x 14) ANOVA examined the combined effects of Stimulus Type and Day of the week across the 14 blocks of the day. There were main effects of Stimulus Type, $F(1,7)=47.03$, $p < .001$, and Block, $F(13,91)=10.99$, $p < .001$, as in the previous analysis, but no effect of Day, $p > .05$. The only significant interaction was between Stimulus Type and Block, $F(13,91)=5.38$, $p < .001$, which reflected the finding of the previous analysis.

Response latency

Speed of response was measured inversely as the interval between stimulus presentation and response. Data were only included for correct responses made within two standard deviations of the participant's mean correct response latency. Means and standard deviations for the two types of stimuli in the four conditions are shown in Table 2.

Table 2: Mean (and SD) of response latencies (milliseconds)

	Target	Non-target
Feedback & extrinsic motivation	668 (115)	415 (50)
Feedback, no extrinsic motivation	666 (115)	401 (52)
No feedback & extrinsic motivation	678 (80)	402 (48)
No feedback & no extrinsic motivation	676 (113)	391 (57)

The data were analysed in a four-way (2 x 2 x 2 x 14) ANOVA of the same design as that was used in the analysis of the error data. There were main effects of Stimulus Type, $F(1,7)=137.03$, $p < .001$, and of Block, $F(3.02,21.11)= 5.75$, $p < .001$. Response latency for target stimuli was longer than those for non-targets. Response latency increased across the duration of the day. There was a significant interaction between Stimulus Type and Block, $F(2.61,18.29)=4.37$, $p < .05$. This interaction is illustrated in Figure 2.

As with analysis of error data, in the absence of any interaction effects between experimental condition and Block, a three-way (2 x 4 x 14) ANOVA examined the combined effects of Stimulus Type and Day of the week across the 14 blocks of the day. There were main effects of Stimulus Type, $F(1,7)= 137.03$, $p < .001$, Day, $F(1.93,13.54)=8.50$, $p < .001$, and Block, $F(13,91)= 5.75$, $p < .001$. Post hoc pairwise comparisons indicated that the mean response latency on the fourth day was significantly higher than the mean response latency on each of days one to three. The only significant interaction was between Stimulus Type and Block, $F(13,91)= 4.37$, $p < .001$, which reflected the finding of the previous analysis.

Discussion

As predicted, target responses were slower and less accurate than responses to non-target stimuli. This is consistent with the coexistence of controlled and automated responses, respectively associated with each of the two types of stimuli (Strayer & Kramer, 1994). The predictions of interactions between Stimulus Type and Block were also confirmed, with a vigilance decrement evident in responses to target stimuli, both in terms of error rates and response latencies. Also as predicted, there were effects of feedback and of extrinsic motivation on target error rates, that were absent from spatial error rates. In both cases, the rate of target errors was lower in the 'inclusive' conditions. These two interaction effects are taken as indications that responses to target stimuli were less automated than were responses to non-targets.

The current task is a typical vigilance task. Vigilance task performance is typified by a decline in target detection rates over time, and usually a decline in false report errors. While the current task demonstrated the former, it was not possible to demonstrate the latter due to the extremely low rate of false alarms. Vigilance decrements tend to show rapid recovery of performance following rest breaks. There was evidence of this in the form of a temporary drop in target response latency that occurred in the block immediately after the lunch break.

Levels of cognitive control and the vigilance decrement

We have suggested that responses to target stimuli in the current task are due a relatively controlled form of processing, as opposed to automaticity. This is despite the fact that the relationship between target and correct response was consistently mapped. In accordance with Fisk and Schneider (1981), it is proposed that consistent mapping is not a sufficient condition for the development of automated processing. Previous research suggests that the development of automaticity is maximised by extended practise (e.g. > 2000 trials), high motivation, speed stressed performance and a high ratio of detection to non-detection searches (after Schneider & Shiffrin, 1977; Schneider & Fisk, 1980 a,b,c). There was no evidence in the current data of an improvement in target error rate across the working day, thus confirming previous suggestions that the detection of target stimuli does not benefit from extended practise, when the ratio of targets to non-targets is low (Schneider & Fisk, 1980a). Moreover, analysis of 'long term' practise effects provided no evidence that the vigilance decrement was diminished by practise across the duration of the week. The vigilance decrement in the current experiment was unaffected by the inclusion of either feedback or extrinsic motivation. This is consistent with the view that automation of target responses was not enhanced by the manipulations of feedback and motivation. The current findings suggest that of the conditions for automaticity specified by Schneider and colleagues, a high ratio of detection to non-detection searches is a necessary condition for the establishment of a high degree of automated processing.

In conclusion, we concur with Fisk and Schneider (1981), that if an operator is to work in an environment where the ratio of target to non-target presentations is low, then simply exposing them to that task as a method of training them for the job is inappropriate. It remains to be seen whether training operators in an environment comprising high target to non-target ratios would be a more effective method of avoiding the deterioration in performance that is associated with the vigilance decrement.

References

Davies, D.R. & Parasuraman, R. 1982, *The Psychology of Vigilance.* (Academic Press, London)

Fisk, A.D. & Schneider, W. 1981, Control and automatic processing during tasks requiring sustained attention: A new approach to vigilance. *Human Factors, 23*(6), 737-750.

Hasher, L. & Zacks, R.T. 1979, Automatic and effortful processes in memory. *Journal of Experimental Psychology: General,* **108**, 356-388.

Schneider W. & Shiffrin R.M. 1977, Controlled and automatic human information processing: 1. Detection, search, and attention *Psychological Review,* **84**, 1-66.

Schneider, W. & Fisk, A.D. 1980a, Automatic and controlled processing in visual search, can it be done without cost? Champaign, IL: University of Illinois, Human Attention Research Lab, Report 8002.

Schneider, W. & Fisk, A.D. 1980b, A degree of consistent training and the development of automatic processing. Champaign, IL: University of Illinois, Human Attention Research Lab, Report 8005.

Schneider, W. & Fisk, A.D. 1980c, Visual search improves with detection searches, declines with non-detection searches. Champaign, IL: University of Illinois, Human Attention Research Lab, Report 8004.

Shiffrin R.M. & Schneider W. 1977, Controlled and automatic human information processing: II. Perceptual learning, automatic attending, and a general theory. *Psychological Review,* **84**, 127-190.

Strayer, D.L. & Kramer, A.F. 1994, Strategy and automaticity: I. Basic findings and conceptual framework. *Journal of Experimental Psychology: Learning, Memory and Cognition,* **20**(2), 318-341.

TEMPORAL VARIATION IN HUMAN PERFORMANCE DUE TO TIME ON TASK

Huw Gibson, Ted Megaw and Ed Milne

Industrial Ergonomics Group,
School of Manufacturing and Mechanical Engineering,
The University of Birmingham,
Birmingham, B15 2TT

This paper begins by reviewing field studies which have reported how performance changes over time-on-task. In particular, the review concentrates on those studies which have reported that performance is relatively poor sometime between the second and fourth hours of the shift, a finding referred to as the 2-4 hour shift phenomenon. Although possibly reflecting exposure factors, it was felt that this phenomenon warranted closer scrutiny under more controlled conditions. For this reason a pilot laboratory study was designed to see the effects of 8 hours of continuous performance on two tasks, one a simple detection task, the other a simple problem solving task. While the results demonstrated some significant changes in performance over time for both tasks, the results could not be taken as providing strong support for the 2-4 hour shift phenomenon.

Introduction

Some of the earliest research into the effects of prolonged work on performance was reported by the Industrial Fatigue Board (later known as the Industrial Research Board) in the 1920s and 1930s. A number of industrial tasks were investigated including inserting hooks into lamps, chocolate packing and tobacco weighing (Wyatt and Fraser, 1929; Wyatt and Langdon, 1937) and inspection tasks including cartridge examining and tile sorting (Wyatt and Langdon, 1932). The results from these studies failed to demonstrate any consistent effect of the time-into-shift (time-on-task) on performance measures such as output and rejection rates. Not only were there large differences in the daily output profiles between operators, many of the studies were restricted to short shifts of four hours continuous work. Despite these shortcomings, there has been a tendency to conclude that, with day shift work at least, performance improves over the first hour reflecting a warm-up period and then begins to decline towards the lunch break, increases again after the lunch break before declining again towards the end of the shift. Sometimes, a small period of improvement is found right at the end of the shift (the so called end-spurt effect). In general the results have been interpreted in terms of the effects of fatigue, monotony and boredom. Since these early studies, there has been a continuing interest in the effects of prolonged work on performance and accident rates, particularly in relation to various driving tasks. The majority of studies into prolonged periods of work have been concerned with time-of-day effects in relation to shift work (Folkard and Monk, 1979). Unfortunately, in a majority of these studies, the data have not been analysed in terms of time-on-task effects.

However, with the recent introduction of the practice of extending working shifts from 8 to 12 hours, there has been a resurgence of interest into the effects of continuous periods of work on performance, albeit regarding the last hours of the work shift. During the course of this research, a potentially interesting finding has emerged from the literature which has been termed the 2-4 hours shift phenomenon. This phenomenon reflects the finding that performance is often relatively poor sometime between the second and fourth hours of the work shift. In other words, performance deteriorates during the initial hours of the work shift followed by a period of improvement throughout the remainder of the shift period. The following lists the results from those studies which demonstrate the 2-4 hour phenomenon.

Verhaegen and Ryckaert (1986). This study was concerned with the performance of train drivers to yellow signals with trains installed with a vigilance monitoring system to overcome the limitations of the dead man's pedal. Although the results showed no circadian effects, a very significant increase in the frequency of delayed reactions was found during the third hour of the shift for those drivers on the early morning shift (starting around 0500 hours).

van der Flier and Schoonman (1988). This is one of the few studies that was deliberately carried out to investigate time-on-task effects on train drivers' performance. The results provide evidence of an increase in SPADs (Signals Passed At Danger) during the second and third hours into the shift for all shifts

Pokorny et al. (1987). An analysis of the accident rates of bus drivers demonstrated a peak during the third and forth hours of the driving period for those working on the early shift. Interesting was the finding that the results for those working a split shift which started the same time as the early shift did not show the equivalent increase in the accident rate over the first three hours.

Leigh et al. (1990). Combining data from all shifts, an increase in non-fatal lost-time injuries was found for both underground and surface Australian mine workers up to the fourth hour of work followed by a decrease for the remainder of the shift period.

Wagner (1988). Data from the night shift workers involved with taconite mining operations showed a marked increase in the number of accidents in the 2 to 4 hour period with a peak at 2 hours. However, this result was only found for younger workers (<29 years) and was totally absent for older workers (>40 years old).

Lauridsen and Tonnesen (1990). Injury data from the Norwegian offshore drilling industry revealed that the highest rate of injuries occurred between the second and fourth hours of both the 12 hour work shifts for the drilling crews, with a greater number of injuries reported by the day shift.

Kovacs and Varga (1996). Accident statistics from the Hungarian bauxite and coal miners yielded greater accident rates for the morning shift with a peak occurring at the fourth hour into the shift.

Ong et al. (1987). Accident data from iron and Steel workers in Singapore demonstrated higher accident rates for the day shift with a peak during the third hour of work. For the other two shifts the rates were higher at the beginning of the shift period.

<u>Novak et al. (1990).</u> The injury rates for workers in chemical manufacturing plants in Texas were highest for the day shift who exhibited a peak at the fourth hour into the shift.

It is important to appreciate that the above represent those studies which support the existence of the 2-4 hour shift phenomenon but that there are numerous other reports which fail to confirm the phenomenon. For example, Ankerstedt (1994) has made an initial analysis of the data held by the Swedish Occupational Injury Information System for the year 1990/1991 involving some 160,000 accidents and has failed to find any influence of time-on-task until duty time exceeds 9 hours.

Naturally, there are obvious shortcomings of the field studies summarised above. The most important of these is that it has been difficult, if not impossible, to take account of exposure effects. It is intuitive to expect that the opportunity to commit errors is likely to occur sometime in the middle of a shift period with workload generally being relative lighter at the beginning and end of the work period. Additionally, some of the results may be confounded by circadian effects although, interestingly, many of the studies mentioned above fail to report the expected circadian effects, again possibly due to the lack of control for exposure factors. Surprisingly, there appears to be an absence of any laboratory studies, where exposure factors have been controlled for, to confirm the presence or absence of a 2-4 hour phenomenon. While there are numerous studies concerned with time-of-day effects, these have only required the intermittent evaluation of performance, with participants being tested for short periods of time every two or three hours.

In this context, we have started a research programme to look at the effects of time-on-task under laboratory conditions where participants are required to perform tasks for, as near as is reasonably possible, 8 hours continuos periods. We decided for the initial pilot experiment reported in this paper to select two tasks which required different information processing demands. The first of these was a simple signal detection task requiring skilled based behaviour (a psychomotor skill) while the other was a task requiring rule based behaviour and placing demands on working memory. No specific hypotheses were generated for this study although it was felt that the 2-4 hour shift phenomenon may result from changes in performance strategy over shift time reflecting individual risk management. In this case the relationship between performance and time-on-task may be different depending on the task requirements.

Method

A mixed experimental design was used. Time-on-task was the within-participants factor, data being collected for each participant for 8 consecutive hourly periods. Task type was the between-participants factor, subjects being allocated to one of two task types. The computerised versions of the tasks were developed for the experiment using Microsoft Visual Basic 4.0.

Signal detection task

The signal detection task required participants to continuously monitor the computer screen for two different signals. They had to press the space bar when a target signal was presented (the letter "c") and ignore the non-target signals (the letter "o"). Both targets and non-target signals were presented in 8.5 point Arial font and were black on a grey background. Signals were presented for 10 ms. Duration between signals was randomised between 3 and 30 sec. so that a signal was presented on average once every 15 sec.. The sequence of targets and non-targets was also randomised with 25% of signals being targets. Signals were presented at the same location at the centre of the screen. Participants were instructed to press the space bar as quickly as possible when a target signal was presented.

Broadbent task

This task was based on the one developed by Broadbent (1977). The task requires participants to input two numbers. An algorithm, which the participants are not told, produces two output numbers from the inputs. Participants are required to match these outputs with two target numbers which are generated randomly by the computer. The input numbers required to solve the solution were always positive integers between 1 and 300 and participants were informed of this. The two equations used to transform participants' inputs to outputs were:

$$\text{output-1} = (220 \times \text{input-1}) + (80 \times \text{input-2})$$
$$\text{output-2} = (4.5 \times \text{input-2}) - (2 \times \text{input-1})$$

Inputting two numbers and comparing the resulting outputs with the targets was termed an "attempt". Participants required a number of attempts to match their outputs with the targets. Once outputs were matched with the targets a "trial" was completed. A new set of targets was automatically generated by the computer and a new trial began. Participants were instructed to reach the goals as quickly as possible and with as few attempts as possible.

Experimental Procedure

Twelve participants were allocated at random to one of the two tasks. They were trained for 2 hours on the signal detection task and 3 hours on the Broadbent task. On the following day participants undertook the experiment. They carried out the assigned task continuously for 45 minutes per hour over eight hours. All participants started at 9.00 a.m. and finished at 5.00 p.m.. Participants had a break for 15 minutes each hour with an extended break of half an hour for lunch. The lunch break was at either 11.45 or 1.45 with 3 participants from each task in each condition.

Results

Tables 1 and 2 give the results for the key performance measures collected for each task. The between-task analyses are not presented in this paper.

Table 1. Summary of results for Signal Detection Task

Measure	PERIOD							
	1	**2**	**3**	**4**	**5**	**6**	**7**	**8**
Hit p.	0.89	0.90	0.88	0.87	0.91	0.87	0.87	0.79
False alarm p.*	0.02	0.01	0.03	0.01	0.02	0.03	0.02	0.03
Mean rt, sec	0.67	0.68	0.68	0.69	0.70	0.69	0.71	0.72
A-prime*	0.97	0.97	0.96	0.96	0.97	0.96	0.96	0.93
R-S Ratio	0.94	0.92	0.99	0.90	0.98	0.95	0.96	0.87

* Significant variation identified using repeated measures ANOVA, $p < 0.05$.

Description of measures

Hit probability (p.) - probability of correctly detecting the target
False alarm probability (p.) - probability of identifying a non-target as a target
Mean reaction time (rt)- time elapsing between signal presentation and response
A-prime - a non-parametric measure of participants' sensitivity to signals.
R-S ratio - a non-parametric measure of response bias.

Table 2. Summary of results for Broadbent Task

Measure	PERIOD							
	1	**2**	**3**	**4**	**5**	**6**	**7**	**8**
Mean trial duration, sec.*	183.1	142.6	133.3	148.0	139.9	145.2	137.1	144.0
Mean attempt duration, sec.*	11.7	10.6	10.3	10.3	9.8	9.8	9.7	9.2
Attempts per trial	23.6	17.8	17.2	19.2	19.2	19.7	19.3	20.2
Errors per attempt	0.038	0.034	0.030	0.037	0.038	0.033	0.039	0.037

*   Significant variation identified using repeated measures ANOVA, $p < 0.05$.

Description of measures

Mean trial duration - mean number of sec taken to match targets to outputs
Mean attempt duration - mean number of sec taken to input two numbers and review the resulting output.
Mean attempts per trial - mean number of attempts taken to successfully complete a trial.
Errors per attempt - errors recorded include miskeying and deletions.

Discussion

The results from both experiments demonstrate some significant changes in performance as a function of time-on-task. However, the trends when looked at individually or in combination do not offer any convincing support for the 2-4 hour shift phenomenon. A possible reason for this lack of support is that individually neither task is sufficiently demanding to provide an opportunity for participants to alter their performance strategies significantly over time. Therefore, in the next experiment it is planned to have participants perform both task types concurrently to see whether this will reveal changes in risk management strategies over time.

References

Akerstedt, A. 1994. Work injuries and time of day - national data. In T. Akerstedt and G. Kecklund (eds.), *Proceedings of the conference on Work Stress and Accidents*, Stockholm Karolinska Institute, Stress Research Report No 248, p.106.

van der Flier and Schoonman, W. 1988. Railway signals passed at danger: situational and personal factors underlying stop signal abuse. *Applied Ergonomics*, **19**, 135-141.

Broadbent, D.E. (1977) Levels, hierarchies and locus of control, *Quarterly Journal of Experimental Psychology*, **29**, 181-201.

Folkard, S. and Monk, T.H. 1979. Shiftwork and performance, *Human Factors*, **21**, 483-492.

Kovacs, S. and Varga, J. 1996. Physiological responses of workloads, shift system and operation with different technologies in Hungarian bauxite and coal miners. In R.J. Koubek and W. Karwowski (eds.), *Manufacturing Agility and Hybrid Automation - I*, Louisville, Kentucky: IEA Press, pp.449-452.

Lauridsen, O. and Tonnesen, T. 1990. Injuries related to the aspects of shift working: a comparison of different offshore arrangements. *Journal of Occupational Accidents*, **12**, 167-176.

Leigh, J., Mulder, H.B., Want, G.V., Farnsworth, N.P. and Morgan, G.G. 1990. Personal and environmental factors in coal mining accidents. *Journal of Occupational Accidents*, **13**, 233-350.

Novak, R.D., Smolensky, M.H., Fairchild, E.J. and Reves, R.R. 1990. Shiftwork and industrial injuries at a chemical plant in south-east Texas. *Chronobiology International*, **7**, 155-164.

Ong, C.N., Phoon, W.O., Iskandarn, N. and Chia, K.S. 1987. Shiftwork and work injuries an iron and steel mill. *Applied Ergonomics*, **18**, 51-56.

Pokorny, M.L.I., Blom, D.H.J. and van Leewen, P. 1987. Shifts, duration of work and accident risk of bus drivers. *Ergonomics*, **30**, 61-88.

Verhaegen, P.K. and Ryckaert, R.W. 1986. Vigilance of train drivers. *In Proceedings of the 30th Annual Meeting of the Human Factors Society*, pp.403-407.

Wagner, J.A. 1988. Shiftwork and safety: a review of the literature and recent research results. In F. Aghazadeh (ed.), *Trends in Ergonomics/Human Factors V*, Amsterdam: Elsevier, pp.591-600.

Wyatt, S. and Fraser, J.A. 1929. *The Effects of Monotony in Work*. Industrial Research Board, Report No 56, London: HMSO.

Wyatt, S. and Langdon, J.N. 1932. *Inspection Processes in Industry*. Industrial Research Board, Report No 63, London: HMSO.

Wyatt, S. and Langdon, J.N. 1937. *Fatigue and Boredom in Repetitive.*Work Industrial Research Board, Report No 77, London: HMSO.

INDUSTRIAL ERGONOMICS

THE APPLICATION OF RAPID PROTOTYPING TECHNOLOGIES FOR INDUSTRIAL DESIGN AND ERGONOMIC EVALUATION

Mark Evans

Lecturer,
Department of Design and Technology,
Loughborough University,
Loughborough, Leics. LE11 3TU

Despite the advent of rapid prototyping technologies, they are virtually unused during the industrial design phase of New Product Development (NPD) where form and user interface are defined. Due to the close relationship between industrial design and ergonomics, this paper explores the potential for ergonomists to exploit this technology.

Case study material from the ergonomic evaluation and industrial design of a nylon-line garden trimmer is used to illustrate the advantages of using rapid prototyping. Two distinct rapid prototyping technologies are applied to produce physical models - Stereolithograpphy and Laminated Object Manufacture. The characteristics of these technologies are discussed, and contributions to ergonomic evaluation identified.

The paper concludes that a change in methodology is required by industrial designers to make full use of rapid prototyping and allow ergonomists the potential for enhanced product evaluation.

Introduction

During programmes of new product development (NPD), industrial designers and ergonomists make significant use of two and three dimensional models for the evaluation of design proposals. The variety of model types available enables evaluation to take place throughout the design and development process, and range from simple sketches to working prototypes.

As the sophistication of computer aided design software increased during the 1980's, it became possible to generate three dimensional virtual models within the computer environment. Whilst these models made a major contribution to the engineering development of products, apart from being accurate visualisations they had little relevance to industrial designers or ergonomists as they could not be held.

Towards the end of the 1980's, rapid prototyping started to emerge from an experimental technology into a commercially viable means of converting virtual computer models into physical objects. It did this by converting the computer models into a series of layers, and then re-creating the layers by hardening or cutting materials and bonding them together. The principal systems to become commercially available were Laminated Object Manufacture, Stereolithography, Fused Deposition Modelling, Selective Laser Sintering, and Solider.

As the availability of rapid prototyping increases, there is considerable potential for it to be effectively integrated into the working practices of ergonomists and industrial designers. To illustrate such potential, this paper examines the nature of conventional industrial design practice, and contrasts this with a strategy developed to exploit rapid prototyping and ergonomic evaluation. A conventional strategy of industrial design is briefly identified, followed by an approach to integrate rapid prototyping using computer aided industrial design (CAID). The latter uses case study material from the design of a nylon-line garden trimmer.

Conventional Strategy

Prior to the advent of CAID and rapid prototyping, industrial designers employed a range of techniques to originate, evaluate, manipulate, and present design ideas. Indeed, these techniques are still very much in use as it is the minority of industrial designers that have access to CAID systems.

Unfortunately, professional ergonomists rarely support industrial designers during NPD, indeed the norm tends to be for them to be used to evaluate products only when they have been in production for some time.

The following methodology identifies the opportunities within a conventional industrial design strategy for 2D and 3d model evaluation by an ergonomists:

Idea Generation
Using 2D and 3D sketching to originate designs. Despite being relatively crude, the 3D sketch models allow basic ergonomic evaluation.

Concept Presentation
Design proposals carefully illustrated (usually in marker) for formal presentation. The ergonomic evaluation of 2D renderings tends to be limited to such issues as conspicuity and basic form issues.

Working Prototype
Whilst working prototypes perform in a similar way to the production item, they do not have the correct appearance. As working representations, ergonomists can determine potential conflict between the performance of the product, and capabilities of the user.

Visual Model
Having selected a design concept, 3D visual models are produced as non-working representations of the product. As they appear no different from the intended production item, they are particularly useful for ergonomic evaluation, enabling such issues as grip, control layout, conspicuity, and reach issues to be assessed.

Visual Prototype
Prior to production, the functional capabilities are integrated with the visual and ergonomic features to produce a hand built visual prototype. These are the highest level of physical model and allow detailed ergonomic evaluation. Unfortunately, as they tend to be hand built they are notoriously expensive and have relatively long lead times.

Rapid Prototyping Strategy

The following methodology for the integration of rapid prototyping technologies and ergonomic evaluation was derived during the design of a nylon line garden trimmer. It illustrates how ergonomists and industrial designers can exploit CAID and rapid prototyping to enhance professional practice.

User trials
Prior to any design activity, user trials were undertaken to provide feedback on the features of products currently in production. A series of tasks were carried out in a moveable frame that could be re-positioned onto uncut grass once a product had been tested. The results enabled an ergonomic wish-list to be produced for the revised design, and highlighted a major problem in terms of fore-arm discomfort after relatively short periods of use.

Idea Generation

Despite the use of a CAID system, no replacement has yet been found for the spontaneity and speed of 2D and 3D sketch models. If the forms of the product are relatively simple, there may be the opportunity for some CAID modelling, but generally this phase of the design process remains unchanged. However, using ergonomic feedback from the user trials ensured a closer match between user needs and design concepts.

Computer Presentation

To present the design concepts and allow rapid prototyping, the surfaces of the concepts were modelled on the CAID system. The speed of the modelling depends on the sophistication of the system and skill of the operator, but appears to take around 30% longer than using conventional marker techniques.

Despite the increase in modelling time, the presentation output was far superior. The renderings were geometrically accurate and extremely realistic.

Whilst it was not possible to hold the virtual computer model, real-time rotation on-screen gave the impression of having the product in front of you. For the relatively sophisticated forms of the trimmer, this proved an invaluable aid to understanding the ergonomics features of the hand-grip in particular.

Concept User Trials

Gaining user feedback on the design proposal at this stage helped reduce the potential for reiteration at a later stage.

Using a foam sketch model (produced during idea generation), computer renderings (produced during computer presentation), and the working prototype, users were able to comment on hand grip, appearance and performance. Whilst the three model types represented specific product attributes, users had no problems in viewing them as a complete product. These trials provided valuable feedback on handle dimensions and orientation.

Rapid Prototyping

Within a conventional strategy of NPD, engineering drawings would be required to produce firstly the visual model (non-working), and secondly the visual prototype (fully working). There is also a degree of replication involved in producing a visual model, and then the visual prototype which combines appearance with function.

Rapid prototyping removed the need for engineering drawings as the process utilised the CAID models produced during computer presentation. The surfaces were converted into rapid prototyping files (STL files) and down-loaded to the selected rapid prototyping system.

Stereolithography and Laminated Object Manufacture were used to produce the components for the nylon line trimmer. Whilst there was no significant difference in costs, these processes use quite different techniques to build the models.

Stereolithography uses a laser to harden a thin layer of photo-curable epoxy resin which forms a slice of the model. Once the layer is complete, it is lowered, and the hardening of the next slice started. The model is built up from a series of layers that have a slight stepped appearance but are completely fused together. A Schematic diagram of the Stereolithography process can be seen in Figure 1.. There were a total of eight components for the line trimmer, which needed a light sanding to remove the stepping.

Laminated Object Manufacture again uses a laser, but in this process it cuts through a layer of paper to form a slice of the model. The next slice of the model is formed by bonding another layer of paper onto the previous and cutting out the next slice. Despite being produced from paper, the Laminated Object Manufacture model had the appearance and working properties of wood.

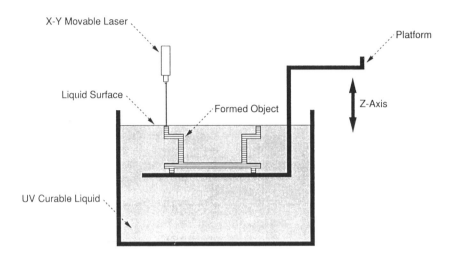

Figure 1. The Stereolithography process

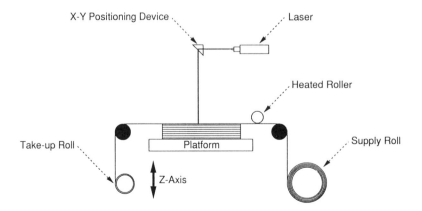

Figure 2. The Laminated Object Manufacture Process

During the assembly of the working rapid prototypes, problems were experienced with the Laminated Object Manufacturing model. All of the components contained relatively thin walls (average 3mm), and in one area went down to 1mm. This proved too thin for the process, and de-lamination occurred in several places. As the components deteriorated this model was abandoned as it was not feasible to dedicate the considerable amounts of time required to produce the high definition model required for evaluation.

The Stereolithography appeared to have no problems coping with the walls, and only a few minor modifications were required to add the components to make the trimmer work. As the Stereolithography rapid prototype had the visual appearance of the design proposal, there was no need to produce a non-working visual model. This removed significant cost and modelmaking time.

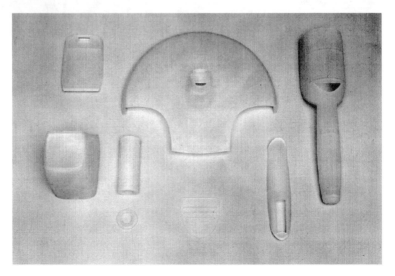

Figure 3. Stereolithography components prior to painting and assembly

Figure 4. Fully working rapid prototype

Rapid Prototype User Trials

As the rapid prototypes were produced overnight, and the finishing and build took only four days, a fully working visual prototype was ready for user trials approximately two months ahead of one produced under a conventional strategy.

The strength of the epoxy resin rapid prototypes was quite adequate for the user trials, and the results gave confidence in the proposal. Only one major design change was required to the angle of the handle.

Conclusions

As the cost of CAID systems decreases, and the availability of rapid prototyping increases, the integration of these technologies is set to transform the working practices of industrial designers. In terms of performance, lead times are set to be dramatically reduced in line with many other activities within the field of NPD.

A key requirement appears to be to select an appropriate rapid prototyping technology for the application. Whilst Laminated Object Manufacture is ideal for models with relatively thick walls (such as engine castings), it appears limited for use where injection moulded components are modelled.

Whilst there will be the opportunity to put products into production quicker, there is also the potential to design poor products faster. It is the authors belief that the improvements in design output must be married with professional ergonomic support during NPD. A rapid prototyping strategy facilitates a range of 2D, 3D and virtual models for ergonomic evaluation at a level previously unknown in NPD. By utilising these approaches to modelling the industrial designer and ergonomist are in a position to reshape the nature of NPD and improve the user's perception of manufactured products.

AN ERGONOMIC MODEL PROPOSED FOR MULTIPURPOSE USE OF ROBOTS IN FLEXIBLE MANUFACTURING SYSTEMS

Ercüment N. DİZDAR[1], Mustafa KURT[2], Abdullah ŞİŞMAN[3]

[1]*I. E. Dept., K.K.Ü., Kırıkkale, Turkey*
tel: +90 318 357 24 58 / 120,   e-mail: dizdar@kku.edu.tr   fax: +90 318 357 24 59

[2]*I. E. Dept., G. Ü., Ankara, Turkey*
tel: +90 312 210 23 10 / 28 54,   e-mail: dizdar@kku.edu.tr   fax: +90 312 230 84 34

[3]*M. E. Dept., N.Ü., Niğde, Turkey*
tel: +90 388 225 01 15,   e-mail: asisman@alp.nigde.edu.tr   fax: +90 388 225 01 12

As the need in rapidness, sensitivity and flexibility of the production increase, the importance of robot automation in production system also increases. In order to automate the production, companies should buy a large number of expensive robots which brings the high financial cost problem. However, this problem can be solved by multipurpose use of robots. In this study made under the guidelines of ergonomic science which presents intelligent and effective approaches to the problems in production systems, a model is developed to show the feasibility of multipurpose use of available robots in flexible manufacturing systems.

Introduction

In the highly competitional world market, companies can only withstand by reaching to a rapid, flexible and low-cost production performance, since the changes in demands forces the companies to keep their production lines renewable.

So they can either get many robots having different hardware, in which companies meet both financial and locational problems on production lines. Or, alternatively, they can use their available robots multifunctional. In this case, the manager has two alternatives. He will either buy many application softwares for one robot or develop an ergonomic software model that will enable him to use his available robot for more than one task.

The way of buying many application software requires considerable financial support, because of the expensive softwares sold by the suppliers with additional payment.

If the available robot can be made multipurposed by means of an ergonomic software model, those troubles will be overcome. Automation supported by the developed software model also solves most of the problems in production.

If there is a limitation in flexible production, this limit is certainly a disadvantage for production. In this case, system should either be designed for a limited product range in which every demand that is not supplied will result in customer lost or more flexibility is added to the system.

In this study, adding high flexibility to robots makes their usages efficient.

Flexible Manufacturing Systems (FMS)

FMS is the latest level of automation along an evolutionary road to achieve more productivity and flexibility from manufacturing equipment. An FMS is a collection of production equipment logically organized under a host computer and physically connected by a central transport system. The object of the FMS is to simultaneously manufacture a mix of workpiece types while being flexible enough to sequentially manufacture different workpiece type mixes without costly, time consuming, change-over requirements between mixes.

Increased competition since the late 1960s has changed the manufacturing environment for end-user products which were traditionally in the mass-production market. The result of such a competitive market is that the products' and their pieceparts' life cycle are decreasing. Suppliers to the market of equipment containing machined pieceparts are finding that their customers require increased flexibility from them while they will not accept higher costs for this requirement. The machining systems of today must provide not only flexibility but also high productivity to provide low-cost goods. For this reason, use of FMS becomes inevitable in our today's industry.

Multipurpose Use of Robots in FMS
In the subject of automation, in order to control and increase the production capacity, high effort has been made by production and design engineers. In this automation process, it is a very difficult problem to select the proper robot in the lines for performing different manipulations for each different workpieces. Ergonomic robot selection, i.e. determination of proper type and number of robots to be used is an optimization problem. Also, it is known that the required time and expense of planning and istallation of them is increasing everyday. If the production system is desired to be capable of flexible manufacturing, it is a must to select the multipurpose robots. A robot can become a multipurpose robot by making suitable modifications on its software.

Depending on software and hardware supports, performing various functions is the most important feature of robots. For example, a robot equipped for welding in a company, must be able to be used for palletizing in any other. Robot manufacturer provides this multifunctionality by preparing a second or a third program option. This, of course, means an additional payment for customers.

The programmability in robots is an important feature which means the capability of changing and controlling robot motion. Reprogrammability is another feature of the robots that it distinguishes them from the single-purpose automatic machines. This feature provides an important advantage to robots for their multipurpose usage. Any available robot can easily be reprogrammed with a small change by using the programming codes of control system. The apparatus and other equipment of robot for different jobs are defined separately.

If an ergonomic software model (ESM) algorithm can be developed, companies will meet their multipurpose robot usage by making their available robots perform other operations without buying expensive softwares. In this study, a software interface model is developed for one robot to perform more than one function.

Ergonomic Software Model (ESM)

In this study, a modular software interface is developed by using the programming codes of operating system of the available robot. In programming, sub-programming method is preferred in order to ease its usage by any person who has a medium-level knowledge about command set.

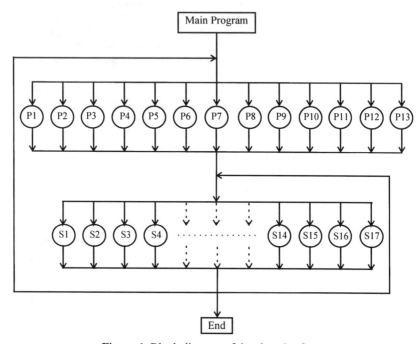

Figure 1. Block diagram of developed software

This program (ESM) consists of one main and thirty sub-programs (Table 1). The modular structure of the program (Figure 1) enables the operator who has a medium-level knowledge about command set to easily understand the software. An additional software is also prepared for user to teach the handling and putting places of workpieces.

The developed software can also perform all the standard palletizing options of material handling robots of the same category without requiring Position Variable Input Function. Moreover, the software performs multiplication and division functions by using standard addition and subtraction functions without requiring any optional software. For this purpose, Cartesian-Coordinate Input Function is employed. The position difference between each workpiece location is performed by step-by-step changing the Work Coordinate System Positioning Input.

A company which uses this philosophy of software interface can meet its multifunctional robot need without buying expensive softwares. As it is performed in this study, a company can use its welding robot in, for example, palletizing also.

Table 1. Table of symbols in block diagram of ESI

Procedure Name		Pallet Operation Options	
P1	Pallet A - Heap First	S1	Apply A-B-C-B-A Motion
P2	Depallet A - Heap First	S2	Calculate P_n
P3	Pallet A - Grid Heap First	S3	Move to P_n
P4	Depallet A - Grid Heap First	S4	Move to Q_{ijk}
P5	Pallet - E - Heap First	S5	Multiplication
P6	Depallet - E - Heap First	S6	Initialization of Dummy Registers
P7	Pallet F - Different Stack Pallet	S7	Move to A at the End
P8	Pallet A - Stack First	S8	Division
P9	Depallet A - Stack First	S9	Preparation for Grid Types
P10	Pallet A - Grid Stack First	S10	Slide X -and Y- Axis
P11	Depallet A - Grid Stack First	S11	General Initialization of registers
P12	Pallet - E - Stack First	S12	Move to Q_{111}
P13	Depallet - E - Stack First	S13	Move to Q_{nms}
		S14	Height Calculation
		S15	Input Data
		S16	Bottom Point File
		S17	Approach Pattern File

The method developed can be applied to almost all the robots used in the industry. In this way, without buying expensive softwares, it will be possible to use a single-purposed industrial robot in any other task too.

The applicability of this model is shown by a study performed at the Middle East Technical University (METU), CAD/CAM Robotics Center. In this study, a software interface model is developed in order to use a FANUC R-G2 controlled FANUC Arc Mate Sr Arc Welding robot as a palletizing robot. In the developed method, in order to reach this aim, we made use of reprogrammability feature of robots.

Application of ESI

FANUC R-G2 Controller robot programming techniques requires knowledge of low level programming language, and a good knowledge of robot's own instruction set, and, of course, theoretical background of robot motion. FANUC R-G2 Controller robot's programming language is not a high-level programming language, rather it is an instruction set coding with mnemonics. So, we can consider it as a low level programming language. As for the developed software, we entirely discussed many used methods for the robot programming. But it was unable to use general techniques of structured programming because of the incapability's of the controller of the ARC Mate Sr robot which has installed in CAD-CAM Robotics Center of METU such as lack of IF ... ELSEIF ... ELSE ... ENDIF like block structure, or LOOP like structure. In addition to unstructured programming facilities, because Robot has no position variable, Multiplication, Division and several important options, we were forced to use work-coordinate-settings to move robot arm from one point to another and are forced to develop multiplication and division sub-programs. These obligations have occupied main tackle for developing software. Note that Robot's sensitivity depends on these mentioned division and multiplication sub-programs. Thus Code developed as unstructured, but modular such that all of the work is managed by a main program. Several sub-programs are used to multiply, divide, put or grip the work pieces, etc.,. One who has a knowledge of instruction set of robot can easily read the code, and understand how it works. For the users, there are separate sub-programs to teach points where the work pieces will be put or gripped, and an input data program. These manipulations can be done by reading User's Guide section.

Conclusion And Discussion

As mentioned previously, the most valuable advantage of robots is in their multifunctional usage. Manufacturers wish to use their robot in more than one function. However, this does not become true during production.

In this study made under the guidelines of ergonomic whose purpose is to increase productivity and takes the production as a combination of many systems in consideration, a new perspective is given such that the robots can be multifunctionally used by a small amount of payment. By such an approach, companies can easily add flexibility to their manufacturing system without buying an additional software.

In this study made under the guidelines of ergonomic science which takes the manufacturing as a combination of many systems into account and tries to increase the productivity, a model developed for flexible use of robots is identified. In this way, user companies can easily add more flexibility to their automation without buying additional softwares.

Robots should be flexible in order to be modified for a second task. In this method developed, we have made use of the reprogrammability feature of robot in order to reach this aim.

The companies using this software interface algorithm can meet their multifunctional robot needs by making their robot capable of performing other tasks without buying expensive softwares.

Robot usage is a hopeful progress to eliminate the last handicap in front the computer integrated manufacturing (CIM). But the flexibility problem has not been solved completely yet. In this study, it's aimed to be of help to similar studies in this area.

This type of problems also show the necessity of application of industrial engineering techniques. Especially, it is now clear that the manufacturing systems should be controlled by ergonomic approaches.

In this study presented, an ergonomic software model is developed for multipurpose use of robots which take part in a flexible manufacturing system.

The concept of flexible manufacturing has been around for over twenty years, applications have until recently been expensive and unique. Developments over the last five years, however, have made widespread applications of flexible manufacturing systems achievable. In this paper, it is intended to show how to increase both the flexibility and the productivity of their production systems.

Acknowledgment

The authors wish to acknowledge to Dr. Yalçın Erol, Dr. Osman Şadi Özkul İlker Keskinkılıç and Serkan Çak for their valuable support.

References

Askin, R. G. and Standridge, C. R. 1993, *Modeling and Analysis of Manufacturing Systems* (John Wiley & Sons, Canada), 125-157.
Blume, C. 1986, Programming Languages for Industrial Robots (Springer Verlag).

Browne, J. Harhen, J. Shıvnan 1990, *Production Management Systems, A CIM Perspective* (Addison-Wesley, Great Britain) 1-143.

Groover, M. P. Weiss, M. Nagel, R. N. Odrey, N. G. 1986, *Industrial Robotics* (McGraw Hill).

Dizdar, E. N. Şişman, A. Kurt, M. and Keskinkılıç, İ. 1997, An ergonomic software interface approach for multifunctional use of industrial robots, *1ˢᵗ International Conference on Engineering Design and Automation*, (Bankok, Thailand).

Hallam, R. and Hodges, 1990, *Industrial Robotics*, (Heinemann Newnes).

Kee, D. Jung E. S. and Chang, S. R. 1994, Man-machine interface model for ergonomic design, *Proceedings of the 16ᵗʰ Annual Conference. on Computer and Industrial Engineering*, (Ashikaga, Japan) 365-368.

Parrish, D. 1993, *Flexible Manufacturing*, (Butteerworth-Heinemann, Cambridge).

Seng-Yuh L. Yuau-Tay, C. Chao-Wei, C. and Menq-Jiun, W. 1995, Computer integration manufacture system designed/manufacture for education purpose, *Proceedings of the 1995 International IEEE/IAS Conference on Industrial Automation Control Emerging Technologies*, (Taipei, Taiwan), 231-235.

Şişman, A. 1994, *Palletizing Software Developed for Multifunctional use of an Industrial Robot*, Graduate School of Natural and Applied Sciences of METU in Partial Fulfillment for the Degree of Master Science ın Mechanical Engineering, (METU, Ankara, Turkey) 1-85.

Şişman, A. Dizdar, E. N. Kemal A. and Gürses C. 1997, Flexible using of robots in computer integrated manufacturing (CIM) systems, *1ˢᵗ International Conference on Engineering Design and Automation*, (Bankok, Thailand).

Şişman, A. Dizdar, A. Balkan, T. 1997, Bilgisayar entegreli üretimde robotların çok fonksiyonlu kullanımı için bir ara yazılım algoritması, *1. Makina Mühendisliği Kongresi, MAMKON'97* (İstanbul, Türkiye).

Westkamper, E. 1986, Increase in flexibility and productivity with computer integrated and automated manufacturing, *Proceedings of FMS5, edited by K. Rathmill*, (IFS Publications, UK) 121-126.

RISK AND ERROR

HUMAN ERROR IDENTIFICATION:
COMPARING THE PERFORMANCE OF SHERPA AND TAFEI

Neville Stanton

Department of Psychology
University of Southampton
Highfield
Southampton
SO17 1BJ

This paper introduces the problems associated with evaluating Human Error Identification (HEI) techniques. In particular, appraisal criteria which rely upon user opinion, face validity and utilisation are questioned. Instead, a quantitative approach, similar to that used in psychometric is proposed. This is demonstrated on two evaluation studies: one assessing Systematic Human Error Reduction and Prediction Approach (SHERPA) and the other assessing Task Analysis For Error Identification (TAFEI). The results of these studies are compared and are remarkably similar, giving some confidence in the use of these approaches. It is suggested that all HEI techniques should be subjected to equally rigorous scrutiny.

Introduction to human error

Human error is an emotive topic. Psychologists and Ergonomists have been investigating the origins and causes of human error since the dawn of the discipline (Reason, 1990). Traditional approaches suggested that error was an individual phenomenon, the individual who appears responsible for the error. Indeed, so-called 'Freudian slips' were treated as the unwitting revelation of intention: errors revealed what a person was really thinking but did not wish to disclose. More recently, error research in the cognitive tradition has concentrated upon classifying errors within taxonomies and determining underlying psychological mechanisms (Senders & Moray, 1991). The taxonomic approaches by Norman (1988) and Reason (1990) have led to the classification of errors into different forms, e.g. capture errors, description errors, data driven errors, association activation errors and loss of activation errors. Reason (1990) and Wickens (1992) identify psychological mechanisms implied in error causation, for example the failure of memory retrieval mechanisms in lapses, poor perception and decision making in mistakes and motor execution problems in slips. Taxonomies offer an explanation of what has happened, whereas consideration of psychological mechanisms offer an explanation of why it has happened. Reason (1990), in particular, has argued that we need to consider the activities of the individual if we are able to consider what may go wrong. This approach does not conceive of errors as unpredictable events, rather as wholly predictable based upon an analysis of an individual's activities.

Since the late 1970's much effort has been put into the development of techniques to predict human error based upon the fortunes, and misfortunes, of the nuclear industry. Despite this development, of many techniques are poorly documented and there is little in the way of validation studies in the published literature.

Predicting human error

Most HEI techniques work in a similar manner (Kirwan, 1992, a). First they require the work activity to be analyse, normally using a derivative of task analysis. Second, errors associated with those activities are predicted. Kirwan (1992b) conducted a comparative study of six potential HEI techniques. For this study he developed eight criteria on which to compare the approaches. In his study, Kirwan recruited 15 HEI analysts (three per technique, excluding group discussion). Four genuine incidents from the nuclear industry were used as a problem to focus the analysts effort. This is the main strength of the study, providing a high level of ecological or face validity. The aim of the was to see if the analysts could have predicted the incidents if the techniques had been used. All the analysts took less than two hours to complete the study. The results for the performance of the techniques were presented by Kirwan as both subjective judgements (i.e.: low, medium and high) and rankings (i.e. worst and best). No statistical analysis was reported in the study, this is likely to be due to methodological limitations of the study (i.e. the small number of participants employed in the study). From the available techniques, SHERPA achieved the highest overall rankings and Kirwan recommends a combination of expert judgement together with the SHERPA technique as the best approach.

A study by Baber & Stanton (1996) aimed to test the hypothesis that the SHERPA technique made valid predictions of human errors in a more rigorous manner. In order to do this, Baber & Stanton compared predictions made by an expert user of SHERPA with errors reported by an observer. The strength of this latter study over Kirwan's is that it reports the use of the method in detail as well as the error predictions made using SHERPA. Baber & Stanton's study focuses upon errors made during ticket purchasing on the London Underground, for which they sampled over 300 transactions during a non-continuous 24 hour period. Baber and Stanton argue that the sample was large enough as 90% of the error types were observed within 20 transactions and after 75 transactions no new error types were observed. From the study, SHERPA produced 12 error types associated with ticket purchase, nine of which were observed to occur. Baber & Stanton used a formula based upon Signal Detection Theory (Macmillan & Creelman, 1991) to determine the sensitivity of SHERPA in predicting errors. Their analysis indicated that SHERPA produces an acceptable level of validity when used by a expert analyst. There are, however, a two main criticisms that could be aimed at this study. First, the number of participants in the study was very low, in fact only two SHERPA analysts were used. Second, the analysts were experts in the use of the technique, no attempt was made to study performance whilst acquiring expertise in the use of the technique.

Research issues for human error identification

The development of HEI techniques could benefit from the approaches used in establishing psychometric techniques as two recent reviews demonstrate (Bartram et al, 1992, Bartram et al, 1995). The methodological concerns may be applied to the entire field of ergonomics methods. There are a number of issues that need to be addressed in the analysis of human error identification techniques. Some of the judgements for these criteria developed by Kirwan (1992, b) could be deceptive justifications of a techniques effectiveness, as they may be based upon:

- User opinion
- Face validity
- Utilisation of the technique.

User opinion is suspect because of three main reasons. First it assumes that the user is a good judge of what makes an effective technique. Second, user opinion is based on previous experience, and unless there is a high degree of homogeneity of experience, opinions may vary widely. Third, judgements may be obtained from an unrepresentative sample. Both Kirwan (1992, b) and Baber & Stanton's (1996) studies used very small samples. Face validity is suspect because a HEI technique might not be able to predict errors just because it looks as though it might, which is certainly true in the domain of psychometrics (Cook, 1988). Finally, utilisation of one particular technique over another might be more to do with familiarity of the analyst than representing greater confidence in the predictive validity of the technique. Therefore more rigorous criteria need to be developed.

Shackel (1990) proposed a definition of *usability* comprising effectiveness (i.e. level of performance: in the case of HEI techniques this could be measured in terms of reliability and validity), learnability (i.e. the amount of training and time taken to achieve the defined level of effectiveness) and attitude (i.e. the associated costs and satisfaction). These criteria together with those from Kirwan (1992, b) and the field of psychometrics (Cronbach, 1984; Aiken, 1985) could be used to assess HEI techniques (and other ergonomics methods) in a systematic and quantifiable manner.

Systematic Human Error Reduction and Prediction Approach (SHERPA)

SHERPA (Embrey, 1986) uses Hierarchical Task Analysis (HTA: Annett *et al.* 1971) together with an error taxonomy to identify credible errors associated with a sequence of human activity. In essence the SHERPA technique works by indicating which error modes are credible for each task step in turn, based upon an analysis of work activity. This indication is based upon the judgement of the analyst, and requires input from a subject matters expert to be realistic.

The process begins with the analysis of work activities, using Hierarchical Task Analysis. HTA (Annett, Duncan, Stammers & Gray, 1971) is based upon the notion that task performance can be expressed in terms of a hierarchy of goals (what the person is seeking to achieve), operations (the activities executed to achieve the goals) and plans (the sequence in which the operations are executed). Then each task step from the bottom level of the analysis is taken in turn. First each task step is classified into a type from the taxonomy, into one of the following types:

- Action (e.g. pressing a button, pulling a switch, opening a door)
- Retrieval (e.g. getting information from a screen or manual)
- Checking (e.g. conducting a procedural check)

- Selection (e.g. choosing one alternative over another)
- Information communication (e.g. talking to another party)

This classification of the task step then leads the analyst to consider credible error modes associated with that activity. From this classification the associated error modes are considered. For each credible error (i.e. those judged by a subject matter expert to be possible) a description of the form that the error would take is given. The consequence of the error on system needs to be determined next, as this has implications for the criticality of the error. The last four steps consider the possibility for error recovery, the ordinal probability of the error, its criticality and potential remedies.

Task Analysis For Error Identification (TAFEI)

TAFEI (Baber & Stanton, 1994; Stanton & Baber, 1996) explicitly analyses the *interaction* between people and machines. TAFEI analysis is concerned with task-based scenarios. This is done by mapping human activity onto machine states. TAFEI analysis consists of three principal components: Hierarchical Task Analysis (HTA), State-Space Diagrams (SSDs which are loosely based on finite state machines: Angel & Bekey, 1968) and Transition Matrices (TM). HTA provides a description of human activity, SSD provides a description of machine activity and TM provides a mechanism for determining potential erroneous activity through the interaction of the human and the device. In a similar manner to Newell & Simon (1972), legal and illegal operators (called *transitions* in the TAFEI methodology) are identified.

In brief, the TAFEI methodology is as follows. First, the system to be addressed needs to be defined. Next, the human activities and machine states are described in separate analyses. The basic building blocks are hierarchical task analysis (describing human activity) and state space diagrams (describing machine activity). These two types of analysis are then combined to produce the TAFEI description of human-machine interaction. It is worth pointing out that the state space diagram also has the potential to contain information about hazards or by-products associated with particular states. However this resource has not been addressed here because the focus of this paper is upon prediction of performance errors. From the TAFEI diagram, a transition matrix is compiled and each transition is scrutinised. Each transition is classified as 'impossible' (i.e. the transition cannot be performed), 'illegal' (the transition can be performed but it does not lead to the desired outcome) or 'legal' (the transition can be performed and is consistent with the description of error-free activity provided by the HTA), until all transitions have been analysed. Finally, 'illegal' transitions are addressed in turn as potential errors, to consider changes that may be introduced.

Comparing the performance of SHERPA and TAFEI

A comparison of the results from two studies will be used to indicate the performance of SHERPA and TAFEI. The SHERPA study is reported in full by Stanton & Stevenage (1996) and the TAFEI study is reported in full by Stanton & Baber (in prep). Both studies take an empirical approach to the assessment of reliability and validity of HEI methods.

The SHERPA study employed 25 undergraduates to undertake a SHERPA analysis of the task steps involved in the purchase of an item of confectionery from a vending machine. Following a period of instruction, participants undertook the

SHERPA analysis on three separate occasions. This was done to test the reliability of the approach (i.e. the consistency of the analysis over time). Validity was examined by comparing predicted errors with observed errors. The analysis was based upon a Sensitivity Index (SI) from the signal detection paradigm reported by Baber & Stanton (1996).

Similarly, the TAFEI study employed 36 undergraduates to undertake a TAFEI analysis of the confectionery vending machine interaction. Following a period of instruction, participants undertook the SHERPA analysis on three separate occasions. Again, analysis of the reliability and validity of the approach was undertaken.

The sensitivity for the two techniques on the three occasions is reported in table one below.

	SHERPA mean SI	SHERPA SD	TAFEI mean SI	TAFEI SD
Time 1	0.76	0.1	0.73	0.1
Time 2	0.74	0.1	0.78	0.1
Time 3	0.73	0.1	0.79	0.1

Table 1. Means and standard deviations for the Sensitivity Index of SHERPA and TAFEI

As table one shows, there is a good deal of similarity in the sensitivity of the two approaches. This confirms a previous study reported by Baber & Stanton (1996). The reliability of the two techniques over the three occasions is reported in table two below.

	SHERPA rho	SHERPA p-value	TAFEI rho	TAFEI p-value
Time 1 to 2	0.65	0.001	0.46	0.005
Time 1 to 3	0.32	0.05	0.36	0.05
Time 2 to 3	0.39	0.05	0.67	0.001

Table 2. Reliability coefficients and probability values for SHERPA and TAFEI

The results suggest that the TAFEI approach appears to improve reliability over time (time 2 to time 3) whereas SHERPA shows quite good levels of reliability initially (time 1 to time 2).

Conclusions

In conclusion, the results are generally supportive of both SHERPA and TAFEI. They indicate that novices are able to acquire the approaches with relative ease and reach acceptable levels of performance within a reasonable amount of time. Comparable levels of sensitivity are achieved and both techniques look relatively stable over time. This is quite encouraging, and it shows that HEI techniques can be evaluated quantitatively.

Human error is a complex phenomenon, and is certainly far from being completely understood. Yet in attempting to predict the forms in which these in which these complex behaviours will manifest themselves armed only with a classification systems and a description of the human and machine activities it is amazing what can be achieved. Despite the gaps in our knowledge and the simplicity of the techniques, the performance of the analysts appears surprisingly good. This offers an optimistic view of the future for human error identification techniques. However, before approaches can be recommended more rigorous studies are needed.

References

Aitkin, L. R. (1985) Psychological Testing and Assessment. Allyn & Bacon: Boston.

Angel, E. S. & Bekey, G. A. (1968) Adaptive finite state models of manual control systems. IEEE Transactions on Man-Machine Systems, March, 15-29.

Annett, J.; Duncan, K. D.; Stammers, R. B. & Gray, M. J. (1971) Task Analysis. Training Information No. 6. HMSO: London.

Baber, C. & Stanton, N. A. 1994, Task analysis for error identification: a methodology for designing error-tolerant consumer products, *Ergonomics*, 37 (11), 1923-1941.

Baber, C. & Stanton, N. A. (in press) A comparison of observational data with two human error identification techniques. Applied Ergonomics.

Bartram, D.; Lindley, P.; Foster, J. & Marshall, L. (1992) Review of Psychometric Tests (Level A) for Assessment in Vocational Training. BPS Books: Leicester.

Bartram, D.; Anderson, N.; Kellett, D.; Lindley, P. & Robertson, I. (1995) Review of Personality Assessment Instruments (Level B) for use in Occupational Settings. BPS Books: Leicester.

Cook, M. (1988) Personnel Selection and Productivity. Wiley: Chichester.

Cronbach, L. J. (1984) Essentials of Psychological Testing. Harper & Row: New York.

Embrey, D. E. (1986) SHERPA: A systematic human error reduction and prediction approach. Paper presented at the International Meeting on Advances in Nuclear Power Systems, Knoxville, Tennessee.

Kirwan, B. (1992a) Human error identification in human reliability assessment. Part 1: overview of approaches. Applied Ergonomics, 23 pp. 299-318.

Kirwan, B. (1992b) Human error identification in human reliability assessment. Part 2: detailed comparison of techniques. Applied Ergonomics, 23 pp. 371-381.

Macmillan, N. A. & Creelman, C. D. (1991) Detection Theory: a user's guide. Cambridge University Press: Cambridge.

Newell, A. & Simon, H. A. (1972) Human Problem Solving. Prentice Hall: Englewood Cliffs, NJ.

Norman, D. A. (1988) The Psychology of Everday Things. Basic Books: New York.

Reason, J. (1990) Human Error. Cambridge University Press: Cambridge.

Senders, J. W. & Moray, N. P. (1991) Human Error. LEA: Hillsdale, NJ.

Shackel, B. (1990) Human factors and usability. In: Preece, J. & Keller, L. (eds) Human-Computer Interaction. Prentice-Hall: Hemel Hempstead.

Stanton, N. A. & Baber, C. (1996) A systems approach to human error identification. Safety Science, 22, pp. 215-228.

Stanton, N. A. & Baber, C. (in prep) Training TAFEI to novices. Manuscript in preparation.

Stanton, N. A. & Stevenage (1996) Learning to predict human error: issues of acceptability, reliability and validity. Submitted to Ergonomics.

Wickens, C. D. (1992) Engineering Psychology and Human Performance. Harper Collins: New York.

FACTORS AFFECTING POSTAL DELIVERY OFFICE SAFETY PERFORMANCE

T.A. Bentley and R.A. Haslam

Health and Safety Ergonomics Unit
Department of Human Sciences
Loughborough University
Loughborough , Leics. LE11 3TU

This paper presents results from a preliminary investigation into the role of supervisor safety practices in postal delivery office safety performance. Supervisor safety practices which may promote improved safety performance were determined from an analysis of contributory supervisor factors for delivery falls, and a survey of management and safety personnel. 'Desirable' safety practices identified included thorough accident investigation, hazard control activities, and use of special practices during severe weather. Further research is considering the use of these practices by supervisors of high and low accident rate postal delivery offices.

Introduction

Accident incidence rates for postal delivery staff working at different delivery offices vary considerably, despite most delivery staff undertaking similar tasks. These differences in safety performance are still observed when the effects of geographical region (effect of weather conditions), delivery area (eg rural/urban), size of delivery office workforce (staff in post), and time supervisor is in post, are taken into account. One explanation for differences in safety performance may be supervisor safety practice.

A number of studies have considered the effectiveness of safety programs and management and supervisor safety practices in occupational safety performance. Many of these studies analysed safety practices common to companies with outstanding safety performance, or compared practices of companies with high injury rates with those having low rates (eg Cohen, 1977; Simonds and Shafai-Sahrai, 1977; Smith et al, 1978; Cohen and Cleveland, 1983; Chew, 1988). Smith et al (1978) notes the following factors were found to be important to successful safety program performance in more than one such study: strong management concern for safety; full-time safety director who reports to top management; frequent use of safety promotions; frequent use of accident investigations; formal training of employees and supervisors; frequent positive contacts between employees and supervisors; stable workforce (older and more experienced employees/ married employees). These and other more recent studies (eg Simard and Marchand, 1994) highlight the importance of manager and supervisor safety practice in occupational safety. The objective of the present study was to identify supervisor safety practices which are effective in promoting good safety performance among postal delivery staff. Two methods were used to determine 'desirable' supervisor safety practices: an analysis of supervisor factors identified as contributory to delivery falls, and a survey of management and safety personnel.

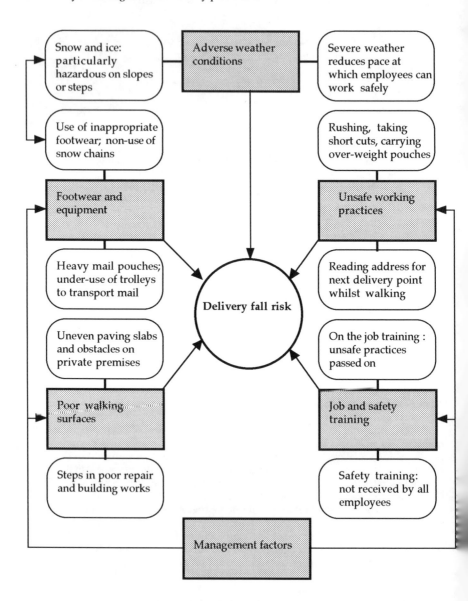

Figure 1. Summary diagram of risk factors for delivery fall accidents

Previous research considering the role of supervisor factors in delivery fall accidents

Delivery falls are the largest cause of accident and lost time for Royal Mail delivery staff. Previous research (Bentley and Haslam, 1996a and 1996b), involving analysis of accident data and reports, and delivery fall follow-up interviews, has found supervisor factors underlie a notable proportion of these accidents. For example, many accidents occurred because employees were attempting to work normally in adverse weather conditions. In addition, an analysis of accident reports suggested accident investigations by office supervisors were often too brief for identification of contributory factors and suitable preventive measures.

A survey of delivery staff, using focus groups and questionnaires, supported these findings, suggesting safety practices employed by office supervisors can affect delivery staff safety performance. Examples of supervisor factors associated with delivery fall accidents include failure to adopt measures to protect delivery staff during periods of adverse weather, failure to respond to employees' reports of hazards on delivery walks, and overlooking employee use of unsafe practice such as carrying more than one delivery pouch. Figure 1 summarises the major risk factors for delivery falls as identified from this research.

Delivery staff argued supervisor safety practices which were most effective in promoting good safety performance among delivery staff included additional time for deliveries during adverse weather, efforts to remove hazards and unsafe conditions from their walks (eg dangerous dogs; holes in pavements; etc), replacement of worn footwear, and lightening delivery loads by providing more 'drop points' where additional mail pouches can be picked up by delivery staff en route.

This research provided an index of 'desirable' supervisor safety practices designed to affect delivery staff safety performance through the control of unsafe conditions and acts connected with delivery accidents. In addition, these practices can be expected to have a positive affect on delivery office safety climate, as employees perceive safety activities of management as reflecting the business' commitment to safety.

Survey of management and safety personnel to identify supervisor safety-related practices which may be effective in improving delivery office safety performance

A survey of Royal Mail management and safety personnel was undertaken to determine safety practices currently used by office supervisors, practices considered to be most effective in promoting improved safety performance, and factors affecting supervisors' use of 'desirable' safety practices. Interviews with senior delivery managers and office supervisors were undertaken to identify practices currently employed, and barriers and incentives which influence the use of safety practices. A discussion session with senior safety personnel considered the areas of safety in which the office supervisor might most effectively play a role in improving office safety performance.

Using data produced from this survey and from the research described above, a list of 'desirable' supervisor safety practices and activities was drawn up (table 1). Some specific examples of 'desirable' supervisor safety practices are listed under the headings accident investigation and follow-up prevention, severe weather practice; safety communications, hazard control and equipment management.

Table 1. 'Desirable' supervisor safety-related practice

Supervisor safety practice	Specific examples of 'desirable' practice
Accident investigation and follow-up prevention activity	Supervisor personally investigates all reported accidents Accident investigation is thorough: * accident-involved employee is interviewed along with witnesses; * supervisor visits site of accident * supervisor identifies all contributing factors; Supervisor takes necessary follow-up preventive action (eg letters to Council or householder to have a hazard removed) Supervisor shares outcome of investigation and follow-up prevention with staff
Severe weather practice	Supervisor takes action to protect delivery staff from increased risk of injury during periods of severe weather (eg withhold staff until conditions improve; allow more time for deliveries; provide transport assistance; heavy mail items not taken out by foot; equipment for adverse weather supplied; special briefings)
Safety communications	Daily contact between supervisors and staff on safety and other job matters. Regular safety-focused team-briefings Display of safety campaign materials Informal safety advice Supervisor operates an open door policy Involvement of senior management (eg safety tours), and high priority of safety in meetings between senior management and office supervisors
Hazard control	Supervisor encourages reporting of hazards Hazards are recorded appropriately (ie hazard cards, 'walklogs') Supervisor takes action to remove 'avoidable' environmental hazards (eg writes to/visits Council or householder), or makes alternative delivery arrangements Supervisor undertakes regular office safety tours Supervisor shares the outcome of hazard reporting and removal efforts with staff

(continued)

Table 1. continued

Equipment management	Supervisor replaces footwear as required (eg due to damage or wear)
	Snow chains and other protective equipment available to delivery staff
	Supervisor undertakes pouch weighing exercises and arranges additional pouch 'drop points' en route where necessary
	Supervisor encourages staff to use alternative delivery methods to manual pouch carriage where possible (ie trolleys/cycles)

Factors affecting current use of safety practices by office supervisors

Factors which affect the use of the above safety practices were considered in the survey and are listed below:

* supervisor's knowledge of 'best safety practice'
* supervisor's training in areas of safety management
* time constraints
* budget constraints
* non-compatibility with quality considerations
* non-compatibility with business policies

Supervisor safety practices in 'high' and 'low' accident rate delivery offices

These preliminary findings are being used in the design of a more detailed study which is to consider the role of activities presented in table 1 and other management related factors in delivery office safety performance. A sample of 'matched' 'high' and 'low' accident rate postal delivery offices are being surveyed, using employee questionnaires and interviews with office supervisors.

Research questions are:

1. Do supervisor safety practices used in 'low' accident rate offices differ from those in 'high' accident rate offices ?

2. What safety practices do supervisors perceive as effective in accident prevention, and how do they understand their role in office safety performance ?

3. Are any of the following factors related to delivery office safety performance ?

* use of unsafe working practices
* delivery method used (eg walk and carry pouch, pouch trolley, pedal cycle)
* training received (job and safety)
* workload factors

The results of this and previous research will be used to determine possible safety intervention measures to reduce the incidence of accidents occurring during the delivery of mail.   As part of this process, 'safety intervention' focus groups will be held with employees, management and safety personnel.   Within these groups risk factors and possible solutions will be discussed, along with ideas for intervention design and implementation.

Acknowledgements

The authors wish to acknowledge the support of the Royal Mail, and in particular wish to thank Mike Dixon, Ian Cooper, Cynthia Yeates, Howard Kilroy, and John Leaviss for their assistance with this work.

References

Bentley, T. and Haslam, R. 1996 (a), Contributory risk factors for falls occurring during the delivery of mail, In: *Contemporary Ergonomics 1996* (Edited by: S .A. Robertson (Taylor and Francis, London), 189-194.

Bentley, T. and Haslam, R. 1996 (b), Outdoor falls in postal delivery employees: a systematic analysis of in-house accident data, In: *Advances in applied ergonomics,* (Edited by: A.F. Ozok and G. Salvendy) (USA Publishing: West Lafayette, Indiana) 204-207.

Chew, D. 1988, Effective occupational safety activities: Findings from three Asian developing countries. Int. Labour Review, 127, 111-125

Cohen, A. 1977, Factors in successful occupational safety programs, Journal of Safety Research, 9, 168-178.

Cohen, A and Cleveland, R.J. 1983, Safety practices in record-holding plant, Professional Safety, 9, 26-33.

Simonds, R. and Shafai-Sahrai, Y. 1977, Factors apparently affecting injury frequency in eleven matched pairs of companies, Journal of Safety Research, 9, 120-127.

Smith, M., Cohen, H., Cohen, A., and Cleveland, J. 1978, Characteristics of successful safety programs, Journal of Safety Research, 10, 5-15.

Simard, M. and Marchand, A. 1994, The behaviour of first-line supervisors in accident prevention and effectiveness in occupational safety, Safety Science, 17, 169-185.

OFF-SITE BIOMECHANICAL EVALUATION OF ELECTRICITY LINESMEN TASKS

D May *, R White **, R J Graves * and E M Wright ***

*   *Department of Environmental & Occupational Medicine,*
**   *Department of Biomedical Physics & Bioengineering*
        University Medical School, University of Aberdeen,
        Foresterhill, Aberdeen, AB25 2ZD
***   *OMS, Aberdeen*

Previous work showed overhead electricity linesmen have a high reported incidence of work related musculoskeletal problems. Extreme postures could be adopted when linesmen were working up poles and crimping overhead wires had particular risk features. The present study further investigates the linesmen's overhead work to assess the nature of potential risks. A postal survey of task and risk factors was carried out and a laboratory based biomechanical evaluation of the crimping task undertaken. Linesman reported low back pain particularly when operating the wire crimping tool. Forces in the low back were found to be excessive when the crimping task was simulated in laboratory studies. Risk reduction can be obtained by preventing extreme postures, and re-designing the crimping tool to be better balanced and lighter with support.

Introduction

A study by Graves et al (1966) found that overhead electricity linesmen had a high reported incidence of musculoskeletal problems which could be work related. This study showed that extreme postures could be adopted when linesmen were working up poles and that crimping overhead wires had particular risk features. This present study further investigates the overhead work carried out by the linesmen to assess the nature of potential risks when undertaking this work. In particular the study was designed to use biomechanical techniques to investigate these tasks and to make recommendations on ways to minimise risks.

Methods

A task analysis was carried out on-site to establish how the pole working tasks were performed and hence determine the postures the linesmen adopted. These were compared with the Extreme Posture Checklist (Parker and Imbus, 1992), to determine potential musculoskeletal risk factors. Those tasks which produced extreme postures were analysed using the Rapid Upper Limb Assessment (RULA) technique, to determine the category of risk and whether any change of posture was indicated (McAtamney and Corlett, 1993).

A questionnaire was designed for a postal survey to establish the extent of the musculoskeletal discomfort the linesmen reported and to relate the tasks to the discomfort. Questionnaires were sent out to all 282 linesmen in the company.

It was decided to simulate the crimping task in the laboratory using a single subject study. The simulation was undertaken with a linesman carrying out the crimping task in three different postures to provide an objective means for assessing the potential risk of musculoskeletal discomfort when using the crimping tool under controlled conditions. Tasks undertaken while wearing a full body harness were compared to wearing a waist belt. Biomechanical techniques (i.e. electromyography (EMG), force platform, and a load cell) were used in this study to determine the forces on the body and the electrical activity (EMG) of the muscles of the shoulder and neck.

The linesman stood with both feet on the force platform so that the three orthogonal foot/ground reaction forces could be recorded (Vertical and Horizontal shear forces - lateral and anterior/posterior). The subject was attached to a rigid frame 1 metre directly in front wearing either a full body or waist belt harness. A strain gauged load cell was placed in series with the harness close to the point of attachment in order to measure shear forces at the lumbar-sacral joint. EMG electrodes were attached to the skin over the deltoid muscles (abductors of the arm) and the neck extensor muscles to monitor the electrical activity of these muscles. The EMG signals provided objective information on the level of activity of the muscles.

The wire to be crimped was positioned at head height 1 metre to the left, right or centre of the linesman. The task was carried out for 30 seconds.

Results and Discussion

From the check list, the following tasks were found to have extreme postures: pole climbing; using the Power Auger; pole work using the wrench; lifting up tools via a pulley rope; lifting pole platform; changing insulators using a wrench; tightening wire between two poles; using the crimping tool; hammering actions; using a ratchet; changing an old transformer to a new transformer; earthing the wire; putting a fuse box up on pole; taking barbed wire off and putting it on the pole. Most of the tasks required extreme postures to be adopted for certain joints of the body. Crimping is the only task that has extreme postures for each part of the body. Table 1 shows a summary of the risk factors/extreme postures involved in crimping.

Table 1. Pole working with crimping tool

RISK FACTORS/EXTREME POSTURES FOR CRIMPING
Ankles extreme dorsiflexion
Knees fully extended
Back extended greater than 20°
Elbows above the head
Arms at or above shoulder height
Neck flexed and rotated greater than 20°
Stretching while holding crimping tool

Following the task analysis a RULA assessment was carried out on those tasks which had three or more extreme postures according to the Extreme Posture Checklist. Table 2 shows those tasks which had an Action Level of 3 or 4.

Table 2. RULA Action Levels (AL) of 3 or 4, and number of extreme postures

Task	RULA AL	No. Extreme postures
Pole climbing	3	3
Using Crimper	4	5
Using Power Auger	3	3
Tightening Bolts	3	3
Pole work - wrench	3	3
Tightening wire	3	3
Putting bolts on insulator	4	3
Using the wrench	4	3

One hundred and twenty questionnaires were returned, giving a 43% response rate. From the analysis of the postal survey data it was found that many of the linesmen reported low back, shoulder and neck discomfort (See Figure 1). The data shows that lower back discomfort seems to be the major area of discomfort in respondents, followed by shoulder and neck discomfort. These results are similar to those found by Graves et al (1996), as can be seen from the comparison of percentages in Figure 1 below.

Figure 1 Percentage respondents reporting discomfort in body regions in present study compared to Graves et al (1996)

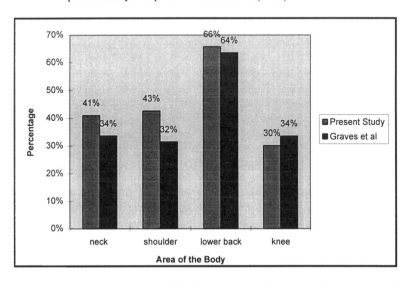

Over sixty eight percent of the linesmen experienced low back discomfort while performing the crimping task. In addition it was found that over ninety percent of the

crimping is done at or above shoulder height. The main task that involved awkward positions of the arms while working was found to be using the crimping task tool (56.7%).

Using a five point rating scale it was found that the main task rated as causing severe discomfort in the lower back, shoulder and neck was crimping tool work. Analysis of this task showed that it produced extreme postures. The crimping operation is used to join two pieces of wire together. This is done by using a hydraulic tool which is pumped to compress the two wires to make them one. There are different varieties of crimps with each type of crimp taking a different amount of time.

Many linesmen thought that the crimping tool was awkward to use, heavy and all the weight was in the head of the tool. Also because the length of the steel arms had been increased they felt more discomfort when crimping, as they had to stretch further.

The linesmen also stated that the full body harness was more restrictive than the waist belt and that the circulation to the hips and legs may be impaired when wearing the harness for long periods. A number of linesmen thought that hoists and pole platforms should be used more frequently. This would reduce the amount of stretching and reaching required when crimping.

Some linesmen stated that the crimping tool placed more strain on the body than any other tool because of stretching, when used at the top of the pole. Policy at present is to use the crimper for many jobs that used to have different tools. There is a new specification to which overhead lines are being built and moving to compressing all connections results in more awkward stretching.

The biomechanical analyses showed that similar forces were found with the full body harness and the waist belt, although slightly higher forces for Anterior- Posterior (AP) horizontal shear forces were found with the waist belt (see Figure 2). These forces were in excess of 30% body weight during the crimping task when the lineman twisted to the left or right. This level of shear force was sustained throughout the task (30 seconds) and is likely to produce musculoskeletal discomfort.

The vertical (compression) forces recorded by the force platform ranged between body weight and 119% body weight. This level of force is not excessive and would not normally cause any discomfort but in the case of the lineman carrying out pole work this force could be concentrated under certain areas of the feet as they are supported on either side of the pole. These pressures could cause compression in the soft tissues and forced dorsiflexion of the foot. The use of appropriate footwear with a thick rigid sole and ankle support can minimise discomfort and injury.

Extreme posture is an important factor in determining the amount of force going through the low back. The waist belt allows the linesmen to take up more extreme postures than the full body harness probably due to the lack of restraint and therefore the AP shear forces were higher in the waist belt. The lateral forces were negligibly small (<4.9% body weight maximum)

The EMG recordings showed the level of activity of the neck extensor and arm abductor muscles during crimping. The arm pumping the crimping tool had considerably higher frequency signals than the arm supporting the tool. The neck extensor muscles also showed low frequency signals and periods when the muscles were quiescent.

When the lineman initially twisted to the right the median frequency of the EMG for the right shoulder (which was supporting the tool) stabilised to about 64 Hz after 10 seconds of operation, it reduced to 19 Hz at 17 seconds and to 13 Hz after 20 seconds. This reduction in median frequency of the EMG signal is an indication of muscle fatigue. For the left shoulder which was operating the lever of the crimping tool showed a similar reduction in frequency in the EMG recording: 43 Hz median frequency at 10 seconds reducing consistently to 33 Hz near the end of the task. The median frequencies measured at the neck also reduced with time. This clearly demonstrates muscle fatigue.

Similar effects were recorded on the EMG traces when the subject twisted to the left except that the patterns for the respective deltoid muscles were reversed. There was less apparent fatiguing of the muscles when the crimping task was carried out in the forward position. Generally the design of the crimping tool and the forces required to operate it will contribute to these fatigue effects.

Figure 2. Comparison of horizontal shear forces (AP) of the full body harness compared to the waist belt

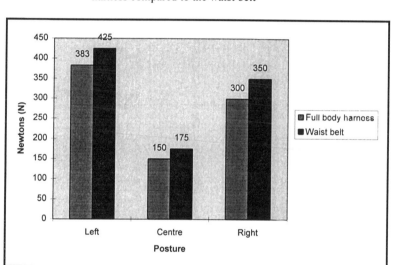

Conclusions

This study has confirmed the earlier work by Graves et al (1996) that there is a high incidence of reported musculoskeletal disorders in the linesman. The detailed task and postural analysis of the pole work has shown that numerous tasks lead to extreme postures and are potentially hazardous. Many linesman complain of low back pain particularly when operating the wire crimping tool.

The laboratory simulation of the crimping task showed that Anterior - Posterior shear forces generated in the low back were found to be excessive and could potentially cause musculoskeletal discomfort. These forces may increase the potential risk of musculoskeletal disorders. The compressive forces however, were not likely to cause musculoskeletal discomfort due to their magnitude, however, prolonged, repetitive loading of these forces of this magnitude could cause discomfort and joint or soft tissue damage.

The EMG traces showed that within twenty seconds of operating the crimping tool the active arm began to fatigue. These results help to identify sources of muscle fatigue and hence which parts of the task need to be modified. In addition it helps prioritise task elements for improvement.

Recommendations have been made to address extreme postures leading to excessive loads on the body and to educate linesmen about musculoskeletal disorders. It has also been suggested that the crimping tool should be re-designed so that it is better balanced and is lighter and supported, for example, by an overhead harness.

The study approach has shown that a combination of on-site analysis and off-site simulation can provide useful objective information on sources of risk. In addition the biomechanical approach provides a useful means of pinpointing degree and sources of risk. It is intended to use these methods to examine task and tool design changes to provide optimum solutions.

References

Graves, R.J. De Cristofano, A. Wright, E. Watt, M. White, R. 1996, Potential musculoskeletal risk factors in electricity distribution linesmen tasks. In: *Contemporary Ergonomics 1996*, London: Taylor and Francis: 215-220.

May, D. 1996, *Identification of overhead electricity linesmen task risk factors and laboratory based biomechanical analysis of selected tasks*, MSc Ergonomics Project Thesis, Department of Environmental and Occupational Medicine, University of Aberdeen: Aberdeen.

McAtamney, L., Corlett, E.N. 1993, RULA: a survey method for the investigation of work related upper limb disorders, *Applied Ergonomics*, **24,** 91-99.

Parker, K.G. Imbus, H.R. 1992, *Cumulative trauma disorders: Current issues and ergonomic solutions: A systems approach*, (Lewis Publishers, USA).

RISK-TAKING BEHAVIOUR OF FORESTRY WORKERS

Simo Salminen*, Tapio Klen** and Kari Ojanen**

Finnish Institute of Occupational Health, Vantaa, Finland
**Kuopio Regional Institute of Occupational Health, Kuopio, Finland*

The risk-taking behaviour of 228 Finnish forestry workers was measured in a postal questionnaire study in 1986. Their risk-taking tendency was not related to accident frequency. Risk-taking behaviour, however, correlated significantly with impulsiveness and neuroticism. The more internal the forestry worker's mean score on the Health Locus of Control scale was, the less likely he was to take risks in the job.

Introduction

The concept of risk is usually defined as a combination of the probability of damage and the seriousness of consequences (Schön, 1980). In addition to potential losses and their significance, is the uncertainty about the realisation of losses, the essential element in the risk (MacCrimmon and Wehrung, 1986). Risk taking is defined as voluntary and conscious exposure to danger (Klen, 1992).

It is a common belief that individual risk taking is an important factor affecting occupational accidents. The different methods in measuring risk taking in different studies render the review of empirical results rather difficult. An experiment with a risk simulator showed that electric workers tending to take higher risks had more accidents at work (Rockwell, 1962). Metal workers involved in accidents made significantly more mistakes and resorted to smaller safety margins than did accident-free workers (Mittenecker, 1962).

Risk taking can be measured by observing the behavior of drivers in traffic situations. Accident-involved drivers were more likely to follow with short headways or to take more risks than accident-free drivers (Evans and Wasielewski, 1982). Higher speeds were observed for drivers with a record of prior accidents (Wasielewski, 1984).

The more accidents the driver has previously caused, the more risky passing manouevres he/she tends to undertake (Risser, 1985).

Risk taking has also been measured by self-assessment with questionnaires. Risk taking did not differ between those truck drivers who were often and seldom involved in accidents (Gumpper and Smith, 1968). The active victims in the accident situations took more risks than the passive victims or accident-free workers (Verhaegen et al., 1985). The risk-taking tendency of construction workers was not found to be related to accidents and safety behaviour (Landeweerd et al., 1990). Risk taking did play a role in 54% of serious occupational accidents, however (Salminen, 1994). On the other hand, risk taking was not related to the collision history of male taxi drivers (Burns and Wilde, 1995).

In seven out of 10 studies there is a positive correlation between risk taking and accident-involvement. In three out of five questionnaire studies there is no relationship between risk taking and accidents. Obviously, there is a need for more detailed study to find the best method to measure risk taking. The aim of this study is to examine the risk-taking behaviour of Finnish forestry workers.

Material and methods

A postal questionnaire was sent to the 450 Finnish loggers in the spring of 1986. In spite of two reminders, only 228 questionnaires (51%) were returned.

The risk taking of the forestry workers was measured on a questionnaire with 25 items. The measure was scored so that the higher the score of the subject, the less the subjects took risks. The internal consistency estimate (Cronbach's alpha) for risk taking was .79.

We measured sociability and neurotism of the subjects with Eysenck's (Eysenck & Eysenck, 1964) Personality Inventory Form C. The impulsive and fatalistic attitude of the subjects was measured in the same question series with six questions like: "Do you feel that you are often at the mercy of fate?". The subjects had five-step scales to answer ("agree completely" to "disagree completely"). The internal consistency estimates (Cronbach's alpha) for sociability were .80, for neurotism .81, for impulsiveness .69, and for fatalism .72.

The subjects filled out the Health Locus of Control Inventory (Wallston et al., 1976). The scale has 11 items of which six measure external and five internal locus of control according to Boyle and Harrison (1981). The reliability (Cronbach's alpha) of the external health locus of control was .76, of internal locus of control .66, and of the whole scale .65. The Accident Locus of Control Scale (Klen, 1992) was a modified version of the Health Locus of Control Inventory, where the term "accident" replaced the term "illness". The internal consistency estimates (Cronbach's alfa) for external accident locus of control were .80, for internal locus of control .61, and for the whole Accident Locus of Control Scale .66.

The occupational accidents of the subjects were checked for the past 7-9 years from the company's records. In the questionnaire the subjects reported their compensation claims from accidents during the past 12 months.

Results

There was no significant difference between accident-free and accident-involved forestry workers in risk taking both with company recorded and self-reported accidents (Table 1). Forestry workers were divided into two risk taking groups by the median. There was no significant difference between risk takers and risk avoiders in accident frequency.

Table 1. Risk taking and accident frequency of forestry workers.

	Risk taking				
Company records	N	Mean	SD	T	p
accident-free	83	91.2	12.7	-0.17	n.s.
accident-involved	107	91.5	12.3		

	Risk taking				
Self-report	N	Mean	SD	T	p
accident-free	172	91.1	12.1	-0.60	n.s.
accident-involved	34	92.5	14.7		

	Accident frequency				
Risk taking group	N	Mean	SD	T	p
risk avoiders	97	0.11	0.14	-1.59	n.s.
risk takers	106	0.14	0.19		

The risk taking tendency correlated significantly with impulsiveness ($r = 0.30$, $p<0.001$) and neuroticism (.18, $p<0.05$). The more impulsive and neurotic the forest worker was, the more likely he was to take risks. However, sociability (-.11) and fatalism (-.02) were not significantly correlated with risk taking.

For both Health Locus of Control and Accident Locus of Control, the more internal the mean score on locus of control of the forest workers, the less likely they were to take risks (Table 2). The correlations were almost the same (for HLC .21 and for ALC .20). The external locus of control was not related to risk taking.

Table 2. Correlations between locus of control measures and risk taking.

Measure	Internal	External
Health locus of control	.21, $p<0.01$	.09, n.s.
Accident locus of control	.20, $p<0.01$	.01, n.s.

Discussion

The risk-taking behaviour of forestry workers was not related to accident frequency. This result is in line with previous studies using questionnaires as measures of risk taking (Gumpper and Smith, 1968; Landeweerd et al., 1990; Burns and Wilde, 1995). One explanation is that our measure is a measure of general risk taking, and not only of the work-related risk-taking of forestry workers.

The more impulsive and neurotic the forest worker was, the more likely he was to take risks generally in life. In very hazardous work environments like the forest, this kind of less prudent person could well run the risk of an accident.

Forestry workers with an internal locus of control tend to avoid risk taking. This result is in line with previous studies (see Salminen and Klen, 1994). However, contrary to previous results, external locus of control was not related to risk taking.

References

Boyle, E.S. and Harrison, B.E. 1981, Factor structure of the Health Locus of Control scale, Journal of Clinical Psychology, 37, 819-824.

Burns, P.C. and Wilde, G.J.S. 1995, Risk taking in male taxi drivers: Relationships among personality, observational data and driver records, Personality and Individual Differences, 18, 267-278.

Evans, L. and Wasielewski, P. 1982, Do accident-involved drivers exhibit riskier everyday driving behavior? Accident Analysis and Prevention, 14, 57-64.

Eysenck, H.J. and Eysenck, S.B.G. 1964, Manual of the Eysenck Personality Inventory, (University of London Press, London)

Gumpper, D.C. and Smith, K.R. 1968, The prediction of individual accident liability with an inventory measuring risk-taking tendency, Traffic Safety Research Review, 12, 50-55.

Klen, T. 1992, Tapaturmariskin arviointi, riskinottotaipumus, persoonallisuus ja työtapaturmat metsureilla [The assessment of accident risk, the risk taking tendency, personality, and occupational accidents of forestry workers]. Unpublished licenciate thesis, (University of Helsinki, Department of Psychology)

Landeweerd, J.A., Urlings, I.J.M., De Jong, A.H.J., Nijhuis, F.J.N. and Bouter, L.M. 1990, Risk taking tendency among construction workers, Journal of Occupational Accidents, 11, 183-196.

MacCrimmon, K.R. and Wehrung, D.A. 1986, Taking risks. The management of uncertainty, (Free Press, New York)

Mittenecker, E. 1962, Methoden und ergebnisse der psychologischen unfallforschung, (Franz Deuticke, Wien)

Risser, R. 1985, Behavior in traffic conflict situations, Accident Analysis and Prevention, 17, 179-197.

Rockwell, T.H. 1962, Some exploratory research on risk acceptance in a man-machine setting, Journal of the American Society of Safety Engineers, december, 23-29.

Salminen, S. 1994, Risk taking and serious occupational accidents, Journal of
    Occupational Health and Safety - Australia and New Zealand, **10**, 267-274.
Salminen, S. and Klen, T. 1994, Accident locus of control and risk taking among
    forestry and construction workers, Perceptual and Motor Skills, **78**, 852-854.
Schön, G. 1980, What is meant by risk? Basic technical views for the initiation
    and applications of safety legislation, Journal of Occupational Accidents, **2**,
    273-281.
Verhaegen, P., Strubbe, J., Vonck, R. and Van Den Abeele, J. 1985, Absenteeism,
    accidents and risk-taking, Journal of Occupational Accidents, **7**, 177-186.
Wallston, B.S., Wallston, K.A., Kaplan, G.D. and Maides, S.A. 1976, Development
    and validation of the Health Locus of Control (HLC) scale, Journal of
    Consulting and Clinical Psychology, **44**, 580-585.
Wasielewski, P. 1984, Speed as a measure of driver risk: Observed speeds versus
    driver and vehicle characteristics, Accident Analysis and Prevention, **16**,
    89-103.

HUMAN ERROR DATA COLLECTION: PRACTICAL METHOD UTILITY AND SOME DATA GENERATION RESULTS

Gurpreet Basra & Barry Kirwan

Industrial Ergonomics Group
School of Mechanical and Manufacturing Engineering
University of Birmingham
Edgbaston, Birmingham, United Kingdom

Over the past eighteen months there have been seven human error data collection exercises associated with the CORE-DATA (Computerised Operator Reliability & Error Database) project. This project aims to produce a database of human error probabilities for supporting Human Reliability Assessments (HRA), risk assessments and ergonomics evaluations. Currently CORE-DATA exists in a fully computerised prototype form. The first part of the paper will present some of the data collected showing the Human Error Probability (HEP) ranges achieved, and the types of data that have been, and will be, incorporated into CORE-DATA. The second part of the paper briefly consider the relative effectiveness, efficiency and utility of four different data collection methods: direct observation; incident review; training record examination; and expert judgement.

1. Introduction

Following an Advisory Committee for the Safety of Nuclear Installations (ACSNI: 1991) report recommending the construction of a human error database, this project was conceived and initiated in 1991. The database exists today in the form of CORE-DATA (Computerised Operator Reliability & Error database). The theoretical developments of CORE-DATA have already been well documented (see Taylor-Adams and Kirwan, 1995). This paper will concentrate on how the system has expanded over the last eighteen months.

There have been seven error data collection projects associated with CORE-DATA, six of which have been industry based, the remaining two being carried out in a laboratory environment. These projects have been conducted in a wide range of industries, in an attempt to extend and broaden CORE-DATA and its usage across industrial sectors. The objective of each study was to collect a range of HEPs. Such data can then support HRA work in hazardous industries.

$$HEP = \frac{\text{Number of errors}}{\text{Number of opportunities for error}}$$

2. Data Collection Studies

Data collection projects have covered the following areas:

- Offshore lifeboat evacuation
- Manufacturing
- Offshore drilling
- Permit to work

- Electricity transmission
- Calculator errors
- NPP emergency scenarios

A selection of data from the seven studies is shown in Table 2. Three of these studies will now be discussed in more detail below.

2.1 Offshore Lifeboat Evacuation

This study was concerned with collecting human error data relating to offshore lifeboat evacuation, via davit-launched (Basra and Kirwan, 1997). These boats are held at the side of the platform by wires or "davits", the boat is released by lowering the boat to sea level, then unhooking the boat using an internal mechanism and driving away to safety. Data were collected using a combination of incident data, observation of training coxswain at a simulator training facility, a search of training records and expert judgement, using trainers as subjects. A total of nineteen HEPs were generated for the davit-launch system, twelve were collected via observation and an additional seven were generated through the expert judgement sessions.

2.2 Manufacturing Environment

Data were collected from four main sections of the plant; two from two different product assembly stations, and two from hardware operation areas. Data from the assembly stations were collected via direct observation. It was only possible to generate two HEPs through this method. One of the main reasons for this, is the fact that the employees were highly skilled individuals, who had been doing the same job everyday for at least five years.

Data from operational areas where tasks included, for example, mixing volatile substances, and casting and moulding, were collected via expert judgement. These tasks were very time consuming and not very repetitive at the time of the study. By employing expert judgement a more diverse range of errors could be investigated.

Finally data from hard-copy data were collected on all aspects of manufacture, across all product types. This search proved to be very useful, and approximately seventy HEPs were generated from this single method of data collection.

2.3 Offshore Drilling

Data were collected using a combination of incident review, direct observation and expert judgement. In this case it was possible to generate quantitative data as well as qualitative data from database searches. The Health and Safety Executive Wells Information System was searched and a number of incidents which were caused by human error were recorded. Observational data were collected from a training simulator. The simulator consisted of operational control panels and a VDU display illustrating the well, drillpipe, and other machinery that was being controlled by the panels, and it was possible to simulate various "kick" sizes and equipment malfunctions. Finally, expert judgement was utilised to quantify errors which were not observed, or recorded from incident review, but could be potentially fatal. Subjects for the expert judgement sessions were made up of simulator trainers and experienced training school consultants.

3. Data Collection Methods

3.1 Direct Observation

Direct observation of trials, both real and simulator based proved to be invaluable, initially as a tool for the development of task analyses, and then as the primary data collection method. Data were collected from real operational environments, training simulators and laboratory controlled simulations. There are a number of pros and cons associated with this method of data collection as discussed below;

1. When collecting observational data there can be problems of obtrusiveness. It could be argued that the presence of an observer would have some effect on the performance, positive or negative, of the observee. This could be brought about by fear of management, observer unfamiliarity, and general uncertainty of the observer's objectives and motives. These fears can be reduced if each subject is allowed firstly to give full permission to be involved in such an activity, and in all the exercises no subject was ever forced into partaking. Secondly, by explaining the exact nature of the work and allowing participants to ask questions, it is usually possible to alleviate such uncertainties. Finally, the general uncomfortableness that some feel when they are being observed, quickly dissolves as the subject gets used to the observer's presence and continues with their tasks normally.

2. Collecting data from simulators proved to be very productive, particularly in the laboratory exercises. It was possible to modify certain performance shaping factors and introduce various stressors, to gain immediate performance feedback. On the other hand, subjects are well aware of the fact that they are performing a task in a simulator, and in this respect it can be argued that subjects do not always react as if they are in a real situation. In terms of error identification, errors that may occur in a real life threatening situation may not be seen, as the subject is not under the same stress. Furthermore, errors that are made in a simulator may not be seen in a real situation as an employee would be more alert in a real emergency.

3. Observational data collection for the generation of error probabilities, requires a number of trials to be observed for the data to be useful and for robust error probability derivation. In some cases it was not possible to watch the required number of trials, in others a good number of trials were observed, but no errors were observed. Some tasks were also very time consuming, and the time between the start of an observation could be hours and even days, so it is therefore necessary to use direct visual observation only for highly repetitive tasks of relatively short duration.

3.2 Hard-copy Data

Hard-copy data refers to, in this case, any company documents that report on accidents, incidents, near-misses and general discrepancies in working practices, as well as more formal accident investigation reports. These data can be used in a qualitative capacity. It can be used firstly to establish what kinds of incidents occur. This is necessary when research is being conducted in an unfamiliar area, the researcher may be initially unaware of the kinds of things that can go wrong. Data collated from incident reports provide a background for the human error analysis stage as well as providing supporting information for later data collection. More specifically this kind of information can be useful in providing data concerned with performance shaping factors. It may also be possible to quantify data, but only if denominator data are available or can be evaluated somehow.

Hard-copy data searches have on the whole been very time consuming without much returns. This can be attributed to a number of reasons. One of the main reasons is that the data encoded in such systems are not recorded in the kind of detail required for a data collection exercise such as is required for CORE-DATA. Furthermore, procedural errors or lapses are generally not recorded in such databases, data are usually concerned with, for example, people falling off ladders. However, accident investigation are typically rich in information and can provide detailed qualitative data.

3.3 Training Record Examination

To date data have been collected from two training establishments. It could therefore be possible, in theory, to collect data from training records, in particular the records of people who do not attain the qualification for which they are training. This kind of search has to date proved to be unfruitful. The main reason for this can be attributed to the fact that these schools do not need to record in explicit detail how a student failed. For example, in the case of the lifeboat evacuation study, a number of students were identified as having failed their coxswains course due to an inability to use the compass. It was not recorded however, whether compass steering problems were caused by over- or under-steering, or whether the compass was being read completely wrong and the coxswain was steering into the platform as opposed to away from it. This level of information is crucial for encoding into CORE-DATA, and for usage in supporting predictive risk assessment.

3.4 Expert Judgement

Expert judgement is a powerful way of quantifying, in a controlled manner, a number of errors for which there are little or no data. In all the data collection exercises a combination of Absolute Probability Judgement (APJ) and Paired Comparisons (PC) were used (Seaver and Stillwell, 1983). When trying to create a database of human error probabilities for risk assessment support, the intention is not to utilise expert judgement at all, since it appears to defeat the object of having a database of 'hard data'. However, expert judgement has been found to have a useful place in the project's data collection activities. Firstly, there are times where no data could be collected, but risk assessment data will be required, and HEPs are needed for a number of installations. In this case, expert judgement techniques can be utilised as a last resort. Preferably, more than one technique of expert judgement will be used. This will then allow the reliability or convergence of the data to be assessed. If it is not convergent, then the data would not be input into CORE-DATA. If convergence does occur, however, then the results will constitute the best estimates until real data are collectable, and would thus be of interest to risk assessors, albeit with the knowledge that the data was 'synthetic'. One of the studies (the NPP data generation study) used expert judgement for a set of tasks that are important to the UK NPP industry, but for which not only are there no data, but there is significant variance in estimates produced by personnel using traditional HRA techniques. This was therefore a study where convergence in the industry was being attempted.

Expert judgement may also be used to fill in data collection gaps, but where real observed data can then be used 'calibrate' the expert judgement. This occurred to an extent in the lifeboat evacuation study cited earlier, where there was some overlap between expert judgement-derived estimates and those derived by observation. The third usage of expert judgement concerns the modification of data to render it more contextually valid. Thus, for example, if data are collected by observing lifeboat evacuations at a training facility, it will be desirable to consider how those data would be influenced by real emergency contexts offshore, i.e. in the context of a fire on the platform in a force six wind and sea state, etc. Expert judgement can be utilised to modify the data to make them more directly usable in risk assessments. Again, as with all usage of expert judgement techniques for CORE-DATA, the final data produced make it very clear as to the role of expert judgement in the data production process.

3.5 Comparison of Techniques

The data collection techniques were compared against five criteria (see Table 1); comprehensiveness in terms of finding all types of errors; consistency of data collection/general approach; usefulness of results for HRA purposes; resource usage; and richness of data in psychologically meaningful terms (enabling the categorising of the error mode, Performance Shaping Factor (PSF), and Psychological Error Mechanism (PEM) for the error).

Essentially the two most practically useful techniques in this projects' data collection activities so far have been direct observation and expert judgement. However, this may be because of poor incident recording systems, which may be unrepresentative of Nuclear Power

Plant (NPP) incident recording system. For example, the nuclear chemical study by Kirwan et al (1990) did successfully generate 36 HEPs from nuclear plant incident and accident records, which in that industry were highly detailed.

Table 1 - Comparison of Techniques

	Comprehensive -ness	Consistency	Usefulness of results	Resources usage	Richness of data
Direct Observation	High	High	Moderate	High	Moderate
Hard-copy data	Moderate	Moderate	Low	Low/Mod.	Low/High
Training record search	Low	Low	Low	Low/Mod.	Low
Expert judgement	High	Mod./High	High	High	High

4. Further Work

Ongoing work will develop the prototype CORE-DATA system and render it into a more usable system within industry, as support for risk assessment work, and other activities that would benefit from the availability of real human error data. CORE-DATA requires development in three main areas:

- **Consolidation of the CORE-DATA system** - consideration of alternative, more robust (quality assurable) software shells; further re-examination of the required functionality for assessors and regulators.
- **Extending the database into key areas** - from discussions with HSE, and from knowledge of the contents of the current hard copy and computerised databases, two key areas for expansion are the transport sector (rail and aviation), and chemical process control.
- **Development of CORE-DATA as an industry resource and service** - raising awareness of the contents and application of the CORE-DATA system, and encouraging industry participation in the usage, continuation and extended development of such a service. Ultimately industry is the largest potential user of such a system, and a series of workshops and demonstrations will be run at regulatory industry locations, and assessors will be able to request data from the CORE-DATA system.

This work will commence in 1997, and will run for approximately two years.

5. References

Advisory Committee on the Safety of Nuclear Installations (1991); Study Group on Human Factors, second report: Human Reliability Assessment - A Critical Overview. Health and Safety Commission, HMSO, London.

Basra, G. and Kirwan, B. (1997) Collection of Offshore Human Error Probability Data. International Journal of Reliability Engineering and System Safety. Paper in press.

Kirwan, B; Martin, B; Rycraft, H; and Smith, A (1990) Human Error Data Collection and Data Generation. Journal of Quality and Reliability Management, vol. 7, 4, 34 - 66.

Taylor-Adams, S.E. and Kirwan, B. (1995) Human Reliability Data Requirements. International Journal of Quality and Reliability Management, 12, 1, 24 - 46.

Seaver, D.A. and Stillwell, W.G. (1983) Procedures for Using Expert Judgement to Estimate Human Error Probabilities in Nuclear Power Plant Operations. NUREG/CR - 2743

Table 2 - Overview of Some of the HEPs Generated

Study	Description of error	HEP	Data Source
Offshore lifeboat evacuation	Before the lifeboat is lowered to sea level, one of the internal checks that the coxswain must perform involves checking the air support system. This check was omitted. If the system is not checked and the crew find out after they have abandoned, there is the possibility of ingress of smoke. Lack of air will cause asphyxiation.	0.028 0.010 0.16 0.070 0.037	APJ (Controlled evacuation scenario) PC (Controlled evacuation scenario) APJ (Severe evacuation scenario) PC (Severe evacuation scenario) Observation
	When the lifeboat is lowered to sea level, it is disengaged from the hooks internally. The coxswain fails to disengage the hooks, and attempts to drive away. If the fall wires are still attached to the boat it will not be able to steer away from the platform. There is a risk the boat may collide with the platform. Evacuation is delayed and there is increased risk to all persons on board.	0.023 0.0036 0.069 0.014 0.019	APJ (Controlled evacuation scenario) PC (Controlled evacuation scenario) APJ (Severe evacuation scenario) PC (Severe evacuation scenario) Observation
Manufacturing	A component has a different profile machined on each end. The operator inadvertently machines the aft end profile on the forward end.	0.00039	Hard-copy data
	One of the mating surfaces of a sub-assembly is not sufficiently coated with adhesive.	0.033	Observation
Electricity transmission	An operator inadvertently opens the wrong circuit breaker.	0.0026	Hard-copy data
NPP	A trainee makes one or more incorrect diagnoses in a set of ten scenarios.	0.16	PC
Calculator errors	A laboratory based simulation found errors whilst performing calculations using handheld calculators. Subjects make an error when performing complex three-step calculations.	0.12	Simulator data
	Subjects make an error when performing simple seven-step calculations	0.20	Simulator data

MUSCULOSKELETAL DISORDERS

ERGONOMIC ASPECTS OF CRIMPING SUB-ASSEMBLY AND QUALITY ASSURANCE

P D Metsios and R J Graves

Department of Environmental & Occupational Medicine,
University Medical School, University of Aberdeen,
Foresterhill, Aberdeen, AB25 2ZD

Components from a sub-assembly process of car harness manufacture had a
high degree of quality control with some rework. A study was undertaken
to examine the ergonomic issues affecting quality using an on-site survey
with a musculoskeletal discomfort questionnaire, discomfort checklists
before and after shift, and defect checklists. Also, workstation and postural
analyses, lighting and knowledge elicitation measures were undertaken.
The survey showed a high percentage of musculoskeletal symptoms in the
neck, shoulders and low back which were associated with risk factors in
task and workstation design. The quality survey showed that 1.665%
defects were detected by the operatives and reworked. Suggestions included
changes to workstations and improved lighting, and organisational changes
such as training, automation, job rotation, and better communications.

Introduction

Manufacturing a car harness depends upon assembling leads and other
components into a harness. The leads are produced with crimped ends in a crimping sub-
assembly process, and appear to have a high degree of quality control. There was some
concern however, that there was little information available on the within crimping
process rework necessary to achieve this level of quality. There was a least some
anecdotal evidence that internal rework was occurring and, some indication of
discomfort and complaints from the workers. The company intended to undertake a new
layout of the area and improve workstation design. This provided the opportunity to
undertake a study to examine the ergonomic issues affecting quality in the crimping sub
assembly process (Metsios, 1996).

The quality of every product depends on product design, process design, work
execution, and inspection. Quality assurance includes activities from design to final test
of a product by setting up standards and procedures to assure that quality is achieved.
Because of the complexity of the electrical harness, the high quality of its components

guarantees the high quality of the final product. Thus, the quality of the leads and their termination is of critical importance. Potential problems even in one lead may cause the whole harness to malfunction. Tracing the problem can be a difficult and time consuming procedure which inevitably imposes financial losses, reduces production and requires additional personnel in order to achieve high standards for the product.

In the context of the continuous improvement that the company has adopted, this project focuses on the identification of ergonomic factors influencing / affecting the quality and productivity in the layout in which the lead crimping process takes place. The ultimate aim of the study reported here was to suggest options for improvement in the job design in this part of the company which may prove useful for the design of a new workplace.

Approach

The study involved a number of activities. These included an on-site survey of a specific work area using the modified Nordic questionnaire (Dickinson et al, 1992), a Discomfort Checklist before and after shift, and recording quality using fault identification checklists. In addition detailed workstation and postural analyses were undertaken, lighting measures taken generally and on the task, and Knowledge Elicitation used to determine task knowledge.

The modified Nordic Questionnaire was supplemented by questions appropriate to the specific work context. It was used to identify the body area and the severity of musculoskeletal complaints. A discomfort checklist was used to obtain additional information on the severity and location of discomfort that the employees were experiencing. The discomfort checklist was adapted from Wilson and Corlett (1995), to identify accurately the most heavily loaded and body parts affected by various task elements. Each worker filled in one checklist at the beginning of the shift and one at the end of the shift for six consecutive working days.

Although, the company keeps records of the outgoing quality of crimped leads, it does not have any records on the quality within the crimping process. Quality checklists and check sheets were constructed to record information on rework, scrap and poor quality, to help identify problems in work design e.g. from machinery breakdown, employees' errors due to fatigue, or discomfort, caused by the nature of the work and poor workstation design. Nearly fifty-five percent of the crimping personnel participated in this part of the study.

The OWAS (Ovako Working Posture Analysing System, see Kant et al, 1990) and RULA (Rapid Upper Limb Assessment, McAtamney and Corlett, 1993) were used in conjunction with video recordings of the tasks to analyse the task and evaluate potential musculoskeletal disorders. In addition detailed dimensions of workstations with different types of machinery were measured in order to evaluate workspace requirements against criteria (see for example, Grandjean, 1988).

Laddering (see McGeorge, and Rugg, 1992) was used to determine a cognitive representation of the task in order to assess if problems could be caused by task misconceptions. This enabled psychological error mechanisms to be identified from a hierarchical task representation. Three operatives with 4 to 7 years crimping experience respectively participated in the knowledge elicitation part of the study.

The task's nature requires precision and detailed examination during inspection of the final product. Since visual performance is dependent on lighting levels, lighting measurements were recorded at 24 out of 43 workstations at the position at which the operatives assemble and inspect the leads, as well as within the presses.

The statistical methods used for the data analysis were frequencies, percentages, Pareto analysis, Pearson r product-moment correlation coefficients ($p \leq .05$), paired samples t-tests (two-tailed, $p \leq .05$). The outgoing quality was taken from the company quality records and the faulty items found (Type II errors) were the basis for a calculation of the Type I (good items rejected) errors as a proportion of the total defects. The Type I errors were impossible to measure otherwise, because the crimps are destroyed when the operatives cut them off in order to replace them. The efficiency of inspection measurements was calculated using the formula: $P_e = PP_2 + (1-P)*(1-P_1)$ (Drury and Fox, 1975).

Findings and Discussion

The task involves fine manipulation of leads as well as manual handling of boxes containing leads (material flow) and components of presses such as applicators and reels. The crimping task consists of six phases. Some additional sub-tasks were also undertaken by the crimping personnel due to system inefficiencies (mainly communication problems) such as cable stripping, trimming copper strands and fitting seals. Illumination was provided by artificial and natural lighting. General illumination levels were supplemented by lights fitted within the machines. The mean general illumination was 466 lux (with a standard deviation (sd) of 242) and the supplementary illumination in the presses 290 lux (sd of 184.8). The mean illumination level in the presses for inspection of the crimped leads was below the guidelines for precision work (IES, 1981) by about 300 lux.

The available sample of employees was 24 out of a potential of 31 due to absence and the response rate was 95.8% of the former (23) with one employee not wishing to respond. The survey results showed a high percentage of reported musculoskeletal symptoms over the last twelve months in the neck shoulders, and low back. Symptoms were mostly reported for the neck (56.5%), shoulders (52.17%), low back (52.2%), wrists (47.74%) and upper back (26.1%). During the study, four workers had sick leave with wrist, back and elbow problems. The workers reported that 100% of their wrist, 26.1% of their low back, 39.1% of their shoulder and 34.8% of their neck symptoms were caused by work. Other reported causes were activities at home, accidents, draughts, and arthritis.

The mean discomfort levels obtained during the 6 days of the study are shown in Figure 1. The discomfort is statistically higher at the end of the shift (two-tailed $t = -2.61$, $df = 16$, $p = .019$). The figure also shows that discomfort increased during the working week with higher levels on Thursday and Friday. The trunk and upper limbs seemed to have more discomfort, mainly the shoulders and the full length region along the spine. Age was moderately correlated with the discomfort (two-tailed $r = 0.5542$, $p = .021$).

Figure 2 shows the difference in defect ratios between the morning and afternoon parts of the shift. The higher value on the fifth day (1/7/1996) was caused by a communication problem. The difference in defect ratios between the two parts of the shift was not statistically significant, except for the highest value ($p = .032$). The proportion of faults was less in the morning than in the afternoon despite the extra hours of work in the morning which increased the likelihood of defects occurring.

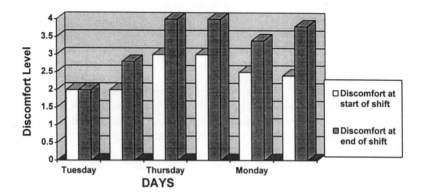

Figure 1. Discomfort levels during the shift and the week.

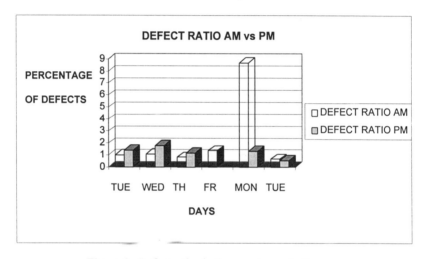

Figure 2. Defect ratios in the morning and afternoon

The types of leads were divided in three generic types; singles, doubles and seals. 58.07% of the production during the quality survey were singles, 32.07% doubles and 9.86% seals. For this reason most defects occurred in single leads followed by doubles and finally seals. Nevertheless, the ratio of faults proportionately to the production was 0.325% for the singles (1.289% taking into account the defectives caused by the communication problem), 0.892% for the doubles and 0.736% for the seals.

Figure 3 demonstrates the primary causes of defects and the proportion of defects derived from them during the study. Human error accounted for 33.7% and the communication problem for 33.57%. Also, the defects caused by the machinery (broken, badly adjusted, lost setting during the crimping process) accounted for 10.61% of the total. The cause under the title "Combination" contains operative and machinery faults which cannot be differentiated.

Rework of the various types of defective leads causes production time to be lost. According to the time measurements obtained by task analysis, the mean production

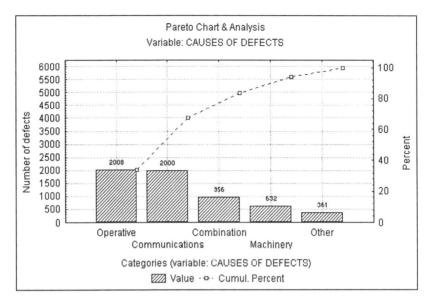

Figure 3. Causes of defects

time lost per day was 2 hours and 20 minutes for the sample of the worker population.

The mean probability of a good item accepted was 0.9942 and the probability of a Type I error 0.0057 during the study. The calculations are based on the quality records of the company and the probabilities of good items being accepted or rejected are calculated mathematically from the defective items accepted and the total production during the study.

The survey highlighted problems in the workplace and the processes of the department where crimping takes place. The combination of workplace characteristics, human factors and inadequate procedures could cause the production of substandard quality leads.

The workstation design and dimensions forced female employees to adopt awkward postures and exert high forces. There was evidence of risk factors associated with workstation design and tasks such as repetitive work with the wrist bent, static loading of the upper limbs, neck and localised mechanical stresses on the elbows and fingers. Other examples included highly repetitive pinch grasping with the wrist bent, and sharp surfaces imposing mechanical stresses on the fingers and elbows (see Hagberg et al, 1995). In addition, some of the loads which the workers handled in relation to the height where manual handling takes place exceeded the Health and Safety Executive (HSE, 1992) manual handling guidelines.

The quality survey showed that 98.333% of the total production conformed to quality standards with 1.665% defectives detected, and corrected immediately by the operatives, with 0.02% undetected and reaching departmental quality control. It appeared the sources of rework faults were the operatives (33.7%), bad intra-departmental communications (33.57%), and the machinery (10.61%).

Recommendations for the improvement of work and the quality of the product included workplace modifications such as; new workstation design, introduction of an adjustable height foot-rest and foot-switch to every workstation, and padded supports on the working surface. Other suggestions were; self-adjusting tables at every workstation for boxes with material, improvements to the height of reel-holders, feeding and the guards of the presses; improved lighting, better flow of material, and machinery maintenance; and substitution of the elastic bands and cutters with ergonomically designed ones.

Organisational changes proposed included; automation of the crimping task, job rotation, improvement of inter/intradepartmental communications, process flow according to an assembly line philosophy, training of the workers, and training of the quality personnel.

The holistic approach recommended by the study would improve many aspects of the crimping process (human resources, organisational, production) and it is estimated that it could reduce the production of defective leads to 1% or less. This would reduce the mean production time lost per day to below the estimated 2 hours and 20 minutes for the sample of the worker population, providing recurring financial benefits.

References

Dickinson, C.E. Campion, K. Foster, A.F. Newman, S.J. O'Rourke, A.M.T. and Thomas, P.G. 1992, Questionnaire development: an examination of the Nordic Musculoskeletal Questionnaire, *Applied Ergonomics*, **23**, 197-201.

Drury, C.G. and Fox, J.G. (Eds.) 1975, *Human reliability in quality control*, Taylor & Francis, London.

Grandjean, E. 1988, *Fitting the task to the man: A textbook of occupational ergonomics*, Taylor & Francis, London.

Hagberg, M. Silverstein, B. Wells, R. Smith, M.J. Hendrick, H.W. Carayon, P. and Perusse, M. 1995, *Work Related Musculoskeletal Disorders (WMSDs): A reference book for prevention*, Taylor & Francis, London.

Health and Safety Executive, 1992, Manual handling Guidance on regulations Manual handling operations regulations 1992. Health and Safety Executive.

IES, 1981, *Lighting handbook*, Kaufman, J.E., (Ed.), Illuminating Engineering Society of North American, New York.

Kant, I. Notermans, J.H.V. Borm, P.J.A. 1990, Observations of the postures in garages using the Ovako Working Posture Analysing System (OWAS) and consequent workload reduction recommendations, Ergonomics, **33**, 2, 209-220.

Metsios, P. 1996, *Ergonomic analysis - redesign of the crimping sub-assembly process within the quality assurance context of a car harness manufacturing industry*, MSc Ergonomics Project Thesis, Department of Environmental and Occupational Medicine, University of Aberdeen: Aberdeen.

McGeorge, P. Rugg, G. 1992, The use of contrived knowledge elicitation techniques, *Expert Systems*, 9,(3) 149-154.

McAtamney, L. Corlett, E.N. 1993, RULA: a survey method for the investigation of work related upper limb disorders, *Applied Ergonomics*, **24**, 91-99.

Wilson, J.R. and Corlett, E.N. 1995, *Evaluation of human work: A practical ergonomics methodology*, Taylor & Francis, London.

THE INFLUENCE OF FLOOR SURFACE ON DISCOMFORT EXPERIENCED BY STANDING WORKERS

Graeme Rainbird and Dr. Corinne Parsons

Ergonomics and Safety Group,
RM Consulting,
Royal Mail Technology Centre,
Wheatstone Road, Dorcan,
Swindon, Wiltshire, SN3 4RD

Royal Mail staff in mail processing centres perform many tasks while standing. Currently, the standard floor surface is a hard vinyl finish, laid on a concrete base. It was proposed that an alternative, hardened concrete finish, be used in new buildings to reduce construction time and costs. There is evidence that floor type can affect the degree of fatigue experienced by standing workers, and it is commonly accepted that fatigue influences productivity. This paper outlines the method, results and conclusions of a comparative trial of staff discomfort levels on the two floor types undertaken prior to acceptance of the new floor. Two types of 'anti-fatigue' mat were also tested. No difference was found between the vinyl and concrete floors. Subjects did report significantly less discomfort when using the mats, compared to the hard floor surfaces ($p < 0.05$).

Introduction

General

A change from the standard floor surface that has been in use within Royal Mail processing centres for over 15 years was proposed to reduce building time and costs in new offices. The current floors consist of hard vinyl tiles laid on a concrete base. The alternative is a hardened concrete finish, laid on setting concrete.

Postal staff working in processing centres are required to perform many tasks while standing. There were concerns that the proposed alternative floor surface would be more uncomfortable to work on and be unacceptable to staff.

Literature review

Long term standing is accepted as a direct cause of pain and discomfort, and it has been demonstrated that local discomfort diminishes as the distance away from the floor increases - in other words the feet hurt more than the ankles which hurt more than the lower legs etc. (Redfern and Chaffin 1988). It has also been shown that the incidence of low back pain is highest in those workers who stand regularly every working day for periods of more than four hours (Magora 1972, cited in Redfern and Chaffin 1988).

It is generally believed that the nature of the floor surface has an impact on staff perceptions of discomfort and fatigue. While the influence of discomfort and fatigue are not simple to isolate or define, it is commonly accepted that increasing levels of fatigue are associated with a loss of efficiency and productivity, and a disinclination for effort.

However, what is not clear is the relative importance of floor type on standing tolerance. Some evidence indicates that floor type has a significant influence on standing workers, and that matting is preferred to concrete floors because it is associated with a significant difference in body comfort (for example Rys and Konz 1988, Rys and Konz 1989, and Hinnen and Konz 1994).

Other studies do not demonstrate a significant difference in fatigue due to floor type; they conclude that the critical factor influencing fatigue for standing workers is the length of the task and that the floor surface effect is relatively insignificant (for example Zhang, Drury and Woolley 1991, and Hansen, Jorgensen and Winkel 1995).

The information available in the literature was considered inadequate to predict with confidence the impact of the alternative floor surface on Royal Mail staff. Comparative trials were therefore undertaken to evaluate the difference between peoples' perceptions of discomfort when working on the current and proposed floor surfaces and to assess the benefits of 'anti-fatigue' matting.

Experimental Method

Location and Set-up
The comparative trial was set up in a Royal Mail warehouse area using simulated tasks and specially prepared floor areas. This allowed greater control than would be possible in an operational environment. Eight areas of concrete floor, $1m^2$ and 50mm deep were laid in wooden trays. The trays were bolted to the floor for stability. Two of each were finished with the following surfaces:

1. current standard vinyl finish
2. alternative hardened concrete finish
3. 'anti-fatigue' matting type 1 (plain surface)
4. 'anti-fatigue' matting type 2 (raised/dimpled surface to stimulate blood circulation)

Work desks were set up at each floor area, at a height of 900mm.

Trial design
Eight subjects, seven male and one female, were hired from an employment agency to take part in the study. Each subject worked for an entire day on each of the floor surfaces in the order shown in the following table:

Table 1. Order of conditions by subjects, for days 1 - 4.

Subject	Vinyl	Concrete	Mat 1	Mat 2
1 & 5	4	2	1	3
2 & 6	3	4	2	1
3 & 7	1	3	4	2
4 & 8	2	1	3	4

Subjects were required to wear the same pair of comfortable shoes throughout the trial. Subjects worked for four periods of 90 minutes with a 15 minute break in the morning and afternoon and 45 mins meal break at mid-day. They performed simulated manual sorting tasks, using playing cards and coins, at the work station allocated to them for the day. The time taken to complete each task cycle was recorded.

Data collection

The literature review indicated that the most reliable method for determining effects of floor surface on fatigue relies on self-reporting of body part discomfort. The approach used was based on body part discomfort scores (Corlett 1990). Each subject was required to complete body part discomfort scales every 30 minutes throughout the four work periods, a total of 16 sets of results per person, each day.

At the end of each day subjects completed a brief questionnaire to record their general opinions of the floor they had been working on that day (for example, its appearance and hardness).

Results

The results reproduced here relate to the body part discomfort scores as these were of most significance in determining the conclusions of the study. Results for subject 5 have been excluded from the analysis because, after reporting minimal discomfort throughout the first three days of the trial, he suffered an injury playing football and reported high levels of discomfort on day 4. Information from other sources, including the end of day questionnaires, is included in the Discussion section.

Overall body part discomfort

Table 2. Overall discomfort scores for each floor type.

Subject	Vinyl	Concrete	Mat 1	Mat 2
1	474	418	241	292
2	225	172	92	104
3	457	451	456	301
4	67	42	57	25
6	373	354	292	485
7	266	363	298	179
8	67	53	24	11
mean	275.6	277.6	208.6	199.6

Overall body part discomfort is calculated as the sum of body part scores.

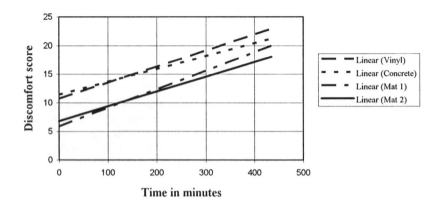

Graph 1. Linear regression for mean discomfort scores over
the working day, for each floor type.

Table 3. Mean collated body part discomfort results

Body area	Vinyl	Concrete	Mat 1	Mat 2
overall body	275.6	277.6	208.6	199.6
feet and ankles	103.0	101.1	84.7	69.6
low back	18.6	16.9	16.6	12.3

Feet and ankle discomfort is calculated as the sum of feet and ankle scores for
subjects working on each floor type. Low back discomfort is calculated as the sum of
low back scores for subjects working on each floor type.

Statistical analysis
        The recorded body part discomfort scores varied widely between individual
subjects (see Table 2.) and consequently an analysis of variance (ANOVA) revealed no
significant differences between conditions ($p < 0.05$). Therefore the data was analysed
using ranked scores rather than raw data. The homogeneity of the ranks of the overall
body part discomfort scores were analysed using the Chi-Squared distribution and were
found to differ significantly ($p < 0.05$). Further analysis using the Wilcox matched
pairs signed ranks test showed a signficant reduction in body part discomfort reported
when subjects worked on the 'anti-fatigue' mats, in comparison to the hard floor
surfaces ($p < 0.025$). No significant difference was demonstrated between the vinyl and
concrete floors, or between the mat 1 and mat 2 'anti-fatigue' mat types.

Discussion

Work rate of subjects
        Subjects performed the same tasks at each stage of each day and their
performance was monitored to ensure that they worked at an approximately consistent

rate. There were variations in work rate, due to: the random element of the simulated sorting tasks (sometimes the task was easier than others); time of day effects; learning and fatigue effects. As the subjects' work rates were generally consistent, it was assumed that the overall productivity was constant for each of the four floor conditions.

Comparison of vinyl and concrete floors

It was proposed that the concrete floor finish might increase staff perceptions of discomfort, and therefore reduce productivity, relative to the vinyl finish. In fact, the overall body part mean score for the concrete finish was almost identical to that of the vinyl finish, (277.6 to 275.6, with the vinyl preferred). There was no statistically significant difference between the two surfaces. All other factors being equal, therefore, the use of the concrete finish would not be expected to be associated with an increase in staff perceptions of discomfort, or a decrement in productivity.

Previous research (Redfern and Chaffin 1988) indicates that the most extreme local discomfort associated with floor type is experienced in the parts of the body closest to the floor. In other words, the scores for the feet and ankles would be expected to show a greater divergence, if there is a difference between floor types. Again, the score for the concrete finish was almost identical to that of the vinyl finish (101.1 to 103.0, with the concrete preferred). Again, there is no significant difference between these body part scores for the two conditions. This supports the initial conclusion that the concrete finish would not be associated with an increase in staff perceptions of discomfort.

Low back pain has been related to long term standing and might be affected by floor type (Kim, Stuart-Buttle and Marras, 1994). However, in this trial, there was little difference between the concrete and the vinyl finishes (16.9 to 18.6, with the concrete preferred). This difference was not statistically significant, again indicating that the introduction of the alternative finish would not affect staff discomfort.

Comparison of 'matting' and 'no matting' floors

Graph 1 shows a clear distinction between the 'matting' and 'no matting conditions', where the matting appears to offer standing workers benefit in terms of reduced discomfort. This difference was shown to be statistically significant ($p < 0.025$). Of the two matting types, mat 2 (with raised finish to stimulate blood circulation) was generally preferred, although the difference was not statistically significant. 'Anti-fatigue' mats may be of benefit to staff who perform standing tasks for extended periods but their use may introduce additional problems if trolley access is required, or where the mats create a tripping hazard.

End of day opinions

Subjects completed end of day questionnaires to determine their general impressions about the different floor finishes on which they had been working. The concrete floor finish scored lower than the vinyl finish for both 'hardness' and 'appearance', although both were perceived as being 'too hard'. These two factors could have a significant influence on staff acceptance of the floor, even though there is no detectable difference in their effects on staff.

Conclusions

There was no overall significant difference between discomfort experienced for staff working on the hardened concrete and the standard vinyl finishes, over a seven and a half hour period. If the hardened concrete finish is laid in new mail processing centres, no increase in perceptions of fatigue, or performance decrement, would be expected.

Staff perceived both the standard vinyl and the hardened concrete floor finishes to be too hard for extended periods of standing work.

The hardened concrete finish was seen to be harder and less attractive than the vinyl finish (even though no effects of this difference in 'hardness' were demonstrated experimentally). These perceptions may prejudice staff against the new floor type.

There was a significant reduction in discomfort reported by subjects when working on anti-fatigue mats compared to the hard floors. Mat 2, with the raised finish was generally preferred to mat 1, although this difference was not statistically significant.

References

Corlett, E.N. 1990, Static muscle loading and the evaluation of posture. In Wilson, J.R. and Corlett, E. N. (eds.), *Evaluation of Human Work: A practical ergonomics methodology*, (Taylor & Francis) 542-570.

Hansen, L. Jorgensen, K. and Winkel, J. 1995, Biomechanical and Microcirculatory Reactions to Walking and Constrained Standing Work According to Softness of Shoe and Floor, In, *2nd International Scientific Conference of Prevention of Work-Related Musculoskeletal Disorders*, (PREMUS 95, Montreal, Canada) 357-359.

Hinnen, P. and Konz, S. 1994, Fatigue Mats. In Aghazadeh, F. (ed.), *Advances in Industrial Ergonomics and Safety VI*, (Taylor & Francis, London) 323-327.

Kim, J.Y. Stuart-Buttle, C. and Marras, W.S. 1994, The Effects of Mats on Back and Leg Fatigue, *Applied Ergonomics*, **25**/1, 29-34.

Redfern, M.S. and Chaffin, D.B. 1988, The Effects of Floor Types on Standing Tolerances in Industry. In Aghazadeh, F. (ed.), *Trends in Ergonomics/Human Factors V*, 401-405.

Rys, M. and Konz, S. 1988, Standing Work: Carpet vs. Concrete. In *Riding the Wave of Innovation. Proceedings of the Human Factors Society 32nd Annual Meeting*, (Anaheim, California) Volume 1, 522-526.

Rys, M. and Konz, S. 1989, An Evaluation of Floor Surfaces. In *Perspectives. Proceedings of the Human Factors Society 33rd Annual Meeting*, (Denver, Colorado) Volume 1, 517-520.

Zhang, L. Drury, C.G. and Woolley, S.M. 1991, Constrained Standing: Evaluating the Foot/Floor Interface, *Ergonomics*, **34**/2, 175-192.

INVESTIGATION INTO THE FACTORS ASSOCIATED WITH SYMPTOMS OF ULDs IN KEYBOARD USERS

Margaret Hanson, Richard Graveling and Peter Donnan

Institute of Occupational Medicine
8 Roxburgh Place, Edinburgh, EH8 9SU, UK

This study aimed to identify the factors associated with symptoms of upper limb disorders (ULDs) among keyboard users. In all, 3503 responders from 4424 selected keyboard users completed a questionnaire concerning discomfort they had experienced in different parts of their upper limb. Groups of cases and controls were defined from these responses. A representative sample of 449 of these subjects (295 cases; 154 controls) took part in a detailed ergonomic evaluation at their workstation. Information was collected on keyboard work; furniture and equipment; postures adopted when keying; physical work environment; psychosocial aspects of work; activities outside work; and personal factors. Analysis of this information allowed some of the significant factors associated with ULDs to be identified. After adjusting for age and gender, the most significant factor associated with symptoms of ULDs was the length of time the subject spent at the keyboard during a week (highly correlated with the length of time spent keying without a break). Other factors found to be significantly associated with ULDs are discussed.

Introduction

Work related upper limb disorders (ULDs) are a major source of ill health and sickness amongst the British working population. The Labour Force Survey (Hodgson et al 1993) showed that upper limb and neck problems including 'Repetitive Strain Injuries' had a combined prevalence of the order of 170,000 cases in 1990. There is evidence to suggest that the extensive use of computer keyboards creates a risk of these disorders developing (eg. English *et al*; 1995, Hopkins, 1990).

Many studies have been conducted which look at the factors associated with (ULDs), but despite the widespread recognition of the multi-factorial nature of musculoskeletal disorders, few have considered comprehensively and simultaneously all the main factors associated with these problems (Hagberg *et al*, 1995). This study aimed to address this by considering, in one study, all major factors that had been shown in previous research to be significantly associated with ULDs in keyboard users.

Identifying those with symptoms of ULDs

A self completed questionnaire was developed concerning discomfort or other adverse symptoms experienced in different parts of the upper limb. The questionnaire also asked about keyboard usage, medical involvement and other personal factors. It was distributed by hand to 4424 randomly selected keyboard users. In all, 3503 completed questionnaires were received (79% response rate). Although not designed as a cross sectional study, the responses do indicate approximately the extent of keyboard users reporting symptoms of ULDs. The responses showed that 55% of the subjects had at some time suffered from some discomfort in their upper limbs. Fourteen percent of all subjects were currently suffering, or had suffered within the last 3 months, from ULDs symptoms which were severe enough for them to seek professional medical advice.

The responses to the questionnaire allowed 6 groups of cases to be classified by anatomical location and pattern of symptoms. These groups were: shoulder disorders; elbow disorders; forearm pain; trigger digit; nerve entrapment syndromes; and tendon related disorders. A group of controls (experiencing no upper limb symptoms, recent or past) was also established.

It must be emphasised that these case groups were not formulated on the basis of clinical diagnoses of ULDs but from self-reported symptoms. However, a small-scale cross-check did indicate a reasonably close agreement between case group classification and ULDs for those cases for whom a clinical diagnosis was available.

Case-control study: Methods

In order to identify the factors that were associated with ULDs, a case-control study was conducted using a representative sample of 449 of these subjects (295 cases; 154 controls). A detailed ergonomic evaluation was carried out at each subject's workstation. The assessment consisted of three parts: a structured interview; a practical period at the keyboard; and completion of a questionnaire concerning psychosocial factors at work. This assessment incorporated all the major factors that had been shown in previous research to be associated with ULDs; these factors were then compared between cases and controls.

A structured interview was conducted covering a wide range of possible causative factors including: keyboard activities and usage; work equipment and furniture; physical work environment; non-keyboard activities at work and outside work; and personal details. The interview lasted approximately 20 minutes. Following this, subjects undertook a typical piece of keyboard work lasting approximately 15 minutes during which their posture was observed and documented. Wrist postures during this time were measured using electrogoniometers (Penny and Giles). The subject's posture and typing style were also videoed. Observations concerning the furniture and equipment were made. Finally, subjects completed the Work Environment Scale (WES) (Moos and

Insel, 1974) concerning the psychosocial aspects of work. This scale has ten subscales covering: involvement; peer cohesion; supervisor support; autonomy; task orientation; work pressure; clarity; control; innovation; and physical comfort.

Case-control study: Results

Analysis of this information allowed some of the significant factors associated with ULDs to be identified for each of the case groups. In light of the number of factors considered (over 90) and the six different case groups defined, a seventh case group was defined and called the 'any syndrome' case group; this group incorporated a positive definition in any of the other case groups. For simplicity in this paper, only this amalgamated case group will be presented and discussed. The results of case-control comparisons of individual factors are presented below. The subsequent section presents the results of a regression analysis comparing all the factors that were found to be significant, including age and gender, in order to determine their relative importance.

Age and gender
Females were more likely to suffer symptoms than males, and increased age was associated with increased prevalence of symptoms for females but not for males

Keyboard use
Although the number of years experience with keyboards was not significantly associated with ULDs, the number of hours keying per week was strongly significant. The length of time at the keyboard without a break was also associated with ULDs and was highly correlated with the total number of hours spent at the keyboard each week, ie. those who spent a high number of hours at the keyboard during the week also had less frequent breaks from the keyboard.

The following job-related factors were associated with ULDs: experiencing difficulty reading the text on the documents or the screen; experiencing frustrating problems with the software; not being able to decide when to take breaks from the keyboard; having a specified rate of keying; and having information presented visually via a document holder. Some of these factors may be associated with ULDs due to the posture they induce, while some will result in stress for the individual. Using a mouse was not shown to be significantly associated with ULDs.

Furniture and equipment
The following furniture factors were associated with ULDs: subject experiencing problems with their chair; not having armrest on the chair; having a footrest; using a document holder; and not having a keyboard that tilted.

Environmental factors
Experiencing disturbing environmental factors (excluding noise or lighting) was found to be associated with ULDs. These environmental factors included extremes of temperature, draughts and odours.

Postural factors

For the 'any syndrome' case group only one of the postural variables was found to be associated with ULDs. This was the tendency for the subject to use particular force when keying. None of the goniometer variables was found to be associated with ULDs.

Psychosocial factors

Of the ten subscales on the WES, having low scores on the task orientation, innovation and physical comfort subscales was found to be associated with ULDs. The task orientation subscale describes the degree of emphasis placed on good planning, efficiency and getting the job done; the innovation subscale describes the degree of emphasis on variety, change and new approaches; while the physical environment subscale describes the extent to which the physical surroundings contribute to a pleasant work environment. The physical environment results support the findings outlined above, ie. that being disturbed by the physical environment was associated with ULDs.

Non-work activities

Being exposed to hand arm vibration was associated with ULDs, as was spending a high number of hours per week in sports or hobbies involving repetitive / forceful upper limb use (eg. knitting, racket sports etc). Having an accident that the subject could relate to the symptoms was associated with ULDs, as was cigarette smoking.

Regression modelling

All factors that had been found to be significantly associated with the 'any syndrome' case group were considered simultaneously in a structured stepwise regression (after adjusting for age and gender) to determine which of these were the most significant. At each stage of the regression the relative importance of the remaining factors was considered. With this analysis, if some factors are highly correlated it is likely that only one of these will appear in the regression model.

The factor most strongly associated with the 'any syndrome' case group was spending a high number of hours keying per week ($p<0.0001$). This was followed by the screen producing flicker ($p=0.0005$); experiencing frustrating problems with the software ($p=0.001$); experiencing difficulty reading text on the documents or screen ($p=0.003$); spending a high number of hours per week in sports or hobbies involving repetitive / forceful upper limb use ($p=0.004$); smoking ($p=0.003$); having a specified keying rate ($p=0.02$); experiencing problems with the chair ($p=0.03$); having a footrest (0.02); and keying while using a hand held telephone ($p=0.01$). The p values given are those in the final model and show significant associations between the designated factors and presence/absence of 'any syndrome', independently of other variables in the model.

Discussion

This paper presents the results for the 'any syndrome' case group which was a combination of 6 specifically defined case groups. This case group therefore contains multiple syndromes of highly different characteristics and hypothesised mechanisms. It should also be noted that the case-control study identified *associations* between risk factors and ULDs, but this does not necessarily mean these risk factors *cause* ULDs.

A number of clinically defined ULDs are regarded as being predominantly associated with aging females. Although the results do support an association between gender and case status, and an association with age in women, the detailed analyses, in particular the regression analysis, showed that many men experienced ULD symptoms and that occupational factors remained significantly associated with case-control status even when the effects of gender and age were allowed for.

Keyboard use

The factor most strongly associated with ULDs was the length of time that the subject spent at the keyboard during the week, with those spending a high number of hours at the keyboard being most likely to suffer ULDs. This factor was also highly correlated with the length of time spent at the keyboard without a break, suggesting that those who spent a high number of hours at the keyboard did not have frequent breaks from the keyboard and were most likely to suffer ULDs.

Furniture and equipment

It is surprising to note that the use of a chair without armrests, use of a footrest and use of a document holder were all found to be associated with ULDs. This appears to go against the usual ergonomic advice that armrests result in poor postures and footrests and document holders help reduce poor postures. However, all these factors are likely to be related to the type of keyboard work being undertaken; those undertaking more intensive keyboard tasks are also likely to have a chair with fewer features (ie. no armrests), to use a document holder (because of the nature of the work), and to use a footrest either because it may be issued automatically to personnel using the keyboard extensively, or because they are aware they require one as they spend long periods at the keyboard. The nature of the work means that personnel with these accessories are likely to spend a high number of hours per week at the keyboard. This is supported by the findings of the regression analysis in which none of these variables were significant after the inclusion of time spent keying.

The use of a mouse was not found to be associated with ULDs. This finding may be due to the fact that during the time of data collection (1992) only 24% of the subjects had a mouse and many of these reported rarely using it.

Postures

The fact that so few of the postural variables were found to be associated with ULDs is surprising, and may be due to the dynamic nature of the work, and the observation period only being a 'snapshot' of the postures adopted. The goniometer measurements of wrist angle also revealed no significant associations. It is thought that this was because during the measurement period the subjects were asked to undertake a typical piece of their keyboard work so that the task would be as realistic as possible. This piece of work usually did not involve them keying 100% of the time, so wrist posture measurements included, in some instances, wrist postures when writing, turning pages, answering the telephone and waiting for the computer to process information.

Psychosocial factors

It is interesting to note that many of the most significant factors were those that may create frustration and stress for the subject: screen producing flicker; experiencing problems with the software; and experiencing difficulty reading the text on the documents or screen. Having a specified rate of keying was also associated with ULDs; it is well recognised that this will put subjects under pressure as well as resulting in a high work rate, both of which may be associated with ULDs. These findings indicate that the psychological stressors of the job contribute significantly to ULDs.

Non-work activities

Non-work activities were also associated with ULDs. This included sports and hobbies undertaken on a weekly basis, which are likely to contribute to ULDs through repetitive movements and forceful exertions - activities that are well recognised as contributing to ULDs (Putz-Anderson, 1988). Smoking cigarettes was also associated with ULDs, which supports the conclusion of Hagberg *et al op cit* who thought it likely that smoking was related to neck and shoulder disorders and carpal tunnel syndrome. Smoking is also an indicator of socio-economic status, and this may be related to jobs which require a high number of hours keying each week.

Conclusions

1.   There is a high prevalence of ULDs among keyboard users, with 14% reporting recent, severe symptoms.
2.   Various factors are associated with the development of ULDs, but the most significant is the length of time spent at the keyboard during the week.
3.   Psychosocial factors and experiencing frustrating problems at work appear to contribute to the development of ULDs as do a number of non-work factors.

References

English, C.J., Maclaren, W.M., Court-Brown, C., Hughes, S.P.F., Porter, R.W., Wallace, W.A., Graves, R.J., Pethick, A.J., Soutar, C.A. 1995, Relations between upper limb soft tissue disorders and repetitive movements at work. *Am.J.Ind.Med.* **27**, 75-90.

Hagberg, M., Silverstein, B., Wells, R., Smith, M.J., Hendrick, H.W., Carayon, P. and Perusse, M. 1995, *Work related musculoskeletal disorders (WMSDs): A reference book for prevention,* Sc. eds: I Kuorinka & L Forcier. (Taylor and Francis, London)

Hodgson, J.T., Jones, J.R., Elliot R.C., and Osman, J. 1993, *Self-reported work-related illness. Results from a trailer questionnaire on the 1990 Labour Force Survey in England and Wales.* (HMSO, London)

Hopkins, A., 1990, Stress, quality of work, and repetitive strain injury in Australia. *Work and Stress,* **4** (2) 129-138.

Moos, R.H. and Insel, P.M. 1974, *The Work Environment Scale,* (Consulting Psychologists Press, Palo Alto, CA)

Putz-Anderson, V. 1988, *Cumulative Trauma Disorders: A manual for musculoskeletal diseases of the upper limbs.* (Taylor and Francis, London).

Acknowledgements

The authors would like to acknowledge and thank all other staff at the IOM who were involved in this research. This work was carried out with funding from the HSE.

REBA (Rapid Entire Body Assessment)
- More than a Postural Assessment Tool

Lynn McAtamney [1] **and Sue Hignett** [2]

[1] Senior Partner, C.O.P.E.,
25 Ashchurch Drive, Wollaton, Nottingham, NG8 2RB.
[2] Ergonomist, Rufford Ward,
Nottingham City Hospital, Nottingham, NG5 1PB

REBA is a Rapid Entire Body Assessment tool which evaluates the risk of musculoskeletal injury, in particular those associated with manual handling operations. The REBA scoring system generates an action level which reflects the level of risk caused by the posture, coupling, loads or forces exerted and muscular activity. REBA will be of particular assistance in a musculoskeletal risk management programme.

Introduction

Risks of damage to muscles and joints (musculoskeletal risks) are found in a wide variety of industries. Although often associated with 'heavy jobs' such as nursing or manual handling operations, musculoskeletal disorders are increasingly found in service industries (HSE, 1994).

In general the risks associated with work related musculoskeletal disorders have been recognised and investigated for some time (Hignett, 1994a, McAtamney, 1994). However it can be difficult to accurately evaluate the impact on the musculoskeletal system from these risk factors.

There are few tools available which combine the knowledge about musculoskeletal risks into a validated method for investigators to use which is also sensitive to risks in a variety of occupations (Wilson and Corlett (1995) provides a detailed description of the current tools).

With the introduction of Risk Management Programmes including the Manual Handling Operations Regulations, 1992 (HSE, 1992) the responsibility for the assessment and prevention of musculoskeletal problems now rests more with managers than the

Occupational Health and Safety Departments. Managers require valid (understandable) information from the ergonomics, health and safety professionals on which to base their risk controls. REBA is an observation tool which has undergone initial validity and reliability trials and will prove useful in meeting the needs of the Risk Manager.

All observation methods involve two (usually contradictory) qualities - generality and accuracy. High generality in an observation method is usually compensated by low accuracy e.g. OWAS (Finnish Institute of Occupational Health, 1992, Hignett 1994b) records a wide range of postures resulting in a low level of detail or NIOSH (1991) equation (Waters *et al.*, 1993) which requires detailed information about specific parameters of the posture to produce highly accurate results with respect to the lifting index, but no general information about the task.

Being an event driven system (Foreman et al 1988) which requires no sophisticated equipment REBA fills the gap between detailed event driven systems e.g. VIRA (Keyserling, 1990) and time driven field methodology, (for example OWAS). OWAS was not designed to record postures which may only be held for a short period of time (e.g. a patient transfer) and the chances are that the postures recorded are not representative of those which cause the highest risk of musculoskeletal problems (Hignett, 1994b).

REBA is more than a postural assessment tool- it has been designed to identify musculoskeletal risks associated with the working posture along with the loads or forces being exerted, coupling and activities associated with the posture being observed. REBA was designed to meet the needs of health care, safety and ergonomics professionals in providing a quick, objective assessment of the musculoskeletal risks to which the worker is exposed as part of a larger workplace assessment.

Development of REBA

REBA was developed to "fill a gap" in the range of field assessment tools which measure the risks resulting from the combination of diverse manual handling operations and associated working postures in heterogeneous tasks. The tool has been designed to evaluate tasks where postures are dynamic, static or where gross changes in position occur.The development of REBA was made using tasks from the electricity and health care industries. REBA is not designed to evaluate sedentary tasks where the loading in principally on the neck, shoulders and upper limbs because RULA was developed for this purpose.

In using REBA a risk rating value is generated which is on a scale of 1-15 and can be compressed into one of five action levels which reflect the magnitude and severity of the exposure. Whilst the body posture diagrams developed for RULA (McAtamney and Corlett,1993) are used as a basis for REBA there are a number of differences between the two methods. These were introduced to fulfil the more general aims of assessing whole body musculoskeletal risks and are described in McAtamney and Hignett, 1995.

Aims

In particular REBA has been designed to:

provide a postural analysis system which is sensitive to musculoskeletal risks in a variety of occupations and task

- divide the body into segments which are coded individually, but with reference to anatomical movement planes
- provide a scoring system for muscle activity caused by static, dynamic, rapidly changing or unstable postures
- reflect the coupling which is important in the handling of loads but may not always be via the hands
- give an action level with an indication of urgency to reduce risks
- require minimal equipment - a pen and paper method

Description

The practitioner undertaking the assessment makes a judgment about which task(s) and posture(s) require assessment and if the assessment is generic in nature what representative sample of staff will need to be observed. After selecting the posture or activity to be assessed there are four steps when using REBA which are:

Step 1. Score the body alignment using the REBA diagrams
Step 2. Measure or estimate the load
Step 3. Categorise the coupling using the REBA codes
Step 4. Score the activity associated with the posture using the REBA codes
          (gross postural change, unstable base, static or repetitive muscle actions)

REBA is based on the RULA system where each body segment is coded with respect to anatomical planes of movement (American Academy of Orthopaedic Surgeons, 1965) with the scoring system adjusted to reflect the biomechanical and postural loading as described in the literature. A score added for specified additional loading, for example, rotation.

The posture scores, load score and coupling score are processed into a single combined risk score using tables which provide a total of 144 combinations. To ensure that risks associated with work activity undertaken in conjunction with the posture being evaluated are included an activity score is added to the combined risk score to give the final score which is on a scale of 1-15. This can then be compared to the action levels which are presented in Table 1 below.

Table 1. REBA Action Levels

Action Score.	Score	Risk level	Action (including further assessment)
0	1	negligible or nil	None necessary
1	2 or 3	low	May be necessary
2	4 to 7	medium	Necessary
3	8 to 10	high	Necessary soon
4	11 to 15	very high	Necessary now

Assessing the Validity and Reliability of REBA

To assess the validity of REBA data have been collected on rated perceived exertion (Borg, 1982) whilst the postures were recorded on photographs or video. This material has then been used to evaluate the REBA system in comparison with OWAS. Where the task is sedentary RULA was also to be administered. Three professionals with qualifications in ergonomics and occupational physiotherapy independently gave a risk rating of 1 to 15 for all 144 posture combinations. This formed the basis of the scoring system which was further refined through discussion of examples from the health care and electricity industries.

The reliability of REBA was then evaluated using a group of 14 experienced ergonomists, physiotherapists, nurses and occupational therapists trained in the method and then tested using photographic examples of tasks with various levels of musculoskeletal loading.

Conclusions

REBA provides a rapid entire body musculoskeletal risk assessment. It assesses the loads on the musculoskeletal system due to posture, coupling, loads or forces and the muscular associated with the task. REBA requires no special equipment. The REBA scoring system generates an action level which reflects the level of risk present when performing the task assessed. The tool can provide useful data when justifying intervention and control measures to be adopted. When changes have been made REBA can be re-administered to demonstrate the change in risk level due to the interventions undertaken. As such REBA is useful as a screening tool, a risk assessment tool and for monitoring the effects of musculoskeletal risk control strategy.

REFERENCES

American Academy of Orthopaedic Surgeons, 1965. *Joint Motion Method of measuring and recording.* Churchill Livingstone, Edinburgh.

Borg, G.A.V., 1982 Psychological bases of perceived exertion. Medicine and Science in Sports and Exercise, **4**. 377-381.

Finnish Institute Of Occupational Health (1992). *Training Publication no: 11. OWAS, a method for the evaluation of postural load during work.* Publication office, Topeliuksenkatu 41 aA, SF 00250 Helsinki, Finland

Foreman, T.K, Davies, J. C., Troup, J.D.G. 1988, A posture and activity classification system using a microcomputer, Int. J. Indust. Ergon, **2**, 285-289.

Hignett, S., 1994a, *Postural analysis of Nursing Work.* MSc Thesis, University of Nottingham, England.

Hignett, S., 1994b, Using computerised OWAS for postural analysis of nursing work, in S. Robertson, (ed) *Contemporary Ergonomics*, (Taylor and Francis, London), 253-258.

Health And Safety Executive (HSE), 1992. Manual Handling - Guidance on Regulations, HMSO, 1992.

Health And Safety Executive (HSE), 1994. *A pain in your workplace?* HSE Books, ISBN 0 7176 0668 6.

Keyserling, W.M.1990, Computer aided posture analysis of the trunk, neck, shoulders and lower extremities, in W. Karwowski, A.M Genaidy, S.S. Asfar, (eds*)* *Computer-Aided Ergonomics, A researchers guide* (Taylor and Francis. London), 261-272.

McAtamney, L. and Corlett, E.N. 1993, RULA: a survey method for the investigation of work-related upper limb disorders, Applied Ergonomics, **24** (2), 91-99.

McAtamney, L. 1994, *The Interaction of Risk Factors Associated with Upper Limb Disorders.* PhD Thesis, University of Nottingham, England.

McAtamney,L. and Hignett, S. 1995, REBA-A Rapid Entire Body Assessment Method for investigating Work Related Musculoskeletal Disorders. in Blewett, V (Ed) Proceedings of the 31st Annual Conference of the Ergonomics Society of Australia.

Wilson, J.R. and Corlett, E.N (1995) *Evaluation of Human Work* ,2nd edn.Taylor and Francis, London. pp 662-714.

FEASIBILITY OF DEVELOPING A DECISION AID FOR INITIAL MEDICAL ASSESSMENT OF ULDs

D T Sinclair,* R J Graves,* M Watt,* B Ratcliffe,** S Doherty***

*   *Department of Environmental & Occupational Medicine,
University Medical School, University of Aberdeen,
Foresterhill, Aberdeen, AB25 2ZD*

**   *233 Wigan Road, Ashton-in Makerfield, Wigan, WN4 9SR*

***   *Health and Safety Executive, Scotland West Area Office,
375 West George Street, Glasgow, G2 4LW*

Anecdotal evidence from a number of sources indicates that there may be difficulties arising from the diagnostic approach adopted by some General Practitioners (GPs) involved in diagnosing Upper Limb Disorders (ULDs), and in particular assessing whether the patient's occupation is a causative factor. Labelling ULDs as "tenosynovitis", or "RSI" may have far reaching consequences to the patient, especially in terms of treatment and legal issues. For this reason a Diagnostic Support Aid covering key points to assist in the diagnosis and management of ULDs was devised by an expert group of occupational physicians, GPs and ergonomists to supplement the existing knowledge and skills of GPs.

Introduction

It has been suggested (see Davis, 1995), that lax diagnostic criteria are frequently used in the initial medical assessment of Upper Limb Disorders (ULDs), and that this leads to incorrect diagnoses and labelling of conditions with "pseudo diagnoses." A survey by Diwaker and Stothard (1995), indicated that labels which may be used inappropriately include "tenosynovitis", and "repetitive strain injury" (RSI), since the diagnostic criteria used by a range of experts to make either diagnosis "varied greatly". Davis (op cit.) suggests that incorrect diagnoses and labelling has far reaching consequences to the patient, both in terms of treatment and legal issues (see also Allen, 1993).

Additional difficulties arise from the use of the term "RSI", since this does not necessarily give an adequate diagnostic description of the nature of the problem, but merely an indication of causation. In addition, the indicated cause may be erroneous since the term "RSI" implies that repetition alone can lead to soft tissue injury, even though there are a number of factors other than repetition which seem likely to increase "risk" (Graves, 1992). According to a number of authors (see Allen, 1993; Diwaker and Stothard, 1995; Hughes, 1990; Macdonald, 1996) use of the term "RSI" or "repetitive strain injury" should be discontinued. The term being used by the Health and Safety Executive to describe soft tissue injuries of the arm/hand is upper limb disorders (ULDs) and this will be used throughout the paper to refer to these types of complaints.

Trained occupational physicians will use an occupational history as well as other knowledge of risk factors to make a diagnosis from their medical assessment. In addition there is information available on procedures which may be of benefit in helping a physician to assess and record his or her findings (see for example Putz-Anderson, 1988; Kuorinka and Forcier, 1995). It has been emphasised that it is important for all appropriate investigations to be undertaken at the outset (Mayou, 1991).

General Practitioners (GPs) are unlikely to have the benefit of the above training and experience, and some type of support aid to help decision making may help them to come to a more considered diagnosis. Providing an aide memoire covering key points to be obtained from the patient's history and carrying out a musculoskeletal examination based upon existing knowledge of the literature would be helpful (D'Auria, 1995). This would maintain awareness of the possibility of work contributing to the symptoms but would help avoid attributing symptoms to the working situation without adequate justification.

It was concluded therefore, that there would be benefit in examining the feasibility of providing an Aid for the Initial Medical Assessment of Upper Limb Disorders (AIMA-ULDs) to act as a decision support aid which would be used by a GP to come to a more considered decision regarding the nature of ULDs and their possible work relatedness. The objective of this work was to develop a simple prototype decision support aid to assist GPs in carrying out an initial medical assessment of Upper Limb Disorders, based on a methodology with structured documentation and aide memoires.

Approach

Two working parties were set up to provide support and guidance in relation to developing the AIMA-ULDs. The first of these was an expert group which was required generally to assist by giving guidance on the form, content and structure of the AIMA-ULDs. This group comprised of an employment medical advisor, a rheumatologist/GP, an occupational physician, and two ergonomists. The second working party was an advisory group which was required to examine the performance and usability of the aid documents, and provide editorial support and operational views in relation to AIMA-ULDs. The advisory group comprised of practising occupational physicians from the following application areas:- local government, manufacturing, service utility, and new technology.

The development of the AIMA-ULDs was carried out in four stages: a literature review to examine the general medical assessment approach of GPs in relation to upper limb disorders and difficulties which can arise; identification of diagnostic criteria of common upper limb disorders; collation of information for assessment by the expert and advisory groups; and development of prototypes under guidance of the groups.

A representation of an appropriate procedure for medical diagnosis of suspected ULDs was adapted from Putz-Anderson (1988), Ranney (1993), and Kuorinka, Forcier (1995). Changes were made to the order of steps involved, and details were added as a result of discussions with the expert and advisory groups.

Findings and Discussion

It was found that the general diagnostic procedure involved: taking the patient history (i.e. history of symptoms and history of patient background), formulation of the differential diagnosis, physical examination, and management of the symptoms/ syndrome. This was examined in some detail but this paper provides only a brief description of the work involved.

Data obtained from the literature included subjective symptoms, objective signs, and potential risk factors associated with a number of syndromes in the following upper limb regions: a) hand/wrist; b) elbow/forearm; and c) neck/shoulder. From the recorded data, initial prototype diagnostic aid documents were formulated for each of the three regions. Two separate options were proposed initially for use as diagnostic aids. The first of these took the form of a set of checklists and the second took the form of flowcharts.

In order to make the aid document as concise and as user friendly as possible, and to target more commonly reported ULDs, the members of the expert and advisory groups decided to concentrate on symptoms of pain in the hand, wrist, and forearm for the final diagnostic aid. (However, it would be possible to develop further aids for other ULDs using a similar proforma.) In addition, it was decided that it would be preferable to produce the diagnostic aid on a single sheet.

Guidance provided by the expert and advisory groups, in relation to aspects of form, content, structure, and usability, was used to design the final prototype draft Diagnostic Support Aid. The aid was incorporated into a draft guidance document, intended to be introduced to GPs in a set of user trials (beyond the remit of this project).

The prototype version of the Diagnostic Support Aid is in two parts. Figure 1 shows a modified version of Part A, the aide memoire for medical assessment of upper limb disorders, which will be printed on an A4 sheet. Part B would be on the reverse and would provide a diagnostic decision aid for symptoms of pain in hand/wrist/forearm. An outline of the latter illustrating the main components only, due to space constraints, is shown in Figure 2. It is proposed that the GP begins the medical assessment using Part A. This will provide points to support taking a patient history. At this point, if symptoms include pain in the hand, wrist or forearm, the GP will proceed to Part B. Otherwise the GP will continue working on the former flowchart, which gives the general procedure for forming a differential diagnosis, conducting a physical examination, and lists management options available to the GP.

For Part B, it is intended that the examiner should proceed through the YES/NO decisions, basing the answers on the information obtained in the patient history. This will allow the examiner to rule out the more easily recognisable syndromes leading to hand/wrist/forearm pain. The less obvious syndromes are given in a lightly shaded area. The patient's background history will then be examined to establish possible causes of the symptoms. To support this process, a list of activities (work or leisure) associated with each syndrome as identified from the literature, was provided after each diagnosis. This is intended to help the examiner to assess whether the patient's complaint is likely or unlikely to be related to activities.

For the less obvious syndromes, the next stage in the flowchart gives criteria to be assessed in a physical examination for confirmation of the diagnosis/diagnoses. However, where symptoms are non-specific, it is suggested that specialist referral may be required to assist with the diagnosis. Such referral will be included within the management of the complaint.

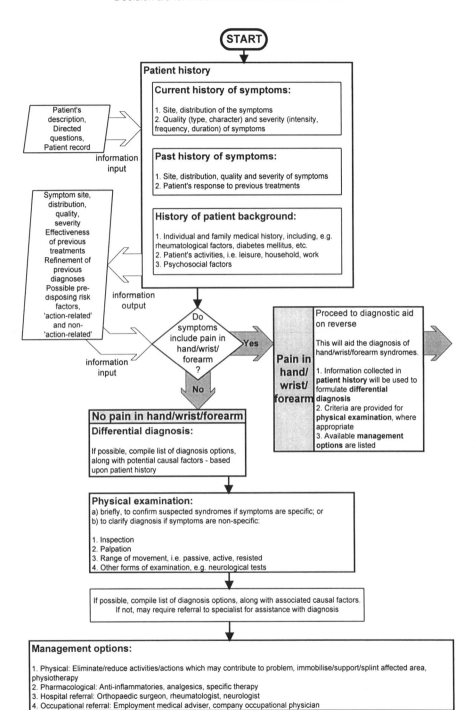

Figure 1 Aide memoire for medical assessment of upper limb disorders (Part A).

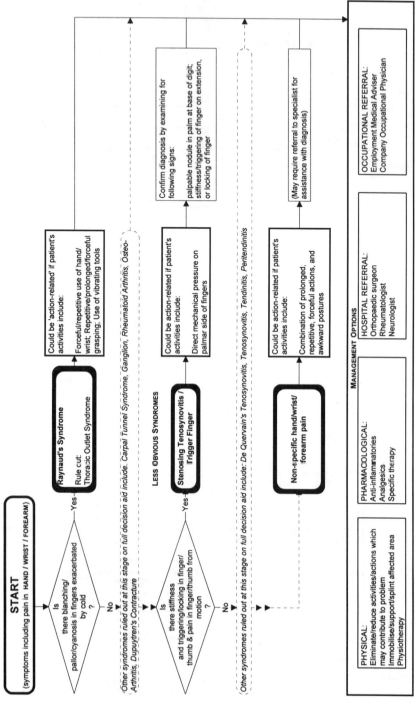

Figure 2   Truncated version of diagnostic decision aid for symptoms of pain in hand/wrist/forearm (Part B).

The final stage in the flowchart gives general prompts to indicate to the GP his/her options to manage the problem. The options are physical, pharmacological, and referral. The latter of these can be subdivided into hospital or occupational referral.

On the advice of the expert and advisory groups the prototype diagnostic aid was designed to include symptoms of pain in the hand, wrist, or forearm region. The fact that it did not cover a wider range of ULD symptoms was dictated by the need to keep the aid document as user friendly as possible. However, there is still some scope to develop flowcharts covering symptoms in other areas of the upper limb (i.e. neck, shoulder, elbow regions).

Now that a prototype has been developed, the next stage will be to conduct a set of user trials with a number of GPs. This would possibly involve an initial pilot study to test for design faults in the diagnostic aid, before re-designing the aid for a wider circulation among GPs. This would be followed by conducting exploratory seminars to test usability of documents and determine training requirements.

Once the user trials of the prototype aid for the hand/wrist/forearm region have been undertaken, it may be appropriate to develop the other flowcharts and produce another prototype aid, perhaps covering symptoms of pain in the neck, shoulder or elbow. This could also be tested and modified by user trials to the endpoint of a full set of diagnostic aids for the initial medical assessment of upper limb disorders.

References

Allen, S. 1993, RSI: Can you claim? *Solicitors Journal, 19 November 1993,* 1152.

D'Auria, D. 1995, Clinical management of Upper Limb Disorders. A perspective from Occupational Medicine. Health and Safety Sponsored Institute of Occupational Health Workshop on Work Related Upper Limb Syndromes - Origins and Management, Birmingham, 10-11 October, 1995.

Davis, T.R.C. 1995, The problems - Early identification, labelling, diagnostic criteria, prevention of progression, interventions and treatment. Health and Safety Sponsored Institute of Occupational Health Workshop on Work Related Upper Limb Syndromes - Origins and Management, Birmingham, 10-11 October, 1995.

Diwaker H.N., Stothard J. 1995, What do doctors mean by tenosynovitis and repetitive strain injury? *Occupational Medicine,* **45** (2), 97-104.

Graves, R.J. 1992, Using ergonomics in engineering design to improve health and safety, *Safety Science,* **15**, 327-349.

Hughes, S. 1990, Repetitive strain injury. How I treat..., *The Practitioner,* **234**, 443-446.

Kuorinka, I., & Forcier, L. (ed.). 1995, Work related musculoskeletal disorders (WMSDs): A reference book for prevention. (Taylor and Francis, London and Washington)

Macdonald, E.B. 1996, Work and disease: an update, *Update,* **52** (5), 249-256.

Mayou, R. 1991, Medically unexplained physical symptoms. *British Medical Journal,* **303** (6802), 534-535.

Putz-Anderson, V. 1988, Cumulative trauma disorders - A manual for musculoskeletal diseases of the upper limbs. (Taylor & Francis, London and Washington)

Ranney, D. 1993, Work-related chronic injuries of the forearm and hand: their specific diagnosis and management. [Review]. *Ergonomics,* **36**, 871-880.

THE DEVELOPMENT OF A PRACTICAL TOOL FOR MUSCULOSKELETAL RISK ASSESSMENT

Guangyan Li and Peter Buckle

Ergonomics Research Unit
Robens Institute
University of Surrey
Guildford, Surrey GU2 5XH UK

An exposure tool is currently under development for health and safety practitioners to use in assessing work-related musculoskeletal risks. Experimental studies were conducted to test the usability and sensitivity of the draft tool. Preliminary results suggest that such a tool is feasible with regard to simplicity, sensitivity and, to some extent, for inter/intra-observer reliability for the tasks evaluated. The need for further improvement of the tool is discussed.

Introduction

Physical exposures to musculoskeletal risks have been described and quantified with a variety of measures, including pen and paper based observations, videotaping and computer aided analysis, direct or instrumental techniques, and approaches to subjective assessment. Literature reviews on some of these methods are available (eg. Kilbom, 1994) and more recent developments of the techniques for physical exposure assessment have been systematically studied by Li and Buckle (1996). Whilst these existing methods have been found useful for research and for the ergonomic evaluation of some real tasks, their usability by health and safety practitioners, for making exposure assessment in many real work situations, has been questioned (Buckle and Li, 1996a). The majority of practitioners studied reported that the exposure assessment methods which are currently available are too technically demanding and time consuming (eg. detailed measurement of angular ranges of segmental movement). Even when the risk factors or hazardous jobs could be identified with some of these tools, the practitioners found it difficult to implement recommendations, or to evaluate whether ergonomic interventions were effective in reducing the potential risks for musculoskeletal disorders.

The objective of the present study was to develop a practical tool for health and safety personnel to use when assessing exposure and the change in exposure to known risk factors for work-related musculoskeletal disorders. The main focus of the tool was to meet the practitioners' needs and to enable its users to evaluate the extent to which interventions at the workplace have reduced exposure. The tool could also be used to advise on the likely impact of a number of alternative ergonomic interventions.

Development of an exposure tool

Studies have been carried out with respect to the development of the exposure tool. These include the review of the knowledge base for the work-related musculoskeletal problems; existing methods for the assessment of physical exposure; interviews with user focus groups; and a questionnaire survey among health and safety personnel. Based on the investigation of the potential users' needs, a preliminary set of usability criteria for the tool was formulated (Buckle and Li, 1996a), with some major concerns that, in addition to its high sensitivity and reliability, the tool should be easy to use, have limited paperwork, be applicable to a wide range of jobs and that the assessment should be completed within a short time (preferably 10-15 minutes).

A draft prototype exposure tool was constructed (Buckle and Li, 1996b). In order to test the usability of the draft tool and understand how it could be improved, tests were conducted by a "practitioners group" to assess several simulated tasks using the draft tool. Some of the preliminary results are reported in this paper.

Experimental studies

Tool users and subjects

Three health & safety personnel participated in the tests acting as the 'tool users' (the 'observers'). They all had some experience in making risk assessment (2.7 years on average, SD=0.82). Six healthy practitioners (3 males and 3 females) acted as 'subjects' performing the simulated tasks. Their mean age was 35.8 years (SD=6.5) and their mean stature was 174.4 cm (SD=7.1).

Simulated tasks

Three simulated jobs were tested and two subjects were randomly assigned to each job. The jobs included manual assembly (bolting), manual material handling (lifting) and VDU work.

The first job, manual assembly, was to assemble bolts onto an horizontal bar which was set at the height corresponding either to the subject's stature (high bar) or to the subject's chest height (low bar, measured 20 cm below the shoulder). In performing the task, the subject picked up a bolt from the box (positioned on the table by the right-hand side) and inserted the bolt into a horizontal hole with the end of the bolt facing downwards. The subject then assembled a nut onto the bolt and tightened it with a spanner.

The subject repeated the same task continuously for 10-15 minutes, and was asked to assume that he/she was being paid for the number of bolts assembled, and the task was to be performed up to 2-4 hours per day. The task was divided into two sessions with either "high bar" or "low bar" assembly for each session. These were randomly presented to each subject and a 10-min rest was allowed between the two sessions (these also applied to the other two jobs as described below).

The second job, load lifting, was also presented as two levels separately to each subject, i.e., lifting from floor and lifting from table. For the lifting-from-floor task, the subject lifted a load (8 kg, placed by the side of the subject) from the floor to a bench located in front of the subject at waist height. For the lifting-from-table task, the subject lifted the load, which was placed on a table at the knee height in front of the subject, to the same bench set at waist height.

The third job was word-processing at VDU. The subjects were required to continuously input a text into the computer through keyboard. Two types of VDU workplace layout were randomly presented to each of the two subjects: one with both the monitor and the script placed on the table surface; the keyboard was set flat on the table surface; a conventional fixed-height chair was used. For the 'modified' VDU workplace the monitor was placed on top of the computer and the script was placed on a file stand; the keyboard was slightly tilted forward; a height-adjustable chair was used and adjusted to the subject's preference.

During the task performance, exposure assessments were conducted by the three observers using the draft tool. Upon receiving the tool, the observers spent about 4-5 minutes reading through a one-page instruction, followed by a further 5-min explanation. It was required that the observers always made their assessments independently. After all the tasks were assessed, the observers completed a questionnaire about the exposure tool.

Results

Figure 1 gives an example on the exposure levels for each item/body part for the bolting tasks. Analysis of variance was conducted to test whether there was a significant change in exposure between the two types of tasks within each job (eg. bolting at high bar and low bar). The results are given in Table 1.

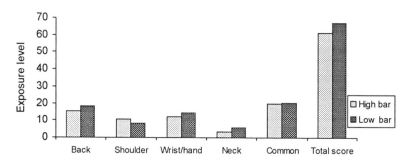

Figure 1. Exposure assessment for bolting tasks (mean values of the three observers)

Table 1. The significance in analysis of variance for the change in exposure between tasks (3 observers, 2 subjects for each task; NS: not significant, $p > 0.05$)

Assessment items	Bolting task	Lifting task	VDU work
Back	$p \leq 0.025$	NS	$p \leq 0.05$
Shoulder	$p \leq 0.001$	NS	NS
Wrist/hand	NS	NS	NS
Neck	$p \leq 0.01$	NS	NS
Common evaluation	NS	$p \leq 0.001$	$p \leq 0.001$
Total score	NS	NS	$p \leq 0.001$

To test whether different observers could reach a similar agreement towards each assessment item, Spearman's coefficients were calculated between each pairs of the observers regarding their assessments for each item, as shown in Table 2.

Table 2. Spearman's correlation coefficients between two observers (matched pairs among 3 observers)

Assessment item	Correlation coefficient	Significance level	No. of observations
Back	r=0.80-0.92	p≤0.01-0.001	12
Shoulder	r=0.62-0.71	p≤0.05	12
Wrist/hand	r=0.10-0.26	NS	12
Neck	r=0.91-0.94	p≤0.001	12
Common evaluation	r=0.78-0.93	p≤0.01-0.001	12
Total score	r=0.80-0.91	p≤0.01-0.001	12

The assessments made by the same observer on different subjects for each job was also analysed by calculating the Spearman's coefficients. The correlation coefficients were r=0.90-0.96 (p≤0.001) for bolting, r=0.92-0.95 (p≤0.001) for lifting and r=0.98 (p≤0.001) for VDU work.

Among the five choices given in the questionnaire regarding the question "how easy or difficult do you find the use of this exposure tool?", (very easy-very difficult), all observers regarded the tool as "easy". The observers also confirmed that they could complete the assessments for each task within 12-15 minutes (including the calculation of the scores).

Discussions

The construction of the exposure tool was based on the current evidence that the physical exposures to musculoskeletal risks are multifactorial (eg. Christmansson, 1994) and that the exposure level is proportional to the level of interactions between different risk factors. Although existing evidence is still very limited regarding the exposure level as a result of the interactions between risk factors, it seems reasonable to assume that the combination of two high-level risk factors is related to a higher level of exposure than the combination of two low-level risk factors. Studies have shown that among the combinations of force (high/low) and repetition (high/low), the combination of high force and high repetitiveness substantially increased the magnitude of association for hand/wrist CTDs (Silverstein et al., 1986). Despite the relationships between physical exposure and the combinations of different risks are still largely unknown, this assumption is considered not to affect the result as long as the same tool is used for exposure assessment.

Sensitivity of the draft exposure tool

One of the most important features of the proposed exposure tool is that the tool should be able to identify the changes in exposure before and after the ergonomic interventions to the same job. The preliminary studies showed that the changes in exposure in some body parts for the three jobs tested could be identified to some extent by this user group. For the bolting tasks, for example, significantly different exposure levels in the back, shoulder and neck were found by the observers (Table 1). Figure 1 shows the higher exposure levels in the back, wrist and neck for the 'low-bar' bolting than that for the 'high-bar' bolting task. Normally, it might be expected that the high-bar bolting would produce higher exposure level than the low-bar bolting. However, it was found during the trials that the subjects performed the high-bar bolting with an almost straight back and neck; while for the low-bar bolting, the subjects were found to bend or twist their back and head/neck from time to time so as to see the bolting point beneath the horizontal bar. As could be expected, the exposure to the shoulder did show a higher level for the high-bar bolting than that for the low-bar bolting (Figure 1).

The exposure levels of the four body parts observed did not significantly change between the two types of lifting tasks (Table 1). A possible reason was that the subjects were well aware of the potential hazards when lifting load from floor, and performed the task with the 'correct posture', i.e., with the back straight and vertical during the lifting.

The assessment of the VDU work also identified a significant exposure changes for the back, common evaluation and the total score, indicating a reduced exposure level for the modified workplace. But the change in exposure to the shoulder, wrist/hand and neck did not reach a significant level (Table 1), suggesting that either the workplace needs further modification, or the tool is not sensitive enough for the VDU tasks tested.

Inter-observer reliability

As shown in Table 2, the assessments between two observers were significantly correlated for all major assessment items except for the wrist/hand. The inter-observer agreement was lower for the shoulder exposure than that for the back and neck. These were in agreement with the previous studies (eg. Baluyut et al., 1995) indicating that observers had difficulty in evaluating the exposure of the upper extremity, particularly the elbow and wrist, while the lower back and neck were easier to evaluate. However, the present results were based on the assumption that the physical exposure is multifactorial, further analysis is needed to understand the inter-observer agreement for each assessment item, eg. back posture.

Intra-observer reliability

The tests showed that the assessments made by the same observer on different subjects for each job were highly correlated, suggesting a high intra-observer reliability of the draft tool regarding its use for these simulated tasks. However, this result should be handled with care because the time interval between the observations was relatively short (within 20 minutes) and the observers might be influenced by his/her previous assessment on the other subject.

The observers' opinion about the exposure tool

Since all three observers rated the tool as 'easy' and they all could make the assessment within 15 minutes, this might have suggested that such type of tool is acceptable in terms of its simplicity and the time needed for conducting the assessment, but the present results were only based on a very small user group and three simulated tasks, further studies are needed to test the tool in more practical situations.

In conclusion, the pilot tests showed that the draft exposure tool shows promise in terms of simplicity and sensitivity; the inter-observer reliability was also acceptable for some of the assessment items. However, since these results were only based on a small user group and some simulated tasks, and the inter-observer reliability was poor especially for the assessment of the wrist/hand, further improvements of the tool are needed as well as testing the tool in real work systems.

Acknowledgements

This work is funded by HSE. The authors wish to thank Professor D. Stubbs for his comments on the draft paper and the practitioners who participated in the pilot study. The support from the members of the Ergonomics Research Unit, Robens Institute is also gratefully acknowledged.

References

Baluyut, R., Genaidy, A.M., Davis, L.S., Shell, R.L. and Simmons, R.J., 1995, Use of visual perception in estimating static postural stresses: magnitudes and sources of errors. Ergonomics, 38,9, 1841-1850.

Buckle, P. and Li, G., 1996a, User needs in exposure assessment for musculoskeletal risk assessment. In: Virtual Proceedings of CybErg 1996: The First International Cyberspace Conference on Ergonomics. (eds.: L. Straker and C. Pollock). URL http://www.curtin.edu.au/conference/cyberg Curtin University of Technology.

Buckle, P. and Li, G., 1996b, A practical tool for the evaluation of the change in exposure to risk of musculoskeletal disorders. Interim report to HSE, Robens Institute, (unpublished).

Christmansson, M., 1994, Repetitive and manual jobs - content and effects in terms of physical stress and work-related musculoskeletal disorders. The International Journal of Human Factors in Manufacturing, 4,3, 281-292.

Kilbom, Å., 1994, Assessment of physical exposure in relation to work-related musculoskeletal disorders - what information can be obtained from systematic observations? Scand J Work Environ Health, 20, special issue: 30-45.

Li, G. and Buckle, P., 1996, Posture based techniques for assessing the exposure to work-related musculoskeletal disorders. (under submission).

Silverstein, B.A., Fine, L.J. and Armstrong, T.J., 1986, Hand wrist cumulative trauma disorders in industry. British Journal of Industrial Medicine, 43, 779-784.

POSTURE

OPERATION OF LOCKS ON BRITISH INLAND WATERWAYS

S Brown and R A Haslam

Health and Safety Ergonomics Unit,
Department of Human Sciences,
Loughborough University,
Loughborough, Leicestershire,
LE11 3TU.

This study examined the manual handling requirements of canal lock operation. Methodology included a search of the manual handling literature; site visits to record operating techniques and differences in equipment and environments; and interviews with employees and canal users. Available force level guidelines are difficult to apply to this dynamic activity and no manual handling studies were found which addressed a user group with such varied ability. Recommendations included: provision and positioning of handles on lock gate beams; stable ground surfaces with foot bracing bricks; and information to users regarding efficient pushing and pulling techniques.

Introduction

Canal boating is a growing leisure activity, one of the attractions being the tradition associated with it. Most inland waterway systems developed during the time of the industrial revolution, almost two hundred years ago. They retain many of the design features of the time, including materials and mechanisms. The structures needed to be strong enough to withstand repeated knocks from loaded barges and the effects of weather, while regularly submerged in water.

The purpose of a canal lock is to allow boats to travel up and down hill, by providing an enclosure where the water level can be raised and lowered. Manual operation involves:-

1. opening and closing the gates to allow boats in and out of the lock
2. opening sluice covers by turning gearing mechanisms with a 'windlass' or winding handle, allowing water to pass from the higher level to the lower level.

Lock gates have extended top beams to counterbalance the weight of the gate and provide leverage for the operator. Pushing and pulling force is applied to the balance beam to open and close the gates. Canal users range from children to septuagenarians.

The aims of the study were to:-
- make an ergonomic assessment of locks and their operation
- identify 'safe' force levels when operating the gates and sluices
- identify potential risks when operating the locks
- identify particular groups of users who may be at risk when operating locks.

The physical demands when operating canal locks depend upon the equipment, the operating environment and the characteristics of the operator. Changes to the gates or sluices would be a long term approach as they are expected to last in excess of twenty years. There appeared to be few obstacles regarding modification of the operating environment. Canal users may be limited by their physique, however, by removing identified hazards and providing necessary information their risk of injury when operating locks should be reduced.

Method

Literature search

A search was undertaken of manual handling literature to identify guidelines for 'safe' force levels relevant to lock operation. Previous manual handling studies were examined to identify criteria which affect the safety and efficiency of subjects when pushing and pulling.

Staff interviews

Waterway's staff were interviewed to gather information from previous studies of locks and to identify variations in the operating equipment. Staff interviewed included a regional operations manager, lock keepers and maintenance staff.

Site visits

Site visits were made to several locks in the region. Due to the historical development of local canal companies there are variations in gate size and gearing mechanisms between locks, however the operating routine is the same. The locks were assessed with regard to their operation and dimensions. The ground surface was checked for secure foot grip and where bracing points were provided for the feet it was noted whether they were in a functional position. Force levels were measured using a force meter.

Canal users

Users were observed operating the locks. Video recording and photographs were taken to analyse the task and postures and techniques used by operators.

Fifty canal users were interviewed using a questionnaire. Details of age, gender and height were collected. Their experience of lock operation was categorised by the number of canal trips they had made over the preceding twelve months. Respondents were asked to identify possible operating difficulties for inexperienced boaters when using locks and to indicate anything they thought would help to reduce or eliminate those problems.

Results

Force levels

Force level guidelines were difficult to apply to this task due to the variation in criteria and the type of subjects used. Many studies address specific industrial problems, the majority of subjects being male, and of working age. However it was possible to summarise from these reports criteria which appeared to affect the efficiency of the pushing and pulling effort of the subjects. Pushing is considered to be more efficient and safer than pulling. Using two hands, the operator is able to exert more force than with one hand (Chaffin & Andersson, 1991). When pushing, it is considerably more efficient to push with the back (Kroemer, 1969).

Operating posture

The height at which force is applied is optimum between shoulder and knuckle height, measured with the arms hanging at the side of the body (Health & Safety Executive, 1992). By leaning forward when pushing and backwards when pulling, body weight can be used to increase the force applied by the operator. The feet must be positioned to allow the operator to lean forwards or backwards (Fig. 1). Provided there is sufficient space, the feet are moved away from the point at which the force is applied when pushing. Pulling requires the feet closer to, and sometimes in front of the point at which the force is applied. Standing with one foot in front of the other will allow the operator to transfer body weight the object begins to move. These postures are identified as the most efficient way of exerting force and are limited by safety and physical comfort.

Figure 1. a) pulling feet together
b) pulling feet apart
c) pushing feet apart

Couplings

Force cannot be exerted without there being an equal and opposite reaction (Grieve, 1993). Operators must have a secure foot grip, a balanced posture and secure handgrip to be able to transmit the forces they are applying to the object being moved (Fothergill, Grieve & Pheasant, 1992). The ground surface must be stable and uncontaminated.

The ratio of horizontal to vertical force acting at the interface of the shoe sole and the ground, gives the coefficient of friction (μ). The ratio should be greater than 0.5 to reduce the risk of slipping when force is exerted. The operator must wear supportive shoes with a non-slip sole. It is important to be able to grip the object securely: if the hands slip then the force will not be transmitted effectively and the operator may be injured. Once the hands and feet are positioned the starting posture of the operator is defined, if there is not sufficient space to take up the required postures the operator becomes less efficient and there is increased risk of injury.

User population

Female boaters were observed to be more likely to operate the locks than their male colleagues. Techniques used to open and close the lock gates included one handed and two handed pushing and pulling, and pushing with the back. Sometimes two people operated the gates, one pushing, one pulling. With most gates it was necessary to start the closing process by pulling, due to the beam being at the edge of the lock. A handle was required for this action but was not always supplied. Those supplied were occasionally poorly located. Beam heights and ground surfaces varied considerably. Together with limited space at a few locks, the operator's posture was often restricted. Provision of bracing points for the feet, in the form of raised bricks, was useful at the sites where they had been well positioned.

Respondents to the questionnaire were representative of the adult user population in age range and height, identified from previous surveys. The gender split was equal. Six respondents, 12% of the sample, were first time canal users. Most problems identified were related to stiffness or heaviness of the mechanisms and not directly to the operating process. Users who appeared to have most problems were those who were of light build or shorter stature. Those with limited experience also had more problems.

Recommendations arising from the manual handling assessment included:
- removal of hazards to the user such as unstable and uneven ground surfaces (Fig. 2)
- provision and positioning of handles on gate beams to accommodate lock users
- installation of raised bricks against which to brace the feet (Fig. 3)
- advice to users regarding the relative efficiency of different operating techniques

Discussion

Operating canal locks in the traditional way is one of the main attractions to users adding variety and interest to their boat journey (Inland Waterways Amenity Advisory Council, 1996). Because of this and because it is a chosen leisure pursuit, the physical demands are accepted as necessary and people do not identify difficulties as problems. Removing the manual handling element of lock operation is not under consideration. Modified gate design and operating environments are suggested to enable safe and efficient use of the locks.

Figure 2. Cobbled surface does not extend to the operating area

Figure 3. Raised bricks do not extend to the edge of the lock for pulling

Identifying 'safe' force limits has proved difficult. The majority of manual handling injuries are associated with lifting, research into pushing and pulling is less extensive. Guidelines are difficult to compare due to the different criteria used in each study.

Encouraging operators to use improved posture during manual handling allows safe and efficient use of effort. Information about the most effective way to push and pull should be available to users. Although the height of the beam is less critical when pushing with the back there should be a handle at a height and in a position which accommodates most users, when pulling the gate away from the edge of the lock. Efforts had been made to improve ground surfaces and provide raised bricks to brace the feet to prevent slipping. At some sites the location of these improvements did not relate to the position of the operator. They were of no functional benefit and in some instances created hazards. Space restriction only seemed to occur where major structures such as road bridges were close to the locks. Although they restricted the operators initial posture when closing the gate, they could be used to brace against for stability.

Enabling operators to adopt stable positions while performing manual handling tasks reduces postural stress and facilitates efficient use of the musculoskeletal system (Ayoub & McDaniel, 1974; Kroemer, 1969). Dynamic activity involves changing body postures (Haslegrave, 1994), and the operating area should allow space to adopt these postures. Efficient transmission of effort from the operator to the object being moved also relies upon stable couplings.

References

Ayoub, M.M. and McDaniel, J.W. 1974, Effects of operator stance on pushing and pulling tasks, AIIE, September, 185-195

Chaffin, D.B. and Andersson , n, G.B.J. 1991, *Occupational Biomechanics*, 2nd edn. (Wiley-Interscience, U.S.A.)

Fothergill, D.M. Grieve, D.W. and Pheasant, S.T. 1992, The influence of some handle designs and handle heights on the strength of the horizontal pulling action, Egonomics, 203-211.

Grieve, D.W. 1983, Slipping due to manual exertion, Ergonomics, **26**(1), 61-72.

Haselgrave, C.M. 1994, What do we mean by a 'working posture'? Ergonomics, 7(4), 781-799.

Health and Safety Executive, 1992, *Manual handling: guidance on regulations*, (H.M.S.O., London.)

Inland Waterways Amenity Advisory Council, 1996, Britain's inland waterways, an ndervalued asset. (Consultative Report, London.)

Kroemer, K.H.E. 1969, Push forces exerted in sixty-five common working positions, Aerospace Medical Research Laboratory. Wright-Patterson Air Force Base, Ohio. AMRL-TR-68-143.

STEREOPHOTOGRAMMETRY - A THREE-DIMENSIONAL POSTURE MEASURING TOOL

Jane E. Mechan
130, Winchester Road,
Urmston,
Manchester.
M41 OUN

Mic. L. Porter
University of Northumbria,
Ellison Place,
Newcastle Upon Tyne.
NE7 8ST

The stereophotogrammetry technique used two photographic still cameras fired simultaneously to record three-dimensional information. This technique was used to record the postures of the upper extremities when operating three different designs of computer keyboard. A 'standard' computer keyboard acted as a control and two 'ergonomic' keyboards were evaluated, with the main objective to ascertain the extent of postural advantage offered by each. In this experiment ulnar abduction and wrist extension (left side of the subject only) were the upper extremity postures considered. The photographic data was compared with the results of the application of the RULA technique.

Introduction

'Stereophotogrammetry' can be applied to the study of posture, either static or dynamic. To be adequate, such a description requires that these body movements be measured in three-dimensions.

Investigations of three-dimensional movements have, historically, used mirrors or three synchronised motion picture cameras to provide second and third views of the subject (Bullock and Harley, 1972). Contemporary ergonomics methodologies use three or more video cameras, together with real time software, to track markers on a subject moving in the camera's field of view. However although the capital price is falling these are at present, costly approaches. Wilson and Corlett (1995) suggest a method for such work in the field could be to interface a posture-recording suit based on light-emitting diodes or strain-gauges.

Bullock and Harley (1972) state that "in some circumstances, one of those being when the instantaneous positions of moving objects are to be measured, photogrammetry affords the most convenient, or perhaps the only, means of measurement".

The conventional keyboard layout has evolved over the past 125 years, but has been implicated in operator musculoskeletal discomfort and injury. This experiment compared this 'control' with two alternative designs that were marketed as able to reduce stresses by keeping the hands in a more natural position. These keyboards are often marketed as 'ergonomic'. The investigation evaluated the effectiveness of a split-design keyboard in reducing the angles of ulnar deviation and wrist extension during operation. Stereophotogrammetry was used as the primary objective measuring tool, but comparisons were made with RULA (McAtamney and Corlett, 1993) assessments.

Method

The procedure for using two cameras to record three-dimensional information is as follows:
- Define X, Y and Z axes of the working area.
- Align the optical axis of one camera with one axis.
- Place calibration marks in the field of view at known X, Y and Z positions.
- Align the optical axis of the second camera in the XZ or YZ plane.
This procedure was applied to the study in the following way:
A fixed height (70cm) table was covered with a grid of 1cm by 1cm squares and the keyboard placed at a position described by the subject, as "comfortable". The axes and origin point were then defined with reference to the keyboard. It is with reference to this point that all measurements would be made. The X axis was defined as running across the working area with respect to the user, the Y axis as running across the table away from the user, and the Z axis as being perpendicular to the plane of the table.

The reference point for each of the axes was defined so that it lay in the field of view of the camera designated to view it. The exact positioning of the origin is not important as the necessary measurements are purely relative and not absolute.

The plan view camera was set up so that its optical axis was aligned with the Z axis as closely as possible to minimise distortion through misalignment. The focal plane of this camera was 148.5cm above the table on which the keyboard was placed. The second camera was then set up so that its optical axis lay in the XY plane, i.e. viewing the edge of the keyboard. The distance between the camera and the nearest edge of the table was measured to be 214cm.

With the cameras set up as described, X, Y and Z co-ordinates can be found for each of the body markers; X and Y being found from the plan view camera and Z being found from the side elevation camera. By obtaining these co-ordinates, the position of each point in space can be determined, and using trigonometry, the relative angles between them calculated. The exact positioning of the points in space is not important, it is the relative positioning of the points to each other is what facilitates calculation of these angles.

The camera positions were devised to give maximum visibility of the body markers with optimal separation, and minimum perspective error. The reader is referred to Paul and Douwes (1993) for further discussion regarding perspective error. Alternative camera positions could have shown both arms, but would have given less separation of the body markers and thus reduced accuracy.

Two Cannon EOS 5 cameras were used, each fitted with a zoom lens and mounted on a tripod. For the plan projection, an angle view finder was attached to the camera. The tripod for this camera was placed on a table 70cm high and to one side, with the camera positioned directly above the working plane. Both cameras used 800ASA film processed by a high street store, throughout. Extra lighting or flash was not desirable since this would have resulted in unwanted reflections, bleaching of the grid squares and could also distract the subject. The cameras were fired simultaneously using an infra red transmitter and frequently checked to ensure that they stayed synchronised.

The Z and Y rulers were constructed from 'Kappa' (foam sandwich) board and covered in centimetre squared paper. Specific areas (rows of squares) were highlighted on each ruler as references, and subsequently at 10cm intervals to aid measurement and later analysis. The rulers height and length were sufficient to ensure they were viewed by the appropriate camera and covered the relevant body markers on the subject. The X ruler was part of the grid used to cover the table.

The body markers used were standard 13mm diameter fluorescent orange discs. They were placed on the subject to illustrate the angles concerned. As a control, they were positioned so that when the arm and hand were in a straight line so were the body markers.

The subject sat on an adjustable "typists" chair with casters on a five spoke base and foot rest was also provided. The subject was helped to adjust the chair and was given the opportunity to undertake a familiarisation exercise. The material to be entered was held on an adjustable stand adjacent to the screen. The general arrangements can be seen in figures 1.1 and 1.2.

Results

The photographs were referenced to ensure that the corresponding plan and side elevations were not mixed and so that the sequence of photographs was correct. The photographs were referenced by subject number, keyboard and photograph number. The same reference was used for each pair of photographs (side and plan elevations). In total 390 photographs were produced for the five subjects. The data for the fifth subject was eliminated since the elbow body marker was not visible. The co-ordinates were taken from the photographs by hand from the centre of the body markers and recorded in tables.

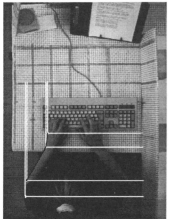 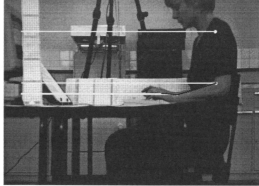

Figure 1.1 (plan) **and 1.2** (elevation): Example photographs illustrating how the X, Y and Z co-ordinates were taken from the photographs.

Figures 2.1 and 2.2 outline the formulae into which the derived X, Y and Z co-ordinates were entered to obtain the angles of ulnar deviation and wrist extension produced by the operation of each of the keyboards. The formulae do not take into account perspective. Any perspective error present was deemed negligible due to the layout used.

$$d_1 = \sqrt{\left(x_1 - x_2\right)^2 + \left(y_1 - y_2\right)^2}$$

$$d_2 = \sqrt{\left(x_2 - x_3\right)^2 + \left(y_2 - y_3\right)^2}$$

$$Cos\theta = \frac{x_1 x_2 + y_1 y_2 - x_2^2 - y_2^2 - x_1 x_3 - y_1 y_3 + x_2 x_3 + y_2 y_3}{d_1 d_2}$$

Figure 2.1 Formulae used to derive the Abduction angles. (Creasy, 1996)

Where:

d_1 = the distance between the knuckle and wrist body markers (plan photographs).
d_2 = the distance between the wrist and elbow body markers (plan photographs).
θ = the angle of abduction
x, y = the co-ordinates taken from the photographs (plan view)

Subscripts:

1 = Knuckle body marker co-ordinates.
2 = Wrist body marker co-ordinates.
3 = Elbow body marker co-ordinates.

$$d_1 = \sqrt{\left(y_1 - y_2\right)^2 + \left(z_1 - z_2\right)^2}$$

$$d_2 = \sqrt{\left(y_2 - y_3\right)^2 + \left(z_2 - z_3\right)^2}$$

$$Cos\theta = \frac{y_1 y_2 + z_1 z_2 - y_2^2 - z_2^2 - y_1 y_3 - z_1 z_3 + y_2 y_3 + z_2 z_3}{d_1 d_2}$$

Figure 2.2 Formulae used to derive the Wrist extension angles. (Creasy, 1996)

Where:

d_1 = the distance between the knuckle and wrist body markers (side photographs).
d_2 = the distance between the wrist and elbow body markers (side photographs).
θ = the angle of wrist extension
y = the co-ordinates taken from the photographs (plan view)
z = the co-ordinates taken from the photographs (side view)

Subscripts:

1 = Knuckle body marker co-ordinates.
2 = Wrist body marker co-ordinates.
3 = Elbow body marker co-ordinates.

Errata: Table 1. in Mechan and Porter

Due to transcription errors, which occurred when the data from the original project report was prepared for "Contemporary Ergonomics 1997", the whole table should be replaced. The paper's conclusions are not affected. We regret these errors, the confusion and inconvenience caused.

Table 1. A summary of the results from the keyboard stereophotogrammetry assessments

Subject No.	Keyboard No.	Mean Wrist Abduction Angle (degrees)	Mean Wrist Extension Angle (degrees)
1	1	18.29	26.24
2	1	26.37	4.49
3	1	20.59	16.22
4	1	6.31	9.21
1	2	20.91	27.24
2	2	24.80	6.33
3	2	17.37	22.34
4	2	11.38	6.79
1	3	25.54	15.48
2	3	5.09	5.27
3	3	12.38	15.65
4	3	2.68	5.74

Table 1. A summary of the results from the keyboard stereophotogrammetry assessments

Subject No.	Keyboard No.	Mean Wrist Abduction Angle (degrees)	Mean Wrist Extension Angle (degrees)
1	1	18.29	26.24
2	1	21.91	27.24
3	1	25.54	15.48
4	1	26.37	4.49
1	2	24.8	6.33
2	2	5.09	5.27
3	2	20.59	16.22
4	2	17.37	22.34
1	3	12.38	15.65
2	3	6.3	9.21
3	3	11.38	6.79
4	3	2.68	5.74

Where:

Keyboard 1 = Control
Keyboard 2 = Non-split design 'ergonomic' keyboard
Keyboard 3 = Split design 'ergonomic' keyboard

The main objective of the research was to examine the extent to which postural advantages were offered by different designs of computer keyboard. Through the stereophotogrammetry, it was revealed that the split design keyboard decreased the angle of ulnar abduction from about 20° to 10° and produced a 29% decrease in the angle of wrist extension. By alleviating ulnar deviation and wrist extension to such an extent, the split design keyboard will reduce the static muscle work of the upper extremities, hence lowering the risk of complications in the forearms and hands. The rows of keys on a standard keyboard are spaced evenly over a distance of 21cm for both hands. However the separation of the elbows (with the arms comfortably by the sides) is so much greater than the width of the keyboard that the forearms must turn inwards across the front of the body when operating. This position will further aggravate ulnar deviation. Where the shoulder or upper torso width of the operator is larger than the opening angle of the fixed angle split keyboard, the opening is not sufficient to alter the angle of ulnar deviation. This was highlighted by the results.

The RULA (McAtamney and Corlett, 1993) methodology was also applied to the study, and results confirmed those of the stereophotogrammetry. The working posture scores produced by the wrist showed the greatest difference between the keyboards. A score of one (wrist in the neutral position) was only produced for the operation of the split design keyboard, for one of the subjects. A score of two was awarded where the wrist was 0-15° in either flexion or extension, with no additional radial or ulnar deviation. This score was most frequently awarded to the split design keyboard. A score of three was awarded for 15° or more in either flexion or extension, with no additional radial or ulnar deviation. This score was most frequently awarded to the standard keyboards and was only awarded to the split design keyboard in one instance. The maximum score of four (as score three but with additional radial or ulnar deviation) was most frequently awarded to the two non-split design keyboards, and was not awarded to the split design keyboard. The working posture scores of three and four produced for the wrist illustrate a movement range with more extreme postures indicating an increasing presence of risk factors causing load on the structures on the wrist. Both the stereophotogrammetry and the RULA assessments reveal that for wrist postures when operating a computer keyboard, the introduction of the split design keyboard can be seen as a successful modification to the workstation.

Conclusion

The technique stereophotogrammetry was reliable, precise and valid when applied to the study of keyboard operating posture in three-dimensions, yielding accurate results from a small sample. The technique is expensive in terms of resource and time, and hence may be less suitable for large scale studies, when the high cost of more capital intensive techniques can be justified. Furthermore, due to the volume and positioning of equipment demanded by the technique, it is likely only to be suitable for a laboratory based study, or a highly controlled field study.

References
Bullock, M.I. and Harley, I.A. 1972, The measurement of three-dimensional body movements by the use of photogrammetry, *Ergonomics*, **15** (3), 309-322.
Creasy, C. 1996, The Department of Mathematics and Statistics. The University of Northumbria at Newcastle, Newcastle Upon Tyne. Personal Communication.
McAtamney, L. and Corlett, E.N. 1993, RULA: A survey method for the investigation of work-related upper limb disorders, *Applied Ergonomics*. **24** (2), 91-99.
Paul, J.A. and Douwes, M. 1993, Two-dimensional photographic posture recording and description: a validity study. *Applied Ergonomics*, **24** (2), 83-90.
Wilson, J.R. and Corlett, E.N. 1995, *Evaluation of Human Work*, 2nd Edition. (Taylor & Francis, London).

Acknowledgement

The work reported in this paper was carried out while Jane Mechan was a student on the University of Northumbria's Applied Consumer Sciences Degree.

MANUAL HANDLING

FUZZY MODELING OF
LOAD HEAVINESS IN THE WORKPLACE

Ashraf M. Genaidy, Waldemar Karwowski,* Doran M. Christensen
Costas Vogiatzis, Nancy Deraiseh and Angelique Prins

Industrial Engineering Program
University of Cincinnati, Cincinnati, Ohio 45221-0116, U.S.A.

Center for Industrial Ergonomics
J.B. Scientific School
University of Louisville, Louisville, KY 40292, U.S.A.

This study was conducted on experienced workers in the package delivery industry to estimate the amounts of load which correspond to various levels of load heaviness. The distribution of loads within each heaviness level was developed using fuzzy sets theory. It was found that the "23 kg" load defined by NIOSH represents a "somewhat heavy" load. Also, the "40 kg" load limit may be classified as a "very heavy" load. Also, one should be more careful in the interpretation of statistical norms for human perception of load handling.

Introduction

One of the work practices frequently taught to employees is to estimate the heaviness of load before it is actually handled. If the load is perceived as too heavy, then one should ask for help. However, limited information can be found in the ergonomics literature about what a person perceives as a heavy load. This study was conducted to fill this gap by estimating the amounts of load which correspond to various levels of load heaviness by experienced load handlers.

Methods

Forty experienced employees participated in this study. Each employee was asked to estimate the most representative amount of load which can be assigned to various levels of load heaviness as defined according to the Borg Scale. The information was collected in the form of a survey during working hours.

A mixed design was used for the statistical analysis. The distribution of amount of load handled for each heaviness level was modelled using a modification of the three-phase method described by Li and Yen (1995).

Results and Discussion

Figure 1 reveals that there were no differences between male and female data for the negligible level up to somewhat heavy. There was a steady increase in the differences between male and female data from heavy to maximum levels (only for very heavy, very very heavy and maximum levels at the 5% level). Figures 2 and 3 present the fuzzy distributions for the various levels of load heaviness.

Figure 4 shows general agreement between our results and those obtained by others for the light range of loads (i.e., very light, light and moderate). However, this study demonstrated higher values for the heavy loads (i.e., somewhat heavy, heavy, and very heavy) particularly in comparison with the findings of Karwowski (1988, 1991) deribed from inexperienced college students.

The 23 kg defined by the U.S. National Institute for Occupational Safety and Health is a somewhat heavy load based on the analysis of load distribution (corresponding to a 1.0 certainty factor). Also, the 40 kg considered in the 1981 NIOSH guidelines can be classified as a very heavy load.

A moderate level of load heaviness (14 kg) corresponds to 85% of the worker population data in the published literature. A somewhat heavy load (i.e., 23 kg) can be handled by about 43% of the worker population. A 32 kg, which is perceived as a "heavy" load, is equivalent to 1% of the worker population. These results further indicate the following: 1) a very heavy level of load or heavier can be handled by less than 1% of the worker population; and 2) a light level of load or lighter is handled by over 99% of the worker population.

The findings of this study can be used by ergonomists as part of awareness programs on safe work practices. Also, the determination of distribution functions for the linguistic levels of ergonomic risk factors can assist in establishing rule-based risk assessment systems.

References

Karwowski, W., 1988, Perception of load heaviness by males, in: *Manual Material Handling: Understanding and Preventing Back Trauma* (American Industrial Hygiene Association, Akron, OH), 9-14.

Karwowski, W., 1991, Psychophysical acceptability and perception of load heaviness by females, Ergonomics, **34**, 487-496.

Li, H.X. and Yen, V.C., 1995, *Fuzzy Sets and Fuzzy Decision-Making*, (CRC Press, Boca Raton).

Luczak, H. and Ge, S., 1988, Fuzzy modelling of relations between physical weight and perceived heaviness: The effect of size - weight illusion in industrial lifting tasks, Ergonomics, **32**, 823-837.

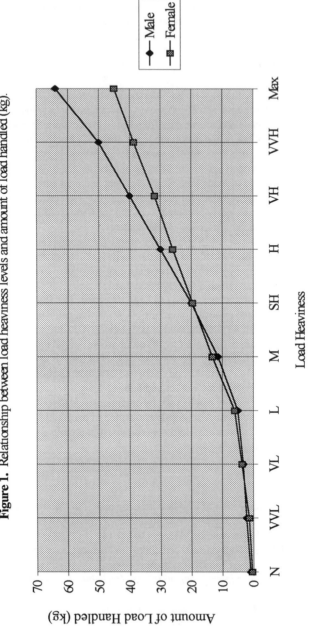

Figure 1. Relationship between load heaviness levels and amount of load handled (kg).

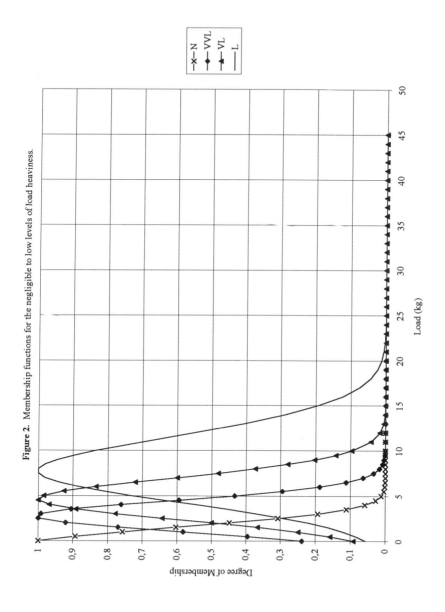

Figure 2. Membership functions for the negligible to low levels of load heaviness.

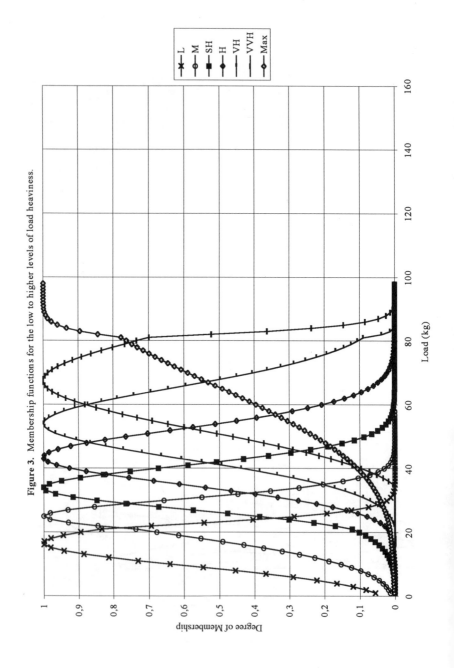

Figure 3. Membership functions for the low to higher levels of load heaviness.

Figure 4. Comparison of present study results and previous findings.

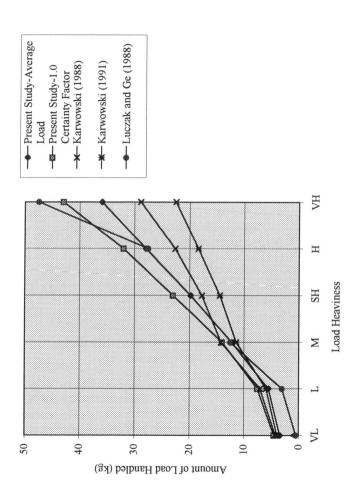

ERGONOMIC ANALYSIS OF MANUAL MATERIAL HANDLING IN CHINESE RESTAURANT KITCHENS

Simon S. Yeung, Roddy A. Ferguson, Zoe K. Siu [1] and Raymond Y. Lee [2]

[1] Dept. of Rehabilitation Sciences
The Hong Kong Polytechnic University
Hung Hom, Hong Kong
[2] School of Physiotherapy
The University of Sydney
Sydney, Australia

This study aimed at addressing the manual material handling injuries in the Chinese restaurant kitchens in Hong Kong. The approach included an ergonomic work health questionnaire to determine the workers physical profile, work nature, workers' perception and satisfaction level of their environment. This was then followed by a series of ergonomic work site analysis to evaluate the physical design of the kitchen area, the weight, nature and location of the load that the workers have to handle and posture analysis of different tasks. The work site analysis revealed that the tasks required in Chinese cooking differs from Western cooking in which lots of pan-fry activities are involved. Based on the findings of the questionnaire and the work site analysis, prevention strategy and risk control are discussed.

Introduction

Manual material handling injury is the second most common occupational accident in Hong Kong. The accident rates in the catering industry are high and account for over one quarter of all industrial accidents. The three most common injuries are manual material handling injuries (22.6%), hand tools injuries (23.6%) and injuries caused by hot or corrosive substances (23.2%) (Hong Kong Monthly Digest of Statistics, August 1995). This high injury rate has always been the concern of employer, employee and occupational health care professional. While Manual Material Handling regulations have not been implemented in Hong Kong, it is expected an ergonomic approach in the evaluation of the work environment should assist in the identification of the causes and hence prevention of manual handling injuries in this industry.

Method

The establishments and persons engaged in the restaurants and hotels in Hong Kong ranged from 1-4 people (2,830 establishments, 7,503 workers employed) to over

1,000 workers (2 establishments, 2,243 workers employed) (Manpower survey of the Hotel and Catering industries, 1991). The research team set out to sample a wide range of establishments and letters, seeking for the arrangement of the site visits were sent out to restaurants with small, medium and large establishment. However, most of the employers were reluctant to arrange the site visit. This possibly related to hygiene conditions and the sensitive nature of the food preparation process. Throughout the period of this project, the research team were able to successfully perform a site visit to 7 restaurants.

The approach adopted in this study included an ergonomic work health questionnaire and a series of ergonomic work site analysis.

Ergonomic work health questionnaire

An ergonomic work health questionnaire has been constructed to review the work health status in this group of workers. It consists of six parts which include the workers physical profile, work nature, injuries statistics, workers' perception and satisfaction level of their environment and a modified Nordic musculoskeletal discomfort survey. A similar type of questionnaire had been given to the railway track maintenance workers, and office workers in Hong Kong (Yeung 1996a, 1996b). Owning to the literacy level of this group of workers and to ensure a good response rate, the questionnaire was applied by way of interview. This was then followed by a series of ergonomic work site analysis.

Ergonomic work site analysis

The work site analysis aimed to identify different tasks involved in Chinese kitchen and evaluate the physical dimension of the workplace, the weight, nature and location of the load that the workers have to handle and the frequent work postures of the workers.

The measurement of the physical dimension of the workplace includes the height, depth and aisles space of the cooking range, height and depth of the working benches, preparation table and dim sum working table, the height and levels of shelves and steam cabinets and the height of the dish washing machine and or working table and shelves.

The research team considered that the weight of the equipment, materials and the frequency of the handling could be important factors in the cause of the frequent low back discomfort. The weight of the most commonly used equipment, utensils and the bulk items were assessed.

The Ovako Work Posture Analysing System (OWAS) (Karhu *et al*, 1977) was used to estimate the frequency distribution of varies body postures when performing different tasks. Video recording was taken during each site visit. The working postures were then subsequently observed by two members of the team. The observation was recorded at 20s. interval and the result were entered on a modified OWAS data collection form. Any discrepancies in the postures would be reviewed again to obtain a consistent working posture.

Results

There are totally 102 workers surveyed. The results of the questionnaire revealed a high incidence of injury rate of the workers surveyed. Manual material handling injury constitutes 24% of all the injuries. The steamer, barbecue cook and the butcher constitutes the highest incidence of injuries. The main concern of the work environment reflected from this group of workers include the weight of the material that they have to lift or handle, the noise level, ventilation system, hot environment as well as the slippery of the floor surface.

The musculoskeletal discomfort survey indicated that the kitchen workers had a high incidence of musculoskeletal discomfort. All the dish delivery workers complaints of one or more than one sites of discomfort. Lower and upper back are the commonest site of complaints in all the posts.

Work Site Analysis

The research team considered the main posts in the kitchen area as defined by the manpower survey of the Hotel and Catering industries (Manpower survey of the Hotel and Catering industries, 1991), the work activities within a kitchen area and classified the cook, steamer and pantry cook as three distinct activities within the cooking area. In the cleaning section, kitchen helper is defined as a worker involved in the cleaning and washing of the cooking and serving utensils. The posture analysis revealed that the Chinese cook differs from the Western cook in which lots of pan-fry activities are involved. The working postures of all the posts, except the dish washers who worked without a proper washing station, as determined by OWAS indicated all fell within the no action category (Karhu *et al.* 1977).

Physical Dimension of the Workplace

The measurement of the physical dimension of the work place includes the height, width, length and depth of the cooking range, working benches, preparation table, oven, steamer and the dish washing area. In general, the dimension does not match with the anthropometric data of the workers.

Weight of the Equipment and Materials

Most of the utensils are stored at the shelves underneath the working table or the filling shelves. The bulk items were mostly packed up on a pallet at the store room, on the floor and on the shelf. The weight of a wok ranged from 2.27-6.21 kg., an empty large soup bowl with holder weighs up to 6.42 kg. The bulky items include bulk bag of rice, sugar, potatoes and onion. These weigh 51, 30, 30 and 27 kg. respectively.

Discussion

The main concern of the work environment indicated by the group of workers studied include the weight of the material that they have to lift or handle, the ventilation system, the hot environment and the slippery floor surface. These areas need to be considered in any recommendations concerning the kitchen environment of this industry. It has to pointed out that the kitchens visited probably represent the better work environment of the catering industry. It is expected that the work environment of the smaller operations are much less satisfactory than these. The past history revealed a high incidence of injury of those workers surveyed. The lower back is the third most frequent reported injury (9%). When the injury site is classified according to post, the steamer, barbecue cook and the butcher constitutes the highest incidence of injuries. This is probably related to the work nature of the post and the improper workstation design.

Work Tasks and Postures

The Chinese cook differs from the Western cook in which lots of pan-fry activities are involved. The manoeuvre of the wok is a frequent source of complaint from the cook. In all the restaurants that have visited, the working bench sat behind the cooking range. The cook had to turn and bring the wok onto the working bench for the dish preparation. When the space between the cooking range and the working bench was too narrow, the cook usually had to twist his back instead of moving his feet to turn his body. The space between the cooking range and the working bench that had measured ranged from 66-80 cm. There should be enough room for the cook to step back and turn around onto the working bench. Posture analysis revealed that only 12% of the time the cook performed his tasks with a straight back and twisting posture. There is risk if this is combined with heavy lifting activities (e.g. when the cook is holding a large size of wok with gelatinous rice).

For the steamer, the OWAS analysis revealed no activities that need immediate corrective action. However, the complaining rate of the subjective musculoskeletal discomfort of the steamers are 71% and the back (50%) and shoulder (25%) are the commonest sites of complaints. It has to be noted that the height of the third level of the steam cabinet (165-193 cm) is too high for the average workers. The workers would have difficulties in putting the dish in and taking it out of the cabinet. This will induce discomfort to the shoulders and the back muscles. The situation would be more hazardous by the high temperature and steam within the working area.

The working posture of the barbecue cook involves a fair portion of back bent posture (16.7%) and above shoulder activities. The barbecue cook also rates a high incidence of musculoskeletal discomfort complaints (86%). The lower back (43%) is the commonest site of complaints. This relates well with the actual activities of the workers. The back bending posture could be minimized by provision of low stool when performing the pig roasting task and the cook has to encouraged to use a stable stool when performing the duck roasting task.

Sixty-three percent of the butchers reported sites of musculoskeletal discomfort. The frequent complaint sites are shoulders (33%), low back (20%) and the neck (13%). This probably related to the height of the working table which may impose additional strain when working.

The working postures characteristic of the two different work place differ significantly. Those working without a proper washing area, frequently used a back bent (41.2%), straight and twist (5.9%) and bent and twist (11.8%) posture. These are above the value of the OWAS scale and corrective action is needed. On the contrary, the back bent and twist movement are significantly less in the sites where proper washing equipment is provided. It is essential to ensure that proper washing equipment can be provided in the kitchen area.

Physical Dimension of the Work place

The main focus of the physical dimension of the work place include the height, width, length and depth of the cooking range, working benches, preparation table, oven, steamer and the dish washing area. The effect of working height in the facilities of the kitchens on the loading of musculoskeletal systems and frequency of complaints had been studied by Pekkarinen and Anttonen (1988). They showed that the symptoms were associated with the raised position of the upper limbs caused by working surfaces which were too high. The mismatch of the work spaces design with the anthropometric data in the present study may partly explain the high incident of the musculoskeletal complaints in this group of workers. Grandjean (1988) had made good recommendation for the work-surface heights for standing workers. In essence, the work surface height for heavy work should be below elbow level. For light work, it could be adjusted to around elbow level and for precision work, above elbow height or better still, to be performed in the sitting position.

Recommendations and Conclusion

While skill training and education in manual material handling is a common strategic in the prevention of manual lifting injuries in the local workforces, It is suggested that manual material handling guidelines, recommendation of work place design and risk assessment should be another dimension in the prevention of manual handling injuries in this industry. Based on the findings of this study, it is recommended that the guidelines should include areas of improvement in work environment and work station design. This should include a recommendation for the optimum working dimension for the Chinese restaurant kitchen based on local anthropometric data, the physical environment and the weight of the equipment and materials.

With the introduction of the Manual Handling Operation (MHO) Regulations 1992 in Great Britain (HSE,1992), a similar regulation will be enforced in Hong Kong in the near future. It is suggested that a risk assessment procedure in the manual material handling should be provided and adopted for the catering industry.

Acknowledgments

This study is funded by a research grant from the Occupational Safety and Health Council, Hong Kong. The research team wish to extend their gratitude to the management and the workers of the participating restaurants who kindly allowed us to perform the site visit and the work health questionnaire.

References:

Grandjean E. (1988) *Fitting the task to the man: A textbook of occupational Ergonomics, 4th edition,* Taylor and Francis pp. 36-81

Health and Safety Executive (1992) Manual Handling Guidance on Regulations L23, Health and Safety Executive, UK

Hong Kong Monthly Digest of Statistics Aug. 1995 pp. 23

Karhu O., Kansi P. and Kurinka I. (1977) Correcting working posture in industry: a practical method for analysis, Applied Ergonomics **8**: 199-201

Manpower Survey report on the catering industry (1991) The Hotel, Catering and Tourism Training Board of Vocational Training Council, Hong Kong pp. 94-110

Pekkarinen A. and Anttonen H. (1988) The effect of working height on the loading of the muscular and skeletal systems in the kitchens of workplace canteens. Applied Ergonomics 19(4): 305-308

Yeung S.S. (1996a) Ergonomics consultancy report to Occupational Safety and Health Council: Ergonomics work site analysis to the railway track maintenance workers of Kowloon Canton Railway Cooperation. Occupational Safety and Health Council, Hong Kong.

Yeung S.S. (1996b) Ergonomics consultancy report to Occupational Safety and Health Council: An office ergonomics report to the Australian Consulate Office of Hong Kong. Occupational Safety and Health Council, Hong Kong.

CYLINDER TROLLEY PUSHING TASKS

O O Okunribido and C M Haslegrave

Institute for Occupational Ergonomics
University of Nottingham
Nottingham, UK

The stresses in the use of one very common type of handling equipment - the cylinder trolley - were investigated, comparing three different orientations of handle mounted at three different heights (all within the range found on existing commercially available trolleys). Pushing forces required to start and to move forward were measured when carrying gas cylinders weighing 19kg and 37kg. Posture was recorded, measuring the wrist and elbow angles and the trolley tilt angle which the subjects chose while pushing the trolley. A biomechanical model was used to calculate the resulting stresses on the spine and arm joints. High moments were found to occur at the elbow during the starting phase, and the force needed to be exerted increased with the angle of the handle to the vertical. Handle height (length from the base) and handle angle also influenced the trolley tilt angle during movement and this had consequences for the perceptions of the stability of the trolley.

Introduction

In response to the high incidence of worker injury associated with traditional manual handling, there has been an increase in the industrial use of material assist devices "MHDs" towards limiting the workers' efforts required in direct hand pushing and pulling (Lee et al, 1991; Woldstad and Chaffin, 1994).

Although hand pushing and pulling have been studied to a considerable extent (Ayoub and McDaniel, 1974; Resnick and Chaffin, 1995; 1996), there is very little yet in the literature on two wheeled trolleys and trucks, such as are used to move gas cylinders (hereafter referred to as cylinder trolleys). One study found that many users of cylinder trolleys in industry find them unstable (the trolley not adequately supporting the weight of the load), difficult to manoeuvre and needing high hand forces to start (Mack et al, 1995). The trolleys appear still to be poorly designed.

In using MHDs, high hand forces are transmitted through the body, particularly affecting the lower back area, and awkward postures that are maintained create body strains which may lead to development of musculoskeletal disorders (Woldstad and Chaffin, 1994; Corlett and Clark, 1995).

The literature suggests that high hand forces would be likely to arise in the use of cylinder trolleys when the weight of the load is not balanced, when the weight of the empty trolley is high compared with that of the load, when the size and type of wheels are not suitable or when the handles are not well located in height (distance from the ground) and direction (angle relative to the vertical), and that poor postures would be a consequence of faulty handle location (Drury et al, 1975; Datta et al, 1978; Resnick and Chaffin, 1996). Since it is through the handles that operators directly interact with the trolley, guidance on design changes to reduce the required hand forces and improve working posture can best be achieved by investigation of different handle locations relative to the physical stresses and other factors associated with using the trolley.

Two experiments were conducted, one a starting trial and the other a movement trial. The aim of the start trial was to identify the magnitude of the forces which had to be exerted and the stresses arising from these, and to investigate the effects of handle angle and height on these. The aim of the movement trial was to identify the aspects that affect perceptions of trolley stability and again to investigate the effects of handle height and angle on these.

Methods

Subjects

In the start trials, eight healthy male subjects, aged between 22 and 32 years (26.3 ± 3.62) were involved. Their stature measured 178.25 ± 5.72 cm, and their weight measured 74.4 ± 8.60 kg. In the movement trials, ten healthy adults aged between 22 and 32 years (25.1 ± 2.85), seven males and three females, served as subjects. Their stature measured 176.55 ± 6.121 cm, and their weight measured 74.3 ± 8.53 kg. Most subjects were novices in using cylinder trolleys, but all had pushed other types of trolleys or barrows.

Apparatus

An experimental trolley, modified to accommodate portable data collecting instruments and to facilitate changes in handle height settings, was used with two representative loads, one a 37kg empty oxygen gas cylinder and the other a 19kg mock-up of the same height and diameter, ensuring the centre of gravity was in a realistic place. In designing the experiments, three different handles were constructed. Each handle could be detached from the trolley, had two separate grip ends pointing backwards at a unique angle to the vertical, and was similar in form to handles found on commercially available trolleys. Figure 1 shows the experimental set-up with the extremes of the range of experimental conditions.

A Kistler forceplate was used to determine foot forces in two axes, horizontal, (Fy) and vertical, (Fz). It was considered that the foot forces would be equal (inversely) to the hand forces when corrected for body weight. Penny & Giles twin axis M180 goniometers were used to determine maximum wrist angles (ulnar/radial deviation and flexion/extension) and left elbow flexion. Trolley tilt angle from the horizontal during movement was measured directly from video recordings.

Lower back compression force and moments at the joints were determined with a three-dimensional biomechanics program requiring inputs of recorded posture, external force, stature and weight of subjects.

Questionnaires were used to obtain subjects' judgments of effort and stability in the movement trials.

Figure 1: Extremes of the range of experimental conditions

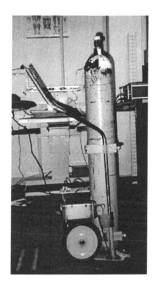 

Design

Nine experimental treatments were involved in the start trials, three heights of handle (1.0 m, 1.1 m, 1.2 m) and three angles of handle relative to the vertical through the grip end of the handle (35°, 50°, 70°), which all the eight subjects performed. The cylinder weight was 37kg. In the movement trials, ten subjects were tested with three heights (length from the base) of handle (1.0 m, 1.1 m, 1.2 m) and two angles of handle (35° and 70°) as well as two loads (19kg and 37kg), twelve experimental treatments which all the ten subjects performed.

Procedures

For the start trials, the trolley was set up in front of the forceplate. At a signal, the subject stepped on to the forceplate and maintained an erect posture with hands down by their sides for about five seconds in order to register a forceplate reading for body weight. They then proceeded to push the trolley foward, applying a force to tilt the handle down and move away from the forceplate, taking two steps. The subject was instructed to adopt natural postures while performing this task.

For movement trials, the trolley was set up behind a line defining the start point of the move. Subjects stood behind the trolley with elbow fully extended in order to provide a reference for the recorded elbow flexion angles. At a signal, they relaxed, started and moved the trolley down a designated 5 metre long path. At the end of the path, they stopped the trolley and returned it to upright position. Subjects then kept their elbow fully extended for five seconds (in order to confirm the initial reference angle) before completing a short questionnaire. Elbow flexion angle was recorded continuously from start to end of movement down the path.

Results

A summary of the principal results obtained from the start trials are given in Tables 1 and Table 2. Figure 2 presents the main results from the movement trials.

In the start trials, subjects generally considered the effort required to be heavy. The maximum force which had to be exerted vertically was in the region of 290N on average, although some subjects exerted forces as high as 550N. The maximum horizontal component of the force was 82N on average. The resulting biomechanical loads on the spine were not particularly high, as can be seen from tables 1 and 2.

Table 1: Trolley start trials - Maximum foot forces and resulting spinal loads(N)

Vertical Force Applied, F_z MEAN (SD)	Handle Height 1.0 m	Handle Height 1.1 m	Handle Height 1.2 m
35° Handle	225.6 (41.78)	197.5 (27.93)	223.1 (34.71)
50° Handle	248.9 (30.06)	244.1 (46.13)	236.4 (63.56)
70° Handle	288.6 (114.13)	250.4 (69.53)	236.1 (35.34)
Horizontal Force Applied, F_y MEAN (SD)			
35° Handle	53.5 (39.50)	65.3 (31.02)	81.8 (54.09)
50° Handle	69.5 (33.31)	49.6 (35.61)	64.0 (34.65)
70° Handle	58.3 (28.41)	71.0 (34.92)	72.9 (29.83)
Spinal Compression Force at L3/L4 MEAN (SD)			
35° Handle	447.7 (343.49)	602.5 (322.15)	504.5 (328.11)
50° Handle	771.5 (268.30)	735.6 (326.17)	502.2 (414.17)
70° Handle	683.4 (371.40)	817.9 (601.28)	546.7 (401.76)
Spinal Shear Force MEAN (SD)			
35° Handle	108.1 (63.27)	121.7 (53.94)	145.1 (79.18)
50° Handle	151.7 (46.70)	106.6 (71.40)	122.6 (60.77)
70° Handle	121.2 (38.98)	135.1 (64.30)	141.0 (58.30)

Table 2: Trolley start trials - Maximum joint moments (Nm) at elbow, shoulder and lumbar spine L3/L4

Elbow Extension Moment MEAN (SD)	Handle Height 1m	Handle Height 1.1m	Handle Height 1.2m
35° Handle	38.6 (22.60)	50.6 (20.90)	75.1 (18.32)
50° Handle	59.5 (10.85)	73.4 (20.97)	73.2 (21.57)
70° Handle	52.0 (29.00)	74.7 (30.12)	75.7 (19.54)
Elbow Adduction Moment MEAN (SD)			
35° Handle	17.1 (9.00)	18.6 (10.89)	26.4 (10.73)
50° Handle	15.0 (5.71)	22.5 (8.47)	33.5 (17.99)
70° Handle	16.6 (6.93)	20.2 (8.65)	27.9 (10.36)
Shoulder Extension Moment MEAN (SD)			
35° Handle	40.9 (26.93)	45.0 (23.96)	52.5 (26.13)
50° Handle	57.0 (21.74)	55.6 (21.86)	49.9 (00.00)
70° Handle	56.5 (24.20)	65.6 (38.24)	51.6 (26.08)
Shoulder Adduction Moment MEAN (SD)			
35° Handle	14.9 (9.46)	14.7 (7.34)	17.4 (10.85)
50° Handle	13.4 (5.95)	18.0 (5.13)	22.9 (12.79)
70° Handle	10.1 (8.32)	16.6 (5.73)	19.6 (9.07)
Lower Back Flexion Moment MEAN (SD)			
35° Handle	38.4 (31.83)	44.4 (25.23)	41.9 (27.50)
50° Handle	53.6 (27.97)	63.1 (29.69)	42.0 (39.62)
70° Handle	57.0 (36.89)	70.7 (54.86)	45.4 (32.16)

There was considerable variability between subjects and the highest compression load impressed on the spine was 1750N. The loads on the joints of the arm were, however high with mean joint moments at the elbow up to 76Nm in extension and 34Nm in adduction. The mean moments at the shoulder joint were as high as 66Nm in extension and 23Nm in adduction.

From analysis of variance, handle angle was found to have a significant effect on the vertical force which had to be applied and on elbow extension moment ($p<0.05$). Handle height was found to have a significant effect on elbow extension moment and elbow adduction moment ($p<0.05$). Wrist loads were not calculated but considerable wrist extension and ulnar deviation were observed to occur with all the subjects.

In the movement trials, the female subjects had some difficulty getting the trolley moving from rest with the 37kg load, and they all first took a backwards step. None of the male subjects needed to do this. Elbow flexion angle for all the subjects and across all the conditions ranged from 0° - 86° and trolley tilt angle ranged from 55° - 78°. However, on average, elbow flexion angle during movement with the trolley was around 45.2° when the handle was at 1.2 metres and 17.0° when the handle was at 1.0 metres. Corresponding average values for trolley tilt angle were around 62.7° and 70.4°. There was considerable variability observed between the subjects in the elbow flexion angles. From analysis of variance, handle height (length from base) and handle angle were both found to have significant effect on the trolley tilt angle ($p<0.05$), while handle height alone was found to have a significantly effect on elbow flexion angle ($p<0.05$).

Discussions and Conclusions

From the results of this study, it is apparent that high force has to be exerted during the start phase when using a cylinder trolley, with stresses on the wrists, elbows, and shoulders. The forces at the hands tend to be higher than those reported for some other assist devices, for example overhead hoists and articulating arm systems (Woldstad and Chaffin, 1994; Resnick and Chaffin, 1996) and the moments at the elbows appear to be considerable.

Figure 2: Results of movement trials

(a) 35° Handle   (b) 70° handle

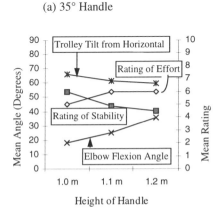

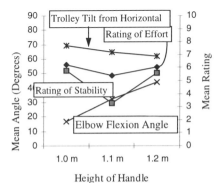

Handle angle has an influence on the vertical force which has to be exerted to pull the handles downwards and free the wheels to start moving forward. The force is lower when the handle angle is more upright. The load moments at the elbow joints are influenced by both handle angle and handle height. The lowest elbow joint moments were found when the trolley handle angle was 35° to the vertical and height was 1.0 m.

When moving and pushing the trolley forward, the effort required is lower than at the start phase but the handle design has a significant effect on the need to balance the weight of the cylinder and thus on the perception of stability of the load being pushed. Subjects tended to tilt the trolley more towards the horizontal when the handles were long and they also reported that the trolley felt less stable. Elbow flexion increased with handle height (length from the base). The observed effects were very similar for large (37kg) and small (19kg) cylinders. The effect of handle angle was not great (as seen from comparing the results in Figures 2(a) and 2(b)).

The trials have shown that proper attention to handle design for cylinder trolleys has potential for reducing both the force that must be exerted in the starting task and the effort during movement.

References

Ayoub, M.M.; McDaniel, J.W. (1974): Effects of operator stance on pushing and pushing tasks, *AIIE Transactions*, Vol. 6, No. 3, pp. 185-195.

Corlett, E.N; Clark, T.S (1995): *The ergonomics of work spaces and machines: A design manual*, Taylor & Francis, London, 2nd Edn.

Datta, S.R; Chatterjee, B.B; Roy, B.N; (1978): The energy cost of rickshaw pulling, *Ergonomics*, Vol. 21, No. 11, pp. 879-886.

Drury, C.G; Barnes, R.E; Daniels, E.B. (1975), Pedestrian operated vehicles in hospitals, *Proceedings of the 26th Annual Conference of the American Institute of Industrial Engineers*, Washington, 20-23 May, pp. 184-191.

Lee, K.S; Chaffin, D.B; Herrin, G.D; Walker, A.M. (1991): Effect of handle height on lower-back loading in cart pushing and pulling, *Applied Ergonomics*, Vol. 22, No. 2, pp. 117-123.

Mack, K; Haslegrave, C.M; Gray, M.I. (1995): Usability of manual handling aids for transporting materials, *Applied Ergonomics*, Vol 26, No. 5, pp. 353-364.

Resnick, M.L; Chaffin, D.B. (1995): An ergonomic evaluation of handle height and load in maximal and submaximal cart pushing, *Applied Ergonomics*, Vol. 26, No.3, pp. 173-178.

Resnick, M.L; Chaffin, D.B. (1996): Kinematics, kinetics and psychophysical perceptions in symmetric and twisting pushing and pulling tasks, *Human Factors*, Vol. 38, No. 1, pp. 114-129.

Woldstad, J.C.; Chaffin, D.B. (1994): Dynamic push and pull forces while using a manual material handling assist device, *IIE Transactions*, Vol. 26, No. 3, pp. 77-88.

HCl

HUMAN FACTORS ENGINEERING IN HUMAN-COMPUTER INTERFACE DESIGN

Daniel L. Welch

Consultant in Human Factors Engineering
4307 Havard Street
Silver Spring, MD 20906 USA

Appropriate and effective human-computer interface (HCI) development methodologies must be based within the framework of total system development. It is possible to survey existing, successful system, software, and HCI development methodologies and distill the essential elements of effective development process from them. This paper presents the results of such a review and distillation, providing examples of criteria for evaluating the "good practice" of an HCI development program in the following areas: General HCI Design Guidance, Task Relatedness of Requirements, General HCI Design Feature Selection, Guidelines for Detailed HCI Design, Analysis for Detailed HCI Design, and HCI Evaluations.

Introduction

The design and development of a human-computer interface (HCI) does not occur in a vacuum; rather, HCIs are developed as part of a larger system intended to carry out some function(s) or accomplish some goal(s). Therefore, methodologies intended to design and develop an effective HCI as part of a total system must be based within the framework of the total system development. In order to determine what program elements contribute to effective HCI design and development, it is possible to survey existing, successful system development methodologies and distill the essential elements of development from them. This paper attempts that survey and distillation.

A number of well defined methodologies can be identified at three levels of development process; the system level, the software level, and the HCI level. Documents which present and describe such successful methodologies are presented in Table 1. From these methodologies, a number of common program elements can be identified and these elements can be organized into nine conceptual groups:

- guidance for the HCI design process,
- the task nature of HCI requirements,
- the selection of HCI design features,
- analysis in detailed HCI design,
- the scope of the HCI design effort,
- generally desirable HCI characteristics,
- detailed HCI design guidance,
- HCI evaluations, and
- HCI design documentation.

Table 1. Well defined development methodologies, at three levels of specificity.

Total System Development Level
DoD-Level Acquisition Policy and Process
  • DoD Directive 5000.1, Defense Acquisition
  • DoD Regulation 5000.2-R, Mandatory Procedures for Major Defense
    Acquisition Programs (MDAPs) and Major Automated Information
    System (MAIS) Acquisition Programs
US Navy Computer Resource Acquisition Management
  • SPAWARINST 5200.22A, Mission Critical Computer Resources
    Acquisition Management
  • SPAWARINST 5200.22-M, Computer Resources Life Cycle
    Management Guide
Software Development Level
  • DoD Instruction 5000.2, Defense Acquisition Management Policies and
    Procedures, Part 6, Section D - Computer Resources
  • DoD Standard 498, Software Development and Documentation
  • SPAWARINST 5200.22-M, Chapter 3, Software Development Cycle, and
    Chapter 4, Software Development Approaches
  • System Engineering Management Guide Software Development Process
HCI Development Level
  • Mil-Hdbk 761A, Human Engineering Guidelines for Management
    Information Systems
  • NASA DSTL-92-007, Human Computer Interface Guidelines
  • HCI IDEA For High Definition (HDS) Application Tool (HI-HAT)
    Operator-Machine Interface (OMI) Design Process
  • International Electrotechnical Commission International Standard IEC 94,
    Design for Control Rooms for Nuclear Power Plants
  • The IBM-Hawthorne Usability Process
  • Multi-Level Flow Modeling
  • The University College London Recruiting Method
  • The Computer-Aided Process Engineering Center Method
  • The Hewlett-Packard "Champion" Method

These elements can be used to identify features contributing to effective HCI design process. The remainder of this paper presents a listing and brief discussion of the detected common elements of effective HCI design and development from six of the nine groups above.

General HCI Design Process Guidance

General guidance describing and defining the HCI design effort should be organized and documented to support standardized and consistent use by members of the HCI design team. Specific guidance should be available for defining task-related HCI design, performing general and detailed HCI design, performing HCI evaluation, and documenting HCI design. In addition, the following aspects should be present.
   1. The guidance should take a systems engineering approach, emphasizing the integration of the HCI design effort into the development of the larger software system and the total system. The HCI, as part of the overall HFE effort, should be definitively integrated into the overall systems engineering planning and effort, and should be represented as such in the HCI Design Process Guidance.
   2. The guidance should emphasize an iterative approach to design, in which an initial concept is developed, evaluated, modified based on the results of

evaluation, and re-evaluated until final acceptance and implementation. Iterative design of the HCI should be closely linked to the iterative development of the larger system and make use of early feedback from representative users. Rapid prototyping of HCI elements should be developed early in the design process and tested with actual users. The design should then be enhanced in small, incremental stages which incorporate user feedback throughout the design cycle.

3. The guidance should strongly stress that final HCI design must be directly connected to the identified functions and requirements of the total system and to each level of decomposition. The "forcing function" for decisions regarding HCI design must be its ability to support and enable the human to fulfill assigned functions and tasks. Decisions made regarding HCI design must be based primarily on this consideration and the results of such decisions, and the resulting HCI design, must be demonstrably traceable to such functional and requirement issues.

Task-Relatedness of HCI Requirements

This criteria indicates that the HCI requirements to support human functions and tasks must be identified, as part of the design process. This implicitly recognizes the systems nature of HCI design and the iterative nature of the development process. Function allocation is not a one-step process, resulting in function distributions determined early in the development process and etched in stone. Rather, function allocation is part of the iterative development process and allocations can be the object of trade studies at any time during the development cycle. If the outcome of a trade study results in the re-allocation of a given function, associated human tasks may (or probably will) change, along with the HCI design elements required to carry out those tasks. While initial requirements can be determined for design, these requirements can change and already completed designs may need to be modified to reflect those changes and support modified task requirements. Therefore, Task-Related HCI Requirements should be evaluated in terms of the following elements.

1. HCI design process guidance must reflect the need to explicitly link HCI design to task requirements. This should be a major consideration in process guidance and should reflect both the need to link (and document) design to task requirements and the need to respond to changes in task requirements (based on function reallocation or task reassignment) by (a) providing for the design of new HCI elements based on new tasks and (b) reviewing accomplished HCI designs to ensure that they are still appropriate to the new requirements.

2. A predecessor system analysis should be performed to identify high driver tasks. Existing systems similar to the developing system can provide indications of tasks that are critical in that they: (a) push the limits of human capability; (b) involve a personnel safety issue; (c) will have an adverse effect on system reliability, efficiency, effectiveness, or safety if not accomplished properly; or (d) are subject to promising improvements in operating efficiency. Tasks identified as critical through comparability analysis should be given special consideration in the HCI Functional Description.

3. An HCI Functional Description (or specification) should be used to define the requirements that the HCI must satisfy and the capabilities the HCI must possess. The functions, tasks, performance requirements and operating experience established in development activities provide the basis for defining an HCI functional description. That functional description can then form the documentation basis to guide the process of (a) selecting general HCI design features (b) developing task required display elements, and (c) developing detailed design concepts.

4. There should be patent traceability of requirements to design. Requirements allocated from the system specification involving HCI should be directly traceable to the HCI Functional Description and from there to design

elements of the HCI. There should be documentation indicating the decision making process by which requirements were translated into specific design elements such as the inclusion of a particular display or controls, or the nature of the display or control.

General HCI Design Feature Selection

This criterion essentially corresponds to the preliminary design stage of the general system development model. The main concern during preliminary design is the translation of requirements into broad system elements. This process may involve decomposing functions into sub-functions, which in turn can be associated with more precise aspects of the system and thus aid designers in developing detailed designs.

1. The selection process should directly link human functions and tasks to each General Design Feature under consideration. Beyond achieving the broadly desirable HCI characteristics, General Design Features must first support the task-related requirements established through system and task analysis. Initial inclusion of a feature should be directly and explicitly linked to those task-related requirements. This linkage is imperative in order to evaluate and determine the most appropriate design alternative for each feature.

2. More than one design alternative should be proposed for each General Design Feature. It is difficult to determine the most effective design alternative for a particular purpose (i.e., satisfaction of associated task-related requirements and desirable HCI characteristics) if only one alternative is proposed. As the number of initial design alternatives increases, the more creative can be the final solution.

3. Design alternatives for each General Design Feature should be subjected to head-to-head evaluation. Operating experience and literature analysis, engineering evaluations and experiments, and benchmark evaluations are of little value in determining the relative merit of General Design Features if those features are evaluated only in relation to themselves and are not compared to each other. Trade studies, employing these and other methods as necessary, should be employed to directly compare specific feature design alternatives, in light of the appropriate evaluation criteria established above.

Guidelines for Detailed HCI Design

This criterion indicates that the design effort should employ tailored HFE guidelines during the detailed design of the selected general HCI features. The criterion also indicates that the guidance should be documented into a guidance document. Guidelines for detailed HCI design should be evaluated on the following elements.

1. A formal, written guidance document should be available to those who are developing the HCI. Traditional HCI human factors guidelines, such as those found in MIL-HDBK-761A and NASA DSTL-92-007, should be employed during the detailed design of the HCI. Guidelines to be employed should be clearly identified, the rationale for their specific selection should be discussed, and the set should be tailored to fit the unique requirements of the specific application. The method(s) by which raw guidelines are translated into the HCI design specification should be described.

2. The guidance document should be appropriately tailored to the current design effort. Simply amalgamating General HCI design guidelines is inefficient and may actually negatively impact design. Tailored guidance should be based directly on (and be traceable to) comparability analysis results and the HCI functional description developed previously.

3. The guidance document should address both static and dynamic design characteristics. During the HCI design process, guidance will address differing aspects of the HCI and should include, as a minimum: (a) functional guidance,

addressing the capabilities the HCI must possess and the requirements the HCI design concept must satisfy; (b) design specifications, containing explicit requirements for detailed design elements; and (c) interaction guidance, describing how user transactions with the system are to proceed.

4. The guidance document should be appropriate for its intended users in terms of scope, depth, and complexity. Guidance must be appropriate for intended users, i.e., the personnel who will do the detailed HCI design. That may be HCI specialists or it may be other engineering personnel with responsibility for HCI design (though, hopefully, with some HFE experience and knowledge). Much current HFE guidance is written with HFE specialists in mind and must be re-expressed to be effectively communicated to non-HFE personnel.

Analysis for Detailed HCI Design

This criterion indicates that analysis methods, such as operating experience and literature reviews, tradeoff studies, engineering evaluations and experiments, and benchmark evaluations, should be employed during detailed design in cases where problems, issues that are not well defined by guidelines, or conflicting guidelines are encountered, or for design details. Mockups, models, and dynamic simulations and HCI prototypes are mentioned as examples.

1. The goals of design analyses must be adequately and effectively expressed and documented. Analysis for analysis sake is of little value; every analytic effort should have a well defined question in mind and that question should be directly linked to the system design. Both the overall approach (literature review, engineering experiment, benchmark evaluation) and the implementing technique (mockup, prototype, computer simulation) should be appropriate to the analytic question.

2. The choice of independent variables must be appropriate and the choice governed by a formal selection technique. The design of any study first consists of choosing appropriate levels of all appropriate independent variables. A formal selection technique provides a means to determine which factors can affect system performance and what to do with each factor.

3. The choice of dependent variables must be appropriate. Human performance in complex systems is multivariate by nature. For that reason, design analyses should avoid examining individual dependent variables and should simultaneously examine indications of efficiency, workload, secondary workload, error proneness, etc. Selected dependent variables should be demonstrably appropriate in terms of the criteria of validity, reliability and sensitivity to the issue.

4. The proper degree of fidelity must be available to the analysis. The use of mockups, models, rapid prototypes and dynamic simulation raises the issue of required fidelity. In a developmental situation, it is not always possible to completely simulate the abuilding system, for the simple reason that it may not be entirely designed yet. However, for many analysis applications, complete fidelity to the total system may not be necessary and may, in fact, be unnecessarily expensive in terms of dollars, time and effort. It is therefore important that decisions regarding analytic fidelity be based on, appropriate to, and traceable to the objectives of the analysis.

5. There should be an objective method for developing tradeoff criteria weighting factors. The outcome of a trade study can be manipulated easily, simply by careful selection of criterion weights (and even more simply by their ex post facto assignment). This can be avoided by using a technique that presents all trade criteria to a number of subject matter experts for pair-wise comparison. The normalized results of these comparisons can then be subjected to eigenvector analysis to determine the relative weightings of all components within the set of criteria.

HCI Evaluations

This criterion requires on-going evaluation of the HCI to ensure acceptability in terms of guidance and functional design. It is noted that special consideration should be given to unique or safety-related HCI elements, and that identified issues or problems should be analyzed as part of problem resolution. The entire HCI should be evaluated to insure appropriate information and control availability and all evaluation activities should be documented.

The distinction between testing as part of design, and evaluation occurring after design should be clearly articulated. A similarly clear articulation should be made between the results of analyses conducted during detailed design, HCI functional evaluation, and user acceptance evaluation. Analyses are directed towards specific design details, issues or problems and imply the search for design solutions. Evaluations are generally directed toward complete HCI elements or the total system HCI and are concerned with usability issues (both user acceptance and functional issues). Identified usability problems are then subjected to analysis and the results employed to correct the problem and enhance usability. Assessment of HCI evaluation efforts should consider the following elements.

1. Scheduled evaluation reviews should be programmed into the design and development effort. "On-going evaluation" does not imply an informal or ad hoc approach. Formal developmental testing of HCI elements should occur shortly after the "completion" of the detailed design of an HCI element (realizing that detailed design is a cyclic design - evaluate process). Formal evaluation reviews, in which the status of the developing HCI is examined in detail, should be scheduled and conducted at appropriate intervals.

2. Rapid prototyping should be employed to permit evaluation of HCI elements prior to design finalization. Rapid prototyping is an approach in which a model of the HCI element is developed quickly, and should require less extensive design, review and documentation than normal. The stated goal of the use of rapid prototyping for evaluation should be to encourage an effective interaction between the user and the designer, in order to produce an HCI element that both meets functional design requirements and achieves user acceptance.

3. HCI evaluation should be based on (or at least heavily include) user acceptance testing. Adherence to HCI guidance and meeting system functional requirements do not ensure usability. User acceptance tests are conducted to ensure that the HCI design approach and elements meet user expectations while meeting acceptance criteria for complexity, ease of learning, retention, transparency, directiveness and workload

Conclusions

The process of translating requirements into hardware and software elements of the HCI is the act of *design*. While extensive guidance and often rigid procedures and techniques are available for the process of requirements definition and analysis, function specification and analysis, job and task analysis, and test and evaluation, little, if any, real guidance is available regarding the *process of design*, the creative act. Since the actual process of design concept creation is difficult to define, the optimization of HCI design process must draw on elements and aspects of the system development process in general, which are commonly accepted as essential to an effective effort.

This paper has presented a number of design and development elements which are commonly applied in successful system, software, and HCI developments. Their application in the systematic HCI design process should enhance the process itself and aid in the optimization of the resulting human-computer interface.

INERT KNOWLEDGE AND DISPLAY DESIGN

Fiona Sturrock and Barry Kirwan[*]

Industrial Ergonomics Group
School of Manufacturing & Mechanical Engineering,
University of Birmingham,
Birmingham, B15 2TT. United Kingdom.

This study aimed at analysing a nuclear power plant scenario in terms of the total knowledge required to diagnose it and also the knowledge/information that was available via the interface. Six types of knowledge were used to categorise such information/knowledge. The results revealed that although the scenario was a classic training scenario the interface did not fully support the operator's knowledge requirements in order for a successful diagnosis to take place. Recommendations for improved interface design were made on the basis of the required 'type of knowledge' for the disturbance.

Introduction

The VDU interface is becoming the primary source of information concerning the status of the process, especially in advanced technological environments, such as recent nuclear power plants. The design of this interface, in process control, will therefore largely determine whether the operator can access and utilise the appropriate information. This is of particular importance during disturbances when the operator must interrogate the system, and aggregate and integrate often incomplete, ambiguous and continuously changing evidence in order to generate a diagnosis of the situation (Vicente, 1995b, 1989; Woods *et al*, 1994; Rasmussen and Vicente, 1989). The diagnostician (the operator) therefore must remain cognitively flexible in order to modify the diagnosis of the situation with the continually accruing evidence.

At present interface displays are designed according to what designers believe is important, and based on system diagrams and functional information, with some input from the operators. From this operator involvement side, however, it is a fairly unstructured approach for which there are very few detailed guidelines. If the interface is of poor design it could potentially 'hide' vital pieces of diagnostically relevant information. Alternatively the presentation of information may not 'trigger' the utilisation of important knowledge, i.e. knowledge that is relevant and appropriate for the

---

[*] Currently Head of Human Factors at ATMDC, NATS, UK

circumstances, and as a result the operator's diagnostic processes will be impaired or hindered. In either case, as far as the joint cognitive system (operator and machine) is concerned, it is inert (Woods *et al*, 1991).

What is placed on displays, due to the huge amount of system information and facts available, is inevitably only part of the total information or knowledge available, and other information must reside in other locations, such as procedures and manuals, and in the operators' minds. Although such an arrangement may be acceptable for routine conditions it may not be so for abnormal situations. During an abnormal or novel disturbance there is a danger that key information is not retrieved or triggered from the operator's memory, or there may not be time or inclination to access procedures, and such procedures may be difficult to navigate through, to find the right information, or simply may not contain the required information. Therefore there are important questions of what to display, what to leave potentially inert, and how to reactivate inert knowledge. These questions have been explored in the work reported in this paper for a particular nuclear power plant scenario (Steam Generator Tube Rupture). It has been found useful to clarify the knowledge elicited into knowledge types, and to relate certain of these types to types of VDU display formats.

Types Of Knowledge (TOK)

A brief description of each of the six knowledge types used in this study is given in Table 1 below. The reader is directed to Sturrock and Kirwan (1996) for a more detailed account of the different 'types of knowledge', which are still undergoing development and validation.

Sturrock and Kirwan (1996) found that the 'types of knowledge' were a useful method with which to categorise the cognitive behaviour of nuclear power plant (NPP) operators during diagnoses of simulated emergency events. The remainder of this paper applies the 'types of knowledge' approach being developed to one of the most important failure scenarios in nuclear power, namely Steam Generator Tube Rupture, to see what insights can be gained on interface representation of required knowledge.

Table 1. Description of knowledge types

Knowledge Type	Characteristics
1	Textbook theory; no practical experience/practice; general NPP knowledge, including functional information and goals of the process.
2	More thorough understanding of the process; practical experience; general fault knowledge; understanding of the logical inter-relationships of the plant systems.
3	Detailed mechanical (functional) knowledge; knowledge of cause and effects; normative knowledge (may also be held by designers and maintenance personnel).
4	Maximum integration of systems, i.e. how changes in one system affect another system.
5	Contextual knowledge; detailed plant layout knowledge; useful for common cause failures.
6	Tactical knowledge; knowledge of specific faults and multiple faults; problem solving skills and realisation of prompt and delayed consequences.

Analysis of the Steam Generator Tube Rupture (SGTR) scenario

A steam generator tube rupture scenario in a nuclear power plant, if undetected, could lead to the radioactive contamination of the secondary side of the nuclear process, which could threaten the lives of the plant personnel and eventually the environment. In Pressurised Water Reactors (PWR) there are four steam generators each containing hundreds of small tubes. If one of these tubes breaks, for what ever reason, then the radioactive water from the primary side contaminates the 'clean' secondary side coolant. If the leakage is small this can be a very tricky scenario to diagnose because the secondary side regulatory system compensates for the leak, hence masking the effect of the leakage.

The first stage of the analysis took the form of an in-depth and exhaustive scenario exploration and knowledge elicitation using both the relative abundance of literature on this scenario (compared with other NPP scenarios) and a very experienced PWR operator, and also a full-scale PWR simulator at the OECD Halden Reactor Project, Norway. A combination of several knowledge elicitation techniques (interviews, walk and talk-throughs, verbal protocols, eye movement tracking) were used as no one technique was considered adequate for the analysis of this complex scenario.

From this data collection it was possible to categorise the knowledge requirements of the task according to what 'type of knowledge' the operator would require in order to diagnose the disturbance and the way in which the interface represented the necessary information as a function of the total information potentially available to the operator. Therefore it was possible to determine what information was inert, i.e. information $_{(total)}$ - information $_{(available)}$ = information $_{(inert)}$. As well as defining knowledge that is inert because it is missing, certain information will be inert because of low salience or because of low relevance to the apparent content of the scenario. Additionally some knowledge may be inert because it is presented at the wrong level, e.g. too detailed or not detailed enough. The operator may require more detailed mechanical or geographical knowledge, or may require an overview display (relating to 'types of knowledge' 3, 5, and 2 respectively). The 'type of knowledge' categorisation system was used to focus on some of these latter aspects.

Results of the SGTR Analysis

Table 2 shows the results from some of the analysis. Due to the restrictions of the length of the paper only a selection of the analysis is shown. Table 2 shows four of the main symptoms required by operators in order to diagnose the SGTR scenario and how such symptoms are presented on the interface (if at all) and whether such symptoms are considered, by the operator, to be salient and relevant in the perceived context of the disturbance. The information presented by the interface was also categorised according to what 'type of knowledge' (TOK) is being represented. Table 2 also shows what knowledge the operator requires in order to acknowledge each symptom and what 'type of knowledge' this can be categorised as. The 'inert' column contains the information/knowledge that the operator requires but which is not displayed on the interface. This is classed as inert knowledge as it is expected that the majority of licensed operators will have the required knowledge but because it is not 'triggered', via cues from the interface, it remains unused (or inert) within the context of the disturbance.

... scenario

Key symptoms for diagnosis	What knowledge does this require?	Visible on interface?	Relevance/ salience of symptoms	Inert knowledge	Interface re-design concepts
Fluctuations in flow through secondary side (SS) regulating valve & closing of feedwater (FW) valve	Knowledge of what flow rates should be (TOK 3). Knowledge that valve is a regulatory valve which will automatically alter to keep flow constant in SS, i.e. closes in response to increased inventory in SS (TOK 3). Key symptom to diagnose leak from primary side to SS (TOK 4).	Shown as regulating valve closing as an increasing percentage value (TOK 2 and 3). Trends formats also available to show increased flow in secondary side (TOK 2 and 3).	Low salience (no alarm) High relevance	Not obvious that anything is changing. Knowledge of inter-relationships, e.g. why is the valve closing, what effect this will have on other parts of the plant (TOK 4).	Function key which highlights all changing or changed parameters during a specified time period. Component or parameter flashing to indicate changes. Alarm message.
Increasing steam generator (SG) level	Knowledge that steam output will remain constant due to the manifold, even if one SG is increasing (TOK 4). Knowledge that if there is an increase in SG level and the regulatory valve cannot cope then the leakage must be large (TOK 4).	Shown as increasing digital flow value in damaged SG (not shown graphically) (TOK 2). All four SGs are on one format which makes comparison of flow rates easier (TOK 2).	Low salience (no alarm until it is too late) Medium relevance	Not obvious that the level is increasing as output remains constant (TOK 4). Knowledge of inter-relationships, i.e. what increasing SG level affects in the plant (TOK 4).	Function key which highlights all changing parameters during a specified time span. Flashing steam generators to indicate change. Highlighting the increased input despite the constant output.
Radiation alarm in secondary side	Knowledge that this distinguishes a SGTR from a Loss Of Coolant Accident (LOCA) (TOK 5 and 6).	Shown as a text alarm (TOK 3). Appears as a radiation value on the display (TOK 3).	High salience (alarm sounds, but takes some time) High relevance	Unless operator reads alarm it may be missed and as it is presented on a different format, i.e. remote from the actual fault and it may not trigger the appropriate responses, or knowledge from memory (TOK 5).	Radiation symbol (flashing or static) on the display in the location of the radiation detection as well as a text alarm, i.e. a visual cue of radiation and it's location.
Absence of increased pressure inside containment	Knowledge that a key symptom of a LOCA is increased pressure inside the containment, and a key symptom of a SGTR is the absence of such an alarm and symptom (TOK 6). Knowledge that if the leak is inside the containment then eventually the pressure will increase, although this may be gradual, especially if the leak is a small one (TOK 6).	Conspicuous by its absence	Low salience High relevance	This will be the same as the knowledge required as there is nothing to make them look at the containment pressure (TOK 5). It is on a display format that is remote from the main system parameters.	Highlighting changing or unchanging parameters. Function key which collates all major parameter changes, i.e. a memory aid which may trigger a diagnosis from long term memory. Suggestions of fault symptom patterns and checklists of parameters to check for each fault pattern. Fault symptom matrices.

The final column in Table 2 gives some suggestions of how the interface could be re-designed in order to support the knowledge requirements of the operator. Such suggestions are aimed at triggering the required knowledge from memory, i.e. reactivating the inert knowledge within the context of the disturbance.

The SGTR scenario is a classic fault with operators being trained to recognise the symptoms and to take corrective actions. However, the analysis of the scenario revealed that the knowledge or information required in order to diagnose the scenario is not optimally supported by the interface, e.g. many of the main symptoms of the scenario are of low salience, despite their relatively high relevance to the context of the disturbance. If an operator's knowledge remains untriggered or inactivated (from long term memory due to the design of the interface) for a classic scenario it is likely that the interface may fail to reactivate the knowledge required for other scenarios. This may be particularly problematic for novel or abnormal disturbances.

Using the above type of analysis, i.e. where the knowledge requirements of the task are taken into account, it is possible to identify potentially inert knowledge that may result in an unsuccessful diagnosis of the scenario, and then to make potential interface design recommendations.

The major problem areas, and those where inert knowledge may affect diagnosis, are likely to be those areas where the type of knowledge represented by the display does not match that required by the operators. Table 3 comprises candidate 'type of knowledge' recommendations for the design of more supportive interfaces, suggesting that different 'types of knowledge' may require different formats of information representation.

Table 3. Interface redesign recommendations associated with types of knowledge

TOK	Interface design recommendations
1	Function displays, e.g. mass or energy transformations
2	Overview displays
3	More detail, e.g. on pump mechanics
4	Need for more integrative displays showing system inter-relationships
5	Geographical, topographical, or *contextual* information
6	Highlighting heuristic inter-relationships such as common mode failures; fault symptom matrices

Benefits of 'Type Of Knowledge' for Interface Design

Designing interface displays for complex and dynamic processes generally follows engineering conventions, with little structured input from the users of the interface, the operators. In addition there are very few guidelines available to aid designers of such interfaces, which tell the designers how to represent different functional information in an integrated way for flexible and adaptive use. As a result designers tend to design a flexible general purpose display which requires that the user can adapt his/her cognitive and physical behaviours depending on the context (Woods et al, 1994; Vicente, 1995a).

The process of identifying, and categorising the knowledge requirements, according to the six types of knowledge in Table 1, potentially results in a more structured approach to display design which places more emphasis on providing operators with more usable information. Figure 1 shows a proposed process for informing display design, based on TOK analysis.

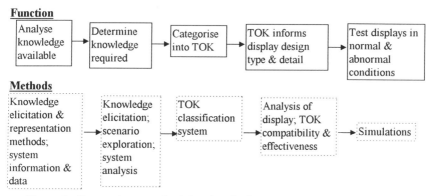

Figure 1. Proposed process for informing display design based on TOK analysis

Summary

The results of this analysis, briefly reported in this paper, show that there is potential utility in categorising the knowledge required to diagnose a NPP scenario in order to structure the content of the displays. The analysis found that displays for the SGTR scenario were not fully supportive of the knowledge required in order to diagnose such a disturbance. The next stage of the research is to consider in more detail the different display types available, and relate these to the types of knowledge. Once such relationships have been established, new displays for the Steam Generator Tube Rupture scenario can be developed and tested in comparison to the traditionally-derived displays. The end goal of the project is to develop a task analysis tool, based on types of knowledge analysis, which can be used prospectively for determining interface requirements for operator diagnostic support and also for generating recommendations for display design.

References

Sturrock, F., and Kirwan, B., 1996, Mapping knowledge utilisation by nuclear power plant operators in complex scenarios, *Proceedings of the Ergonomics Society's 1996 Annual Conference* (Taylor & Francis), 165-170.

Rasmussen, J., and Vicente, K. J., 1989, Coping with human errors through system design: Implications for ecological interface design, *International Journal of Man-Machine Systems*, **31**, 517-534.

Vicente, K. J., 1995a, Ecological interface design: A research overview. In T. B. Sheridan (Ed.), Analysis, Design, and Evaluation of Man-Machine Systems, Pergamon Press.

Vicente, K. J., 1995b, Supporting operator problem solving through ecological interface design, *IEEE Transaction on Systems, Man, and Cybernetics*, **25**, 529-545.

Woods, D. D., 1991, The cognitive engineering of problem representations. In G. S. R. Weir and J. L. Alty (Eds.), Human-Computer Interaction and Complex Systems, (Academic Press. London) 169-188.

Woods, D. D., et al, 1994, Cognitive System Factors, State of the Art Report, Behind Human Error: Cognitive systems, computers and hindsight, CSERIAC 94-01.

Acknowledgements: The authors would like to thank both the HRP staff and the operator for their co-operation, time and enthusiasm. A particular thanks to Erik Hollnagel.

Disclaimer: The opinions expressed in the paper are those of the authors and do not necessarily reflect those of their respective organisations nor the HRP.

SIMULATION AS A TOOL FOR ENGINEERING EDUCATION - AN INTEGRATION APPROACH

Kai Cheng
Department of Engineering, Glasgow Caledonian University
Cowcaddens Road, Glasgow G4 0BA, U.K.

W.Brian Rowe and S.S. Douglas
School of Engineering and Technology Management, Liverpool John Moores
University, Byrom Street, Liverpool L3 3AF, U.K.

In this paper a novel approach is proposed on applying simulation to computer assisted learning and teaching in engineering education. The approach concerns with the integration of simulation, hypermedia and artificial intelligence (AI) technologies with the conventional education techniques such as lectures, group working, individual study, tutorials and laboratory experiments. The approach is presented in detail through the development of a simulation library with its hypermedia-based course guides for control engineering. The design of the media was focused on the interactive manipulation, comprehensive exploration and progress feedback which are essential for engineering subject in particular. The cognition and human-computer interaction issues involved in the development of the library and course guides are also explored.

Introduction

Simulations have been proved an effective tool for engineers to solve a variety of problems in engineering systems and processes. It has been suggested that simulation, hypermedia and artificial intelligence will enrich higher education in the 21st century (Sculley, 1989). The potential of using simulations in engineering education is increasingly being recognized and explored by educationalists from engineering disciplines (Smith and Pollard, 1986; Magin and Reizes, 1990; Eaton, 1992). Using one or more realistic simulations, students can investigate an engineering system or process that would be too large, complex, expensive, or hazardous to be operated in the university laboratories. However, comprehensive research need undertaking on how designing simulations and using them in courses in the light of enhancing the quality, effectiveness, flexibility, efficiency and productivity of engineering undergraduate provision.

The main purpose of this paper is to explore the educational advantages of simulation technology for engineering application. The relationship between simulation based media and teaching/learning is discussed with particular reference to supporting a variety of teaching/learning activities using simulation based courseware. The teaching and learning trials have demonstrated that a well designed simulation based courseware, with support from techniques of artificial intelligence (AI) and hypermedia, can provide great potential enhancement to teaching/learning practice of an engineering subject.

Engineering education and simulation

Engineering education

There are five traditional types of teaching and learning activities in engineering

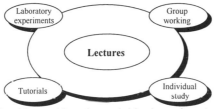

Figure 1. Teaching and learning activities in engineering education

education as shown in Figure 1. These activities are lectures, laboratory experiments, tutorials, group working and individual study. Lectures are usually regarded as the main activity with enhancement provided by other activities. However, the traditional viewpoint of engineering lecturers is currently challenged in engineering education. The situation is being changed for the following reasons (Dixon, 1991; Dinsdale, 1991):

- Students have little or no engineering background knowledge and experience.
- Staff-student ratio is worsening.
- Increasing numbers of students enter with weak analytical and mathematical abilities.
- Emphasis on the 'knowledge' content of subjects rather than on the development
  of skills and deep understanding is recognized as inadequate.
- Staff have less time to deliver lessons.
- Fewer hours are available for student contact.

Use of technology-based teaching and learning, such as simulations, multimedia coursewares and AI based learning tools, may be an appropriate approach to address the problems described above. A computer based learning (CBL) environment can be enriched through the provision of a variety of multimedia style materials such as text, graphics, image, animation, simulation, video and audio. Materials for these five teaching and learning activities can be incorporated into one technology-based CBL environment which has the potential to offer:

- a learning environment in which students are actively involved in the learning process and through which a deeper understanding of the principles, concepts and practice may be achieved.
- a shift from predominantly passive, lecture-centered delivery to active student-centered learning.

• different learning rates.
• provision for diverse backgrounds by access to wide-based learning materials..
• a way round artificial boundaries between disciplines and subjects.
• a 'safe' environment through animation/simulation modes to explore hazardous or expensive experiment activities.
• more enthusiasm in staff and students through the environment-learner interaction. Enthusiasm is a crucial ingredient for successful learning.

Simulation

A simulation is a computer program which incorporates a mathematical or logic model of an engineering system or process, allowing the user to specify the values of one or more system parameters and, following computation, to examine the resulting values of other system parameters. As shown in Figure 2, a simulation can be a realistic tool for engineers to bridge the gap between the real world, physical and mathematical modeling,

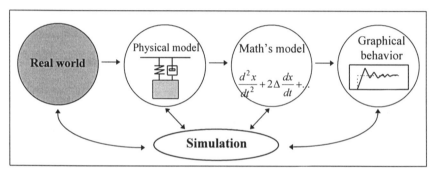

Figure 2. A simulation used to solve a real world problem

and the results from the world. In the education context, simulations can be well suited to play a complementary role for all of the following educational activities:
• extension of laboratory activities especially these are too expensive or too hazardous to be made available in the laboratory.
• support for lectures and tutorials with the numerical or graphical presentation of elegant solutions on the simulated systems or processes.
• teaching and learning for individual or group students through manipulating and deep exploring with simulations.

Interaction with a simulation-based CBL courseware may result in a better understanding of the engineering principles taught because of its flexibility of nonlinear and interactive access to information and ability to foster exploration of relevant information in different media. Simulations that combines abstract and conceptual learning help to create an intense, content-rich learning opportunity grounded in experience and reality.

Simulated animation is also a simulation technique well used in engineering courseware. A simulated animation sequence can supply a learner with information concerning with physical movement, changes over time, motions within three dimensions which are essential in the representation of engineering objects and processes. In the education context, simulated animations can:

• immediately identify what objects are available.
• clearly show the orientation of actions on these objects.
• seamlessly link learner input and animated demonstration or response.

Given the current number of high quality simulation tools, it is becoming increasingly easy to develop CBL coursewares including simulated animation sequences.

Exemplars of using simulation

The authors have developed several simulation based coursewares including significant simulations and animation sequences (Cheng, 1993; Cheng et al, 1996). Each courseware uses text information as a structural spine as shown in Figure 3. A 'book' paradigm is superimposed on the 'card' paradigm to assist in navigation. The text is divided into chapters and cards which correspond to pages. Other medium information, such as graphics, images, animation and simulation, is added on as required. These added media are authored in hypermedia node-link techniques and incorporated with the text information to enhance the knowledge representation and transfer. The text information is based on the lecturing notes. Other media, especially, simulations and animations are programmed to provide learners with scope to undertake a deep interactive study.

Figure 4 illustrates a simulation operation copied from an engineering dynamics & control courseware developed. As shown in the figure, the simulation concerns a first order water-tank system. A learner can vary the variables such as the system 'Cooling Factor', 'Water Volume' and 'Power' and select the heating input modes which include 'impulse', 'step', 'ramp' and 'sinusoidal'. Then the learner can manipulate the simulation as set up and see the simulated temperature outputs both graphically and numerically. The learner can also access the 'theory' and 'assessment' modules inside course guides shown at the bottom screen. The 'theory' module includes specific background knowledge about the simulated system. The 'assessment' module is programmed using AI techniques and used for a learner to receive feedback on his or her progress during and after a learning session. The interactive manipulation can be undertaken and repeated

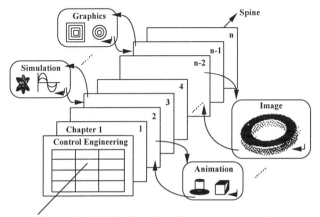

Figure 3. A technology based courseware structure

as a learner requested until the learner receives a satisfied learning outcome. The simulation and course guides are implemented respectively using LabVIEW and HyperCard.

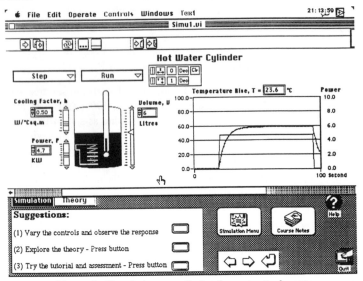

Figure 4. Simulation on a first order water-tank system

Table 1. Some of the guidance for designing a simulation based courseware

Illustration of physical situation:
• can vary main variables
• achieve simplest presentation
• be easily understandable
• eliminate clutter
• standardize variable controls, e.g. buttons and slides
Presentation of mathematical modeling:
• focus attentation at one point
• eliminate text where possible
• highlight parameters consistently with physical situation
• achieve simplest presentation
• be easily understandable
Provision of graphical output and learner interaction:
• complement modeling
• reflect effect of manipulations on the system
• highlight key input/ output parameters
• achieve simplest presentation
• be easily understandable
• achieve simplest presentation
• add new understanding
• eliminate clutter

Discussions on using simulation

The approach used by the authors to the design of simulation based courseware focuses on the aspects of concepts understanding, physical modeling, simulated presentation, and learning outcomes in engineering teaching and learning. Table 1 lists some of the guidance for designing a simulation based courseware. The guidance is oriented from the authors' first hand practices on courseware development and research.

Concluding remarks

In this paper, the relevance of simulation with computer based learning is explored concerned with the learning process and outcome in engineering education in particular. Guidelines are formulated for the design and use of simulations in a technology based courseware. The exploration appears using simulation is particular appropriate to enhance the learning activities such as lectures, laboratory experiments, tutorials, group working and individual study which are essential in engineering education.

References

Cheng, K. and Rowe, W.B. 1993, A hypermedia-based tutoring system for education in mechanical engineering, Proceedings of The 30th International MATADOR Conference, UMIST, UK, 31 March - 1 April 1993, pp.277-284.

Cheng, K., Rowe, W.B. and Douglas, S.S. 1996, A developer based approach to applying hypermedia in engineering education, Proceedings of the 13th International Conference on Technology and Education, New Orleans, Louisiana, U.S.A., 17-20 March 1996, Vol. 1, pp.248-251.

Dinsdale, J. 1991, Engineering design education, Annals of the CIRP, Vol.40/2, pp.595-601.

Dixon, J.R. 1991, New goals for engineering education, Mechanical Engineering, March, pp.56-62.

Eaton, J.K. 1992, Computer-based, self-guided instruction in laboratory data acquisition and control, Proceedings of the 22nd Annual Conference on Frontiers in Education, 11-14 November 1992, Nashville, Tennessee, U.S.A., pp.809-813.

Magin, D.J. and Reizes, J.A. 1990, Computer simulation of laboratory experiments: an unrealised potential, Computer and Education, Vol. 14 No. 3, pp.263-270.

Milheim, W.D. 1993, How to use animation in computer assisted learning, British Journal of Educational Technology, Vol. 24 No. 3, pp.171-178.

Sculley, J. 1989, The relationship between business and higher education: a perspective on the 21st century, Communications of the ACM, Vol. 32 No. 9, pp.1056-1061.

DISCLOSING DIFFERING INTERPRETATIONS OF USER NEEDS: AN APPLICATION OF SOFT SYSTEMS METHODOLOGY

Kieran Duignan

InterFaces, 84 Alderton Road, Croydon, CR0 6HJ, UK

M. Andrew Life

*Ergonomics & HCI Unit, University College London,
26 Bedford Way, London WC1H 0AP*

Soft systems methodology was the tool chosen to address an unstructured problem of eliciting user requirements of an Information Technology system in an Occupational Health and Safety Unit in a large public sector organisation. The methodology is characterised. Its application is described up to the point of disclosure of contrasting patterns of thinking which support alternative perspectives on user requirements. The value of such an analysis is briefly discussed and some implications for use of the methodology in a human-computer interaction context are considered.

Introduction

Soft Systems Methodology (SSM; Checkland, 1984) originated outside the domain of information technology (IT) as a tool to support organisational change. Patching (1990) very briefly describes its application to exploring user requirements of an IT system in the Social Services department of a County Council in the U.K. To explore the potential of the method for wider use in human-computer interaction (HCI), it was applied to elicit interpretations of user needs, so that differences in assumptions of policymakers are brought out clearly.

The approach of 'Soft' Systems Methodology

Systems thinking generally was characterised by Checkland (op.cit.) as an epistemology which, when applied to human activity, is based on four basic characteristics of a system: emergence, hierarchy, communication and control. Together these characteristics endow any system, including the human activity systems of an organisation, with both structure and dynamism. Checkland chose the expression 'soft systems' to highlight the relevance of the methodology he developed to tackling real-world problems in which known-to-be-desirable ends *cannot be taken for granted*. This he contrasted with a 'hard systems' methodology, appropriate for tackling real-world problems in which an end-to-be-achieved *can* in practice be taken as given, as the basis for engineering a system to achieve a stated objective.

The model of human functioning implicit in SSM is one in which individuals and groups negotiate their interpretations of reality in the process of solving a problem, while these multiple interpretations at the same time constitute social and organisational reality. As a result, interventions of an analyst are not limited to those of a primarily technical kind. The task of an analyst applying SSM calls for deploying each of the five styles of intervention identified by Blake and Mouton (1976): acceptant, catalytic and confrontational as well as technical i.e. prescriptive and didactic (to do with theory and principles). The interventionist is actively required to draw on each of these styles of intervention to ensure adequate characterisation of similarities and differences in interpretations of reality by different groups in the problem solving process.

The philosophy guiding SSM is one of 'Action Research': it is intended to bring about change in the problem situation while the participants learn from the process of change. The 'Action' orientation of SSM implies that the appropriate criterion of evaluation is of a reflexive nature. The criterion of evaluation appropriate to the methodology is not the production of 'objective' knowledge, such as experimental or survey data, for knowledge gained using SSM relies not on formal proof but on demonstration of a 'weak sufficiency relationship'. In practice, the criterion of evaluation lies in the extent to which the people involved acknowledge that they have learned, either explicitly or through actions of some kind, which actually implement changes.

In relation to HCI, SSM is useful in addressing what Malone (1985) describes as 'the organizational interface', referring to social structures and interactions required for elaborating and negotiating policy on information management and the role of IT in an organisation. Hosking and Morley (1991) emphasise that at this level of description of an organisation, units and departments are the product of assumptions and of social negotiation in the real world of social behaviour; that is to say, they constitute 'desired ends which cannot be taken for granted', the subject matter SSM was designed to address. The organisational interface, therefore, is the primary plane of a SSM intervention.

Application of the complete soft systems methodology is a 7-stage process. Stage 1 is a facilitative stage, during which participants share their perceptions of the issue of interest and Stage 7 is one of action to implement any changes agreed. Each stage from 2 to 6 has a defined output. This paper reports on stages 1 through 4 of an application.

SSM supplies two sets of 'rules' to support appropriate choices and uses of interventions: 'constitutive' and 'strategic' (Naughton, 1977). 'Constitutive' rules, Checkland regarded as essential to the application of SSM, 'as a linked whole, as an enquiring system' (op.cit.). A set of 'strategic' rules, intended as guides to decisions, especially during stages 1 and 2, are useful but less integral than the 'constitutive' rules to the methodology's process of problem-resolution.

During Stages 1 and 2, the interventionist guides expression of the issue of concern from a relatively unstructured representation to a relatively structured one, expressed in what are called 'rich pictures'. These pictures may includes diagrams, sketches, quotations or other symbols which represent the diversity of views of informants, based on data from conversations with individuals or groups.

The interventionist's tasks at stages 3 and 4 are to analyse data gathered in 'rich pictures' of informants' views, with a view to abstracting significant conceptual models that underpin their interpretations. In accordance with the constitutive rules, during stage the analyst constructs two of more 'root definitions' of the system being explored. Here, the analyst's task is a didactic intervention of a creative kind. By contributing a concise definition of a human activity system which captures a particular view of it, the intervention should offer the client system a means of acquiring a reliable perspective. Another constitutive rule of SSM prescribes that a 'well-formulated' root definition is sufficiently substantial that it represents the essentials of a dynamic system in terms of six elements: Customers, Actors (participants), the Transformation process of inputs into outputs, Weltanschauung (a German term for 'viewpoint'), Owners of the system and the Environment. There is a hierarchy of importance amongst the six elements of a root definition. Checkland (op.cit.) contends that it is both *the transformation* - the conversion of some input into an output - stated in a root definition which affects the construction of the model, and *the viewpoint* which makes any particular human activity system a 'meaningful' one to consider.

Adequate characterisation of contrasting viewpoints, 'Weltanschauungen', in root definitions is vital. For a well-phrased expression of a Weltanschauung conveys the image or model of the world held by the actor(s) or participants in a system. In practical terms, it is contrasting Weltanschauungen that the interventionist has to take pains to uncover, patiently and with as high a degree of fidelity as possible to the integrity of all informants. Regarding interpretation of user needs, the points of interest are that 'actors', 'customers' and 'owners' may each have different viewpoints of a system and that the task of elucidating these differences accurately has an affective quality.

The output of stage 4 also includes providing the client system with sufficient data at stage 5 to compare the conceptual models with the rich picture of stage 2, so that they may be in a fair position to recognise options for deliberation at stage 6. So, Checkland (op.cit.) proposes comparing each conceptual model with a formal systems model and with one or more other model's relevant to the domain of enquiry. After the comparison at stage 5 and the deliberation and action planning at stage 6, the interventionist's task is complete: it is for the client system to move on to implement action plans as stage 7 of the methodology.

Applying SSM in a problem situation

The manager of the Occupational Health and Safety (OHS) Unit in a large metropolitan local authority reported that the level of computerised assistance available provided inadequate support for professional tasks of the Unit. Professional work of the OHS Unit involves educating line management and councillors, about their legal responsibilities and providing them with advice when difficulties arise. Frequently, this work requires detailed analysis of records of individual employees, their occupational and medical history and patterns of absence from the workplace. In view of the ratio between the staffing complement of the OHS Unit (6) and the number of employees (about 10,000), adequate computerised support was required for this kind of analysis.

The OHS Unit is a constituent part of the Personnel Department of the local authority; the manager of the OHS Unit - a Consultant in Occupational Health and Safety (COHS) - reports to the Personnel Director, who has primary control of resources in all sections of the Personnel Department. Despite sustained representations by the Unit manager, problems of computer support in the OHS Unit have not featured clearly in the annual plans for the Department of which the OHS Unit was part. They were not clearly addressed by a joint Personnel/I.T. working party which recently reviewed the provision of computerised support in the Personnel Department.

The primary research objective was to elicit user requirements for adapting the organisational interface to better support the professional tasks of the OHS Unit. The COHS recognised that the enquiry had the character of 'intervention' in the sense that the primary tasks of an interventionist are to generate valid information, to help the client system make informed and responsible choices, and to develop internal commitment to these choices (Agyris, 1970). Respecting any concern of informants about how the research intervention represented their views, it was agreed that each of the main informants should receive a copy of the research report.

During stage 1, the interventionist used a checklist derived from the model of a formal system outlined by Checkland (op.cit.) to gather data from staff with relevant experience during stages 1 and 2 of the methodology. Individual interviews were conducted with each of the staff in the OHS Unit; two senior staff in the I.T. Department; the two Senior Personnel Officers with responsibilities for IT. Four members of the Corporate Safety Liaison Group, including the COHS, contributed to a group discussion about the support of IT for safety management in the organisation. The Personnel Director, who controlled the budget and I.T. strategy constraining the OHS Unit, was not available to contribute to data-gathering.

At this juncture, the interventionist characterised the 'real-world problem' as 'how to make effective, efficient, robust and satisfying use of I.T. in the OHS Unit', recognising that each of the problem-owners might perceive it differently. He classified data in terms of the methodology's definitions of the concepts of 'structure', 'process' and 'climate' and represented them in two 'rich pictures' of the health and safety services and of I.T. systems in relation to these services. In this way, 'IT-related issues' reported by informants were fleshed out in terms of performance problems.

Applying stages 3 and 4 of the methodology, the interventionist gradually formed the opinion that two alternative 'human activity system' versions of the OHS Unit were relevant to resolving the problem situation, namely the OHS Unit as
*   a supplier of professional services;
*   an administrative part of the Personnel Department, within its managerial control.
He drafted root definitions of the OHS Unit for each of these versions, fleshing out the essentials of a dynamic system for each. The differences in 'Weltanschauungen' illustrate the contrast between the alternative root definitions, as summarised in Table 1.

The tone, tenor and orientation of the viewpoints expressed in the contrasting root definitions illustrate the central point of this paper. Different policymakers, as stakeholders within the organisation, operated on the basis of contrasting interpretations of needs of users. By eliciting these interpretations in terms of root definitions at stage 3, the interventionist was in a position to generate correspondingly contrasting conceptual models at stage 4; within SSM, these models in turn could serve as a structure for comparison and deliberation at later stages of the application. To test the adequacy of each conceptual model, it was compared to the formal systems model and to an Interactive Worksystems model of human-computer interaction (Dowell and Long, 1989); this had the merit of suggesting enhancements to the formulation of each conceptual model. At stage 4, the interventionist also decomposed each conceptual model to the level where information-handling by I.T. support was made explicit; this was the point at which disclosure of key differences between 'administrative/managerial' and 'professional' interpretations of user requirements of I.T. support for the OHS Unit became clearly evident. On the one hand, OH staff were preoccupied with professional goals of providing high quality healthcare and safety management services as well as reports and advice to managers; on the other hand, the managerial priorities were primarily to do

Table 1. The 'Weltanschauung' element of the root definitions

Professional version	*Administrative, managerial* version
A pro-active approach to occupational health and safety is necessary, founded on comprehensive assessments of risk, accident prevention, incident and accident reporting, counter-hazard protection and in educating managers and employees at all levels about OHS. It includes a commitment to developing 'best practice' approaches to OHS by line managers as a strategic objective of the OHS Unit. Methods include a clear Mission Statement, up-to-date Quality Indicators, systematic analysis of feedback from clients and formal, proceduralised attention to human factors objectives in the use of IT.	As workplaces present risks to physical and psychological health and safety of employees, line managers must be informed and developed to take appropriate action about OHS matters to ensure compliance with their legal responsibilities. Up-to-date Quality Indicators within the OHS Unit are *not* required. A strategy for Information Technology does *not* require systematic, formal updating in the light of changing demands on the OHS Unit or its practices; it does *not* require formal attention to explicit usability objectives in the use of IT.

with legal obligations and financial control. This disclosure enabled 'problem solvers' in the client system to view the data in the 'rich pictures' from fresh perspectives, with different orientations, and to approach the contrast at stage 5, between the 'rich pictures' and the conceptual models of their realities, with a view to identifying new options.

Discussion

Two tentative conclusions about the usefulness of SSM emerge from this case study. Firstly, the open-ended character of SSM offers scope for fertile ways of framing user requirements within the socio-technical system of an IT application. The variety of interventions required to apply SSM with fidelity to interpretations of different users affords clear acknowledgement of the diversity of needs of users as a basis for system redesign. As a result, disciplined use of SSM restrains pre-emptive definitions of user requirements. Secondly, a distinctive contribution of SSM lies in its efficacy at the plane of the organisational interface, providing options for addressing elusive problems to support redesign. The case study suggests that this may be particularly the case where a history of substantial commitment on the part of senior management to optimising usability aspects of IT is not apparent.

Three issues about applying SSM came to light in this case study. Firstly, when the task of the interventionist is to clarify assumptions of different groups within an organisation, it is reasonable to propose that disclosure of the needs can be managed mainly through interactions with each group or with both groups. In the instance reported, it is possible that the researcher's lack of access to informants *as an inter-disciplinary group* including IT providers and users, may have limited the range of perspectives which contributed to the construction of 'rich pictures' of the problem situation at stage 2. Perhaps the addition of a 'strategic' rule to the methodology might usefully specify data-gathering on group and inter-group level levels. Secondly, extant procedural guidance of SSM to deploy 'conventional' methods of enquiry for eliciting a full picture of their requirements from users (Patching, op.cit.) is insufficient. In the system explored, the intervention might have been more fruitful with more informative stimuli to guide dialogue with informants, for example through the support of scenarios about system usability. Thirdly, the methodology does not make sufficiently clear quite how versatile the intervention style required of someone who uses SSM as a research method may have to be, in order to respond coherently to 'actors' at different stages of an intervention. For applying SSM involves didactic application of a set of definitions of terms which characterised the approach, such as 'root definition', 'structure' and 'process'; this constraint requires the interventionist and actors to adopt linguistic categories supplied by the methodology. Yet, practical exigencies are also likely to require the interventionist to adopt an acceptant style of intervention with informants who wish to assert their view of their problem situation in their own words; alternatively, a catalytic style is required in other circumstances. Accordingly, more complex cognitive strategies than the methodology at present acknowledges are required of the analyst. Acknowledgment of tensions between the didactic and the acceptant or catalytic dimensions of the methodology would help the analyst accommodate tensions arising from such demands.

Conclusion

SSM was applied to elicit interpretations of user needs of an Information Technology system in an Occupational Health and Safety Unit in a large public sector organisation. This methodology throws some useful light on differences in constructions of user requirements by different groups of participants in the socio-technical system supporting the work of the Unit. Effective use of SSM is mediated by the diversity of intervention styles it calls for. Contrasts between these intervention styles as well as the significance of inter-group communications call for complex cognitive strategies on the part of an interventionist; they also perhaps suggest that additional 'rules' within SSM would be useful, at least in HCI applications.

Acknowledgments

The first author thanks the Engineering and Physical Sciences Research Council for the grant which supported this research project. Both authors thank the staff of the unnamed local authority who contributed their time and information to it.

References

Agyris, C., (1970). *Intervention Theory and Method.* (Addison-Wesley)

Blake, R.B. and Mouton, J., (1976). *Consultation.* (Addison-Wesley)

Checkland, P., (1984),. *Systems Thinking, Systems Practice.* (John Wiley & Sons Ltd.)

Dowell, J. and Long, J., (1989). Towards a conception for an engineering discipline of human factors, *Ergonomics.* **2**, no. 11, 1513-1535.

Hosking, D.-M. and Morley, I. (1991), *A Social Psychology of Organizing.* (Harvester Wheatsheaf)

Malone, T. W., (1985). Designing organizational interfaces. In L. Borman & W. Curtis. *Human factors in computer systems - II. Proceedings of the CHI '85 conference.* North-Holland.

Naughton, J. (1977). *The Checkland Methodology - A Reader's Guide.* (John Wiley and Sons Ltd.)

Patching, D. (1990). *Practical Soft Systems Analysis.* (Pitman Publishing)

HUMAN-FACTORS KNOWLEDGE REPOSITORY FOR DESIGNING ACCESSIBLE USER INTERFACES

Demosthenes Akoumianakis and Constantine Stephanidis

Institute of Computer Science
Foundation for Research and Technology - Hellas
Science and Technology Park of Crete, P.O. Box 1385
Heraklion, Crete, GR-71110 GREECE

This paper reports on the development and use of a human factors knowledge repository for building accessible and usable user interfaces. Two issues are explored and discussed in detail, namely the underlying knowledge representation for consolidating human factors knowledge, and the computer-designer collaboration for the propagation of such knowledge to the design of accessible and usable user interfaces. Additionally, the paper presents recent experience in using such a human factors knowledge repository in the context of a European collaborative research and development project.

Introduction

When reviewing the current practices, methods and tools by which user interfaces are designed, it becomes evident that ergonomic principles are frequently neglected (Bevan and Mcleod, 1994). This is partly due to the fact that such knowledge is not easily exploitable by user interface designers (Tezlaff and Schwartz, 1991), but also due to the view that guidelines and style guide documents are inefficient ways of communicating human factors knowledge to the designer (Bevan et al., 1994; Lowgren and Nordqvist, 1992). The appreciation of these shortcomings has led many researchers to develop novel methodologies, techniques and tools towards the application of human factors expertise in the design of user interfaces of interactive systems and services. These are classified under three main headings, namely non-metric methods, metric methods, and software tools for early design and evaluation of user interfaces. The non-metric methods category comprises a collection of low-end and relatively cheap methods for usability evaluation and assessment which do not require measurement of any type. Representative techniques include *heuristic evaluation methods, cognitive walk-through, guidelines and checklists, design guides* and *cognitive and analytical models* (e.g. GOMS family of models). Metric methods stems from the principle that user-based evaluation can feed the design process with usability problems. An indicative instrument facilitating this line of work is the MUSiC toolset, developed in the context of the MUSiC project of the ESPRIT-II Programme of the European Commission (Bevan et al., 1994). The Music toolset provides a collection of methods used to evaluate relatively stable interface prototypes and it is best suited for evaluations relatively late in the product's life cycle. These include the *Performance Measurement Method (PMM)*, the *Diagnostic recorder for Usability Measurement (DRUM)*, the *User Perceived Quality (SUMI)* which is a technique for subjective usability evaluation and *Cognitive workload measures*.

Finally, software tools for design assistance, guidance and evaluation, usually involve the development of knowledge based systems for the evaluation of a user interface and its subsequent refinement by the designer. The objective is to consolidate and encapsulate human factors expertise into an appropriate knowledge base which augments the design and implementation environment of a high level user interface development system, such as a User Interface Management System (UIMS), thus bringing human factors support closer to the design process. Indicative systems which have been developed to pursue this line of work include the KRI/AG system (Lowgren et al., 1992), various types of critiquing systems and components (i.e. Malinowski and Nakakoji, 1995), and tools for working with guidelines such as the EXPOSE system (Gorny, 1995), SIERRA (Vanderdonckt, 1995) and IDA (Reiterer, 1995).

This paper describes work which follows a slightly different paradigm. The main objective of the present work is to investigate the feasibility and demonstrate the applicability of software components in the form of explicit, organisation-wide, human factors knowledge repositories, contributing to the design of accessible user interfaces. The primary role of such repositories is to encapsulate an organisation's evolving design experience, design wisdom and rationale, and to ensure availability of this body of knowledge across different departments of an industrial organisation, thus reducing the gap between design and development. As a result, the current work fosters an *organisational memory* approach towards *experience-based usability guidelines*.

Consolidating and propagating human factors knowledge

A human factors knowledge repository is a software component which provides a reusable pool of knowledge on the ergonomic design of accessible user interfaces. It has been implemented as a knowledge server which may be interrogated by designers or client applications. An interrogation cycle involves passing data to the server regarding the *context* of an interaction, and subsequently, the delivery of *design recommendations* that can be directly embedded into the implementation of a user interface. In the current implementation, the context of an interaction is characterised by four primitives, namely the underlying interaction metaphor used to implement the user interface, the dialogue states of the interactive application, the range of interaction object classes used in a particular dialogue state and finally, the attributes of these object classes which characterise the physical level of the interaction.

In the present work, binding design recommendations to context entails that the scope and interpretation of such recommendations are bound to particular interaction metaphors, task contexts and abstract object classes. It is precisely this context-specific interpretation as well as the non-trivial range of interaction elements reflected in such recommendations that differentiates them from default parameters, as used in current graphical user interface environments and makes their derivation a knowledge-intensive task.

Metaphors may be either embedded in the User Interface (e.g. menus as interaction objects follow the "restaurant" metaphor) or may characterise the properties and the attitude of the overall interaction environment (e.g. the desktop metaphor presents the user with an interaction environment based on sheets of papers called windows, folders). The interactive environment of a metaphor is realised by specific user interface development toolkits. Thus, for example, OSF/Motif and MS-Windows support a particular embodiment of the visual desktop metaphor. Different interaction metaphors may be facilitated either through the enhancement of existing development toolkits or by developing new ones. For instance, an enhancement of the interactive environment may be facilitated by embedding in the toolkit automatic scanning facilities for interaction object classes (Stephanidis, 1995). Alternatively, new interaction metaphors may be realised through novel toolkits. An example of the latter case is reported in (Savidis and Stephanidis, 1995), where Commonkit is used to support non-visual interaction based on a non-visual embodiment of the Rooms metaphor.

Depending on the choice of metaphor, the designer will typically have to deal with different object classes and alternative interactive behaviour. Consequently, design recommendations at the level of the metaphor are required so that an interface may exhibit the attitude and characteristics of that metaphor. This behaviour is established through a collection (i.e. library) of object classes which exhibit the required semantic, syntactic and lexical behaviour. For the purposes of the present work, we have dealt with two interaction metaphors, namely the visual desktop as embodied in Windows'95, enhanced with automatic scanning facilities and a non-visual metaphor suitable for non-visual interaction with hypermedia systems (ACCESS, 1996).

The final component of the context of a design recommendation is the notion of a task context. In the present work, a task context is considered as a characterisation of "dialogue states". Dialogue states are indicators of what the user interface is doing at any point in time, either as a result of specific task requirements or due to user preference. Task contexts are identified as part of task analysis in the context of a design activity. It is important to underline that the task contexts defined by the designer reflect dialogue states in *envisioned tasks* (i.e. tasks performed through the interface being developed) as opposed to existing tasks. Moreover, task contexts encompass a degree of *abstraction* which entails that the same task context may be performed differently by different users, or the interface may exhibit differentiated behaviour in the same task context depending on the current user. With regards to implementation, task contexts are *polymorphic*. In other words, the same task context may be mapped onto several physical interface components. Consider for example, a hypothetical situation where the designer wishes to design the dialogue through which a user will review a video. The designer will typically try to map higher level dialogue components such as for example the video component to specific interface instances, exhibiting similar cognitive principles (e.g. in the design of Figure 1 the interface instances depicted in right hand-side provide alternative embodiments for containment; one uses the toolbar the other a pop-down menu to convey containment). Consequently, the design space involves decisions at different levels such as those illustrated on the diagram of Figure 1. In this diagram, we assume that through task analysis the designer observes that several users may have to engage in video review and these users do not perform the task in uniform

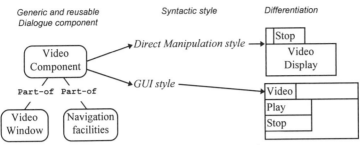

Figure 1: Polymorphism in realising task contexts

manner. In other words, some of them prefer a direct manipulation style while others are used to a more conventional menu-like dialogue syntax. Moreover, within the same category of users (i.e. those with a preference for direct manipulation) there are variations. Such variations may be due to preferences for specific syntactic styles (e.g. function-object versus object-function) or interaction modes (explicit versus implicit activation). Given these observations, the designer decides to identify an abstract task context called **Video-review** whose description is going to be populated in such a way that it suits different usage patterns. Such knowledge descriptions are formally expressed in a knowledge representation formalism, called *task context scheme*.

The above primitives give rise to a design recommendations space which is contextually defined by a five-tuple relation:

< Metaphor, State, Object, Attribute, Assignment >

where Metaphor is the interaction metaphor (as embedded in a particular user interface development toolkit), State is a dialogue state or task context (identified and specified during the design phase), Object is the object class, Attribute is the object attribute to which the recommendation applies and Assignment is the recommendation.

Embedding design recommendations into user interface implementation

Having reviewed the format and interpretation of design recommendations, we now address the issue of their generation and propagation to user interface implementation. With regards to generation, a tool called USE-IT has been developed which automatically reasons about design options and criteria and automatically compiles maximally preferred recommendations, based on user, task and platform (i.e. development tool) constraints. USE-IT has been described in detail elsewhere (Akoumianakis and Stephanidis, in press) and therefore it is not further elaborated in this paper.

With regards to propagation of design recommendations, the main implication following from the above discussion is that the user interface development system should be equipped with the appropriate mechanisms to interpret and apply the derived design recommendations in the context of particular dialogue states (locality of decision). Through such a localisation of a design recommendation, it is made possible to practically support a wide range of syntactic and lexical differentiation of the interactive behaviour of an interface. In the ACCESS unified interface development platform, this is supported as follows. When an interaction object in constructed using a particular toolkit, it is assigned to a particular dialogue state or application-specific task context. The Application Programming Interface (API) of these toolkits provides a dedicated function which allows the developer to utilize externally generated design recommendations such as the above, automatically. Although, the details in which this is achieved are beyond the scope of this paper, it should be mentioned, that such toolkits have been developed (ACCESS, 1996) and used by the ACCESS consortium partners to develop user interfaces in selected application domains.

Applications and Experience

In the context of the ACCESS project, the concept of the Human Factors Knowledge Repository has been experimentally validated in two application domains, namely, the development of an educational hypermedia application accessible by blind users and the development of two interpersonal communication aids for language-cognitive and speech-motor impaired users. In compiling the knowledge base of the server, the objective was merely the collection of a minimum set of design recommendations which would subsequently grow as the organisation's experience grows towards experience-based usability recommendations, thus the facilitation of an *organisational memory* approach (Henninger et al., 1995) towards the development of a reusable knowledge repository of experience-based guidelines or "living design memory". In what follows, we briefly review the design recommendations generated for the non-visual application, to illustrate some of the principles already discussed. The example describes recommendations generated and delivered through the server, for two object classes, namely *NonVisual_TextReviewer* and *NonVisual_Button* and one task context, namely *Bookmark.* USE-IT automatically generated a set of design recommendations in the form 6-tuple clauses introduced later. This collection when fed to the Human Factors Knowledge repository delivers toolkit-specific recommendations that can be automatically interpreted by the toolkit through one or more interrogation cycles. An extract the knowledge delivered by the Human Factors Knowledge Repository for the example being described is as follows:

```
NonVisual 372
    Bookmark
        Bookmark NonVisual_TextReviewer input_device joystick
        Bookmark NonVisual_TextReviewer joyNavigation 1
        Bookmark NonVisual_TextReviewer kbdNavigation 0
        Bookmark NonVisual_TextReviewer touchNavigation 0
        Bookmark NonVisual_TextReviewer presentInBraille_Lines 2
        Bookmark NonVisual_TextReviewer presentInBraille_Cells 40
        Bookmark NonVisual_TextReviewer presentInBraille_Policy Options on Braille
        Bookmark NonVisual_TextReviewer gestures 0
        Bookmark NonVisual_TextReviewer sayClass 1
```

The first line, in the above specification, is a counter giving a total of the recommendations generated for a particular interaction metaphor. In this case, it is specified that 372 recommendations were compiled in total. Following this line, the recommendations are listed by task context identifier (i.e. Bookmark). It can be seen that the compiled recommendations cover several attributes of lexical interaction as supported by the target toolkit. Thus, the selected device is joystick (from a range including touch tablet, keyboard and joystick). Similarly, joystick navigation within the non-visual space is enabled. With regards to output, the output device selected in Braille followed by recommendations detailing certain parameters of the attributes output_device and output_Technique.

Concluding remarks

This paper has briefly presented the development of a software platform for generating and consolidating ergonomic design recommendations into reusable and expandable human factors knowledge repository. The underlying tool environment provides facilities for depositing new knowledge as well as the incremental growth of the knowledge server towards experience-based usability guidelines (i.e. living design memory). Currently, the approach described and the tools developed have been experimentally tested in two application domains. In the future, several enhancements of the basic functionality of the server are foreseen including explicit support for design rationale and design critiquing.

Acknowledgement

Part of this work has been carried out in the context of the ACCESS project (TP1001) funded by the TIDE Programme of the Commission of the European Commission, DG XIII.

References

ACCESS Consortium, 1996, *Progress Report No. 3 on the Development of User Interface Development Tools*, (available from the authors).

Akoumianakis, D., Stephanidis, C., in print, Knowledge Based Support for user adapted interaction design, to appear in *Experts Systems with Applications*, **12**(1), 1997.

Bevan, N. Macleod, N., 1994, Usability measurement in Context, *Behaviour and Information Technology*, **13**(1&2), 132-145.

Gorny, P., 1995, EXPOSE: An HCI-Counselling tool for User Interface Design, Conference *Proceedings of INTERACT'95*, 297-304.

Henninger, S., Heynes, K., Reith, M., 1995, A Framework for Developing Experience-Based Usability Guidelines, *DIS'95 Conf. Proceedings*, (ACM, New York), 43-53.

Lowgren, J., Nordqvist, T., 1992, Knowledge-based Evaluation as design support for graphical user interfaces, *CHI '92 Conference Proceedings*, (ACM Press, New York), 181-187.

Malinowski, U., Nakakoji, K., 1995, Using Computational Critics to facilitate Long-Term Collaboration in User Interface Design, *CHI'95 Conference Proceedings*, (ACM Press, New York), 385-392.

Tetzlaff, L., Schwartz, D., 1991, The use of guidelines in Interface Design, *CHI'91 Conference Proceedings*, (ACM Press, New York), 329-333.

Reiterer, H., 1995, IDA: A design environment for ergonomic user interfaces, *Conference Proceedings of INTERACT'95*, 305-310.

Savidis, A., Stephanidis, C., 1995, Building Non-Visual Interaction through the development of the Rooms metaphor, *Companion of CHI '95*, (ACM Press, New York), 244-245.

Stephanidis, C. (1995): Towards User Interfaces for All: Some Critical Issues, *HCI International '95 Conference Proceedings*, (Elsevier, Amsterdam), 137-143.

Vanderdonct, J., 1995: Accessing guidelines information with SIERRA, Conference Proceedings of INTERACT'95, 311-3-16.

SUPPORT SYSTEM CONFIGURATIONS: DESIGN CONSIDERATIONS FOR THE FACTORY OF THE FUTURE

Anne Dickens, Chris Baber and Nick Quick

Industrial Ergonomics Group,
School of Manufacturing and Mechanical Engineering,
University of Birmingham,
Birmingham, B15 2TT

Factory configurations revolve around three concurrent design axes: sociotechnical systems design, manufacturing systems design and support system design. This study concentrates upon the latter of the three design axes, support systems. The research aims to investigate the effect that the configuration (or organisation) of support systems has upon the overall effectiveness of the manufacturing system. Initial results reveal that poor design of communication structures, systems architecture and physical layouts are adversely affect manufacturing performance. The results also lead us to dispute the existence of 'true' cells.

Introduction

Support systems are at once 'indirect' elements of manufacturing and essential contributors to the success of the overall process (Dickens and Baber, 1996). The role of support systems is to enable the manufacturing process to function as efficiently as possible and encompasses functions such as logistics, design, engineering, maintenance and marketing, etc. During this project, support systems were considered in the context of cellular manufacturing companies. Franks, Loftus and Wood (1993) describe cells as that part of a manufacturing system which has crisp boundaries having the minimum material and information flows across them. This represents the consensus within the manufacturing community; that cells are independent, autonomous units. The majority of manufacturing companies in Britain, i.e. 66%, currently use cells (Ingersoll Engineers, 1990), ensuring that the results remain highly applicable.

Factories consist of three separate but interacting systems, as depicted in Figure 1. Sociotechnical aspects of the factory encompass job design, management structure and supervision methods whereas engineering aspects focus on the physical process of manufacturing. This physical process includes cell design, equipment selection, line balancing and control systems. The third and final aspect is that of support systems. As previously explained, support systems include the functions performed to support manufacturing. Any relationships between these three separate systems is represented by the shaded 'overlap' area shown in Figure 1.

In the early 20th century, the factory system would have looked substantially different, with the absence of the sociotechnical aspect and virtually no if any, overlap between engineering and support systems. This is because every employee had their own defined role, from which no deviation was allowed (Taylor, 1907). In the factories of today, team work and empowerment are key factors in the success of cells, both of which demand a significant degree of overlap between the three systems shown in the diagram below.

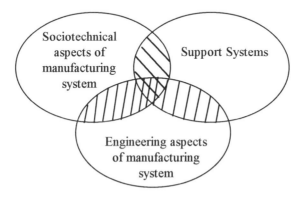

Figure 1. The three sub-systems of a factory.

The hypothesis tested in this paper is that support systems are not interacting effectively with manufacturing aspects and sociotechnical aspects. The following sections describe research into support systems and discusses the role of communication, and support system configurations of the two main types of cell used in industry: process and product cells.

Data Collection

Scope

The companies targeted were cellular manufacturing companies, and 53 contacts were obtained using the 'snowballing' technique (Oppenheim, 1992). These companies formed the sample for the primary research. Comparative studies in the area of cells (Wemmerlov and Hyer, 1995; Ingersoll Engineers, 1990) were also considered in order to provide a benchmark for the results.

Methodology

A postal survey was used to collect the primary data. The survey was designed to compile three main types of data; descriptive, relational and attitudinal. The first, descriptive, was designed to supply general information on cellular companies which could be used in comparison with the Wemmerlov and Hyer (1995) study. Relational data was intended to show any relationships that existed between support systems and the success of cells, whilst the attitudinal data recorded personal judgements about the success or failure of support systems within the company. The points discussed in this paper only relate to descriptive data. The target respondents were senior managers, directors or managing directors to ensure a high quality of general factory information.

Results

Returned Surveys

The response rate for the 53 companies who were targeted to complete the questionnaire was 49% and is expected to improve upon completion of a reminder programme. Inter-rater and intra-rater reliability using Pearson's product-moment correlation coefficient were both found to be significant at the 0.05% level.

Descriptive data

The descriptive data obtained from the postal survey can be split into several distinct categories:

• Types of companies that use cells: The vast majority of the companies using cells were classified as SME's (small to medium enterprises) having an average of 367 employees, and an annual turnover of 31.1 million pounds.

• Types of cell used: 62% of companies used cells in conjunction with other methods of manufacture, there was a 1:1 ratio of product and process cells. The average number of both product and process cells used within a company is 4. In terms of labour content, process cells have an average of 8 direct employees and 1 indirect employee, whereas product cells have 11 and 1 respectively.

• Direction of decision making: 54% of companies employ senior managers to make any manufacturing decisions, whereas 46% of companies prefer cell employees to have the responsibility.

• Communication methodology: the type of communication methodology used was recorded in three scenarios; within cells, within support systems (or offices) and finally, between cells and support systems. In all instances, verbal communication was by far the most frequently used method of communication, i.e., in excess of 90% in all cases.

• Support system configurations: from previous research (Dickens and Baber, 1996), it was discovered that support systems could be configured in one of four ways. For every product or process cell recorded in the survey by the sample companies, a type of configuration was identified. The figures given below may appear high because there are eleven different support 'functions' (e.g. a function could be engineering, marketing, logistics etc.) associated with each configuration. These configurations are given overleaf in Table 1.

Table 1. Distribution of support systems for process and product cells

Configuration of support → Type of cell ↓	Cell based & responsible for supporting 1 cell	Centrally based & dedicated to a few specific cells	Centrally based and directly supporting all cells	Centrally based with little direct cell contact
Process cells	38	24	118	97
Product cells	47	40	135	88

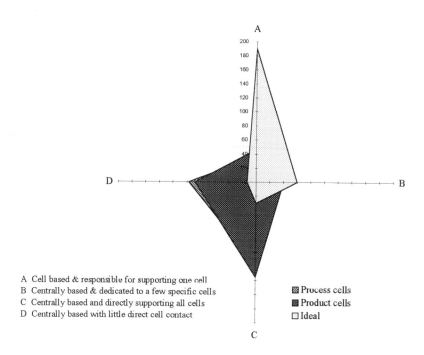

A Cell based & responsible for supporting one cell
B Centrally based & dedicated to a few specific cells
C Centrally based and directly supporting all cells
D Centrally based with little direct cell contact

■ Process cells
■ Product cells
□ Ideal

Figure 2. Configurations of support systems

Using the figures given in Table 1, it is possible to plot a radar chart which in effect shows a 'blueprint' of support system configurations. Superimposed upon the process and product cell configurations, is a third area named 'ideal'. The implications of this third area will be discussed below, alongside the other results.

Discussion

The results in Figure 2, contributes to dispelling the idea that a 'true' cell exists. They also go some way to explaining why support systems have the potential to restrict the throughput of the manufacturing function (Dickens and Baber, 1996). As with many manufacturing problems, the cause seems to stem from a combination of several distinct, but interacting areas.

The majority of companies only partially use cells in conjunction with other manufacturing systems. Cells require a very distinct form of support in that all the relevant functions should be based at trackside, rather than the remote environment of a central office. If other types of manufacturing system are also used within the factory, this type of support could prove to be highly unsuitable. Due to the fact that there is very little research in the area, the company could well decide, whether consciously or not, to retain centralised support structures. This notion is borne out by the results obtained for the support system configurations. When examining Figure 2, it can be clearly seen that the 'ideal' support system configuration, based upon Franks' et al. (1993) definition of cells, is diametrically opposite to the 'actual' process and product cell configurations. Two additional elements can also be drawn upon to interpret this graph; the first being that engineering research is progressing at a faster rate than it's complementary systems such as support, i.e. the companies are able to successfully implement the physical manufacturing systems but not the support, and second point is that without the correct support systems in place, cells are downgraded to a mere change in the shopfloor layout, disputing the fact that 'true' cells exist in the majority of companies.

One of the key philosophies of the cellular approach is the empowerment of shopfloor employees, yet the survey shows that the majority of companies are still relying upon senior management to make manufacturing decisions. This suggests a reluctance to relinquish control which again questions the existence of true cells.

Recent developments in the area of flexible manufacturing systems (FMS's) tend to suggest that there is a general trend towards the high technology factory, and therefore an increase in more up-to-date communication methods such as email and electronic data interchange (EDI). The survey results dispelled this idea with companies having face to face verbal communication as their primary source of information exchange. When combining this with the fact that the majority of support for cells is remotely based, the potential for 'noise' in general communications is enormous. To demonstrate this fact, eleven identified and remotely based support functions could have any number verbal inputs into a cell at any given time. This alongside the existing communications structure within the cell, has the potential to cause considerable disruption to the cell output (or product). In support of Franks et al. (1993), if the information passing the cell boundaries was minimised, the 'noise' factor could be significantly decreased, leading to minimum disruption of the cell output.

One of the major studies in this area conducted by Wemmerlov and Hyer (1995), concentrate upon the cell engineering aspect, providing very little data which can be compared with the results in this survey. They also neglect to discuss communication and support methodology, which has already been proved to have a potentially negative affect upon manufacturing processes. This paper is intended to bridge this particular research gap.

Conclusions

The cellular environment has the potential to gain increased efficiency through the restructuring of the support system configurations. This is demonstrated by the following points:

• The majority of support systems are remotely based rather than within the cell, facilitating communications 'noise'.

• Communication methodology is predominantly verbal, and although this would not present a problem if support was cell based, it currently has the potential to induce confusion, misinterpretation and disruption within the cells.

• Management must develop the issue of empowerment within cells, rather than retaining control from a remote environment.
The above points were highlighted in the survey, but in isolation, they provide very little aid to a company wishing to improve their support structures. In response to this, a framework needs to be develop to encompass a coherent company development model.

References

Dickens, A. and Baber, C. 1996, Can support systems withstand the demands of factory 2000?, *Proceedings of the twelfth international conference on CAD / CAM robotics and factories of the future conference* (Middlesex University Press), 562-567.

Franks, I.T., Loftus, M., and Wood, N.T.A. 1993, Attributes of a discrete cell control system, *Computer Integrated Manufacturing Systems*, 6 (3), 176-184.

Ingersoll Engineers 1990, Competitive manufacturing - the quiet revolution, *A Survey of implementation and performance across British manufacturing industry.* (Ingersoll Engineers).

Oppenheim, A.N. 1992, *Questionnaire design, interviewing and attitude measurement.* (Pinter Publishers, New York).

Taylor, F.W. 1907 On the art of cutting metals, *Trans. American Society of Mechanical Engineers*, 28, 31-350.

Wemmerlov, U. and Hyer, N.L. 1995, Cellular manufacturing in the U.S. industry: a survey of users. In Moodie, C., Uzsoy, R. and Yih, Y. (eds), *Manufacturing cells - a systems engineering view*, (Taylor and Francis, London) 1-24.

USER REQUIREMENTS AND NETWORKED APPLICATIONS

Andrée Woodcock and Stephen A.R. Scrivener

Design Research Centre,
Derby University,
Mackworth Road,
Derby, DE22 3BL

Systems are now emerging which allow users to work together locally and when separated by distance. In building Computer Supported Co-operative Work (CSCW) systems, developers need to consider the activity to be supported, the way people need to work together, and hci standards and issues. In moving these systems out of the laboratory into the real world factors such as the amount of available and affordable bandwidth required by the user present additional problems. The work reported here relates to preliminary results from user trials run over a local Asynchronous Transfer Mode (ATM) network at Queen's University Belfast. The main objective of this part of the work was to evaluate the perceived performance of three applications (FINS, Scribble and ShowMe) running over different bandwidth configurations in a series of realistic user trials.

Introduction

The work reported here is part of that being conducted by the FashionNet project funded under the EU TEN-IBC (TransEuropean Network - Integrated Broadband Communication) programme. A primary aim of the project is to evaluate the technical and task performance of multimedia networked applications (MNAs) across an experimental pan-European network. In the case of the FashionNet project this involved connections between sites in Belfast, Dublin, Lisbon and Stuttgart. Users at these different locations interacted with each other via and supported by MNAs to complete prespecified tasks. Prior to the onset of the international trials, local trials were conducted at QUB to test the experimental procedures for studying the effects of bandwidth performance on the perceptions of system usability under real tasks. This paper outlines the experimental protocol employed and initial results from the local trial.

The Domain and the Task Context

As the name suggests, FashionNet is concerned with the potential and use of MNAs in the fashion industry. Fashion design utilises diverse media in a wide variety of tasks. Designers, for example, create images of designs that are discussed and modified during meetings with other designers, managers, buyers, and clients. Designers receive inspiration for their designs and obtain design materials and components from a variety of sources. A typical design scenario might involve an initial search for materials and inspiration, followed by design and design appraisal. Frequently, these activities are realised collaboratively with design colleagues or mediated through staged group meetings. We were interested in investigating whether MNAs could support these kinds of task and interactions between participants separated by distance.

The applications

A toolkit (workstation) allowing fashion designers to work co-operatively on a number of 'real world' design tasks or scenarios was configured by bringing together the following three independently developed applications:

a)   FINS - Fashion Information Service - a multimedia database developed for the clothing and textile industry. It provides information to assist in sourcing (e.g. of yarns, fabrics, trimmings, packaging suppliers, design services, machinery, transportation and distribution for goods), the identification of retail buyers seeking new suppliers, and for linking manufacturers. It also provides access to forecasting information, allows companies to sample more accurately and earlier in the season, and includes reports from international trade fairs conferences and exhibitions. Visual material from company catalogues, swatches, shade cards and photographic information is also provided.
b)   Scribble - a shared drawing package which allows designers, separated by distance, to develop and annotate designs, either new designs or earlier design imported into the system from the designers' personal data bases. It also allows image file transfer between remote locations.
c)   ShowMe is a suite of video conferencing tools enabling users at different sites to interact with each other over audio-visual channels. It can be used for multipoint conferences (two or more end users), for one-way conference presentations (either as sender or receiver), and it can also be used in "Multicast" mode when there are many users in a conference. FashionNet is one of the first projects using ShowMe over a wide area ATM network. (ATM technology is the base upon which all future "SuperHighways" will be built).

An experimental procedure based around realistic design tasks or scenarios had been developed during a previous project, Marshall, Scrivener and Woodcock (1995) and B2004 Final Report. For this project, four realistic design scenarios were developed which required fashion designers to use all three applications in the sourcing and exploration of ideas, through to collaborative design and the presentation of final ideas. An example of one such scenario is given below:

"Design a co-ordinated winter costume (trousers and top) which will appeal to 8-11 year old boys. It should reflect the latest trends, be durable, and allow them to pursue their normal outdoor activities (eg football, running, cycling)."

To complete this scenario designers have to source for ideas in FINS eg from the fashion shows, other designers, choose fabrics, and then sketch out trousers and top (in Scribble), based on their findings. ShowMe allows the pair of designers to exchange their ideas and co-ordinate their activities.

Methodology

All local trials took place between two rooms in the Department of Electrical and Electronic Engineering at Queen's University Belfast, where European ATM network conditions could be simulated. The users were eight students from the BA Foundation Course at the University of Ulster Belfast Art College. These were grouped into 4 pairs. Each group undertook all four design scenarios, under four different bandwidth conditions (1,2,4,8Mbsec). Show-Me and Scribble ran as shared applications, FINS ran separately on the two user machines. Each session lasted for approximately one hour. A similar experimental procedure has been adopted for the international trials. Restricting the bandwidth creates a bottleneck of cells during complicated activities. These are rejected or passed on later. The lower the bandwidth the more cells are rejected, especially during complicated tasks such as joint drawing and image transfer. The scenarios had been constructed to ensure that all tasks (e.g. talking, drawing, image transfer) occurred under each bandwidth. The relationship between the tasks and bandwidth activity could then be determined.

It was hypothesized that the applications would run better over the higher, unrestricted bandwidth conditions, and that users would notice and be able to quantify this difference in their answers to questionnaires. The following measurements were taken either preceding, during or anteceding the actual trial:

a)   Measurement of pro computer attitudinal scores (before and after participation in each trial) after Badagliacco (1990). If the applications performed well it was hypothesized that pre and post attitudinal scores would remain constant through the trials.
b)   Usability questionnaires relating to each application completed after each of the trials, in order to ascertain whether there had been any noticeable effect on certain facets of system usability which could be attributed to bandwidth differences (e.g. speed, response rate, video and audio quality).
c)   All sessions were video recorded and subsequently subjected to a task and breakdown analysis Urquijo, Scrivener and Palmen (1993). In this type of analysis only verbal breakdowns need to be fully transcribed. These have been rated firstly in terms of their severity: firstly, speech breakdown (a verbal expression of dissatisfaction to self or another); secondly, user stuck -verbal/action request for assistance from system or trial administrator. Although subjects were trained in the use of the applications prior to the sessions intervention was necessary (e.g. when users forgot how to perform certain commands); thirdly, system crash, most usually a non recoverable error. As these were development systems, running over reduced bandwidths, this situation did arise.

Each breakdown was further analysed to determine the time it occurred during the session (pre, beginning, middle or end), it's length and the application it related to (FINS, Scribble, ShowMe). Breakdowns which occurred in consecutive 10 second intervals were treated as a single event. Each incident was classified as relating to Task (the scenario),

User (the partnership), Tool (the application/feature of the application) or Environment (bandwidth configuration).

d) OpenView/Foreview, a network management and monitoring tool, was used to record the amount of traffic generated during each trial scenario in terms of the number of cells received, transmitted and rejected. These could then be related back to user actions.

e) Cost benefit assessment, Eason (1988), was conducted after all the trials to look at anticipated issues relating to the introduction of the technologies

f) Evaluation of Scribble and FINS running under optimum conditions. This was based on the questionnaire developed by Ravden and Johnson (1989). Previous releases of the applications had been subjected to a similar analysis in the earlier project , B2004.

Results

The evaluation of Scribble and FINS running under optimum conditions, highlighted a number of areas where system improvements could be made. These results themselves will again be fed back to the system developers. In terms of the analysis of bandwidth requirements, an application which rates poorly in terms of usability when run over optimum conditions will probably fair even worse when the bandwidth is restricted. Regarding usability over restricted bandwidths, as was hypothesized, the teleconferencing system (ShowMe) and Scribble were rated least favourably in the lower bandwidth conditions (1 and 2Mbsec). Although FINS ran independently of bandwidth condition, images had to be exported from it into Scribble. This was difficult for subjects and sometimes resulted in system crashes (especially on the lower bandwidth conditions). The perceived usability of Scribble and FINS therefore co-varied.

The pro attitudinal computer scores dropped slightly during the trials, regardless of the order of bandwidth presentation. The post trials scores were usually more negative than the pre trial ones. This was especially true of the 1 and 2 Mbsec bandwidth conditions.

For the breakdown analysis, a representative subject pair was selected from each experimental condition. It was hypothesized that firstly as the bandwidth is decreased, the number of breakdowns attributable to the system will increase: secondly, that as users were trained in the use of the systems, the 'user stuck' figure should remain fairly stable throughout the trials: thirdly, the nature of the speech breakdowns would change as a result of the bandwidth configuration. These results are summarised in Table 1 below.

Hypothesis 1 was upheld, more system breakdowns occurred in the lower bandwidths. They were also of greater duration (18 minutes as opposed to 90 seconds). For example in 1Mbsec condition - despite attempts by the system administrators - ShowMe remained unusable with poor sound quality, bad echo, subjects could only hear each other when they shouted through the door! In 2Mbsec bandwidth condition, the trial ended with both machines crashing. The users spent more time on Scribble, but had to wait 5 minutes for images to be transferred, during which time they were unsure as to whether they had performed the operation correctly or whether the system had crashed. In the higher bandwidths subjects were able to spend more time concentrating in the task, and seemed more confident overall.

Hypothesis 2 was upheld. Users experienced the same sort of problems on all the conditions, despite having been trained in the use of the systems e.g. they found navigating in FINS difficult, could not remember how to cut and paste images, or how to draw and delete with Scribble. These problems were exacerbated by the lack of help facilities in both systems. These confirmed/exemplified issues raised in the usability evaluation.

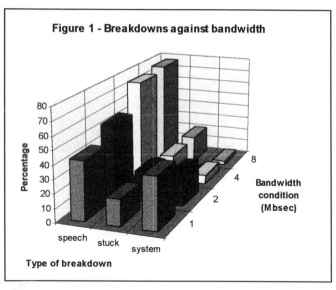

Hypothesis 3 - the number of speech breakdowns increased with increased bandwidth. This could be attributable to the following reasons: firstly, in the higher bandwidths the trials went on longer; secondly, it was easier for subjects to engage in conversation in the higher bandwidth conditions; thirdly, there was a qualitative as well as quantitative change in the nature of the speech breakdowns, they became more task oriented, rather than just asking for repetitions.

The cost benefit assessment showed that major issues in the introduction of this technology would be inter and intra-organisational benefits resulting from increased communication, concern over security and confidentiality (which is a major issue in the Fashion Industry), retraining and recognition of new skills.

Conclusions

All the measures used - pro computer scores, breakdown analysis, and usability measures - indicated that from a users perspective the 1 and 2 Mbsec conditions were most unsatisfactory. The bandwidth data showed that more cells were rejected in the 1, 2 and 4 Mbsec conditions. None were rejected in 8 Mbsec condition. In the local trials it was not been possible to determine whether usability was significantly better under 4 or 8 Mbsec condition. In the next stage, the international trials, similar scenarios will be used with 2, 4, and 8 Mbsec conditions and fashion designers.

Discussion

The scenarios have shown that the type of applications, if fully developed, could support distance working. The video analysis has shown that the present configuration of the toolkit is not optimum for database searching. Much of the users time was spent trying to co-ordinate their actions, they remained unsure as to the progress made by the other partner, and where information was located. Ideally the users needed a window on the other system, so they could refer to the screens being accessed by the partner. One pair changed the configuration of the equipment so that the camera pointed at the other screen rather than at their partners face. Good quality audio was essential.

The results from the local trials have shown that the battery of methods employed in the local trials do yield useful results concerning perceptions of usability. The use of video and breakdown analysis has provided results not captured in the questionnaires. The tendency, at least in the local trials, was for subjects to rate applications more favourably than they sometimes merited. Although partly computerised, the method of recording, transcribing and interpreting breakdowns still remains time consuming and subjective. During the analysis of the international trials, the procedures will be refined in order to provide a robust, accessible methodology which can be employed in the assessment of systems supporting co-operative working.

In its wider context, the methodology employed by the project involves bringing fashion designers and retailers (from small and large enterprises) into the laboratory to test and experience new technology as part of user trials or international workshops held at each of the sites. The technology can be classified as new - because it is still some years before broadband communication will be widely available, at low cost, to the fashion industry. In taking this approach, users have been exposed to immature applications (in terms of their content and functionality) and untested networks, Marshall et al (1995). In all cases they have been able to look beyond the limitations of the applications and networking problems to consider and comment on the long term implications the technology will have for them at a personal and organisational level.

Acknowledgements

1   The work was funded by the EU under the TEN-IBC. The user trials were conducted by Roisin Donnelly (of the University of Ulster-Jordanstown) at the Department of Electrical and Electronic Engineering, Queen's University Belfast. Further information relating to the consortium can be found on http;//dougal.deby.ac.uk/fnet/, or by contacting the authors.
2   FINS - Fashion Information Service was developed at Nottingham Trent University in collaboration with Nottinghamshire County Council.
3   Show - Me is a proprietary system developed by Sun Microsystems.
4   Scribble was developed by Trinity College Dublin and ADETTI (Portugal).

References

B2004 Final Project Report, submitted to the EU in March 1995, available from author.
Badagliacco, J.M., 1990, Gender and race differences in computing attitudes and experience, Social Science Computer Review, **8**,1, 42-63.
Eason, K. D. 1988, *Information Technology and Organisational Change*, Taylor and Francis, London.
Marshall, A., Scrivener, S.A.R. and Woodcock, A, 1995, FashionNet/Temin - Experiences with Distributed Multimedia for the Fashion Industry, paper presented at Broadband Islands Conference.
Ravden, S and Johnson G, 1989, *Evaluating usability of human-computer interfaces*, Ellis Horwood, London.
Urquijo,S.P., Scrivener, S.A.R. & Palmen, H., 1993: 'The Use of Breakdown Analysis in Synchronous CSCW System Design', *Proc. of the Third European Conference on Computer-Supported Co-operated Work*, Milan, September, 281-293.

THE EFFECT OF COLOUR ON THE RECALL OF COMPUTER PRESENTED INFORMATION

Jeanette E. Rollinson and Paul T. Sowden[1]

*Department of Psychology, University of Surrey,
Guildford, Surrey. GU2 5XH*

This research examined whether presentation colour influenced the recall of computer presented information. Subjects were required to read text passages and subsequently recall them both immediately after reading them and 24 hours later. Four different presentation colours and both directions of luminance contrast were compared. Results indicated that contrary to previous findings reading speed was not affected by presentation colour. However, surprisingly recall was better for text presented in colours familiar to the subjects from their use in computer packages within the company from which the sample was drawn. Direction of luminance contrast did not affect recall or reading speed, but subjects expressed a preference for positive luminance contrast displays.

Introduction

Until the 1980s colour played little part in man-machine interfaces, with computer displays being produced on monochrome screens. Since the mid-1980s rapid advances in computer technology have allowed colour to become readily available on computer screens. The majority of research on the effects of introducing colour into displays has been concerned with the effect of colour on speed of performance, error rates and clarity on the display interface. For instance, research has indicated that chromatic contrast between background and foreground colours, in the absence of any luminance contrast, results in poorer reaction times (Widdel and Post, 1992); that identification of alphanumeric characters is poorer when they are presented in blue (e.g. Shurtleff, 1980) due to the lower number of short wavelength receptors in the human eye; that legibility of text on VDUs is

---

[1] Correspondence should be directed to Paul Sowden at the address above

impaired by low levels of luminance contrast (Timmers, Van Nes and Blommaert, 1980).

An interesting study for illustrative purposes was conducted by Matthews (1987) who found that subjects were able to identify more spelling errors and read more pages of a computer based task when the text was presented in medium wavelength colours on a black background (e.g. yellow on black, green on black). This finding could suggest that performance is better for text presented in yellow or green. However, Matthews points out that a likely interpretation of these results is that, since yellow and green have high luminance values in VDU displays compared with colours such as red and blue, the differences may have been a function of luminance contrast rather than the specific display colours, a suggestion in agreement with other research (e.g. Radl, 1980).

Matthew's (1987) suggestion highlights a major problem with a great deal of research in this area. Colours can be described on a number of dimensions and there are many formal ways of doing so. One such way is using the Munsell system which describes colours on three dimensions, hue (colour to the lay-person i.e. red, blue etc.), chroma (the saturation or intensity of the colour) and value (the luminance of the colour). Most studies claim to vary colour (meaning hue), but in fact simultaneously vary the saturation and luminance as well, thus confounding the interpretation of results as illustrated so well by Matthew's experiment. The present study avoids this limitation of earlier work by only varying hue, whilst controlling for variations in luminance and saturation.

Whilst a great deal of research has considered the effects of colour on legibility and reaction time, little work has considered the effect of colour on retention of information. The findings of early work comparing retention of information between colour and black and white television were equivocal (e.g. Reich and Meisner, 1976 vs. Farley and Grant, 1976). However, previous research has indicated that arousal level affects memory (e.g. Hockey, 1979) and further some researchers present evidence to suggest that different colours have different arousing properties (e.g. Wilson, 1966). Thus, there may be reason to expect an effect of presentation colour on retention of information. The present study investigates this possibility further.

Finally, whilst it has been established, as described above, that performance is generally enhanced by high levels of luminance contrast, the findings of research have also been equivocal with respect to whether the direction of this luminance contrast (i.e. whether text is lighter or darker than the background) has any impact on performance, and there has been no investigation of the effect of direction of luminance contrast on the retention of information. Consequently, the present study also examines whether direction of luminance contrast influences retention of information.

Method

Sample, apparatus and materials

There were 48 subjects (26 female and 22 male) all employees of a West London company. Their mean age was 37.4 years and they were all regular (min.

eight hrs. per week) and experienced computer users (min. of one year). All subjects had normal or corrected to normal colour vision.

The subjects were presented with two passages of text taken from the Logical Memory Test of the Wechsler Memory Scale-Revised (WMS-R). The passages were presented by an IBM compatible PC on a 15 inch monitor (NEC Multisync 4FGE) in 80x25 text mode, double spaced and in standard ASCII format. The monitor was calibrated such that colours from the Munsell Computer Book of Colour, V22 could be selected and displayed. Four chromatic colours were chosen that varied only in hue (value and chroma were the same for all colours). In addition, an achromatic grey was chosen of lower luminance (value) than the chromatic colours. The Munsell co-ordinates for the colours are shown in table one.

Table 1. Munsell Colour Co-ordinates

Colour Label	Hue	Value	Chroma
Blue	7.5 B	8	8
Green	7.5 G	8	8
Yellow	7.5 Y	8	8
Red	7.5 R	8	8
Grey	7.5 B,G,Y,R	3	0

An eight item evaluative questionnaire was constructed for subjects to complete after viewing each text passage to measure subjects' preferences for the various colour combinations (for instance, there were items on eye strain, pleasantness and liking). Cronbach's Alpha over seven of the eight items was 0.80 indicating that the reliability was sufficient for the scores on these seven items to be summed to provide an overall 'preference' scale score. In addition, subjects completed a single item after viewing both passages that asked them whether they preferred the positive or the negative luminance contrast display (see below), or had no preference.

Design and procedure

The text was presented by pairing one of the four chromatic colours with the achromatic grey. The subjects were divided into four groups. Each group saw text presented in only one of the chromatic colour/grey combinations. One of the passages was presented with the text in a chromatic colour on a grey background (positive luminance contrast) and the other was presented with the reverse arrangement (negative luminance contrast). The order of passage presentation and the direction of luminance contrast used was fully counterbalanced.

The subjects completed the experiment in a darkened room. After reading preliminary instructions the first text passage was displayed. Subjects read through the passage and then pressed a key. The passage was then replaced by a blank screen. The computer recorded the time taken to read the passage. The subject was then required to write down as much of the passage as he/she could recall and to complete an evaluative questionnaire for that colour/grey combination. The subject then pressed a key to display the second passage and the procedure repeated as for the first passage. The subjects were again required to write down as much of the two passages as they could recall 24 hours later.

Results

Subjects' recall was scored as instructed in the WMS-R manual. To ensure reliable scoring two individuals scored each passage (Cohen's Kappa=0.78). The maximum score for each passage was 25.

The impact of hue and direction of luminance contrast on reading speed
A two way analysis of variance (colour (4) — blue, green, yellow, red; direction of luminance contrast (2) — positive, negative), with repeated measures on direction of luminance contrast, examined whether there were any differences in subjects' reading speeds. There were no significant main effects and no significant interaction.

The impact of hue and direction of luminance contrast on recall of information
A three way analysis of variance (colour (4) — blue, green, yellow, red; direction of luminance contrast (2) — positive, negative; time (2) — immediate recall, delayed recall), with repeated measures on time and direction of luminance contrast, examined whether there were any differences in subjects' recall of the passages. As expected there was a main effect of time (F(1,36)=19.57 p<0.0005). Subjects recalled more information immediately after reading each passage than they did 24 hours later (immediate recall mean = 13.49, delayed recall mean = 11.85). There was also a main effect of colour (F(3,36)=4.58 p=0.008). Mean recall scores and standard deviations are shown for each colour/grey combination and for each time in table two.

Table 2. Mean recall scores and standard deviations (in brackets) for each colour/grey combination at each time

Colour	Immediate recall	Delayed recall
Blue	15.33 (3.54)	13.85 (5.44)
Green	15.62 (4.70)	13.50 (4.19)
Yellow	10.75 (3.78)	9.18 (4.03)
Red	12.25 (4.28)	11.50 (3.79)

As can be seen from table two recall was better, both immediately after reading the passages and 24 hours later, for green/grey and blue/grey displays than for red/grey and yellow/grey.
There was no main effect of direction of luminance contrast and there were no significant two way interactions.

Subjects' preferences
A two way analysis of variance (colour (4) — blue, green, yellow, red; direction of luminance contrast (2) — positive, negative), with repeated measures on

direction of luminance contrast, examined subjects preference scores. There were no main effects and there was no interaction.

However, on the single item measure of preference for direction of luminance contrast subjects indicated a trend for preferring the positive contrast displays (Pearson Chi-square(6)=11.99 p=0.062). Fifty percent of subjects preferred the positive contrast display, compared with only 21% preferring the negative contrast display and 29% indicating no preference.

Discussion

This study was designed to investigate the effect of presentation colour and direction of luminance contrast on the recall of computer presented information. Chromatic colours were carefully selected such that they varied only in hue and not luminance or saturation, thus avoiding some of the problems of earlier research. The results indicated that, in agreement with much previous research (e.g. Bruce and Foster, 1982), direction of luminance contrast had no effect on subjects reading speed or recall of the information, although there was some indication that subjects preferred a positive contrast display. Presentation colour also had no effect on reading speed, but it did affect subjects recall of the information.

The lack of an effect of presentation colour on reading speed agrees with the interpretation of previous research made by others (e.g. Radl, 1980; Matthews, 1987) that it is the high levels of luminance typically used in the display of medium wavelength colours, such as yellow and green, that enhances reading speed for these colours, compared with low and high wavelength colours such as red and blue. Thus, in the present study, by controlling for levels of luminance, the improved performance that has been reported for medium wavelength colours was eliminated.

Perhaps the most surprising result was that presentation colour did in fact affect recall. Information presented in blue/grey or green/grey was recalled better than that presented in red/grey or yellow/grey both immediately after reading the text and 24 hours later. This result does not appear to result from enhancement of memory due to simple changes in arousal level with presentation colour, since Wilson (1966) suggests that red is more arousing than green: that is unless one accepts the Yerkes-Dodson law and argues the present red was too arousing.

However, Kaiser (1985) suggests that psychological factors may moderate the influence of presentation colour on recall. In the present study all of the subjects were drawn from one West London company. During debriefing it was apparent that the subjects were more familiar with the green and blue colours used in the present experiment as they were similar to the colours used in their company computer packages (e.g. cc:Mail). It may be that this familiarity had some effect on arousal levels, or other mechanisms, leading to the enhancement of recall for text presented in those colours. At present this explanation is somewhat circular (higher arousal leads to better memory: we found better memory for green and blue therefore it must be because of higher arousal). To avoid this problem we intend to conduct future studies that provide independent measures of factors such as arousal for each presentation colour.

If the findings of the present study were substantiated and a specific mechanism identified, then this would have important implications for the choice of colours to be used in computer packages. It may be that through careful choice of colours tailored to the individual (or, given pragmatic considerations, at least to 'company colours'), recall of computer presented training materials could be enhanced.

Acknowledgements

We gratefully acknowledge the support of the information resources department of SmithKline Beecham PLC.

References

Bruce, M. and Foster, J.J. 1982, The visibility of colored characters on colored backgrounds in Viewdata displays, *Visible Language*, **XV14**, 382-390.

Farley, F.H. and Grant, A.P. 1976, Arousal and cognition: memory for colour versus black and white multimedia presentation, *The Journal of Psychology*, **94**, 147-150.

Hockey, R. 1979, Stress and the cognitive components of skilled performance. In V. Hamilton and D. M. Warburton (eds.), *Human Stress and Cognition: An Information Processing Approach*, (Wiley, London).

Kaiser, P.K. 1985, Physiological response to color: A critical review, *Color Research and Application*, **9**, 29-36.

Matthews, M.L. 1987, The influence of colour on CRT reading performance and subjective comfort under operational conditions, *Applied Ergonomics*, **18**, 323-328.

Radl, G.W. 1980, Experimental investigations for optimal presentation-mode and colours of symbols on the CRT-screen. In E. Grandjean and E. Vigliani (eds.), *Ergonomic Aspects of Visual Display Terminals*, (Taylor and Francis, London).

Reich, C. and Meisner, A. 1976, A comparison of colour and black and white television as instructional media, *British Journal of Educational Technology*, **7**, 24-35.

Shurtleff, D.A. 1980, *How to Make Displays Legible*, (Human Interface Design, La Mirada, California).

Timmers, H., Van Nes, F.L. and Blommaert, F.J.J. 1980, Visual word recognition as a function of contrast. In E. Grandjean and E. Vigliani (eds.), *Ergonomic Aspects of Visual Display Terminals*, (Taylor and Francis, London).

Widdel, H. and Post, D.L. 1992, *Colour in Electronic Displays*. (Plenum Press, New York).

Wilson, G.D. 1966, Arousal properties of red versus green, *Perceptual and Motor Skills*, **23**, 947-949

AN ATTITUDE-BEHAVIOUR MODEL FOR USER-CENTRED INFORMATION SYSTEMS DEVELOPMENT

Andy Smith and Lynne Dunckley

Department of Computing
University of Luton
Park Square, Luton, Beds, LU1 3JU
Tel: +44 (0) 1582 34111
Email andy.smith@luton.ac.uk

User acceptance is an important determinant in the successful implementation of computerised information systems. User satisfaction with, and acceptability of, new systems is related to individual user *attitude* within the organisational context. Based upon generic theories of attitude and behaviour, and specific theories of user acceptance and user-centred design the authors describe an attitude-behaviour model for the development of user-centred information systems which attempts to describe the full range of factors which effect acceptance and determine behaviour.

Introduction

The ISO define the usability of computer systems in terms of the 'effectiveness, efficiency and satisfaction with which specified users can achieve specified goals' and state further that satisfaction is 'the comfort and acceptability of the system'. Satisfaction and acceptability are closely related to individual user attitudes within the organisational context. An information system may, through inappropriate individual attitudes, elicit user behavioural responses such as resistance and even positive strategies of rejection. In this paper we will review the theories underpinning user attitude, behaviour and resistance, and analyse strategies to maximise individual and organisational acceptance of information systems. Specifically we will propose an attitude-behaviour model, which based upon theories of attitude / behaviour and user-centred design, establishes a framework on which appropriate attitudes and behavioural positions may be developed.

Resistance and rejection are recognised behaviours in relation to information systems and have been studied by a number of authors. Hirschheim and Newman (1988), for example, provide an extensive review of the theory and practice of user resistance. Marcus (1983) describes three broad theories explaining why information systems may come up against resistance. In the *people determined theory*, resistance occurs because of factors internal to the person or group using the system. The second, *system determined theory*, is founded on the assumption that resistance occurs because of factors inherent in the information system itself. Lastly in the *interaction theory* resistance is brought about by an interaction between characteristics related to the people and characteristics related to the system and both a *socio-technical variant* and a *political variant*, can be described.

Attitude and behaviour adoption: the theories

Within what is referred to as *attitude theory*, social psychologists define an attitude as 'a learned and organised collection of beliefs towards an individual, object or situation predisposing the individual to respond in some preferential manner' (Bentler and Speckart, 1981). Social psychologists have classified behaviours as those under volitional control and those which are not. Volitional behaviours include wilful ones which are a direct result of deliberate attempts by the individual to engage in a certain manner. Volitional behaviour has been explained by the theory of *reasoned action*. Reasoned action is a function of two basic determinants, one *personal in nature* and the other reflecting *social influence*. The personal factor is the individual's attitude towards the behaviour; the individual's positive or negative evaluation of performing the behaviour. The second determinant, known as the subjective norm, arises from the social pressure as perceived by the individual to perform, or not perform, the behaviour. In this theory people will perform a behaviour when they evaluate it positively and they believe that others, who they deem to be important, think they should perform it. Problems however occur when the theory is applied to behaviours, such as smoking, which are not under complete volitional control. Various factors and external circumstances will also modify the performance of a behaviour. These concepts led Ajzen to extend the theory of reasoned action to the theory of planned behaviour which postulates three independent determinants of intentions: attitudes, subjective norms and perceived behavioural control.

Bentler and Speckart (1981) appear to largely agree with Ajzen and Fishbein and have integrated a number of similar ideas within a generic *attitude behaviour model* which, based upon prior behaviour, individual social and personal norms, and individual attitude to specific behaviours, attempts to describe how individuals might decide to take up particular behavioural positions. As a basis for our specific attitude-behaviour model for IS development we present in Figure 1 a distillation of the ideas of Ajzen and Fishbein and Bentler and Speckart. The model shows how individuals will form predispositional attitudes based upon prior behavioural patterns (for example in our case reactions to earlier computer implementations), and social and personal norms of behaviour (which are themselves effected by attitude to individual behaviours).

Image theory (Beach, 1990) provides a second, and complementary, model of behaviour adoption and focuses particularly on the decision making process. It views a decision maker as possessing three distinct but related images of his / her situation with which reference is made when any decision is required. These three images are a *value image*, or set of principles, which defines how events should transpire in the light of the decision maker's values, morals ethics etc., a *trajectory image*, or set of goals which are about the kind of changes the decision maker wants for himself, herself or the organisation, and constitutes an agenda for the future, and finally a *strategic image*, which describes the plans and tactics the decision maker has for accomplishing these goals. Any decision, or course of action, is made in the context of these three images. Within the theory two distinct types of decision are identified. Firstly *adoption decisions* are about the adoption, or rejection, of courses of action with reference to the decision maker's value, trajectory and strategic image. Secondly *progress decisions* are about whether a particular plan on the strategic image is producing satisfactory progress towards attainment of its goal. When making a decision the decision maker engages in a process called *framing*, in which a subset of elements from his or her images are identified as being relevant to the decision at hand.

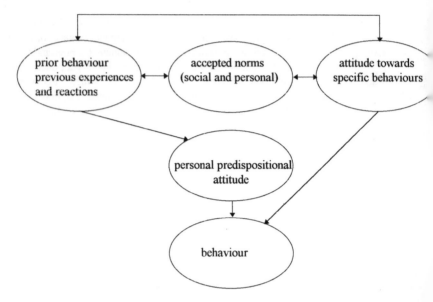

Figure 1. **Determinants of behaviour**

Attitude and behaviour: application to information systems

Individuals who hold negative attitudes are unlikely to make effective and efficient use of information systems. In our terms user behaviour can range from full systems acceptance, through marginal resistance to total rejection. Hirschheim and Newman (1988 expand on the types of behaviour associated with resistance and distinguish between *aggression*, a behaviour which represents an attack (either physically or non-physically) with the intent of injuring or causing harm to the object presenting the problem, *projectior* a behaviour exhibited when the person blames the system for causing difficulties and *avoidance*, which occurs when a person defends himself from the system by avoiding or withholding from it. Through an understanding of the factors which determine attitude an behaviour we can improve the way in which we design the interaction between human and computer and between the system and the organisation. Strategies to overcome, or at least minimise the effects of all three theories lie within the theory of user-centred design (Smit and Dunckley, 1995). In essence a user-centred design process can facilitate the modification of user attitudes and the enhancement of positive response behaviours. Smit and Dunckley (1995) show how approaches to user-centred design can be addressed unde the following broad concepts: *structures, processes,* and *scope*. The categories are:

- *structures* which provide a controlling and / or supporting mechanism to the design and focus on design team structures; structures provide an environment in which use predispositional attitudes can be determined and through which they can be modified.

- *processes* within design which focus around the design methodology adopted and include issues relating to requirement elicitation, communication, implementation and training; processes provide the mechanisms in which system can be developed to meet the specific needs of their users,
- *scope* of the design process which relates to how far the analysis reflects a socio-technical, as opposed to just a technical, solution.

An attitude-behaviour model for user centred IS development

It is clear that a wide number of factors and theories underpinning behaviour in general, and information systems development in practice, are important in determining user attitude and behaviour. In proposing the attitude-behaviour model for information systems development three major elements have been selected and integrated:
- Bentler and Speckart's and Ajzen and Fishbein's models of attitude and behaviour
- Beach's image theory
- Smith and Dunckley's categories of user-centred design.

Figure 2 extends the generic model of Figure 1, applies it specifically to the users of computerised information systems and integrates aspects of image theory. In addition to generic determinants of behaviour, the model shows how user-centred design, through the categories of structures processes and scope, can modify personal predispositional attitudes over the main stages of systems development life cycle. Image theory provides us with a method for analysing how and why users decide on particular behavioural stances (tendency towards acceptance and rejection) in response to new information systems. The model proposes that individual behaviour in response to computerised information systems can be influenced by both the individual's:

- *attitude to individual behaviours,* individuals differ in their likely behaviour patterns, some workers are inherently less likely to react in negative ways,

and
- *modified personal attitude and images,* which is based upon the individual's predispositional attitude but is modified by:
  - *system issues,* the major elements within the socio-technical system are also important, those systems which clearly demonstrate wide will tend to mitigate against the effects of predispositional attitude,
  - *development and implementation issues,* the actual processes adopted during the design stage will effect both the degree of effective participation and the nature of the final product, in addition methods of system changeover and approaches to training and user support are important
  - *user participation,* the degree to which users are able to effectively contribute to design team *structures*

Essentially the user is making an *adoption decision* which is made as a result of a *framing* process on the three *images*. Specifically the *value image* will be significant if any new information systems is seen as being in conflict with the users values, morals and ethics, the *trajectory image* will determine whether the system is either in conflict with the personal goals of the user (for example systems which dehumanise or downgrade work practices) or in synergy with them (the empowerment ideal) and the *strategic image* provides a basis for mapping the functions of the system to currently established plans.

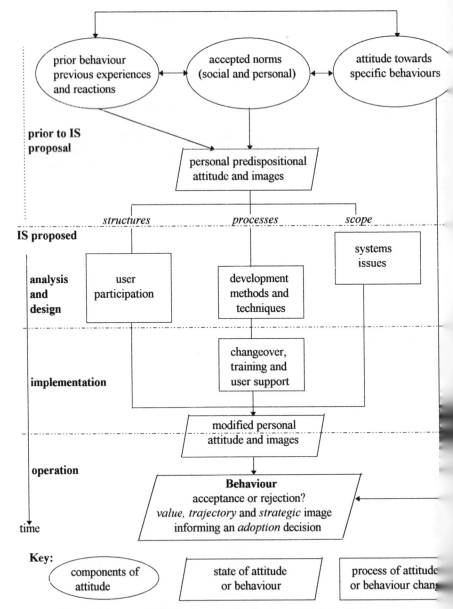

Figure 2. **An attitude-behaviour model for users of information systems**

Case Study

In order to explore and validate the model it will be appropriate to show how it has been applied to a case study. The organisation selected claimed a positive approach to user-centred design based on a central IS project steering group which aimed at a balance of user representatives and technical experts. Systems development was based on structured analysis techniques with some prototyping. The group had agreed to the development of a small PC database for a department which the manager claimed urgently needed a manual record system converted to a database for ad hoc inquiries and to help them cope with a much increased work load. The small database was designed and implemented, the clerical and administrative staff were trained in the use of the system and supporting documentation produced. However when the sixth month review was carried out on the system the developer discovered that it had never been used, no data had been entered and the staff preferred to continue with the existing manual system..

There was considerable conflict in the department between the manager and sections of the staff with both sides blaming eachother for failures. The manager had seen the new computer system as raising his profile in the organisation. The staff rejected the system because they considered that it would not solve the real problems of the department, would only increase their work load while providing no benefit. They perceived their problems arising from their manager's tendency to continually move the goal posts and tinker with the business rules of the department. The staff's prior behaviour and previous experience led to negative attitudes. The social norms also led them to feel that it would be safe for them to reject the system as this might actually lead to more attention to their conflict with the manager and perhaps undermine his position. Reviewing the development, in terms of *structures* there was certainly insufficient genuine user involvement in design where end-users had been compliant rather that actively involved, whilst the adherence to structured analysis techniques *(processes)* had meant that the focus had been on providing a good technical solution. The greatest weakness had been in the *scope* of design where it was acknowledges that although the designers had picked up some undercurrents they considered that the computer system would help the department by introducing disciplined working practices and therefore had not paid attention to the conflicting objectives of different groups in the department or perceived that these non-technical issues would have a significant impact on their work.

References

Ajzen, I. and Fishbein, M. 1980, Understanding attitudes and predicting behaviour (Prentice Hall)

Beach, L. R. 1990, *Image Theory: Decision Making in Personal and Organisational Contexts* (Wiley)

Bentler, P. M. and Speckart, G. 1979, Models of Attitude Behaviour Relations, Psychological Review, 86

Hirschheim, R. and Newman, M. 1988, *Information Systems and User Resistance: theory and practice*, Computer Journal 31 (5)

Marcus, M. L. 1983, Power, Politics and MIS Implementation, In Baecker, R. M. and Buxton, W. A., *Readings in HCI*, (Morgan Kaufman)

Smith, A. and Dunckley, L. 1995, Human factors in software development - current practice relating to user centred design in the UK. *In Human-Computer Interaction*, Proceedings of INTERACT-95 (Chapman and Hall)

USING USER-CENTREDNESS TO DEVELOP A TAXONOMY OF SOFTWARE DEVELOPERS

Andy Smith and Lynne Dunckley

Department of Computing
University of Luton
Park Square, Luton, Beds, LU1 3JU
Tel: +44 (0) 1582 34111
Email andy.smith@luton.ac.uk

In this paper the authors describe how three categories of user-centred design (*structures*, *processes* and *scope*) have been used to develop a template which software developers can use to assess their own approach to user-centred information systems development. Within the template eight distinct Software Developer Types are identified and their characteristic approaches to user involvement are described. Together the eight types constitute a *user-centred taxonomy* of software developers.

Introduction: evaluating user-centred design

IT systems would still appear to be failing because of a lack of effective user engagement. In a previous study (Smith and Dunckley, 1995a) a major survey of UK commercial user organisations was undertaken in order to identify to what extent, and in which ways, genuine user centred design (UCD) principles have been integrated into mainstream commercial IT systems design. From the survey it was possible to identify which particular aspects of UCD are being adopted at a faster rate than others, and to ascertain a number of factors which influence the degree of user-centredness in both UK organisations and IT projects. Through the survey it emerged that although software development managers have a commitment to user involvement, user-centred principles are not being adopted uniformly across the IT industry. There is insufficient take up of tools and techniques to enable active user involvement. This is in line with the previous findings of Hornby and Clegg (1992), and Green (1992) who reveals that while organisations consider the involvement of users to be an important aspect of their development approach in practice few support 'active' involvement.

Evaluating user-centred design

The authors have described (Smith and Dunckley, 1995a) how approaches to user-centred design can be addressed under the following broad concepts: *structures, processes* and *scope*. It is these concepts which have been used to evaluate approaches to UCD. The categories are:

- **structures** which provide a controlling and / or supporting mechanism to the design process: structures include design team structures, project management, and mechanisms for user involvement and communication between the design team and the user community,
- **processes** within design which focus around the design methodology adopted and include issues relating to requirement elicitation, communication and implementation, including approaches to prototyping,
- *scope* of the design process which relates to how far the analysis reflects a socio-technical, as opposed to just a technical, solution: an IT project with only a narrow scope will focus on technical and functional requirements whereas one with a wider scope will address issues such as the allocation of functions between man and machine, the design of work structures and individual jobs and the ways of enhancing job satisfaction within the organisation.

Templates for user-centred design

The findings of the survey highlighted the lack of a shared understanding within the software development community about what constitutes UCD and how individuals might address any weakness in their approach. In response to this the authors have developed two self-assessment templates (Smith and Dunckley, 1995b) which enable individual software developers to evaluate the user-centredness of their approach. The templates has three specific aims:

- to raise awareness within the design community about what constitutes UCD
- to enable individuals and organisations to identify their strengths and weaknesses in respect of UCD
- to provide a sign-posting facility which will direct software developers to the many tools and techniques which are currently available to integrate human and organisational factors within the development process.

Initially two versions of the template were developed. The second version of the template is more comparative and it is this which has led to the development of a taxonomy of software developers. This template adopts an approach similar to that developed by Belbin (1981) for a self-perception inventory for management teams. It allows for a qualitative comparison between the approaches to each of the categories (structures, processes and scope) within UCD. In addition a fourth dimension has been introduced to represent the *intent*, to involve a range of users (whether or not effective structures and processes are adopted to support them).

When completing the template the software developer is presented with 5 sections and within each section is required to distribute a total of ten marks between eight responses to a statement. For each statement, such as shown in Figure 1, a response is provided which might be typical of each of eight Software Developer types (A - H). Each of the software developer types equates positively or negatively to each of the four UCD concepts (intent, structures, processes and scope) as shown in Figure 2.

By recording scores in an analysis sheet the user is able to calculate an index score for each of the developer types and plot these on a User Centred Profile. The UCD Profile generated has the advantage of an effective mechanism for individual comparative self-assessment. The template have been trialled within a large number of UK organisations, some of which has taken part in the earlier survey. The aims of trialling was to elicit views of software developers concerning ease of use and clarity in both the methods for UCD assessment and ways of identifying remedial action. From the results of trialling it emerged that the template was easy to complete and provided mechanisms for clear identification of user-centredness.

For each section below distribute 10 points among the sentences you think best describe your views. The points can be distributed among several sentences; in extreme cases they might be spread amongst all the sentences.

Section 1
From my experience of IT project design and implementation:

(a) I consider the establishment of Local Design Groups with full user involvement to be important in the management of successful projects
(b) I have found that success can only be achieved by involving all potential users in the development process
(c) I have found that successful project design and implementation is best led by central DP/IS departments
(d) I have found that it is not necessary practice to consider the design of jobs for employees
(e) I believe that the best approach to prototyping is to not use it at all
(f) Where user requirements are concerned I know that methods such as GOMS and Hierarchical Task Analysis can be used successfully
(g) I consider that when creating a design team structure it is sufficient to ensure that users are available for requirement elicitation
(h) I consider that when identifying user requirements for a project the future impact on organisational structures is highly important

Figure 1. An Example Section from the Template

A taxonomy of software developers

Although many of the underlying factors influencing UCD relate to the nature of the project itself and the organisation in which it is being implemented, other factors are specific to the individual software developer. This individual element inherent within all software developers is in itself determined by a number of factors. These will include their own personal characteristics (such as their tendency towards being an introvert or extrovert) and their knowledge of the various methods and techniques which can support UCD.

The taxonomy has been developed from the eight software developer types to address the individual element within systems development. In the descriptions below the extreme position is discussed. Developers with a strong bias towards one role will demonstrate the qualities indicated. Two alternative names for each type, together with a characterising statement and a description are provided. In addition the percentage of developers in the trial who were dominant (first or second) in each type. In practice it was found that most software developers show a tendency towards a number of types.

Type   A   **Romantic** (Partnership Seeker)   %: 45
Characteristic statement: *I try to involve as many users as possible*
The Romantic, or Partnership Seeker, has a rose coloured view of the ease / difficulty in involving all types of user. In terms of user involvement the romantic has his / her heart in the right place but may not be effective in user engagement. Where this is the predominant style it is likely that the Romantic is person- rather than task-centred.

Type   E (opposite to A)   **Hermit** (User Avoider)   %: 4
Characteristic statement: *I do not need contact with users*
The Hermit, or User Avoider, inherently likes working on his / her own and will only interact with users on a minimalist basis. For example, even though this type of developer may have a tendency to create and participate in structures and processes designed to promote UCD, in practice they allow barriers to limit effective communication between users and developers.

Type   B   **Nestler** (Partnership Builder)   %: 22
Characteristic statement: *I adopt structures to support the user*
The Nestler, or Partnership Builder, is someone who is able to implement a range of structures which will enable a variety of users to make an effective contribution. The Romantic appears to adopt the Nestler role as a secondary role. The reverse of this, however, may not be true. The results obtained so far indicate that when the Nestler is the predominant role, a range of secondary roles, e.g. technocrat or Explorer is preferred.

Type   F (opposite to B)   **Chancer** (Informal Worker)   %: 0
Characteristic statement: *I do not provide mechanisms which would integrate users*
The Chancer, or Informal Worker, assumes that goodwill towards user involvement on his / her part will be sufficient to provide effective participation. Interestingly this type was identified through observation of project development in practice, but no developer assessed himself as predominantly belonging to this type in the template trial.

Type   C   **Rover** (User Supporter)   %: 2
Characteristic statement: *I use a range of user centred processes*
The Rover, or User Supporter, adopts a range of socio-technical processes to meet the specific individual requirements of users. In selecting the methods, tools and techniques to be used within the project a flexible approach will be used in order to ensure that all needs (technical, human and organisational) are addressed. In the trial this type was rarely a predominant type and this may be due to the lack of adoption of specialised UCD processes.

Type   G (opposite to C)   **Technocrat** (Functional Specialist)   %: 5
Characteristic statement: *I mainly use processes which address technical issues*
The Technocrat, or Functional Specialist, will only adopt techniques which address technical issues. He / she will be inherently inflexible in approach and will follow standard

procedures, for either personal or organisational reasons. This is also rarely a predominant role but features as a secondary role for 'Nestler' types.

Type   D   **Explorer** (Organisational Visionary)   %: 13
Characteristic statement: *My approach is socio-technical*
The Explorer, or Organisational Visionary, will look at all aspects of the organisation, including its mission, values and beliefs, its individual employees and their job satisfaction needs together with functional requirements in order to co-optimise effectiveness between the social and technical systems within the organisation.

Type   H (opposite to D)   **Fixer** (Technical Implementer)   %: 9
Characteristic statement: *My approach is primarily technical*
The Fixer, or Technical Implementer, will take a narrow, technically focused, view of the project. He / she will not be interested in wider human / organisational issues either being unaware of their importance, or believing that they can be sorted out later.

As we have stated, the process of completing the template involves the calculation of an index score for each software developer type. All developers will have one fully shaded region (index score 1.0) on the template (Figure 2). This indicates the role which they most closely adhere to. The degree of shading (magnitude of other index scores in the range 0 to 1) indicates their *relative* adherence to other types. Trialling of the taxonomy and template has indicated that a number of developers are heavily biased towards one type whilst some others show a reasonable spread across two or three types.

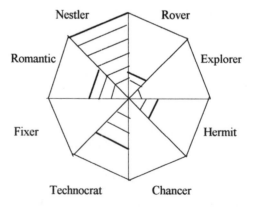

Figure 2. Example User Centred Profile

Discussion and conclusions

Initial trialling indicates that the most frequent developer type is that of the Romantic. This result is supportive of previous work of the authors and other researchers. In their survey the authors found that a large majority of organisations stated that user participation was important to them, although a detailed analysis under the three categories did not support this in practice. Green (1992) found that 98 per cent of organisations considered involvement of users to be important but that under 20 per cent achieved active involvement. The second most frequent developer type was the Nestler. Again this is compatible with the authors previous survey which identified that structures was the must supported UCD category.

An interesting question arises as to why individual developers generate differing profiles. It is possible to speculate on the basis of the template results, and the experience gained through case study work, that some of the differences may be due to organisational culture rather than personality traits. For example the Romantic may be driven by the work culture to perceive UCD as a good idea. The fact that the Nestler tends to create UCD structures which are then used positively may also be the result of a culturally driven perception that this is the correct thing to do, coupled with relatively low resource costs and a probable high payoff. Similarly the Explorer type could be the product of a cultural environment which has sprung from the perceived need to develop hybrid managers whereas the Fixer could be the product of a culture which takes a narrow view of the contribution of Information Systems to the enterprise. In the same way it is not yet clear whether the Hermit is the product of a personality type (e.g. an anxious extrovert who is unhappy working with users) or an organisational culture in which technical experts are expected to know best without unnecessary consultation, which might reduce their technical productivity.

Both the template and the taxonomy suggest that developers have been adopting processes which have functional / technical objectives rather that specific UCD processes. Results so far have not generated a Rover type as predominant. This is unlikely to be the result of organisational culture. It was noticeable that some developers had simply no knowledge of UCD tools and techniques. It may be that whereas the adoption of structured methodologies has been driven by external agents (such as government requirements and the pursuit of ISO9000), with the possible exception of the EU directive on usability, there is no comparable driving force to justify an investment in education and training relating to UCD.

Through wider distribution, implementation and discussion within the software community it is suggested that the template and taxonomy will make a contribution to raising the importance of UCD and to the spread of UCD techniques. Materials are available from the authors at the University of Luton.

References

Belbin, R., 1981, *Management Teams* (Heinemann, London).

Green, P., 1992, Involving Users in Systems Development, Butler Cox Product Enhancement Programme Paper 19 August 1991, (CSC Index, London)

Hornby, P. and Clegg, C., 1992, User Participation in Context: A Case Study in a UK Bank. In Behaviour and IT, 1992, Vol. 11, No 5. (Taylor and Francis, London)

Smith, A. and Dunckley, L., 1995a, Human factors in software development - current practice relating to user centred design in the UK. In Human-Computer Interaction, Proceedings of INTERACT-95, (Chapman and Hall, London)

Smith, A. and Dunckley, L., 1995b, A template to assess user centredness in software quality management. In Software Quality Management 3 - Proceedings of SQM-95, (Computational Mechanics Publications, Southampton)

CLOSE-COUPLED DYNAMIC SYSTEMS
CHALLENGE AND OPPORTUNITY FOR ERGONOMISTS

Hugh David

Eurocontrol Experimental Centre
91222 Bretigny-sur-Orge, France

Computer-control or computer-mediation is becoming increasingly common in large and small-scale systems. Most such systems have evolved from mechanical or electro-mechanical precursors and have been adapted to, rather than designed for, computer operation. Short-term considerations of surface appearance and (sometimes) the lack of time for in-depth analysis have produced computer-based displays that mimic the preceding electric or mechanical systems, neglecting the enormous potential of modern computer systems, and equally neglecting basic ergonomic and cognitive principles, resulting in unnecessary routine strain on the operators, and failure to use their true strengths in emergencies. This paper suggests that the ergonomist is particularly well placed to guide the development of future interfaces.

Introduction

It has been suggested, more or less seriously, that ergonomics is in the process of fission into HCI (Human-computer Interaction) and 'The Rest'. The multiplication of societies, associations and study groups on HCI may well serve to draw off the excess pressure, but I wish to develop in this paper the idea that the basic training and attitudes of the ergonomist are particularly well-adapted to the study of human-computer interaction. Ergonomists should not allow the case to go by default to the electronic engineers, applied psychologists, and 'cognitive scientists' who make much of the running at the moment. Having spent more than a quarter of a century grappling with the ergonomics and human factors of Air Traffic Control, I shall draw most of my examples from that domain, although my message is definitely not confined to that field.

Background

(Although most readers will be familiar with the general processes of ATC, the following selective history is introduced to form a common background for the subsequent argument.)

Air Traffic Control (ATC) is typical of many advanced systems in that it has evolved under continuous pressure for increasing 'efficiency'. It came into being initially from the clear need to avoid aircraft colliding with each other and to settle the order of precedence of take-off and landing. There has always been a tension between ground and air aspects of Air Traffic Control. From the earliest times, rules were devised to govern the behaviour of aircraft in the air. Typically they were derived from the rules of navigation at sea, and make interesting reading today. Ground control was limited to the vicinity of airfields, operating with flags, flares and lamps. As technology developed, air routes were defined. In the United States, ATC routes were actually instituted around 1922, before radio aids were available, using bonfires and flare pots (Perrow 1984). From these, it was a natural transition to air routes marked by radio beacons, initially simple transmitters (NDBs - Non-directional Beacons). Flight Plans were required, and Procedural Control, which required aircraft to report on passing over beacons, and allocated height levels to separate them was introduced as airborne radio became available. (Procedural Control required the Flight Plan to be transferred to 'Strips' so that they could be compared and used to check the progress of the flight. This was done by hand, by the controller or an assistant.)

Radar was introduced after the Second World War, initially around airports only. SSR(Secondary Surveillance Radar), developed from war-time IFF (Identification Friend or Foe) which relied on aircraft-based equipment interrogated by the ground-based radar, to identify and give the height of aircraft came in to use in the early 1950's. NDBs were elaborated to provide distance and direction measurements (VOR-DMEs - Very High Frequency Omni-directional Ranging - Distance Measuring Equipment).

ICAO (the International Civil Aviation Organisation) defined a standardised international language for air traffic control, and rules were laid down about which languages should be used between aircraft and the ground-based ATC system. Standard forms for Flight Plans were agreed, and a specialised teleprinter network was derived to distribute these throughout the world.

Initially, flight plans were sent through the teleprinter network a few hours in advance of the flight, and copied onto strips by hand by assistants. Later, strips were generated automatically from incoming flight plans (which required considerable tightening of standards for their acceptance). Later still, computer files of repetitive flight plans were developed for months ahead, and strips were generated directly at the sectors.

Aircraft navigation was improved and simplified by various technical devices - details are not important here, but the consequence, the disappearance of the specialist navigator from the aircrew, is highly significant. Contemporary modern commercial aircraft navigate using satellite-based location systems which can, in theory, provide positions accurate to metres, compared with the hundred metres accuracy of sophisticated multi-radar systems. Navigation systems are supplied with the flight plan of the aircraft, and this is used by the aircraft to navigate itself from point to point with little or no intervention by the pilots.

The 'Glass Cockpit' introduced initially in the Airbus 320, and adopted by most major aircraft manufacturers, involves a computer-based system to choose what information to display to the pilots, and essentially adapts these displays as the flight progresses, or in response to unexpected phenomena. (Many of the displays in a 'glass cockpit' are, however, imitations of what would have been presented on conventional instruments.) The introduction of the 'glass cockpit' has been accompanied by the disappearance of the 'flight engineer' reducing the aircrew to two. ACAS (Airborne Collision Avoidance System), which is currently being introduced, is an aircraft based system designed to give emergency warning of the approach of other aircraft. Initial versions of ACAS were entirely self-contained, but more recent versions interrogate other aircraft to obtain accurate height measurements.

Currently, ATC expects problems. The traditional response to increased traffic - reduction in sector size - is reaching the limit of the practical, and the airlines are becoming restive about the restrictions imposed on their activities by ATC requirements. Technical developments palliate these problems to some extent - ATC centres are now linked directly for computer-based transfers of aircraft from centre to centre, as well as from sector to sector within a centre. More sophisticated computer-based displays have been simulated (Prosser, David and Clarke, 1991, Graham, Young, Pichancourt, Marsden and Ikiz, 1994) and are already being introduced into existing and updated ATC systems. These displays incorporate new ways of displaying data, such as the C(A)RD (Conflict (And) Risk Display), which warns the controller of potential future conflicts in his sector. More radical solutions are in the air (Dee,1995).

Although some parts of the system (for example, displays, traffic flow management, and flight plan processing) are in the vanguard, others (ground-air communication - Cushing 1994) are, regrettably, straggling to the rear. Although data links have been under consideration for many years, and are actually installed in many modern aircraft for company communications, ATC relies on voice communications, using VHF and even HF frequencies in a manner that will be familiar to anyone who has watched films set in WWII, while passengers may be provided with satellite telephone links that provide clear communication with anywhere on the surface of the earth.

Trends

Air traffic has increased steadily for the last forty years, and shows no sign of levelling off. Overall, the development of ATC has been by the adaptation of devices originally devised for other purposes. (ACAS, a rare exception, was introduced on the orders of the US Congress.)

There has been a steady development of international standards, mainly via ICAO (the International Civil Aviation Organisation)

In recent years computers have become increasingly involved in ground and air operations, and there has been a corresponding reduction in aircrew per aircraft. Ground and air systems have steadily increased their precision. The speed of data transmission has increased , and expectations for efficiency have risen. Ground control staff have not decreased, but have been required to increase the traffic they can handle. (The current ground-based control system is reaching saturation in many areas.)

The civil aviation system is passing from being relatively loosely-coupled, where actions take minutes or hours to propagate through the system, to being 'closely-coupled' where actions taken in one place may have effects within seconds on freight handling, passenger bookings, catering, and security at airports hundreds or thousands of miles away. It is becoming increasingly clear that it must be designed as an integrated whole, for safety, economy and efficiency. It is one of many systems for which this is true.

Design and Re-design

It is rare for any system to be designed from a completely blank sheet. Even where there is no physically pre-existing system, attitudes and traditions are imported with the designers. Most systems are said to have 'evolved', although the implications of Darwinian selection and 'survival of the best adapted' are not necessarily justified. More often, the system is modified by the introduction of some innovation, and the existing elements of the system must be adapted to make use of the innovation. There are some valid arguments in favour of this process, particularly where systems are relatively loosely-coupled, and depend on custom for much of their efficiency. However, a point can be reached where the accumulation of traditions, unexamined assumptions and habits leads to inefficiency and eventually to collapse or catastrophe.

It may be rewarding to attempt a re-design of the system, as if from first principles. Although the resultant system may never be introduced in practice, it will usually bring into focus the problems of the existing system, and may suggest practical actions to resolve them.

System Description

Kirwan and Ainsworth(1992) provide an excellent review of Task Analysis methods for describing systems. Most form models, in words, diagrams, or equations."A theory has only the alternative of being right or wrong. A model has a third possibility: it may be right, but irrelevant" (Eigen 1973).

There is always a temptation to accept a task description in place of task analysis. A task description is a description of the way a task is done at the moment. This need not necessarily be the best way to do the task, particularly where the task has 'evolved' over the years. The painstaking work reported in part in Ammerman, Ardrey, Bergen, Bruce, Fligg, Jones, Kloster, Lenorovitz, Phillips, Reeves and Tischer (1984) provided an extremely detailed quantitative description of the US ATC system at that time (amounting in total to some 12,000 pages). It was used as the basis for the projected AAAS system, which was rejected by the controllers, although controllers had been heavily involved in writing the specification. This system provided a high-quality replication on computer displays of the existing manual system, but made no attempt to add facilities to it, to take advantage of the computer's information organising abilities.

A task analysis for conceptual design may be quite sketchy, (Dee 1996). It should be qualitative, rather than quantitative, since a premature involvement in detail may result in over-specific design. ATC centres and sectors are extremely variable, and it is unwise to become fixated on a specific region.The same will apply to most complex systems.

Task allocation

Task allocation is the second step in the re-design of a system. Each task should be looked at in terms of whether it is suitable for human performance, whether human operators would enjoy performing it, and whether they can maintain the degree of reliability required in the context. Tasks that humans find irritating, impossible or boring (such as monitoring) should be performed by the computer system. There should be a category of tasks that may be performed either by human or computer, preferably in such a way that the human can perform them better than the computer. The characteristics of such tasks are (or should be) well known to ergonomists. Hopkin (1995) provides detailed discussion in the context of ATC.

Overall, the aim should be to define a satisfying, suitably rewarding human task, backed by a reliable, simple, automatic system. This may involve considerable changes in the way tasks are carried out. Dee (1996), for example, suggests that Air Traffic Control is best exercised via a digital data-link, as did Cushing (1994) in view of the proven unreliability of speech communications. This would eliminate the need to insert messages into the system as well as speaking them, would eliminate the inherent ambiguities of voice communications and would make it possible to complete manoeuvres in one order, rather than remembering the state of incomplete ones.

Interface design

The design of the system interface should start from the tasks allocated to the human operator. The interface should provide the information for the tasks the operator will need to perform, and should provide it in a suitable form. There is always a temptation to 'play safe' by providing a copy of whatever display the original system provided, or by providing all the information all the time. Neither of these solutions is adequate in the long run. Many original system displays were constrained by mechanical engineering problems, and provide information in an indigestible form. The existing ATC system requires the controllers to keep considerable amounts of precise information, defining the present positions, nature and intentions of all their current aircraft, in their memories, refreshing it from time to time from their strips and radar - 'the picture'. Controllers experience difficulty in maintaining this 'picture' (Stein and Bailey 1994). Hopkin (1996) discusses the implications for future systems, where it may be necessary for the controller to accept that he cannot continue to maintain 'the picture'. If the controller passes from explicit control of each flight to a 'control by exception' system, where he intervenes only to make changes at the instance of a conflict detection routine, then most aircraft may never come to his attention. (In emergencies, he will require completely different displays. Most emergencies fall into predictable categories - in ATC: fuel shortage, equipment failure, navigation mistakes or inaccuracy. Displays and computer-assisted procedures can be devised to cope with most of these.)

Certain well-known principles apply to displays and controls:.
- Displays and controls should be compatible with the real world.
- Display characteristics should 'fit' the aspects being displayed.
- Displays should acknowledge actions immediately.
- Control sequences should be short, 'complete' and natural.
- Irrelevant information should not be shown.
- Displays should show what the user needs to know in a relevant form.

Vincente, Christoffersen and Hunter (1996) in a response to a critique (Maddox 1996) of Christoffersen,Vincente and Hunter(1996) make the point that interface design should include specifying what needs to be measured not merely the optimal display of what happens to be available. "We also wish to have a say in the specification of interface content, and, thus, design functionality".

At present, software is more flexible than hardware, so that the different types and styles of display, different uses of colour coding or flashing or pulsing can be explored relatively easily, using existing knowledge of human perceptual abilities. Economic realities tend to restrict input devices to a few widely available devices. The ubiquitous 'mouse' and various functionally equivalent devices (trackballs and joysticks), and the standard keyboard, or its close relatives cover most current possibilities, but it would be unwise to suppose this will still be true in five years time.

Basic ergonomic principles, as taught in most undergraduate courses, apply as much to the control of nuclear reactors, ATC systems, chemical plants, and vehicles of all sorts as they do to babies bottles, kettles, washing machines and personal computers.

To quote Vincente et al (1996) again : -

"Human Factors engineers must be at least be copilots in the design process..... This point goes well beyond the confines of the process control domain and the interface design problem......It is relevant to all other domains in which designing for human use is an important consideration..."

Conclusion

It is generally accepted that 'Human-Centred Automation' like 'Motherhood' and 'Apple-Pie' is a Good Thing. What it means in practice is less clear.

I have tried to show in this brief paper that the attitudes and methods of the broad tradition of ergonomics provide the answer.

Acknowledgements

The Author wishes to thank the director of the Eurocontrol Experimental Centre for permission to publish this paper. This paper represents the opinions of the author, and should not be taken to represent Eurocontrol policy.

References

Ammerman H A, Ardrey R S, Bergen, Bruce, Fligg, Jones G W, Kloster G V, Lenorovitz D, Phillips M D, Reeves, Tischer,1984, *Sector Suite Man-Machine Functional Capabilities and Performance Requirements*, CTA Inc.,Englewood Colorado

Briers, S., Reed, J. and Stammers, R.,1996, Research into the human factors desirability of a truly paperless control room at BNFL, in S.A.Robertson (ed.),*Contemporary Ergonomics 1996 :Proceedings of the Annual Conference of the Ergonomics Society , Leicester,* April 1996 (Taylor and Francis London), ISBN 0-7484-0549-6 , pp 470-475

Christoffersen, Klaus, Hunter, Christopher N. and Vincente, Kim J., 1996, A Longitudinal Study of the Effects of Ecological Interface Design on Skill Acquisition, Human Factors,**38(3),**pp 523-541

Cushing, Steven, 1994, *Fatal Words : Communication Clashes and Aircraft Crashes*, University of Chicago Press,ISBN 0-226-13200-5

Dee, T.B., 1996, Ergonomic Re-design of Air Traffic Control for increased Capacity and Reduced Stress, in S.A.Robertson (ed.),*Contemporary Ergonomics 1996 :Proceedings of the Annual Conference of the Ergonomics Society , Leicester,* April 1996 (Taylor and Francis London), ISBN 0-7484-0549-6 , pp 563-568

Eigen, Manfred, 1973, in *The Physicist's Concept of Nature* ,ed Mehra J, Reidel, Dordrecht,The Netherlands.

Graham R V, Young D, Pichancourt I, Marsden A and Ikiz A, 1994, *ODID IV Simulation Report*, EUROCONTROL Experimental Centre Report No. 269, Bretigny, France.

Hopkin, V.D., 1995, *Human Factors in Air Traffic Control*, Taylor and Francis, London, ISBN 0 85066 8239

Kirwan, B. and Ainsworth, L.K., 1992, *A Guide to Task Analysis*, Taylor and Francis, London ISBN 0 7484 0057 5

Maddox, Michael E, Critique of " A Longitudinal Study of the Effects of Ecological Interface Design on Skill Acquisition"by Christoffersen, Hunter and Vincente, 1996,, Human Factors,**38(3),**pp 542-545

Perrow, Charles,1984, *Normal Accidents*, Basic Books, New York, ISBN 0-465-05143-X

Prosser,M., David, H. and Clarke, L., 1991, *ODID III Real-Time Simulation,* EUROCONTROL Experimental Centre Report No. 242 , Bretigny, France

Stein, E S. and Bailey, J, 1994, *The Controller Memory Guide: Concepts from the Field*, DOT/FAA Technical Centre, Atlantic City, USA

Vincente Kim J., Christoffersen, Klaus and Hunter, Christopher N. 1996, Reply to Maddox Critique, Human Factors **38(3)**, pp 546-549

AUTHOR INDEX

SUBJECT INDEX